R. E. LITTELL
2 - 93

CONSTRUCTION AND GEOTECHNICAL METHODS IN FOUNDATION ENGINEERING

McGraw-Hill Series in Construction Engineering and Project Management

Raymond E. Levitt, *Consulting Editor*

Barrie and Paulson: *Professional Construction Management*
Douglas: *Construction Equipment Policy*
Koerner: *Construction and Geotechnical Methods in Foundation Engineering*
Parker and Oglesby: *Methods Improvement for Construction Managers*
Shuttleworth: *Mechanical and Electrical Systems for Construction*

CONSTRUCTION AND GEOTECHNICAL METHODS IN FOUNDATION ENGINEERING

Robert M. Koerner, Ph.D., P.E.

Professor of Civil Engineering
Drexel University

McGraw-Hill Book Company

New York   St. Louis   San Francisco   Auckland   Bogotá   Hamburg
Johannesburg   London   Madrid   Mexico   Montreal   New Delhi
Panama   Paris   São Paulo   Singapore   Sydney   Tokyo   Toronto

This book was set in Times Roman by University Graphics, Inc.
The editors were Kiran Verma and J. W. Maisel;
the production supervisor was Marietta Breitwieser.
The drawings were done by Wellington Studios Ltd.
Halliday Lithograph Corporation was printer and binder.

CONSTRUCTION AND GEOTECHNICAL METHODS IN FOUNDATION ENGINEERING

2 3 4 5 6 7 8 9 0 HALHAL 8 9 8 7 6 5 4

ISBN 0-07-035245-3

Library of Congress Cataloging in Publication Data

Koerner, Robert M., date
    Construction and geotechnical methods in foundation
engineering.

    (McGraw-Hill series in construction engineering and
project management)
    Includes bibliographies and index.
    1. Foundations. I. Title. II. Series.
TA775.K57 1984   624.1'5   83-16261
ISBN 0-07-035245-3

To Paula
"Simply Fantastic"

CONTENTS

The rationale behind the writing of this book is to provide a vehicle for the student of construction engineering to learn and appreciate geotechnical engineering, and for the student of geotechnical engineering to learn and appreciate construction methods. The book is intended to provide for both groups an accurate, representative, logical, and properly sequenced coverage of the subject of foundation engineering in regard to civil engineering structures. The subdivision of material is as follows:

Chapters 1 through 3 cover topics of concern to all persons involved in foundation engineering.

Chapters 4 through 6 have to do with in situ stabilization methods.

Chapters 7 and 8 discuss relatively recent foundation engineering topics.

Chapter 9 covers techniques for monitoring the systems discussed in the rest of the text.

There are several appendixes providing backup information on many of the subjects treated.

In regard to the background helpful to a student reading this book, it is assumed that she or he has had an introductory course on soil and rock materials, behavior and mechanics, as is often offered in the junior year or first-term senior year of civil engineering at most colleges and universities.

It should be possible to cover the material in the text within a period of 15 weeks. To do so, however, would be to cover it less than completely considering the detail given both in the text and in the references at the end of each chapter. There is considerable opportunity for the instructor to deviate from the text or to supplement specific areas as felt to be desirable.

The author is a strong advocate of homework (yes students, even for every class period). An instructor can only provide the framework for a course, offer clarification when needed, and expand on various topics when necessary—then he or she must let go. It is the student's desire and drive to learn that are required for proper comprehension and utilization of the material being studied.

Regarding the problems at the end of each chapter, a few comments are in order:

They generally follow the order in which the material was presented in the chapter.
They vary randomly in their difficulty—some needing only simple, one-word answers, others requiring more complex responses that can even go beyond material presented in the book itself.
Therefore, to answer some of the problems, it will be necessary to do additional reading in other references, e.g., to consult ASTM standards or ASCE Journals.
Also, some of the problems in each chapter involve material presented in earlier chapters.
The more numerically oriented problems are given in Chaps. 1 through 3, while problems of a more subjective nature are presented in Chaps. 4 through 9.

On an overall basis, and in contrast to many other subjects, the author feels that foundation engineering is a beautiful subject. It combines problem recognition, conceptualization, and design with engineering judgment about what is constructable within the framework of the processes and materials available and affordable for a particular project. A difficult task, but one that is worthwhile, challenging, and enjoyable.

Best wishes and full speed ahead.

Robert M. Koerner

CONSTRUCTION AND GEOTECHNICAL METHODS IN FOUNDATION ENGINEERING

SHALLOW FOUNDATIONS

1.1 HISTORY, CONCEPTS, AND TYPES

Lost in the pages of history are the first rational decisions about the value of a "solid" foundation—that, for example, rock is a better foundation material for a structure than is soil and, in turn, dense granular soils are better than high-water-content fine-grained soils to avoid foundation problems after the structure is built. (See Appendix A for soil classification values.) It seems logical to expect that early builders developed empirical procedures which, through a trial and error procedure, evolved into reasonable guidelines to keep failures and settlements within tolerable limits. It also seems logical to expect, in the absence of written communications, that misguided concepts were carried over through many generations and that lessons learned in one area of the world had little chance of reaching distant locations. Undoubtedly, the subject was discussed as formalized training in military and civilian engineering took place in Europe in the early 1700s and then in America in the late 1700s and early 1800s. Real progress into current times took a major step forward, however, when the concept of a footing *grillage* (parallel sets of wood, iron, or steel beams placed perpendicular to one another) to spread column and wall loads was first used in the 1880s in Chicago.[1] The development of reinforced concrete shortly thereafter replaced the grillage concept, providing greater flexibility of design at reduced costs. These foundation methods gave rise to quasiformalized design guidelines such as the following: The foundation area should be proportional to the imposed load, the load (if vertical) should act at the center of the footing, the load (if inclined) should act within the center third of the footing width, etc. True formalization, however, first came about with the 1925 publication by Terzaghi of *Erdbaumechanik*[2] (see also Terzaghi's *Theoretical Soil Mechanics*[3]) and the teaching of soil mechanics and foundation engineering, now known as geotechnical engineering, in the late 1930s and early 1940s in several universities and colleges. This formalization process is still continuing and forms the basis of current methods used to design and construct shallow foundations.

The concept of a shallow foundation for building loads, bridge piers, retaining wall loads, industrial machinery, utility towers, etc., is to provide for the adequate transfer of load to the underlying soil or rock. The usual construction material is reinforced concrete placed at or slightly beneath the ground surface. This feature gives rise to the basic distinctions between shallow and deep foundations in that:

Shallow foundations have a least width generally greater than their depth beneath the ground surface.
The shear strength (not the weight) of the backfill soil above the bottom of the foundation does not contribute to its stability.

When considering a shallow foundation for a given loading system, the design engineer must take into account three areas. They are the following, each of which will be dealt with in detail later in this chapter:

1. Footing placement, which involves its proper siting as to location and depth. Obviously, this requires close interaction with the client or owner and a knowledge of the past use of the site along with detailed information about subsurface soil and rock.
2. Safety against bearing capacity failure, which involves the proper footing design vis-à-vis the strength of the foundation soil or rock. Note that we are not considering the structural strength of the reinforced concrete footing, but of the foundation materials beneath the footing. In this area a knowledge of the geotechnical stratigraphy and individual properties of the soil or rock is essential. See Appendix B for typical soil property values.
3. Tolerable foundation settlements, which involves calculating the anticipated settlement and comparing it to the allowable settlement that the structure will stand. Again, a knowledge of the soil or rock profile and properties are necessary, along with communication of the anticipated settlements to the client or owner. In many cases the time frame in which these settlements will occur is vitally important and must also be calculated or estimated.

Generally, shallow foundations are treated as a separate category within foundation engineering but are subdivided into a number of types according to their size, shape, and general configuration:

1. *Spread footings* (or foundations) are also known as *pad footings, isolated footings,* and *square* or *rectangular footings* and are characterized by having L/B ratios of less than 5. Seen in Fig. 1.1, it is obvious that this type of footing is the

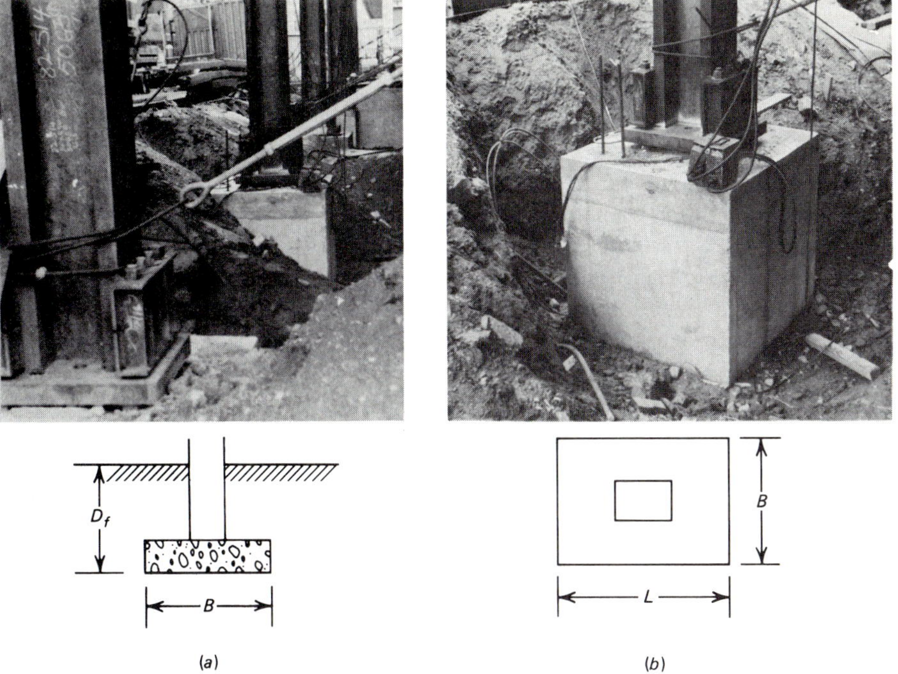

Figure 1.1 Photographs and schematic diagrams of spread footings. (*a*) Elevation view; (*b*) plan view.

most common of all types and represents the minimum in complexity, formwork detail, cost, etc. Yet, it still provides the essential function of spreading the column load to a value compatible with the strength and deformation characteristics of the soil or rock upon which the footing is placed.

2. *Combined footings* (or foundations) are characterized by combining two or more column loads onto one footing. This results in a more even load distribution in the underlying soil or rock, and consequently there is less chance for differential settlement to occur. While these footings are usually rectangular in shape, they can be trapezoidal (to accommodate unequal column loadings or close property lines) or in *strap* form (to accommodate wide column spacings or close property lines); see Fig. 1.2 for details.

3. *Continuous footings* (or foundations) are also known as *strip footings* or *wall footings* and are characterized as having L/B ratios greater than 5. Seen in Fig. 1.3 is that this type of footing carries a number of relatively closely spaced column loads or a continuous wall load in such a way that the load intensity is lowered and relatively uniform as it is transmitted to the soil or rock below. For this type of footing, loads are usually expressed in force per unit length of footing.

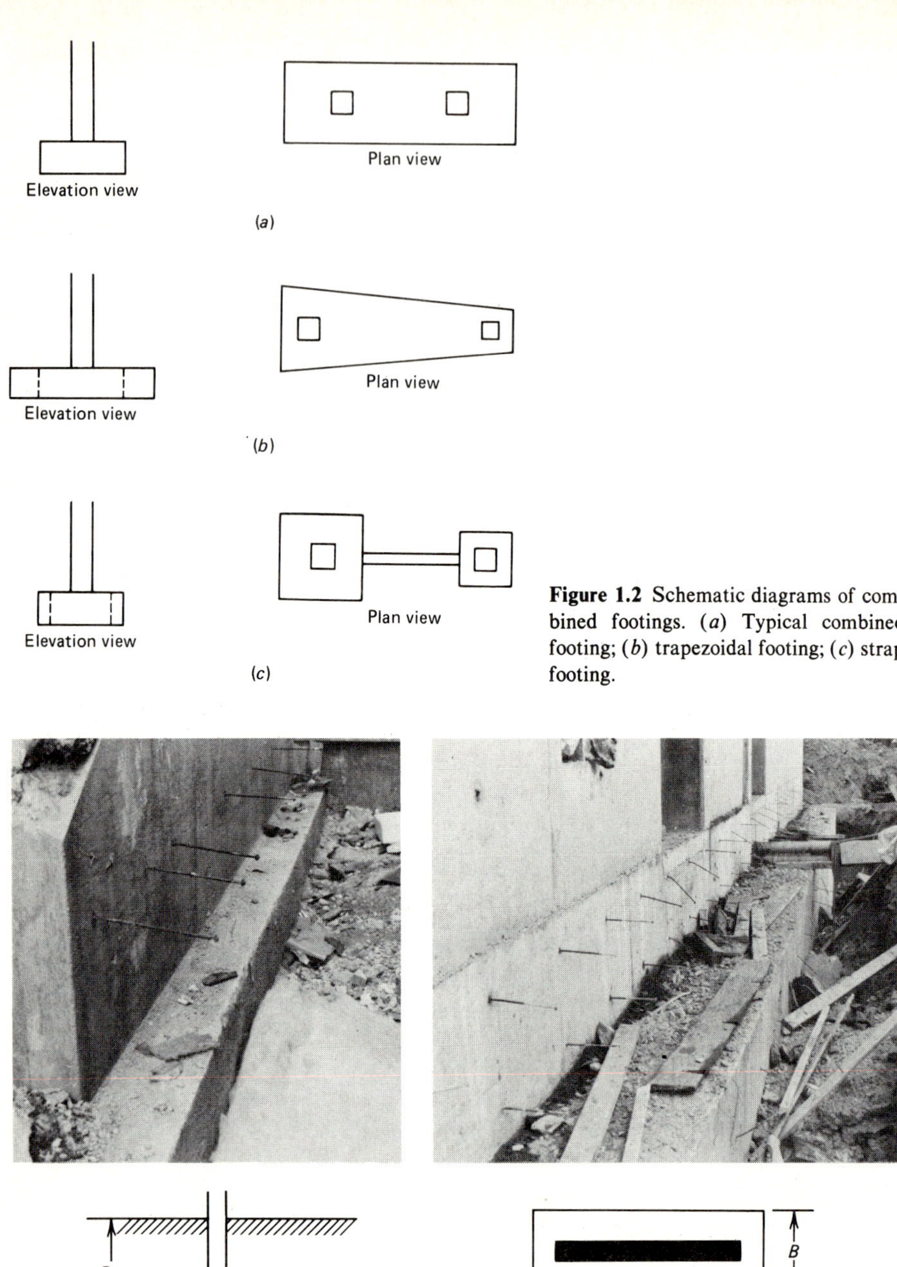

Figure 1.2 Schematic diagrams of combined footings. (*a*) Typical combined footing; (*b*) trapezoidal footing; (*c*) strap footing.

Figure 1.3 Photographs and schematic diagrams of continuous footings.

4. *Mat footings* (or foundations) are also known as *mass footings,* and are characterized by the feature that columns frame into the footing in two directions. These footings can accommodate as few as four columns or as many columns as the building or structure has. They are usually considered when the sum of all individual or combined footings exceeds one-half the total building area (to economize on formwork costs) or for structural stability on poor foundation soils (to avoid differential settlements). In this latter regard the mat footing becomes a very significant structural feature of the building and a number of variations are possible; see Fig. 1.4.

It should be noted that mat footing terminology is often confused with *compensated foundations* (where all, or a part of, the building load is compensated for by excavating the soil and placing the building at a lower than normal depth) or with *raft* or *buoyant foundations* (where the structure itself is made watertight and placed a distance beneath a permanently established water table elevation).

1.2 MISCELLANEOUS DETAILS REGARDING SHALLOW FOUNDATIONS

As with many technical designs, the detailed, and usually analytic, design often occupies our attention to the point where obvious flaws are sometimes overlooked. Indeed, the obvious is as significant in shallow foundation design as it is elsewhere. This section is meant to cover a number of significant topics which should never be dismissed as trivial. So as not to be omitted, they are placed before both Sec. 1.3 on bearing capacity and Sec. 1.5 on settlement analysis.

1.2.1 Freeze-Thaw Zones

Shallow foundations must be placed at a depth beneath the maximum frost penetration in the particular area of concern. Figure 1.5 gives frost penetration contours in the contiguous United States. These depths can be used in the absence of building codes, which usually have a minimum depth restriction for most urban areas. It is important to consider frost since, upon freezing, the pore water in the soil will expand by approximately 10 percent, thereby exerting pressures on adjacent areas or upon the base of footings, floor slabs, walls, etc., founded above. The forces exerted are more than adequate to cause structural damage, as can be seen from the photograph and schematic diagrams shown in Fig. 1.6. Here, frost heave lifted up the wall footings of a one-story industrial building, which placed the wall in compression. The roof structure had its own column supports which extended down to footings approximately 3 ft beneath the wall footings. The wall buckled outward at its weakest point (where the windows frame into the sill area), as was seen by the tension cracks along the outside of the building. It should be noted that in many buildings, water often

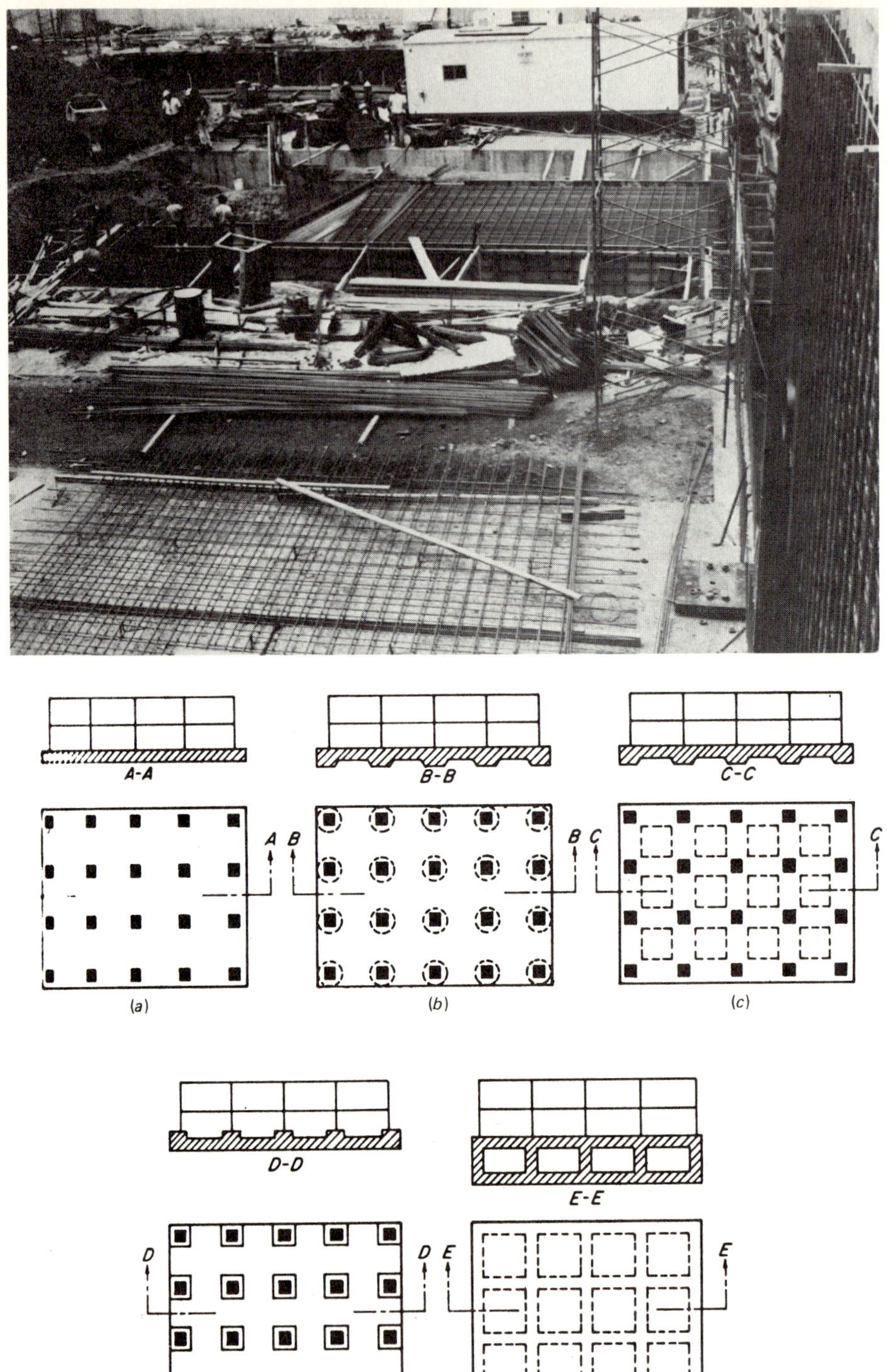

A-A
B-B
C-C
D-D
E-E
A B
B C
C
D D
E E
(a)
(b)
(c)
(d)
(e)

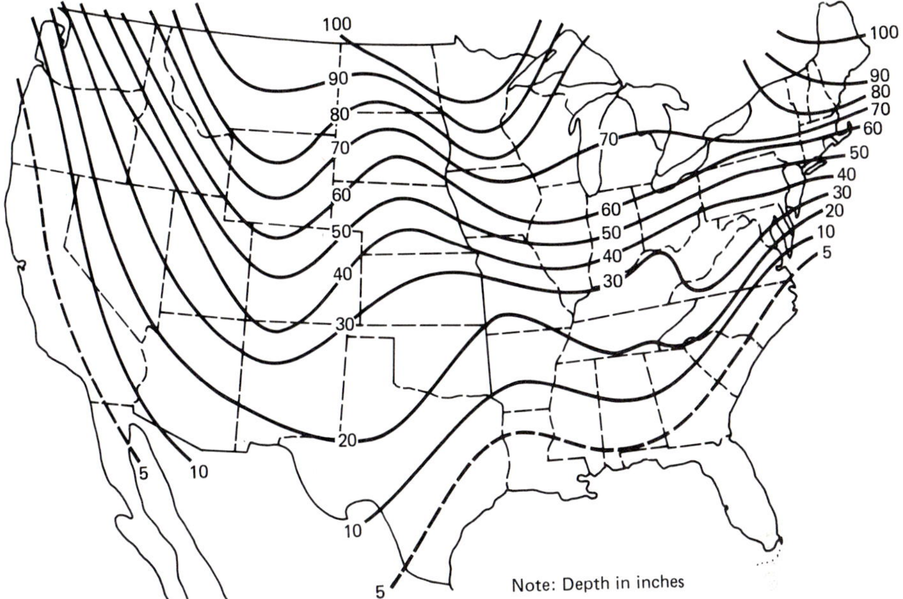

Figure 1.5 Maximum frost penetration in the United States. *(After Sowers and Sowers, Ref. 5.)*

enters the foundation soils, which can freeze in exactly the same way as ground water from natural circumstances.

The idea of placing a foundation beneath the maximum depth of frost penetration has an interesting parallel in building practices necessary to combat conditions in permafrost areas of northern climates. Such areas (Alaska, northern Canada, Greenland, etc.) never have completely ice-free soils but do have periods of partial melting during summers. This thaws the upper portions of the soil from the ground surface downward and any footings founded within this "zone of thaw" will become unstable. Indeed, a relatively thin zone of thawed soil beneath the footing and above frozen ground can easily extrude laterally from beneath the footing, causing large settlements and eventual failure. Thus, footings in permafrost areas should be placed at an elevation beneath the depth of maximum thaw penetration.[6]

1.2.2 Frost Heave

Frost penetration and uplift is often confused with the continuously increasing uplift caused by frost heave. The distinction is easily seen when numbers are put on the

---

Figure 1.4 Schematic diagrams of different types of mat foundations. (*a*) Flat plate; (*b*) plate thickened under columns; (*c*) beam and girder support; (*d*) pedestal support; (*e*) box structure. [*Adapted from Bowles (Ref. 4) and photograph of a mat foundation during construction.*]

(a)

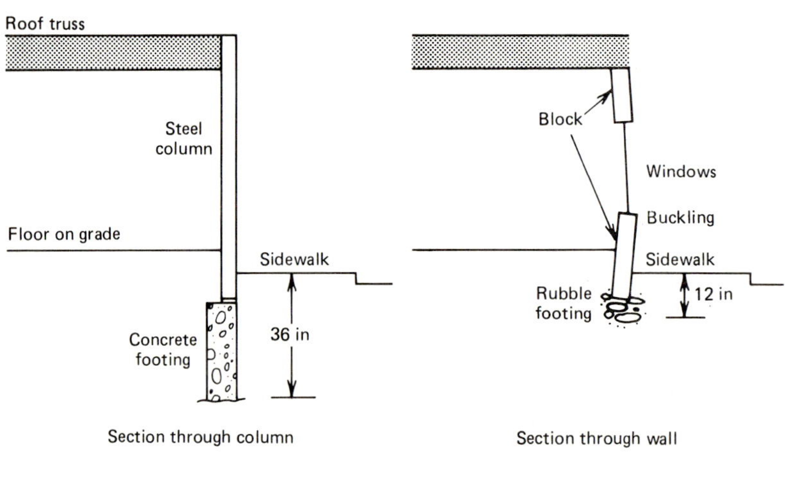

(b)

Figure 1.6 Buckling of a wall due to foundation uplift (footing depth 12 in in a 36-in frost penetration area).

respective processes. For example, frost penetration and uplift in a saturated soil of 0.3 porosity would be approximately 0.10 × 0.3, or 3 percent by volume. Yet, frost heaves of 20 to 30 percent by volume have been measured, clearly indicating that a very different mechanism is occurring. Necessary for frost heave to occur are the following[6,7]:

A permanently established water table
A soil capable of supporting a capillary zone
Freezing temperatures over a time period capable of transferring the water from the
    water table through the capillary zone to the zone where temperatures are below
    freezing

When the above combination of events is present, the water near the top of the capillary zone freezes in lenses which progressively grow, heaving the ground and structures above it. Essentially, there is no limit to the phenomenon. The situation occurs in cold northern regions (e.g., in permafrost areas), but also beneath cold-storage buildings and ice skating rinks and arenas.

The fundamental mechanisms responsible for the continued transmission of water through the capillary zone is an area actively being researched. These mechanisms include the following[8]:

1. *Capillary theory of freezing pores.* This original concept of frost heaving, according to which moisture migrates toward the freezing point as a result of the rise of water in the capillary zone, has not been shown to explain completely, or even generally, the phenomenon.
2. *Pressure head theory.* This theory seems to be applicable only to saturated soils freezing in closed systems.
3. *Film migration theory.* Involving moist soils, this theory seems to be applicable to unfrozen and, with refinements, to frozen soils.
4. *Theory of crystalline forces.* Originally thought to explain the ice crystallation process, and thereby supplement the film migration theory, it does not explain water movement above freezing temperatures, and is not a general theory.
5. *Theory of osmotic pressure.* This theory, which is very useful in regard to partially saturated soil water flow, does not seem well suited to explain such flow under frozen conditions since major water movement then only occurs when rather high salt concentrations are present.
6. *Theory of absorption forces.* Here, suction forces are roughly equated to the net effect of migration forces, but the theory is not fully developed.
7. *Adsorption film theory.* Explaining the mechanism as the result of the combined action of several processes, this theory is the most current. It is thought that the properties of water in films of various thicknesses, which are governed by the influence of active sites on the soil particle surfaces, set the water in motion from thick films, which have more mobile molecules, toward thin films, which have fewer.

8. *Phenomenological models.* These models, based on differential equations of heat and mass transfer, have also enjoyed recent popularity.

It follows from the above brief review that the basic theory of moisture migration in frozen soils is in a very unsettled state. As of yet, no integrated theory has been developed in general form. As far as construction projects are concerned, the fact that the phenomenon exists and an appreciation of its magnitude are sufficient to make one take positive steps to avoid its occurrence.

Remedies include construction methods that break the continuity of the capillary zone (insulation, gravel layers, or geotextiles) or that render the zone impermeable (grouting or cutoff). Unless such positive steps are taken, the damage frost heave produces can be considerable.[6-10]

1.2.3 Ground Water Level

The location of the level of ground water, which can be determined from an initial site survey, is of importance in shallow foundations for a number of reasons. For one thing, the bearing capacity of the soil (as will be seen later) is a function of the unit weight of the soil. When a high water table is encountered, the effective unit weight (saturated unit weight less unit weight of water) becomes significantly less than the total unit weight, which has the effect of decreasing the soil's bearing capacity. A high water table will also cause added construction costs to adequately waterproof basement walls and temporarily dewater the site during construction of the foundations and other below-ground-surface activities.

It should be recognized that the ground water elevation is not a stable, unyielding surface but fluctuates in most areas in a rather dynamic way. Seasonal variations are tied to climatic conditions, while local variations are tied to subsurface nonhomogeneities. The usual preconstruction method to determine the water table level is by using a piezometer and/or water observation pipe at the location of interest. Equilibrium will be reached within hours for granular soils and within days for fine-grained soils. For large areas, a number of locations should be monitored or a suitable nondestructive testing method may be utilized (see Chap. 9).

1.2.4 Scour

Whenever an obstacle to water flow, such as a bridge pier, is placed in a river bed, the increased water velocity around the obstacle causes erosion of the soil adjacent to and below the foundation. The flow regime is extremely complex (see Fig. 1.7), and the resulting scour depth prediction is an area of active research. Variables in the process include the following:

Water velocity
Water turbidity
River width
River depth

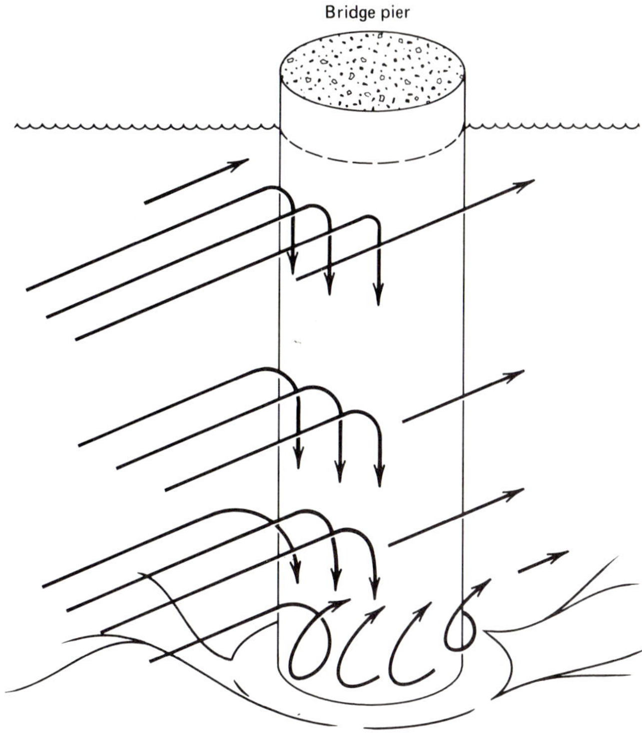

Figure 1.7 Idealization of flow regime near base of circular pier placed in a river. *(From Ref. 11.)*

Pier width
Pier geometry
Pier orientation
Soil properties
Soil density
Soil stratigraphy
Other (?)

Easily seen is that many approaches toward obtaining a unified equation to predict scour depth can be generated. Beginning in 1894, when Engels published the results of model studies, a vast number of formulas have been generated. In 1970, the Federal Highway Administration published a summary of these formulas along with the results of two sample problems using the various predictive methods[11]:

Problem 1. Clear water scour was to be calculated around a pier in water of 69-ft depth with average flow velocity of 2.0 ft/s. The average size of the soil was 0.1 mm. The pier was rectangular, 47 ft wide, and oriented with the river flow. The various methods illustrated predicted scour depths from 0 to 79 ft!

Problem 2. Scour was to be calculated around a river pier in sediment transporting water, for a 50-year storm, in 28 ft of depth at a 3.9-ft/s velocity. The soil had a mean size of 0.5 mm. The pier was rounded, 6 ft 7 in wide, and aligned with the river flow. The various methods used predicted scour depths ranging from 0 to 13.8 ft!

The report states that "it is not possible to recommend a specific method for predicting scour. Judgment must be exercised in selection of a method, and in interpreting the resulting predictions." The author certainly agrees with this finding. The uncertainty in scour depth prediction is also reflected in the answers of various agencies to a questionnaire as to the technique they used to predict scour depth: 46 of 95 agencies replied that "engineering judgment" was used.

At this point in time, judgment in river pier placement (and avoidance if at all possible) or use of protective techniques seem to be justified. In this latter category are the following:

Deep foundations
Physical protection (rip rap, mattresses, etc.)
Flow attenuators (artificial seaweed, nets, etc.)
Upstream barriers

1.2.5 Expansive Soils

While volume change in soils as moisture is added or withdrawn from the soil/water system has been a long established concept in explaining the consistency (Atterberg) limits (see Fig. 1.8), its significance as a problem in its own right is a relatively recent finding. It has been recently estimated that shrinking and swelling soils cause over $2 billion of damage per year to homes, buildings, pavements, utilities, etc.[12] Furthermore, it is estimated that the susceptible soils exist in about 25 percent of the United States and are found in almost every country of the world.

Geologically, there are two basic groups which give rise to volume change susceptible soils[13]:

1. Basic igneous rocks in which the parent feldspar and pyroxene minerals decompose to form montmorillonite
2. Sedimentary rocks that contain montmorillonite and physically break down to form expansive soils

The distribution of these soils is shown in Fig. 1.9. Since the consistency (Atterberg) limits seem to be intrinsically tied to the phenomenon, it seems only natural that they should be used to identify the troublesome soils; see Table 1.1. In instances where this is too generalized an approach, a consolidation test can be modified for quantification. In such a test, a sample would be placed in its undisturbed state and, after

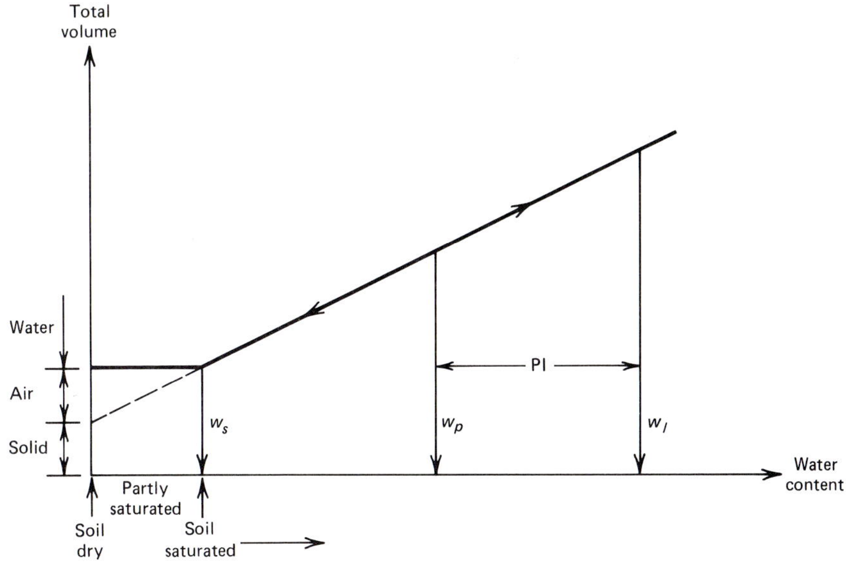

Figure 1.8 Idealized volume change response to moisture variation in fine-grained soils. The consistency (Atterberg) limits are as follows: w_s, shrinkage limit; w_p, plastic limit; w_l, liquid limit; *PI*, plasticity index ($= w_l - w_p$).

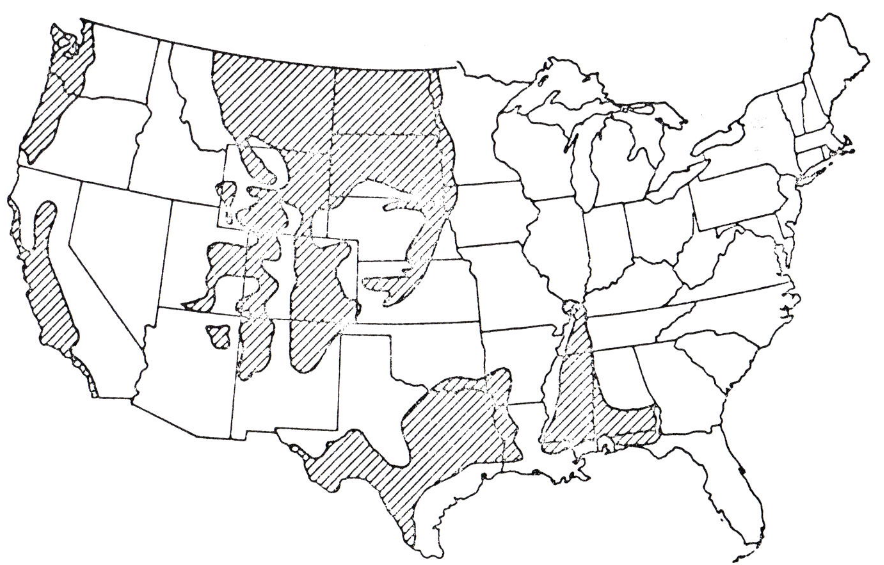

Figure 1.9 General abundance of montmorillonite (expansive soils) in near-outcrop bedrock formations. *(After Chen, Ref. 13.)*

Table 1.1 Volume-change-susceptible soils†

Likelihood of volume change with change in moisture	Plasticity index		Shrinkage limit
	Arid region	Humid region	
Little	0–15	0–30	≥ 12
Little to moderate	15–30	30–50	10–12
Moderate to severe	≥ 30	≥ 50	≤ 10

Note: See Fig. 1.8 for definitions.
†After Sowers (Ref. 1).

the addition of water, the swell with time noted. The test time is rather lengthy, which suggests that additional research into the topic is warranted.

Remedial measures usually center around avoiding construction in the troublesome area altogether, excavating and replacing the susceptible soil with granular soils (i.e., less plastic soils), treating the in situ soils with lime, cement, fly ash, or other injection materials (see Chap. 6), or providing seepage barriers.

1.2.6 Creep-Sensitive Soils

There are numerous situations where long-term movements, i.e., deformation under a constant stress which is generally called *creep,* pose a particular problem for foundations and the structures founded upon them. These situations are often on hillsides which present quite attractive building sites from an aesthetic point of view. Some insight into the behavior and general magnitude of creep, as well as methods of predicting its occurrence, is appropriate.

The general behavior of soil to sustained load is shown in Fig. 1.10. There, in the curve labeled (3), three stages can be identified as primary, secondary, and tertiary, respectively. The limiting situation, at the end of the tertiary stage, is obviously failure with a resulting slide of the soil mass and any structures founded on or near to it. While this physical situation is obvious, a good predictive method to determine which curve the creep will follow, or what position a particular instance of creep is at on a particular curve, is not so obvious. Currently, there are three different approaches toward achieving accurate prediction:

1. Rheologic model analysis, which consists of various combinations of springs and dashpots. The spring constants and dashpot viscosities are usually experimentally obtained.
2. Rate process theory, which predicts a strain rate equation using microscopic deformations. The shear across interparticle bonds and free energy of activation are utilized.
3. An empirical approach using creep test data. This approach will be described more fully here since it is straightforward and gives reasonable response behavior.

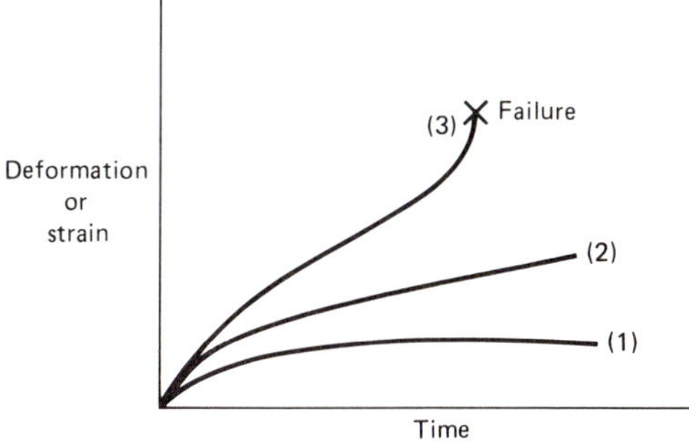

Figure 1.10 General stages of sustained load type of deformation, i.e., soil creep. Notes: (1) Noncreep-sensitive soil, typical of granular soils or soils containing sufficient granular material that the particles are touching and mobilizing effective stress. (2) Constant, or steady-state, creep, typical of clay soils that have some frictional components. (3) Creep-sensitive soil typical of clays in their undrained state.

It has been observed by Singh and Mitchell[14] that creep data, consisting of data sets of creep rate ($\dot\epsilon$) and time (t) for various intensities of deviator stress ($\sigma_1 - \sigma_3 \equiv D$), can be plotted as shown on the two curves illustrated in Fig. 1.11. By using the slopes of these curves, the following equation can be derived for the creep rate:

$$\dot\epsilon_{(t,D)} = Ae^{\alpha D}\left(\frac{t_1}{t}\right)^m \tag{1.1}$$

which, for $t_1 = 1$ min, reduces to

$$\dot\epsilon_{(t,D)} = A\,\frac{e^{\alpha D}}{t^m} \tag{1.2}$$

where $\dot\epsilon_{(t,D)}$ = strain rate at time t and deviator stress D
D = deviator stress
t = time, min
m, α, A = experimentally obtained constants (as shown in Fig. 1.11)
The constants α and D in the above equations carry the units of inverse stress and stress, respectively. If the maximum deviator stress of the soil is known, $\overline\alpha$ and $\overline D$ can be substituted, where $\overline\alpha = \alpha D_{max}$ and $\overline D = D/D_{max}$, thereby making the replacement constants $\overline\alpha$ and $\overline D$ dimensionless. The dimensions of the constant A are usually percent per minute.

Singh and Mitchell propose these relationships for creep stresses from about 30 to 90 percent of the initial soil strength, depending on the susceptibility of the soil to fail in creep rupture. This is certainly within the range of engineering interest for most problems encountered with shallow foundations.

By appropriate integration of the creep rate equation, the corresponding creep, i.e., strain, can be obtained. From Mitchell,[15] the following equations result which are graphically shown in Fig. 1.12:

$$\epsilon = \epsilon_1 + \frac{A}{1-m}\,e^{\overline\alpha\overline D}(t^{1-m} - 1) \qquad \text{for } m \neq 1 \tag{1.3}$$

$$\epsilon = \epsilon_1 + Ae^{\overline\alpha\overline D}\ln t \qquad \text{for } m = 1 \tag{1.4}$$

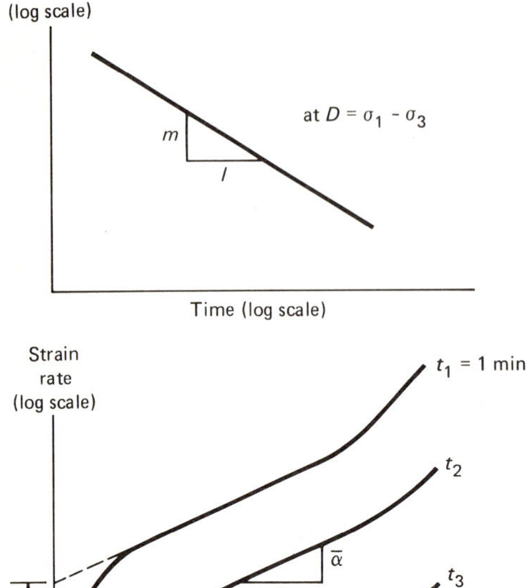

Figure 1.11 Method of plotting creep data to empirically determine the constants m, α, and A. *(After Singh and Mitchell, Ref. 13.)*

Since creep (ϵ) is a more intuitively acceptable design parameter than creep rate ($\dot{\epsilon}$), it is generally considered to be the focal point in a design or analysis situation.

Regarding the constants, m, $\bar{\alpha}$, and A in these equations, Hirst and Mitchell[16] have determined that m is related to clay type and stress level, $\bar{\alpha}$ is inversely related to shear strength and directly related to deviator stress, and A is a function of clay content and type. Some representative values for these constants in different soils are

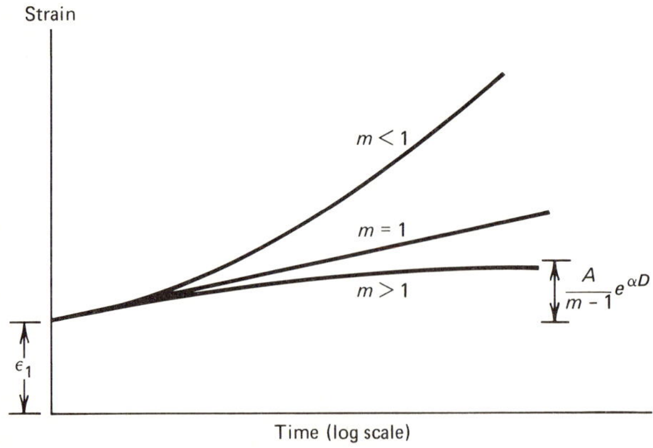

Figure 1.12 Creep curves predicted by Eqs. (1.3) and (1.4). *(After Mitchell, Ref. 14.)*

given in Table 1.2. It should be noted that the actual values for the particular soils being evaluated must be experimentally determined for each specific case.

Thus the creep rate or deformation can be calculated for any time and stress state for the soil in question. The procedure is quite useful in comparing the results to field monitoring measurements taken using standard techniques, e.g., slope inclinometers or extensometers. See Chap. 9.

1.2.7 Ground Subsidence

While ground subsidence caused by a subsurface disturbance has long been recognized as being devastating to structures, the subject has only recently been approached on a unified and concentrated basis.[17-19] Shown in Table 1.3 are some of the various mechanisms along with their range of dimensions and the approximate mobilization time for such mechanisms to occur.[20]

Regarding actual occurrences vis-à-vis a shallow foundation design, the most troublesome subsidence events are small in scale and generally related to human activity. This is primarily because of liquid or solid extraction from the soil or rock in the immediate vicinity. It thus becomes important to postulate the nature and extent of the extraction and then to project the impact of this withdrawal to the ground surface or the structure in question. Figure 1.13 presents the various elements of ground movement due to solids (mineral) extraction and its impact on the ground

Table 1.2 Representative values for constants in creep equations governing fine-grained-soil behavior†

	Creep equation constants		
	m	$\bar{\alpha}\ddagger$	A, %/min
Determination	Slope of log time vs. log strain rate response	Slope of deviator stress vs. log strain rate response	y intercept at zero stress of deviator stress vs. log strain rate response
Primary dependence:	Soil type and stress level	Shear strength and deviator stress	Clay content and soil type
Typical values†:			
San Francisco Bay mud, remolded	0.73	5.4	0.088
San Francisco Bay mud, undisturbed	0.81	4.5	0.30
Kaolinite	0.75	5.6	0.040
Ygnacio Valley clay	0.90	5.0	0.15
Grundite (illite and quartz)	0.88	2.2	0.32
Osaka alluvial clay	1.0	5.5	0.008

†Values provided by J. K. Mitchell, Univ. of Calif., Berkeley, Calif.

‡$\bar{\alpha}$ values, where $\bar{\alpha} = \alpha D_{max}$.

Table 1.3 Subsidence mechanisms along with typical dimensions and time for mobilization†

Mechanism	Dimension, km	Time, years
Crustal isostatic adjustments	1000–100	10,000
Plate tectonics, volcanic	· · ·	100–10
Large-scale water withdrawal	· · ·	100–10
Earthquake	· · ·	10^{-6}–10^{-7}
Fluid withdrawal	100–10	100–10
Liquid or solids extraction	10–1	100–10
Surface loading, consolidation, hydrocompaction	1–0.1	100–10
Vibration densification	1–0.1	10^{-6}–10^{-7}
Compaction	0.1–0.01	0.1

†After Scott, Ref. 20.

surface.[21] This type of physical model has generated many analytic and quasianalytic attempts at predicting ground movement and, in some cases, the time for such movement to occur. Singh[22] categorizes and references both mine subsidence and geofluid (oil and water extraction) subsidence models.

A related type of ground subsidence due to natural causes occurs in areas of Karst topography; see Sec. 1.2.9.

1.2.8 Collapsing Soils (Including Loess)

Collapsing soils (also called *metastable soils*) are those dry or partially saturated soils which undergo significant loss of total volume upon wetting. Obviously, structures founded on such soils will settle in proportion to the amount of volume change that the subsurface soil undergoes. The volume change comes about via a large-scale particle rearrangement from the in situ conditions. Some typical initial soil structures which can easily collapse are shown schematically in Fig. 1.14. Easily seen is that the primary soil structure consists of bulky-sized particles (sands and silts) which are weakly bonded by water, clays, or silts in a relatively unstable manner. Upon wetting, or vibration, collapse and large volume decrease will occur. Such soil situations are widespread, as indicated by a number of review articles on the subject.[24-26]

The phenomenon is typified by a certain type of soil, and its structure, called *loess*. Loess is a wind blown (aeolian) deposit of silt-sized particles occurring in many areas of Europe, Russia, China, and the United States; see Fig. 1.15. The deposits accumulate in areas of high humidity, often near major rivers, causing deposition at depths of 3 to 15 ft but occasionally up to 50 ft. Loess is often characterized by a high void ratio (up to 1.50) and a low density (as little as 70 lb/ft³). The situation can be evaluated quantitatively by the standard consolidation test. Knight[28] suggests consolidation of the undisturbed sample up to 30 lb/in², then saturation for 24 h, followed by the next higher load increment. The void ratio decrease during this incre-

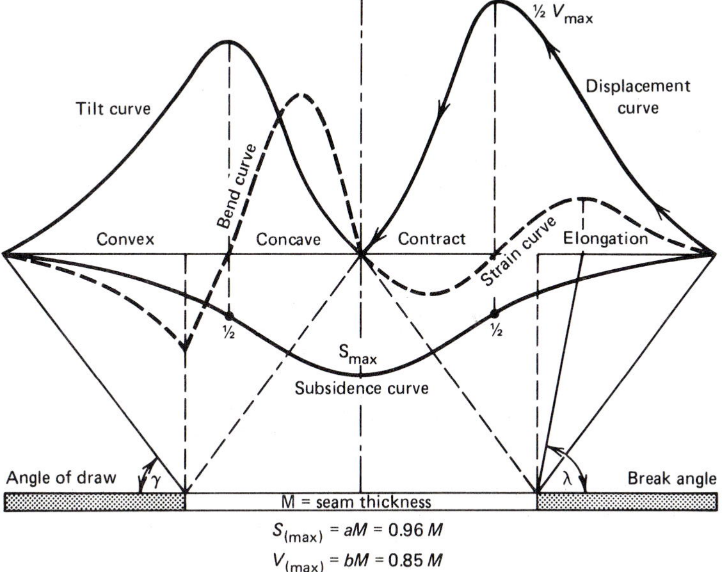

$$S_{(max)} = aM = 0.96\,M$$
$$V_{(max)} = bM = 0.85\,M$$

Elements of ground movement
Horizontal movement (displacement) elongation, contraction
Vertical movement (subsidence) tilt, bend.

The following applies to the separate ground movement elements:

Subsidence	Maximal in the center of the excavation zone
Tilt	First derivative of the subsidence, maximum over the edge of the excavation zone and zero in the center of it
Curvature	Second derivative of the subsidence, zero over the edge of the excavation zone, convex over the marginal zone and concave toward the center of the excavation area
Elongation and contraction	First derivative of the displacement; over the edge of the excavation zone the displacement is maximal and transition from elongation to contraction takes place; over the center of the zone of excavation displacement is zero and no contraction occurs

Figure 1.13 Hypothetical ground movement, i.e., subsidence trough, from solid ore extraction. *(After Pottgens, Ref. 21.)*

ment is indicative of the collapse potential (CP) as follows:

$$CP = \frac{\Delta e_c}{1 + e_0} = \frac{\Delta H_c}{H_0} \tag{1.5}$$

where CP = collapse potential
Δe_c = void ratio change upon wetting
e_0 = initial void ratio

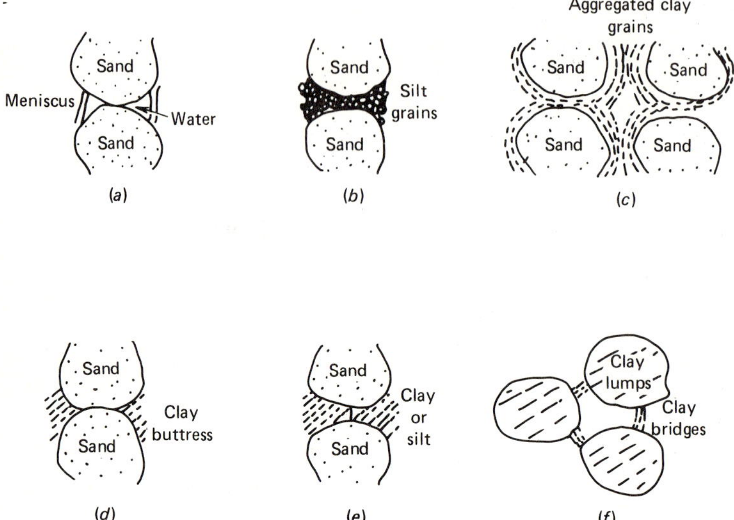

Figure 1.14 Soil structure types which are prone to collapse upon wetting. (*a*) Capillary tension; (*b*) silt bond; (*c*) aggregated clay bond; (*d*) flocculated clay bond; (*e*) mud flow type of separation; (*f*) clay bridge structure. *(After Barden et al., Ref. 23.)*

Figure 1.15 Location of major loess deposits in the United States. *(After Gibbs and Holland, Ref. 27.)*

20

ΔH_c = specimen height change upon wetting

H_0 = initial specimen height

Numerical values are given in Table 1.4.

The usual construction remedies for building sites where loess is encountered are as follows:

1. Excavation and replacement of the soil. For shallow surface deposits, this option is well suited since the soil in itself is not unsuitable, only its structure must be modified. Relative density and unit weight are probably the best control parameters during replacement.
2. Wetting the in situ soil (hydroconsolidation). For thick deposits of collapse-prone soils, prewetting or flooding the soil can be used.[29] Here the unstable structure is purposely broken down before construction begins. The efficiency of this remedy must be verified by in situ tests.
3. Surface densification via vibration. For relatively shallow deposits, vibratory surface rollers can be used; for deeper deposits, deep dynamic compaction methods can be used.
4. In situ densification via vibration. For deep deposits, Vibroflotation or Terra-Probe densification (both vibratory methods) can be used.
5. Deep foundations. As a last resort, deep foundations (piles or caissons) placed through the collapsing soils to firm underlying strata can be used.

Although some design examples of shallow foundation systems founded directly on collapsing soils are available,[26] it is generally not a recommended direction to follow. Thus, some alteration of the in situ soil should be made prior to construction.

1.2.9 Rockhead Formations (Including Karst Formations)

A most important consideration for shallow foundation siting in cases where the separation between soil and rock is very abrupt is the nature of the rockhead geometry, i.e., the shape of the surface of the rock. Rockhead formation is often assumed to be uniform, but this is rarely the case. Some of the more severe, and potentially troublesome, cases are illustrated in Fig. 1.16.[30] Such situations underscore the

Table 1.4 Collapse potential values†

Collapse potential (CP), %	Severity of problem
0–1	No problem
1–5	Moderate trouble
5–10	Trouble
10–20	Severe trouble
>20	Very severe trouble

†After Knight, Ref. 28.

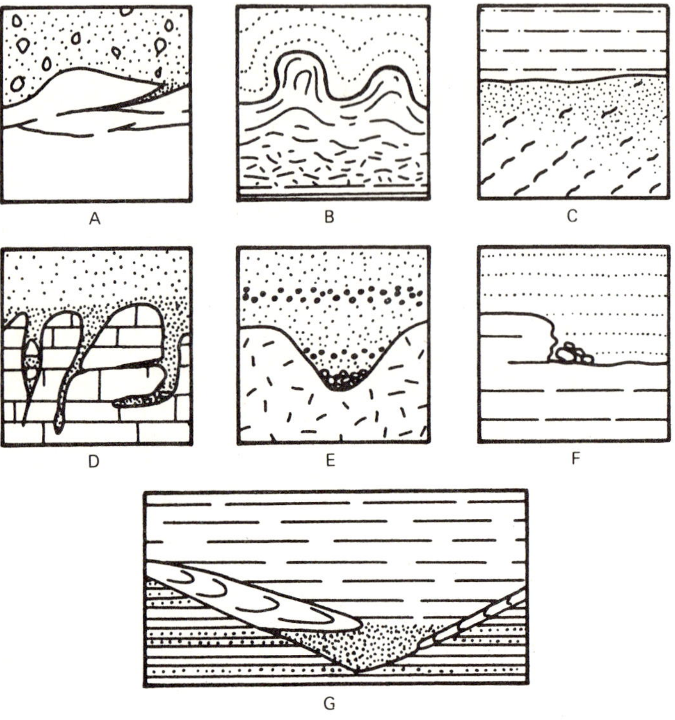

Figure 1.16 Differing rockhead situations. (*a*) Till on rock with glacially induced shears containing till infilling; (*b*) bulged periglacially disturbed rockhead; (*c*) weathered rock below rockhead; (*d*) solution-affected limestone; (*e*) buried channel with alluvial fill; (*f*) buried wind-eroded and undercut rock face; (*g*) buried channel with slope failures on flanks. (*After Knill, Ref. 30.*)

importance of an understanding of the area geologically and a detailed subsurface investigation of the bedrock situation. Nowhere is this more significant than in solution-affected limestone rock, often called *Karst topography.*[31]

Rocks composed of calcium and magnesium carbonates, known generally as *limestone,* comprise over 10 percent of the rocks exposed at the earth's surface.[32] As far as foundations are concerned, there are two significant features of limestones:

They are soluble in water, resulting in deep, randomly oriented channels and cavities.
They are generally covered by nonsoluble soils that hide the voids, which gives a false
impression of their solidity.

Sowers[32] has characterized some typical situations of limestone-activated collapse due to subsidence and raveling mechanisms; see Fig. 1.17.

Usually construction techniques in areas of soluble limestones involve filling exposed voids in the rock with a lean concrete and then constructing a large, heavily reinforced footing. Consideration should be given to combined footings or mat foun-

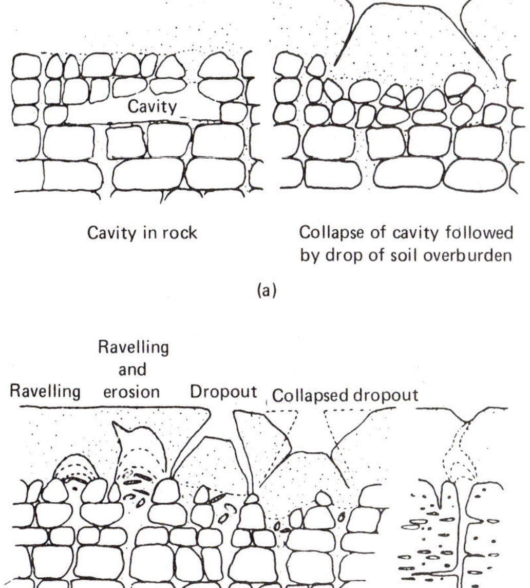

Figure 1.17 Solution limestone situations showing subsidence and raveling collapse mechanisms. *(After Sowers, Ref. 32.)*

dations. For cases where there are large depths of unsuitable overburden over the soluble limestone, and piles or caissons are called for, great care must be exercised. Better quality piles should be considered along with very conservative pile loads. Whatever the foundation design, it is imperative that the final drainage and grading layout of the site be designed so that both subsurface and ground water movement is away from the limestone rock. In that way solution will not occur during the lifetime of the structure.

1.2.10 Underground Facilities

In urban areas many subsurface structures have been installed, have served useful functions in their day, and then have long been forgotten. Included in this category are old water pipes, masonry sewers, old septic tanks, manholes and catch basins, tunnels, etc. All of these must be avoided, or excavated and properly backfilled, if adequate foundations are to be provided for construction projects in the areas where they exist.

Information about the existence and location of these facilities can be found in municipal government archives, with an occasional valuable assist from a local "old timer" who has a good memory. Most municipalities have good records of underground facilities currently in service, which obviously must be investigated.

Because of the cost of borings and the possibility of damaging a service facility by drilling through it, the use of some type of close-in geophysical or nondestructive testing method should be considered. These methods include:

Magnetometer
Metal detector
Ground-probing radar
Very-low-frequency electromagnetic detection
Continuous microwave penetration

They will be discussed more fully in Sec. 9.4.

1.2.11 Adjacent Structures

The proximity of adjacent structures to a given site development scheme always poses some questions. The obvious ones involve new excavations close to and deeper than the foundations of the already existing structures. Various retaining systems for the existing structures will be discussed in Sec. 3.3. More subtle questions arise because of additional stresses being imposed on subsurface strata by the new foundation loads or because of changes in unit weight due to the lowering of ground water in the new excavation. Both actions have been known to cause building distress (cracks, distortions, etc.), particularly in old masonry structures. In such cases a preconstruction building survey of all adjacent structures (by an independent agency) is highly recommended.

1.3 BEARING CAPACITY THEORY

The very essence of design and analysis of shallow foundations is the concept of *bearing capacity*. In this regard, reference is usually made to the strength of the soil (occasionally rock) beneath the footing in resisting a shear stress failure. Such failures have indeed occurred and the consequences are devastating; see Fig. 1.18. The usual design approach taken to avoid such situations is to calculate a factor of safety against failure. Numerically, this is obtained by comparing the soil strength against failure to the average pressure exerted by the footing on the soil. Required, then, is a knowledge of the type of anticipated failure and the appropriate theory for computational purposes.

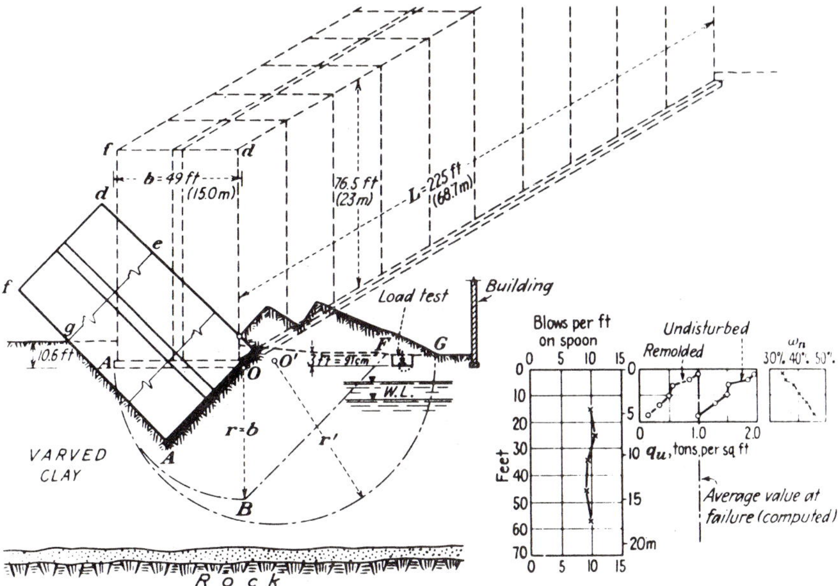

Figure 1.18 General shear-type bearing capacity failure of a set of interconnected grain storage silos. *(After Tschebatorioff, Ref. 33.)*

Regarding the general modes of failure that a shallow foundation can undergo, model tests have shown that three types are possible (see Fig. 1.19):

General shear. Characterized by distinct slip planes, large surface heave, and a well-defined maximum load in a load-versus-settlement curve. It is usually associated with dense granular and overconsolidated cohesive soils.

Local shear. Characterized by slip zones, small surface heave, and a poorly defined maximum load on a load-versus-settlement curve. It is usually associated with intermediate-density granular and normally consolidated cohesive soils.

Punching shear. Characterized by large deformations beneath the footing and little movement adjacent to it. There is no surface heave and no adequately defined load-versus-settlement curve. It is usually associated with loose granular and sensitive cohesive soils.

Most theoretical work has focused on the general shear type of failure. This is rationalized on the basis that the amount of settlement necessary for local or punching shear types of failure to occur is already excessive for most structures, and because analytic solutions for those cases are not yet available.

In regard to the general shear type of failure, the original work was done by Prandtl.[35] However, his theory was applied to a punch forcing its way through a metal plate rather than to a soils problem. It was Terzaghi[3] who recognized its appli-

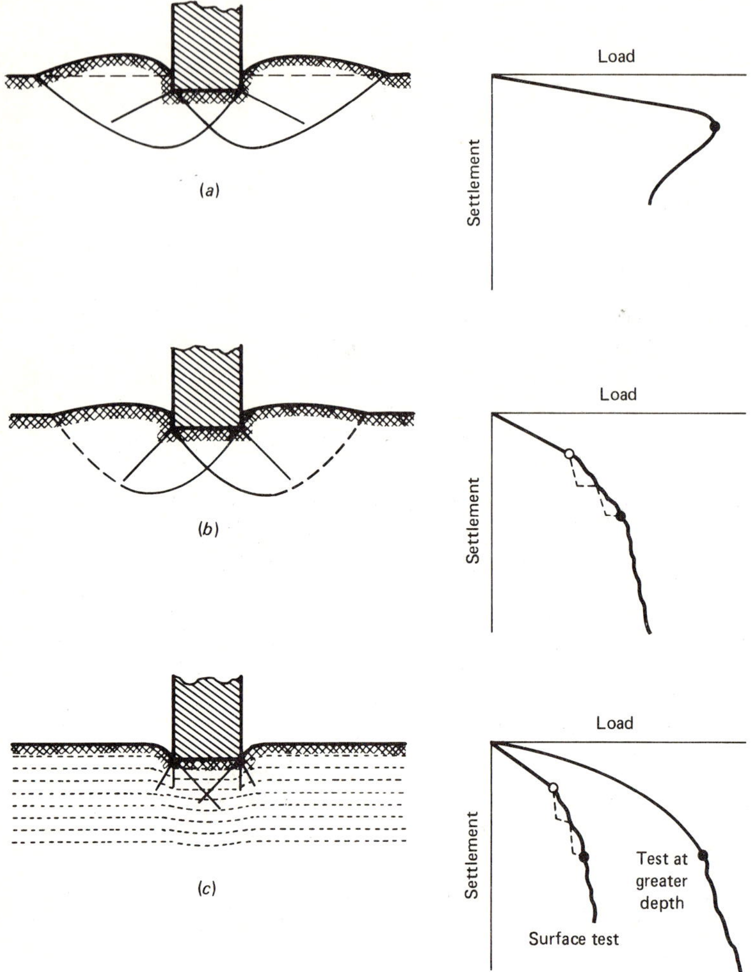

Figure 1.19 General modes of bearing capacity failures. (*a*) General shear; (*b*) local shear; (*c*) punching shear. (*After Vesic, Ref. 34.*)

cability to shallow foundations and modified Prandtl's solution to include the weight of the soil in the configuration shown in Fig. 1.20. By assuming four distinct areas of soil mass within the failure planes, and coupling it with the further assumption that the soil is homogeneous, isotropic, and semi-infinite in its depth, the following equation results as the ultimate bearing capacity of a uniform-contact-pressure, infinitely long, strip footing (i.e., a "plane strain problem" in the terminology of mechanics):

$$q_0 = cN_c + \gamma_2 D_f N_q + \tfrac{1}{2}\gamma_1 B N_\gamma \tag{1.6}$$

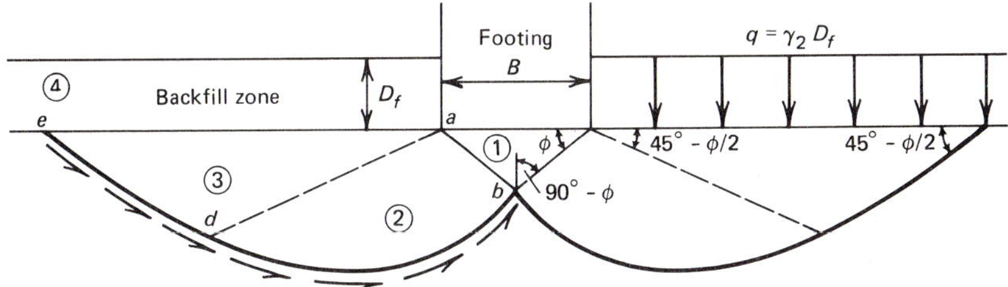

Figure 1.20 General shear bearing capacity scheme used by Terzaghi. Areas within failing soil mass: (1) elastic wedge; (2) passive log spiral area; (3) passive Rankine area; (4) backfill area acting as surcharge but possessing no shear strength. *(After Terzaghi, Ref. 3.)*

where q_0 = bearing capacity of soil

$\quad\quad c$ = cohesion

$\quad\quad \gamma_1$ = unit weight of soil beneath footing

$\quad\quad \gamma_2$ = unit weight of backfill soil

$\quad\quad D_f$ = depth of footing beneath ground surface (some references use $q = \gamma_2 D_f$, where q is the equivalent surcharge load caused by the backfill soil)

$\quad\quad B$ = footing width

$\quad\quad N_q = \tan^2(45° + \phi/2)e^{\pi \tan \phi}$

$\quad\quad N_c = (N_q - 1)\cot \phi$

$\quad\quad N_\gamma = 2(N_q + 1)\tan \phi$

$\quad\quad \phi$ = angle of shearing resistance

In the above formulation the N_c and N_q terms come about from the original solution and are derived in a straightforward manner (see Bazant[36]). The manner of including the weight of the soil (the N_γ term) is not as well defined (see Vesic[37]); Vesic has tabulated these bearing capacity factors for values of ϕ from 0 to 50° and they are given in Table 1.5. They are based on an initial failure surface beginning at an angle $45° + \phi/2$ for the elastic wedge rather than ϕ as shown in Fig. 1.20. The former has been shown experimentally to be a more correct assumption.

The effect of footings which are not in keeping with the plane strain assumption of being infinitely long can be considered by making suitable modifications to each term in Eq. (1.6). These modifications are based on laboratory tests, in comparison with strip footing, and are given in Table 1.6.

The effect of load inclination, i.e., inclination at other than a vertical angle, can be considered by a suitable modification to the general shear bearing capacity equation. Craig[38] presents the following formulation of the problem where α is the angle of load inclination from the vertical and ϕ is the angle of shearing resistance of the soil:

$$i_\gamma = \left(1 - \frac{\alpha}{\phi}\right)^2 \tag{1.7}$$

$$i_c = i_q = \left(1 - \frac{\alpha}{90°}\right)^2 \tag{1.8}$$

Table 1.5 Tabulated bearing capacity factors†

ϕ	N_c	N_q	N_γ	N_q/N_c	$\tan\phi$
0	5.14	1.00	0.00	0.20	0.00
1	5.38	1.09	0.07	0.20	0.02
2	5.63	1.20	0.15	0.21	0.03
3	5.90	1.31	0.24	0.22	0.05
4	6.19	1.43	0.34	0.23	0.07
5	6.49	1.57	0.45	0.24	0.09
6	6.81	1.72	0.57	0.25	0.11
7	7.16	1.88	0.71	0.26	0.12
8	7.53	2.06	0.86	0.27	0.14
9	7.92	2.25	1.03	0.28	0.16
10	8.35	2.47	1.22	0.30	0.18
11	8.80	2.71	1.44	0.31	0.19
12	9.28	2.97	1.69	0.32	0.21
13	9.81	3.26	1.97	0.33	0.23
14	10.37	3.59	2.29	0.35	0.25
15	10.98	3.94	2.65	0.36	0.27
16	11.63	4.34	3.06	0.37	0.29
17	12.34	4.77	3.53	0.39	0.31
18	13.10	5.26	4.07	0.40	0.32
19	13.93	5.80	4.68	0.42	0.34
20	14.83	6.40	5.39	0.43	0.36
21	15.82	7.07	6.20	0.45	0.38
22	16.88	7.82	7.13	0.46	0.40
23	18.05	8.66	8.20	0.48	0.42
24	19.32	9.60	9.44	0.50	0.45
25	20.72	10.66	10.88	0.51	0.47
26	22.25	11.85	12.54	0.53	0.49
27	23.94	13.20	14.47	0.55	0.51
28	25.80	14.72	16.72	0.57	0.53
29	27.86	16.44	19.34	0.59	0.55
30	30.14	18.40	22.40	0.61	0.58
31	32.67	20.63	25.99	0.63	0.60
32	35.49	23.18	30.22	0.65	0.62
33	38.64	26.09	35.19	0.68	0.65
34	42.16	29.44	41.06	0.70	0.67
35	46.12	33.30	48.03	0.72	0.70
36	50.59	37.75	56.31	0.75	0.73
37	55.63	42.92	66.19	0.77	0.75
38	61.35	48.93	78.03	0.80	0.78
39	67.87	55.96	92.25	0.82	0.81
40	75.31	64.20	109.41	0.85	0.84
41	83.86	73.90	130.22	0.88	0.87
42	93.71	85.38	155.55	0.91	0.90
43	105.11	99.02	186.54	0.94	0.93
44	118.37	115.31	224.64	0.97	0.97
45	133.88	134.88	271.76	1.01	1.00

Table 1.5 Tabulated bearing capacity factors†
(*Continued*)

ϕ	N_c	N_q	N_γ	N_q/N_c	$\tan\phi$
46	152.10	158.51	330.35	1.04	1.04
47	173.64	187.21	403.67	1.08	1.07
48	199.26	222.31	496.01	1.12	1.11
49	229.93	265.51	613.16	1.15	1.15
50	266.89	319.07	762.89	1.20	1.19

†After Vesic, Ref. 37.

where these terms are multiplied by the bearing capacity factors N_γ, N_c, and N_q, respectively.

The effect of eccentrically applied loads to a footing is a serious departure from the general design concept that load resultant and footing centroid should be coincidental. As such, severe penalties are imposed on any footing design where this is done. Meyerhof[39] indicates that the effective footing area should be computed using the following reduced footing dimensions:

$$B' = B - 2e_B \tag{1.9}$$

$$L' = L - 2e_L \tag{1.10}$$

$$A' = B'L' \tag{1.11}$$

where B = actual footing width
$\quad B'$ = hypothetical footing width
$\quad e_B$ = load eccentricity in B direction
$\quad L$ = actual footing length
$\quad L'$ = hypothetical footing length
$\quad e_L$ = load eccentricity in L direction
$\quad A'$ = hypothetical reduced footing area

The impact of the above procedure is to place the eccentrically applied load concentric with the hypothetical footing. Calculations easily show that this has a serious impact in footing design.

Table 1.6 Factors to include in calculations of footing shape for bearing capacity†

Footing shape	Factor to be multiplied to bearing capacity term		
	N_c	N_q	N_γ
Continuous	1.00	1.00	1.00
Rectangle	$1 + (B/L)(N_q/N_c)$	$1 + (B/L)\tan\phi$	$1 - 0.4(B/L)$
Square or circle	$1 + (N_q/N_c)$	$1 + \tan\phi$	0.60

†After Vesic, Ref. 37.

For soil conditions which result in a local shear (or worse, a punching shear) condition, a reduced set of shear strength parameters is used and then procedures follow along standard lines. For example, Terzaghi[3] recommends using the general shear theory with the following reduced parameters:

$$c' = \tfrac{2}{3}c \qquad (1.12)$$

$$\tan \phi' = \tfrac{2}{3} \tan \phi \qquad (1.13)$$

where c' and $\tan \phi'$ = reduced parameters

c and $\tan \phi$ = actual test parameters

In such cases, however, footing settlement may well be the controlling feature in the design process.

The value of the bearing capacity calculated in the manner just described is considered to be the ultimate or greatest value in regard to the load the soil can withstand, i.e., it means a factor of safety (FS) of only 1. Obviously, for design purposes this value must be reduced to an allowable bearing capacity in accordance with a number of considerations:

Adequacy of subsurface soil exploration programs
Adequacy of soil testing procedures
Type of structure to be built
Significance of a failure if one should result
Reliability of design load, magnitude, and type

The usual factors of safety to be applied to computed bearing capacity range from 2.0 to 4.0, with a value of 3.0 recommended for general-purpose applications. Thus, for design, one uses the following equation:

$$q_a = \frac{q_0}{\text{FS}} \qquad (1.14)$$

where q_a = allowable bearing capacity

q_0 = ultimate bearing capacity

When one considers the time and expense involved in obtaining a statistically reliable value of allowable bearing capacity in a given situation, it is not surprising that a number of empirical methods are available (see Appendix B). These empirical methods, called *presumptive bearing capacities,* are aimed at evaluating the allowable bearing capacity on the basis of limited information. Quite often they are based on standard penetration test (SPT) values and, as such, reduce the entire design process to selecting a value from a chart or deducing it from a simple equation. In general, such procedures are not recommended for design purposes.

In closing this section, it should also be noted that many urban areas have building codes which limit the allowable bearing capacity to a certain value. These values are generally thought to have been developed on the basis of local experience but their origin is often lost in the past. Whatever the value is, it does represent the maximum permissible value, but it should be checked for reasonableness according to the procedures outlined in this section.

Example 1.1 Determine the ultimate bearing capacity and the allowable bearing capacity based on a factor of safety of 3.0, for the square footing foundation shown on the figure.

SOLUTION Bearing capacity factors:

$$N_c = 23.9 \qquad N_q = 13.2 \qquad N_\gamma = 14.5$$

Shape factors

For N_c:
$$1 + \frac{N_q}{N_c} = 1.55$$

For N_q:
$$1 + \tan \phi = 1.51$$

For N_γ:
$$= 0.60$$

Therefore, from Eq. (1.6) and (1.14),

$$q_0 = (200)(23.9)(1.55) + (115)(2.0)(13.2)(1.51)$$
$$+ (0.5)(122)(4)(14.5)(0.60)$$
$$= 7400 + 4580 + 2120$$
$$= 14{,}100 \text{ lb/ft}^2 \qquad (\text{that is, } Q_0 = q_0 A_{\text{base}} = 225 \text{ kips})$$

$$q_a = \frac{q_0}{\text{FS}}$$
$$= \frac{14{,}100}{3.0}$$
$$= 4700 \text{ lb/ft}^2 \text{ (that is, } Q_a = 75 \text{ kips)}$$

1.4 SHALLOW FOUNDATION DESIGN

Shallow foundation design means that the proper footing dimensions are determined so as not to overstress the soil beneath the footing. To do this, one must know the allowable bearing capacity of the subsurface soil (refer to Sec. 1.3) along with the imposed loads to be carried by the footing. These two values are then integrated into the basic design philosophy that *the resultant of the applied loads be coincident with the centroid of the footing* to arrive at the proper footing dimensions. Obviously, the basic laws of statics must be obeyed during the process. The subsections to follow will give specific cases.

1.4.1 Spread Footings

Spread footings are usually square or rectangular in area and support an individual load due to a column, tower, bearing plate, caisson, or other quasiconcentrated load scheme. Figure 1.21 shows the usual configuration where the design load and allowable soil bearing capacity are known (from the structural and geotechnical engineer, respectively) and the footing thickness must be estimated. Typical thicknesses are 18 to 30 in for light to heavy footing loads. Thus we have the relationship

$$A_{\text{req'd}} = \frac{Q_{\text{design}}}{q_{\text{allow}} - \text{footing weight}} \tag{1.15}$$

Example 1.2 As an example, assume

$$Q_{\text{design}} = 100{,}000 \text{ lb}$$

$$q_{\text{allow}} = 3000 \text{ lb/ft}^2$$

$$d = 30 \text{ in}$$

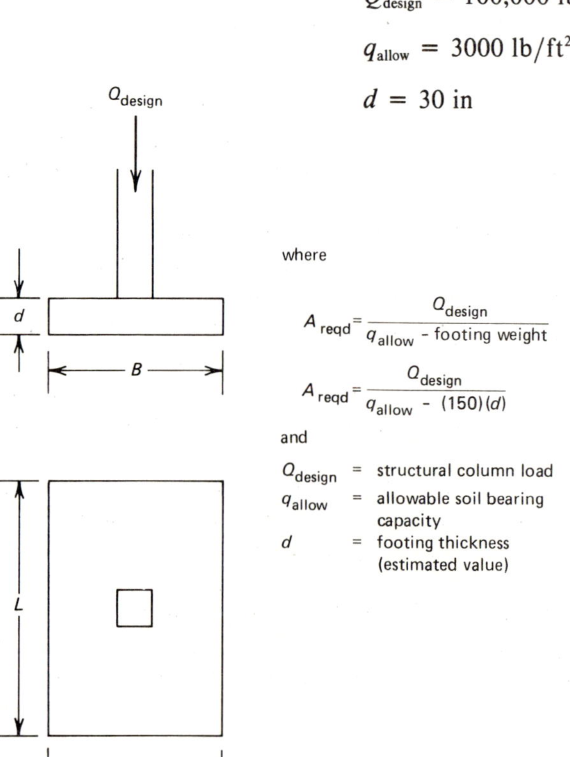

Figure 1.21 Typical spread footing configuration.

Then,

$$BL = \frac{100,000}{3000 - (150)(2.5)} = 38.1 \text{ ft}^2$$

So the possible footing combinations can be made into a schedule as shown below:

If B, ft:	Then L, ft:	And L/B
6.17	6.17	1.00
6.00	6.36	1.06
5.00	7.62	1.5
4.00	9.53	2.5
3.00	12.7	4.2

All of the above are acceptable solutions provided that no constraints, such as property lines, other footings, etc., are imposed on the problem.

1.4.2 Continuous Footings

Continuous footings are found beneath the exterior walls of buildings, retaining walls of various types, some machine foundations, etc., and are characterized by a length-to-width (L/B) ratio greater than 5; see Fig. 1.22. The design philosophy is exactly the same as with spread footings except that the units are in force/unit length.

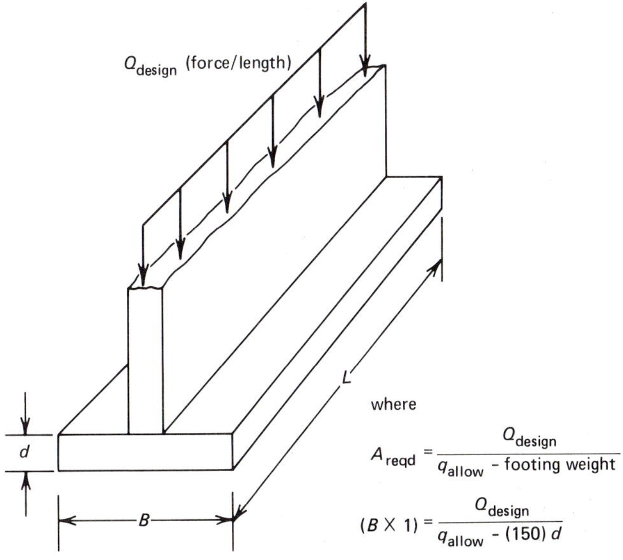

Figure 1.22 Typical continuous footing configuration.

Example 1.3 As an example, assume a structural wall load is 10,000 lb/ft of wall length and the allowable soil bearing capacity is 4000 lb/ft^2 with $d = $ 24 in.

Then, the unknown footing width is

$$B \times 1 = \frac{10,000}{4000 - (150)(2)}$$

$$B = 2.7 \text{ ft}$$

1.4.3 Combined Footings

When a number of structural loads frame into the same footing, it is called a *combined footing*. In this case the load resultant must be calculated and made coincident with the centroid of the footing. The procedure is illustrated in the following examples.

Example 1.4a Given two column loads of 300 and 150 kips, respectively (1 kip = 1000 lb), an allowable soil-bearing capacity of 4000 lb/ft^2 and an assumed footing thickness of 36 in, design a combined footing of width 5.00 ft (see Fig. 1.23).

As before, the required area is determined:

$$A_{\text{req'd}} = \frac{300,000 + 150,000}{4000 - 150(3.0)} = 127 \text{ ft}^2$$

And for a required $B = 5.0$ ft:

$$L_{\text{req'd}} = \frac{127}{5.0} = 25.4 \text{ ft}$$

Now, to properly position this footing, the resultant is determined by taking moments about the 300-kip load:

$$x = \frac{150,000(11.00 + 0.83 + 0.58)}{450,000}$$

$$= 4.1 \text{ ft from 300-kip column}$$

Using one-half of the length (25.4/2 = 12.7 ft), the footing extends 12.7 − 4.1 = 8.6 ft beyond the 300-kip column. See Fig. 1.23 for the final dimensions.

Example 1.4b Given the same problem described in Example 1.4a, except that a property line is 5 in from the edge of the 14-in column, design the footing.

As before, $A_{\text{reqd}} = 127$ ft^2, but now the width cannot be chosen randomly. The length can be determined, however, since the load resultant is 8.3 ft from the 150-kip column. It is 8.3 + 0.42 + 0.58 = 9.3 ft from the property line,

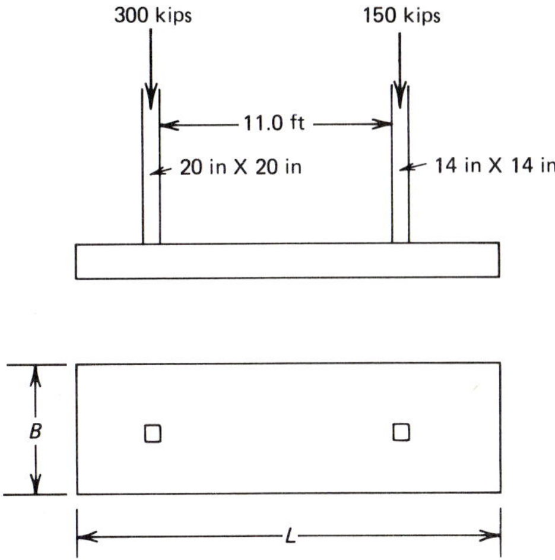

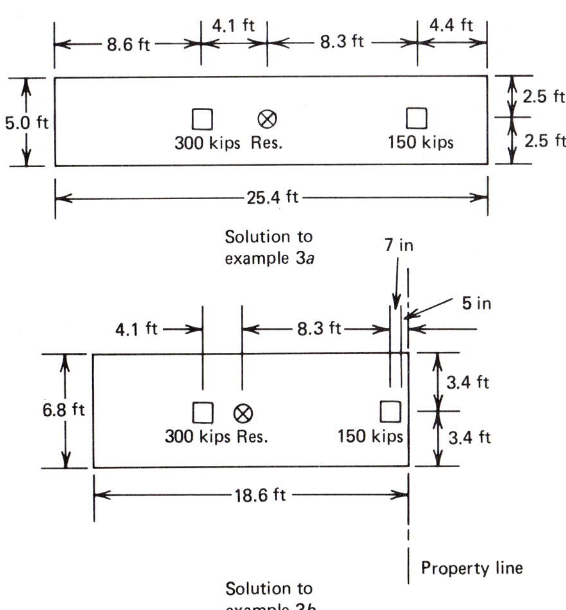

Figure 1.23 Combined footing configuration and design example.

which is now one-half of the footing length. Thus

$$L = 2(9.3) = 18.6 \text{ ft}$$

and

$$B = \frac{A_{\text{req'd}}}{L} = \frac{127}{18.6} = 6.8 \text{ ft}$$

For both of these design examples shear and moment diagrams can now be drawn. The common assumption is that the footing acts as a rigid unit and that the soil pressure exerts a uniform and upward load against the footing. This assumption results in maximum shear values beneath the columns and maximum moment between the columns. Figure 1.24 illustrates the procedure.

It should be noted that there are a number of different styles of combined footings which have been developed over the years to address specific problems. They include:

Trapezoidal footings: For situations where one column is carrying a much higher load than another.

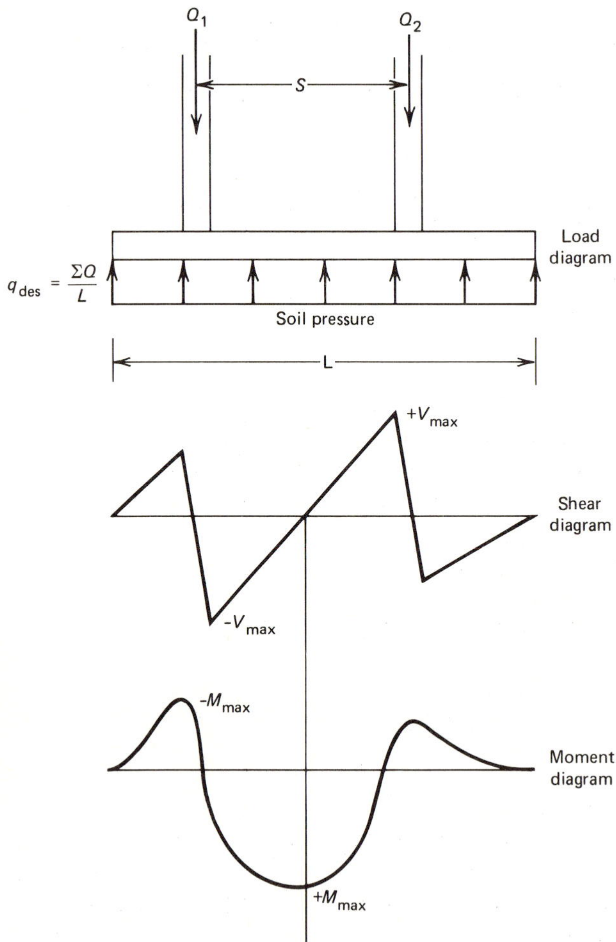

Figure 1.24 Common method used to obtain maximum shear and moment in combined footing design (assumes a rigid footing).

Strap footings: For situations where columns are widely spaced and all that is required for continuity is reinforcement for moment resistance.

Strap footings with void beneath strap: To further reduce the negative moment (see Fig. 1.24), it was common in the past to disturb the soil beneath the strap—not the soil beneath the column footings—to relieve the soil pressure in this area.

1.4.4 Mat Footings

For situations where the site area is more than half covered by individual spread footings, continuous footings, and combined footings, the foundation designer should consider the use of a mat footing. Such a mat footing, or mat foundation, will then cover the entire site (or a major portion of it) and act as an integral unit. The trade-offs are reduction in formwork versus increased concrete and reinforcement, which are obviously site-specific economic decisions. On a more technical basis, however, mat footings should be considered whenever the foundation soil is weak or nonuniform in its composition; recall Secs. 1.2.5 through 1.2.10.

Standard mat foundation design assumes a rigid footing with a uniform soil pressure resisting the structural loads. For the illustration of Fig. 1.25 the soil pres sure is

$$q_{\text{design}} = \frac{\Sigma Q_i + yLd\gamma_c}{y} \tag{1.16}$$

where q_{design} = uniform soil pressure on footing strip in question

ΣQ_i = total column loads in a strip

y, B, L, d = footing dimensions

γ_c = unit weight of reinforced concrete ($\simeq 150$ lb/ft^3)

To obtain a solution, the footing is arbitrarily divided into strips of width y and designing proceeds as with a combined footing. Shears and moments are calculated as though this strip acted independently of the rest of the foundation, as shown schematically in Fig. 1.25. A concept[40] recently introduced in regard to mat footings (and also in regard to combined footings) is the use of posttensioned, prestressed concrete in areas of nonuniform-soil subgrade support, which aids in a better distribution of column and wall loads.

It should be noted, however, that the above-mentioned design procedures are quite conservative, i.e., the calculated footing thickness and required reinforcement will be larger than actually necessary. The alternative method is to consider footing flexibility, which will be discussed in the following section.

1.4.5 Footing Flexibility

The preceding design methods all assumed that the footing was perfectly rigid in resisting the structural loads on top of it and the soil pressures beneath it. In most cases this is not so. It is a particularly poor assumption in regard to firm soils, where loads are widely spaced and unequal and where the footing thickness is relatively small.

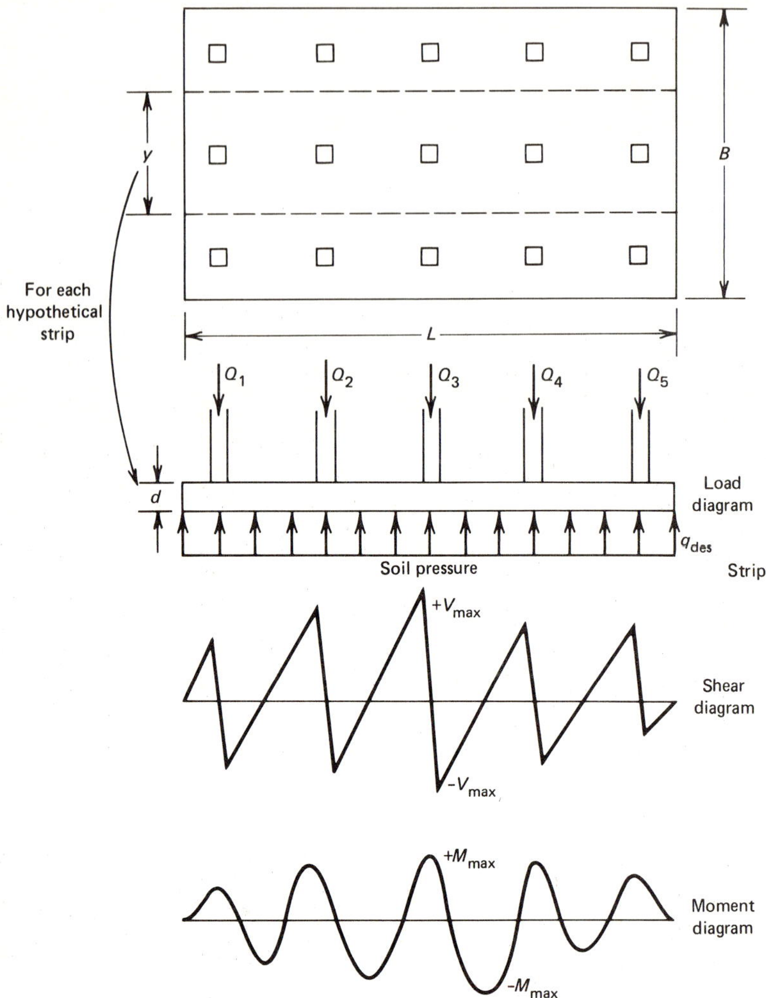

Figure 1.25 Standard mat foundation design scheme which assumes a rigid footing and uniform soil bearing capacity.

In order to determine if, and how, flexible a footing is, there are two approaches which can be followed. One is based on the fundamental characteristic of the governing differential equation upon which the flexibility analysis is based.[41] That approach is as follows:

$$\lambda L = \sqrt{\frac{K_s L^4}{4EI}} \tag{1.17}$$

where $K_s = kB$

k = coefficient of subgrade reaction as determined by a plate bearing test (ASTM—D119, D1196)

B = footing width
L = footing length
E = footing modulus of elasticity
I = footing moment of inertia
λL = dimensionless characteristic, where if $\lambda L \leq \pi/4$, the footing is rigid; if $\pi/4 < \lambda L < \pi$, the footing is somewhat flexible; and if $\lambda L \geq \pi$, the footing is very flexible

The other approach, due to Vesic,[42] subdivides the area into beams ($L/B > 5$) and mats ($L/B < 5$). For footing beams, the flexibility criterion is

$$\alpha = \frac{E_s L^4}{E_c I_c} \tag{1.18}$$

where E_s = modulus of elasticity of soil
L = length of beam
E_c = modulus of elasticity of concrete
$I_c = \frac{1}{12} Bd^3$ = moment of inertia of beam
B = width of beam
d = thickness of beam
α = dimensionless criterion number, where if $\alpha \leq 5$, the footing beam is rigid; if $5 < \alpha < 60$, the footing beam is somewhat flexible; and if $\alpha \geq 60$, the footing beam is very flexible

For footing mats, the flexibility criterion is

$$\bar{\alpha} = \frac{E_s B L^3}{E_c I_c} \tag{1.19}$$

where E_s = modulus of elasticity of soil
B = width of mat
L = length of mat
E_c = modulus of elasticity of concrete
$I_c = \frac{1}{12} Bd^3$ = moment of inertia of mat
d = thickness of mat
$\bar{\alpha}$ = dimensionless criterion number, where if $\bar{\alpha} \leq 1$, the footing mat is rigid; if $1 < \bar{\alpha} < 12$, the footing mat is somewhat flexible; and if $\bar{\alpha} \geq 12$, the footing mat is very flexible

Irrespective of the approach taken to test for footing flexibility, it will be seen that most situations place the footing in the "somewhat flexible" or "very flexible" categories. Thus by using a rigid-design approach, as is customary, we are almost always overdesigning our footings! While this uses extra concrete and reinforcement in footings, the alternative design approach considering flexibility is quite complex and requires a considerable amount of design time. Bowles[4] covers the subject nicely for footing beams (including computer programs) and Timoshenko and Wionowsky-Krieger[43] cover the basic elements of footing mat design.

1.4.6 Compensated and Raft Footings

As noted by Sowers,[1] the concept of excavating a weight of soil approximately equivalent to the weight of the proposed structure was first recognized in the 1700s by the English engineer Thomas Rennie. It was apparently forgotten, then redeveloped earlier in the twentieth century, and today it is practiced frequently. The power of the concept is that those soils which remain beneath the footing level are not stressed any more heavily by the newly constructed structure than they were by the now excavated overburden soils. This procedure avoids the possibility of bearing capacity failures and essentially eliminates postconstruction foundation settlements.

A variation on a fully compensated foundation is a partially compensated one where only a part of the structural load is offset by the weight of excavated soil.

Still another foundation concept is to utilize the buoyant effect of a stationary water table to offset the structural load system. Thus the creation of a "raft" or "floating" foundation. The essential elements of this type of foundation scheme are the following:

A stationary water table near the surface of the ground
A mat foundation which can be made watertight
An auxiliary pumping system within the structure to remove infiltrating water from it during the lifetime of the structure
A temporary wellpoint or seepage cutoff system to allow for the building of the foundation in the "dry"

The essence of the technique rests on the fact that the weight of the structure can be compensated for by the unit weight of water times the depth beneath the minimum elevation of the water table. The most notable success using this type of raft foundation is the Tower Latin America in Mexico City.[44]

The technique can be carried too far, for if light structural loads are placed beneath the water table they can easily become unstable and either rise up in elevation or tilt out of vertical alignment. Thus in some areas, such as those affected by tides, the use of ground anchors to hold down buried storage tanks when they are empty and the water table elevation rises is often necessary (see Sec. 3.5).

1.5 FOOTING SETTLEMENT

As noted earlier in this chapter, the design methodology for shallow foundations is a three-part process. The parts are:

Acquiring adequate knowledge of the general ground and subsurface conditions at, and near, the specific site (see Sec. 1.2)
Providing adequate resistance against bearing capacity failure (see Secs. 1.3 and 1.4)
Achieving tolerable footing settlements (this section)

It should first be recognized that all footings settle. Even footings on rock have small, but measurable, settlements due, if to nothing else, to elastic compression of the parent material. Recognizing this, it becomes the designer's duty to estimate the settlement that will occur, compare it with the allowable settlement for the structure under consideration, and report the findings (which involves conveying their meaning) to the client.

1.5.1 Immediate Settlement

For most of the soil situations encountered in the placement of shallow foundations, settlements can be estimated on the basis of classical elasticity theory.[3] The basic equation is as follows:

$$\rho = qB \frac{1 - \mu^2}{E} I_w \qquad (1.20)$$

where ρ = settlement
     q = contact pressure
     E = modulus of elasticity of soil (see Table 1.7)
     μ = Poisson's ratio of soil (see Table 1.8)
     I_w = influence factor depending upon footing geometry and location (see Table 1.9)

Table 1.7 Typical values for secant modulus of elasticity†

	E_s	
Soil	ksi	kg/cm²
Clay		
Very soft	0.05–0.4	3–30
Soft	0.2–0.6	20–40
Medium	0.6–1.2	45–90
Hard	1–3	70–200
Sandy	4–6	300–425
Glacial fill	1.5–22	100–1600
Loess	2–8	150–600
Sand		
Silty	1–3	50–200
Loose	1.5–3.5	100–250
Dense	7–12	500–1000
Sand and gravel		
Dense	14–28	800–2000
Loose	7–20	500–1400
Shales	20–2000	1400–14,000
Silt	0.3–3	20–200

†After Bowles, Ref. 4.

Table 1.8 Typical values for Poisson's ratio for various materials†

Type of soil	μ
Clay, saturated	0.4–0.5
Clay, unsaturated	0.1–0.3
Sandy clay	0.2–0.3
Silt	0.3–0.35
Sand (dense)	0.2–0.4
Coarse (void ratio = 0.4–0.7)	0.15
Fine-grained (void ratio = 0.4–0.7)	0.25
Rock	0.1–0.4 (depends on type of rock)
Loess	0.1–0.3
Ice	0.36
Concrete	0.15

†After Bowles, Ref. 4.

An adaptation of this analysis can be applied to the rotation of free-standing structures, e.g., to chimneys, towers, etc., as follows:

$$\tan \theta = \frac{Ve}{BL^2}\left(\frac{1 - \mu^2}{E}\right) I_m \tag{1.21a}$$

where θ = angular rotation of the footing

$Ve = M$ = applied moment causing rotation

E, μ = elastic constants (see Tables 1.7 and 1.8)

I_m = influence factor (see Table 1.9)

Table 1.9 Influence factors for footing settlement and rotation calculations†

Shape	Flexible			Rigid	
	Center	Corner	Average	I_w	I_m‡
Circle	1.00	0.64 (edge)	0.85	0.88§	6.0
Square	1.12	0.56	0.95	0.82	3.7
Rectangle:					
L/B = 0.2					2.29
0.5					3.33
1.5	1.36	0.68	1.15	1.06	4.12
2	1.53	0.77	1.30	1.20	4.38
5	2.10	1.05	1.83	1.70	4.82
10	2.54	1.27	2.25	2.10	4.93
100	4.01	2.00	3.69	3.40	5.06

†After Bowles, Ref. 4, and Lee, Ref. 45.

‡Lee, Ref. 45.

§Others have used the value 0.79 = $\pi/4$ for the rigid-footing influence factor for circular footings.

Use of the above equations is generally applicable to granular soils at all moisture contents from dry to saturated, and for fine-grained soils at moisture contents less than 70 percent. For these soils, load application and compression is very rapid. This is because of air permeability being essentially instantaneous and water permeability, in the case of saturated granular soils, being very high. For these reasons, the phenomenon is not time-dependent (practically speaking) and the theory generally overestimates the observed settlements in the field since measurements are often taken after construction has begun and whatever load is in place has already caused settlement. Nevertheless, the calculation is worthwhile to perform at least for an approximate value.

Example 1.5 Given a rigid 4-ft round shallow foundation exerting an average constant pressure of 1200 lb/ft^2 in a dense, silty, sand-soil, determine the immediate settlement:

$$\rho = qB \frac{1 - \mu^2}{E} I_w$$

$$= (1200)(4) \frac{1 - 0.3^2}{7000} (0.88) \left(\frac{1}{12}\right)$$

$$= 0.046 \text{ in}$$

Since footings are rarely placed on the ground surface, and the theory behind Eq. (1.20) was developed on this basis, an empirical correction is appropriate for footings at a depth. Fox[46] proposed such a correction, which has been solved by Bowles[4] and is given as Fig. 1.26. For Example 1.5, if the footing were located 4 ft beneath the ground surface, then D/B would be 1.0, and for $L/B = 1$ and $\mu = 0.3$, the correction factor would be 0.65 and the estimated immediate settlement 0.030 in.

As an alternative to the above method, it is quite common to use either the standard penetration test (described in Sec. 9.2.1) or the cone penetration test (described in Sec. 9.2.3) to obtain a limiting bearing capacity, which has been shown by experience to have a limiting settlement. This limiting settlement for granular soils has been arbitrarily set at 1.0 in.

For standard penetration test data, Teng[47] has proposed:

$$q_a = 0.720(N - 3) \left(\frac{B + 1}{2B}\right)^2 \tag{1.21b}$$

$$K_d = 1 + \frac{0.2D}{B} \tag{1.21c}$$

where q_a = allowable net increase in soil pressure for a maximum settlement of 1.0 in, kips/ft^2

B = footing width, ft

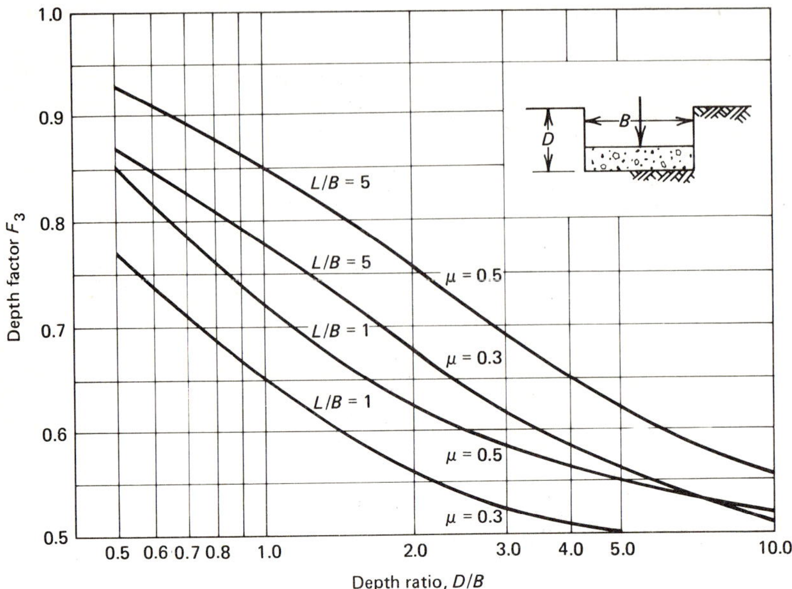

Figure 1.26 Correction factor to apply to immediate settlement calculation for footings located at depth D beneath ground surface. *(After Bowles, Ref. 4.)*

N = standard penetration blow count (blows/foot) corrected for footing embedment, if applicable

K_d = footing embedment correction factor (maximum value = 1.20)

D = depth of bury of footing, ft

For cone penetration data, Meyerhof[48] has a similar technique as follows:

$$q_a = \frac{q_c}{15} \qquad \text{for } B \leq 4 \qquad (1.21d)$$

$$q_a = \frac{q_c}{25} \left(\frac{B+1}{B}\right)^2 \qquad \text{for } B > 4 \qquad (1.21e)$$

where q_a = allowable net increase in soil pressure for a maximum settlement of 1.0 in, kips/ft^2

q_c = cone point resistance, kips/ft^2

B = footing width, ft

In each of these empirical procedures it is the allowable bearing capacity which is the focal point, the assumption being that if this value is within tolerable limits, the resulting settlement should be of no serious consequence.

1.5.2 Time-Dependent Settlement

The settlement of shallow foundations founded on saturated (or nearly saturated) fine-grained silts and clays creates many more problems than does immediate settlement. This is because of both the greater magnitudes of settlement involved and the time-dependent nature of the consolidation phenomenon. Originally developed and formalized by Terzaghi,[2,3] consolidation analysis forms a key role in geotechnical engineering. The process involves first calculating total settlement, then calculating the predicted time over which this settlement will occur. The equations for one-dimensional, vertical drainage consolidation are as follows:

$$\rho = H_1 \frac{C_c}{1 + e_0} \log \frac{p_1 + \Delta p}{p_1} \tag{1.22}$$

$$t = \frac{T_v H_2^2}{c_v} \tag{1.23}$$

where ρ = consolidation settlement

H_1 = thickness of compressible stratum

C_c = compression index (see Fig. 1.27)

e_0 = initial void ratio

$p_1 = \bar{\sigma}_v$ = effective vertical pressure at location where settlement is being calculated

Δp = increment of pressure caused by load at location where settlement is being calculated—usually obtained using Boussinesq or Westergaard elastic stress analysis procedures as found in most geotechnical engineering texts

T_v = dimensionless time factor which is a function of the percent of consolidation U_z, as follows:

U_z	10	20	30	40	50	60	70	80	90
T_v	0.008	0.031	0.071	0.126	0.197	0.287	0.403	0.567	0.848

H_2 = maximum drainage path length (= H_1 for single drainage; = $H_1/2$ for double drainage)

c_v = coefficient of consolidation (see Fig. 1.28)

In utilizing Eqs. (1.22) and (1.23), C_c and c_v are experimentally obtained from laboratory tests on undisturbed samples, p_1 and e_0 are obtained from soil property calculations, H_1 and H_2 from soil boring logs, and Δp from imposed load conditions; T_v is a constant that depends upon the specific percentages of ρ with which one may be concerned in any given case. The following example illustrates the procedure.

Example 1.6 Given a soil profile as shown below containing a normally consolidated Newfoundland silt (see consolidation test results of Fig. 1.28), a struc-

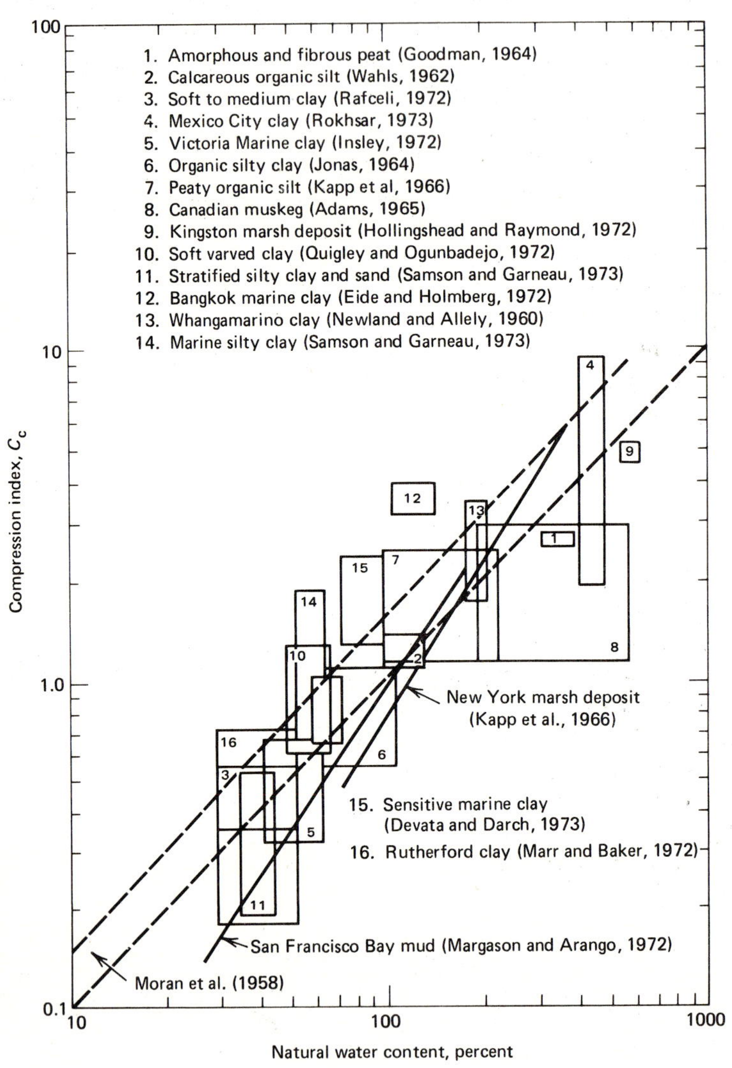

Figure 1.27 Typical value of compression index. *(After Mesri and Rokhsar, Ref. 49.)*

46

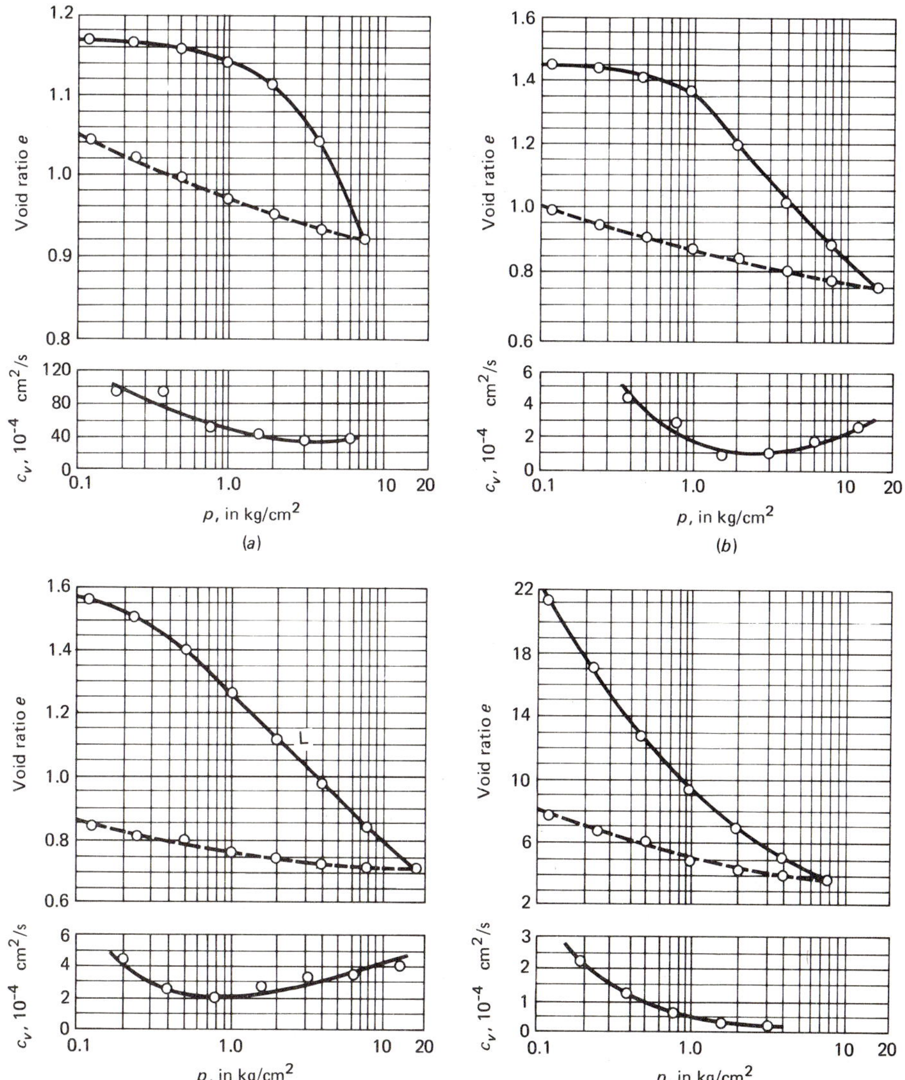

Figure 1.28 Consolidation characteristics of four undisturbed soils. (*a*) Boston blue clay; (*b*) Chicago clay; (*c*) Newfoundland silt; (*d*) Newfoundland peat. (*After Taylor, Ref. 50.*)

tural load of 700 lb/ft^2 at the midpoint of the compressible silt will be mobilized by the surface loading. Determine the settlement-time response under these conditions.

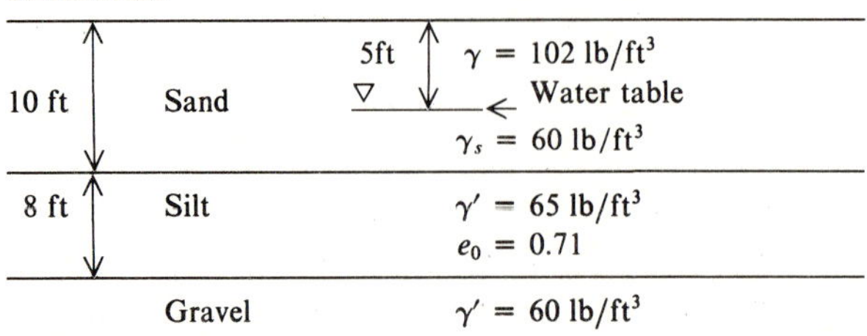

SOLUTION: From Fig. 1.28, $C_c = 0.43$ and $c_v \simeq 2.0 \times 10^{-4}$ cm^2/s. Also $p_1 = \bar{\sigma}_v = 5(102) + 5(65) + 4(60) = 1075$ lb/ft^2 and $\Delta p = 700$ lb/ft^2. So:

$$\rho = H_1 \frac{C_c}{1 + e_0} \log \frac{p_1 + \Delta p}{p_1}$$

$$= 8(12) \frac{0.43}{1 + 0.71} \log \frac{1075 + 700}{1075}$$

$$= 5.26 \text{ in at } 100\% \text{ consolidation}$$

The time for this settlement (double drainage) to occur is obtained as follows:

$$t = \frac{T_v H_2^2}{c_v}$$

$$= T_v \frac{(4 \times 12 \times 2.54)^2}{2.0 \times 10^{-4}} \frac{1}{60 \times 60 \times 24}$$

$$= 860 T_v \text{ days}$$

In tabular form, this time for consolidation can be distributed over the entire settlement range:

Percent consolidation	Actual settlement, in	Time factor T_v	Time for settlement, days
10	0.53	0.008	7
20	1.05	0.031	27
30	1.58	0.071	61
40	2.10	0.126	108
50	2.63	0.197	169
60	3.16	0.287	247
70	3.68	0.403	347
80	4.21	0.567	488
90	4.73	0.848	729
100	5.26	∞	∞

After the excess pore water pressure has been dissipated, some soils (particularly those with high organic content) continue to expel water and settle. This phenomenon is known as *secondary compression* and occurs over very long time periods. It is distinguished from creep deformation (recall Sec. 1.2.6) in that this situation has drainage whereas creep occurs in soils under undrained conditions. An approach toward estimating the magnitude of secondary compression is to use the laboratory consolidation test where deformation is plotted against the log of time well beyond primary consolidation. The slope of the resulting straight line can be used as follows:

$$\frac{\Delta H}{H} = C_s \log_{10} \frac{t_2}{t_1}$$

where $\Delta H = \rho_s =$ amount of secondary compression

$\quad\quad H =$ stratum thickness

$\quad\quad t_1 =$ time for primary consolidation to end, $\simeq t_{90}$

$\quad\quad t_2 =$ any specific later time

$\quad\quad C_s =$ coefficient of secondary compression (typical values are less than 0.001 for overconsolidated clays, 0.005 to 0.02 for normally consolidated clays, 0.03 or higher for sensitive clays and organic soils; see Lambe and Whitman, Ref. 51)

1.5.3 Allowable Settlement

In the past, many attempts have been made to stipulate a maximum or allowable settlement that a given type of structure can be subjected to. An example of one such

Table 1.10 Maximum allowable settlements based on type of movement[†]

Type of movement	Limiting factor	Maximum allowable settlement
Total settlement	Drainage and access	6 to 24 in (15 to 60 cm)
	Probability of differential settlement	
	Masonry walls	1 to 2 in (2.5 to 5 cm)
	Framed buildings	2 to 4 in (5 to 10 cm)
Tilting	Towers, stacks	$0.004B$[‡][§]
	Rolling of trucks, stacking of goods	$0.01S$[‡][§]
	Crane rails	$0.003S$[‡][§]
Curvature	Brick walls in buildings	$0.0005S$ to $0.002S$[‡][§]
	Reinforced concrete building frame	$0.003S$[‡][§]
	Steel building frame, continuous	$0.002S$[‡][§]
	Steel building frame, simple	$0.005S$[‡][§]

[†]After Sowers and Sowers, Ref. 52.

[‡]B is base width; S is column spacing.

[§]Differential settlement in distance B or S.

Table 1.11 Settlement criteria for building damage

Category	Description	Maximum radius of curvature[†]
Architectural	Cracking	1/300
Structural	Strength reduction	1/150
Functional	Impaired use	1/50

[†]Radius of curvature $= \rho/s$, where $\rho =$ settlement and $s =$ column spacing.

attempt, made by G. B. Sowers and G. F. Sowers,[52] is shown in Table 1.10. While there is a certain amount of utility in such guidelines, classification by category seems to be the current research direction. Nowhere is the problem more widespread than in regard to buildings where the categories of damage and their maximum limits have been identified as shown in Table 1.11.

Two sources (Refs. 53 and 54) are available which, basing their data on numerous case histories of building settlements, allow one to estimate maximum allowable settlement (see Table 1.12). Thus, for example to prevent architectural cracking in a building on spread footings in fine-grained soil, the maximum allowable settlement according to Ref. 54 (Grant et al.) would be 1200 (1/300) = 4.0 in. For the same situation in granular soil, the maximum settlement according to the same source would be 600 (1/300) = 2.0 in.

With time and the accumulation of case histories on settlements of other structures in addition to buildings, the procedure will become more refined and provide better insight into the maximum settlements that various structures can withstand.

Note, however, that it is the differential settlement, i.e., the ρ/s values used in Tables 1.11 and 1.12, that is of critical concern. Such differential settlement shows up immediately in wall cracks, windows and doors being stuck open or closed, utility problems, floors and ceilings buckling or pulling apart, etc.

Table 1.12 Maximum allowable building settlements for various foundation types and soil conditions

Foundation type	Soil type	Maximum allowable settlement	
		Skempton and MacDonald[†]	Grant et al.[‡]
Isolated footings or piles	Granular	$600(\rho/s)_{max}$[§]	$600(\rho/s)_{max}$
Isolated footings or piles	Fine-grained	$1000(\rho/s)_{max}$	$1200(\rho/s)_{max}$
Mat foundations	Granular	$750(\rho/s)_{max}$	$750(\rho/s)_{max}$
Mat foundations	Fine-grained	$1250(\rho/s)_{max}$	$1250(\rho/s)_{max}$

[†]Ref. 53.
[‡]Ref. 54.
[§]$\rho =$ settlement, $s =$ column spacing.

PROBLEMS

1.1 Consider the problem of protecting a 5-ft-wide by 50-ft-long bridge pier footing placed in moving water from the potential scour problem that might occur.

(*a*) Sketch some solutions to the problem on the basis of physical protection schemes if the water is a maximum of 5 to 10 ft deep.

(*b*) Sketch some solutions to the problems using flow attenuators or upstream barriers if the water is a maximum of 20 to 30 ft deep.

1.2 Classify soils 1 through 5 according to the unified soil classification system using the values of consistency (Atterberg) limits shown below and rank them from most to least sensitive as far as volume change is concerned for your climate.

Soil	Liquid limit w_l, %	Plastic limit w_p, %	Shrinkage limit w_s, %
1	12	7	14
2	55	21	10
3	54	30	10
4	200	40	8
5	38	14	12

1.3 Calculate the creep strain of a kaolinite clay on the basis of a constant initial creep movement and for a $\overline{D}$ of 25, 50, and 75 percent of D_{max}. Show your resulting three curves on a graph of time (x axis) versus creep strain (y axis).

1.4 On the map of the United States shown in Fig. 1.15, it can be seen that a concentration of loess follows the eastern side of the Mississippi River throughout its course. Explain why this is the situation.

1.5 Determine the allowable bearing capacity for a shallow, 4-ft-wide continuous footing using a factor of safety of 3.0 as a function of depth beneath the ground surface from 0 to 10 ft. Use $\gamma = 122$ lb/ft³, $c = 200$ lb/ft², $\phi = 27°$, and $\gamma(b'$ fill$) = 115$ lb/ft³, and assume general shear failure.

1.6 Determine the allowable bearing capacity for a 4-ft-wide continuous footing using a factor of safety of 3.0 and a depth beneath the ground surface of 2.0 ft for the soil conditions listed below. In all cases backfill density is 115 lb/ft³.

Soil condition	Unit weight, lb/ft³	Cohesion c, lb/ft²	Friction angle ϕ, deg
In situ	115	150	25
Compacted using surface rollers	119	200	30
Compacted using deep dynamic compaction	123	250	36

1.7 (*a*) Repeat Prob. 1.6 on the basis of the footing being rectangular and of size $B = 4$ ft and $L = 20$ ft.

(*b*) Repeat Prob. 1.6 on the basis of the footing being square and of size 4 × 4 ft.

1.8 (*a*) Determine the allowable bearing capacity of a continuous footing 5 ft wide and placed on the ground surface for load angle inclinations varying from vertical to 45° from the vertical. Assume general shear failure and use $c = 500$ lb/ft², $\phi = 30°$, $\gamma = 123$ lb/ft³, and a factor of safety of 2.5. Plot your results on a curve with angle of load inclination as the x axis and allowable bearing capacity as the y axis.

(*b*) Repeat Prob. 1.8*a* for a footing that is placed 3 ft beneath the ground surface and is backfilled with soil weighing 112 lb/ft³.

(*c*) Repeat Prob. 1.8*a* for a footing that is placed 10 ft beneath the ground surface and is backfilled with soil weighing 112 lb/ft^3.

(*d*) Plot the results for parts *b* and *c* on a curve having the same *x* and *y* axes as the curve in part *a*.

1.9 Based on FS = 2.0, what is the allowable bearing capacity of a 5 x 20 ft rectangular footing placed 3.5 ft beneath the ground surface on an ML soil that is best typified by a local shear failure. Design parameters are γ = 115 lb/ft^3, γ(*b′* fill) = 105 lb/ft^3, *c* = 450 lb/ft^2, and ϕ = 28°.

1.10 Of necessity, a vertical load will be applied to a 10 × 15 ft rectangular footing (see Fig. P1.10) in an eccentric manner in both directions. What is the allowable load that it can carry based on FS = 2.0 and parameters D_f = 0 ft, *c* = 600 lb/ft^2, and ϕ = 21°, γ = 118 lb/ft^2.

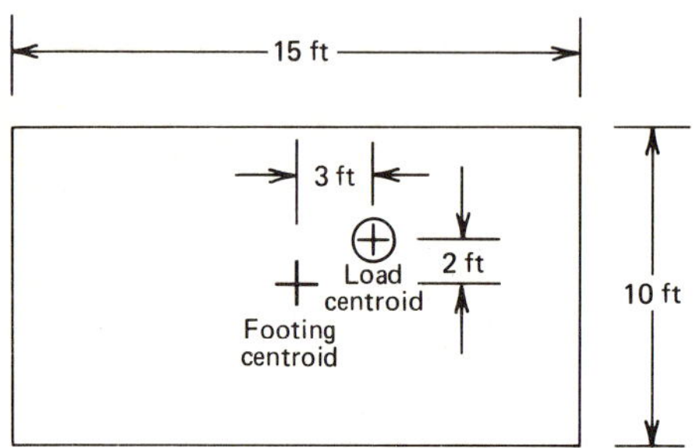

Figure P1.10 Plan view of footing.

1.11 If the ultimate bearing capacity of a soil is 10,000 lb/ft^2 and the FS = 3.0, what square footing size is required to support a 75-kip design load. The footing is estimated to be 24 in thick.

1.12 (*a*) Calculate a footing schedule, i.e., various acceptable *B*/*L* values, for the support of an 80-ton load on soil having an allowable bearing capacity of 4000 lb/ft^2. The footing will be 30 in thick.

(*b*) Repeat Prob. 1.12*a* with allowable bearing capacities of 2000 lb/ft^2, 6000 lb/ft^2, and 8000 lb/ft^2.

(*c*) Plot the results of a *B*/*L* ratio of 1.0, i.e., of a square footing, as *B* versus allowable bearing capacity of the soil.

1.13 (*a*) What will be the required width of a long, concrete masonry wall footing carrying a vertical load of 8 kips/ft on a soil of 3 kips/ft^2 allowable bearing load? The footing weight has already been deducted from the above figure.

(*b*) What would be the width if the load were offset by 9 in from the footing centroid?

1.14 (*a*) Given two column loads of 100 and 150 kips as shown in Fig. P1.14*a* and a soil having an

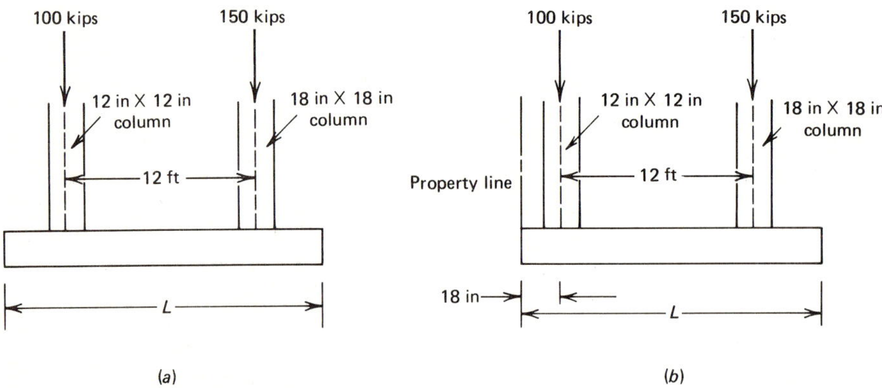

Figure P1.14 (*a*) Width is 4.0 ft; (*b*) property line.

allowable bearing pressure of 3500 lb/ft³ (which includes the footing weight). What length footing is required if the width is 4.0 ft? Also, draw the shear and moment diagrams.

(*b*) What will be the length and width of the footing if it is now restricted by a property line as shown in Fig. P1.14*b*.

1.15. Given three columns in a line, as shown in Fig. P1.15. For a width of 3.5 ft, what is the required length, if the allowable soil bearing pressure is 4500 lb/ft² and the footing thickness is 30 in? Determine the maximum shear and the bending moment by drawing the V and M diagrams.

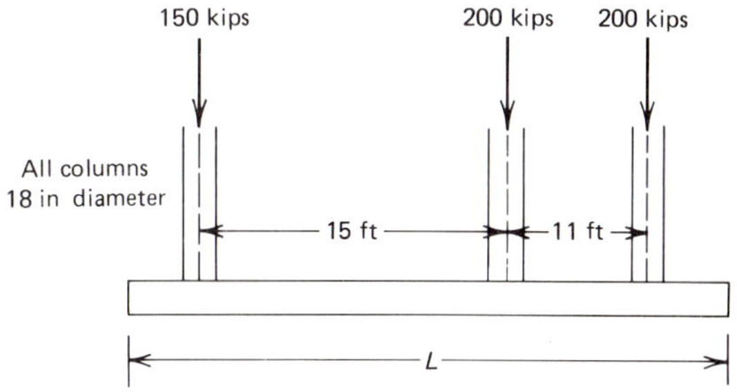

Figure P1.15

1.16 For the mat foundation shown in Fig. 1.25 where Q_1 and Q_5 column lines are 100 tons, Q_2, Q_3, and Q_4 column lines are 225 tons, and all columns are 18 in in diameter and spaced at 17.5-ft centers, draw the shear and moment diagrams (*a*) in the lengthwise (*L*) direction and (*b*) in the widthwise (*B*) direction. Use $L = 85$ ft, $B = 45$ ft, $t = 30$ in, and an allowable soil bearing pressure of 2.1 ton/ft² (includes footing weight).

1.17 For a continuous concrete footing of $L = 30$ ft, $B = 5$ ft, $t = 2$ ft, soil modulus of 1000 lb/in², and concrete modulus of 3×10^6 lb/in², determine if the footing is rigid, somewhat flexible, or very flexible according to Eq. (1.18).

1.18 (*a*) For the same set of conditions given in Prob. 1.17, except that L varies from 25 to 60 ft, plot α versus L/B and label the appropriate areas.

(*b*) Again, for the same set of conditions given in Prob. 1.17, except that this time t varies from 1 to 6 ft, plot α versus t and label the appropriate areas.

(*c*) For both $L = 30$ ft and $t = 2$ ft again as originally, plot the curve α versus E_s and label the appropriate areas when the modulus of the soil varies from 100 to 5000 lb/in².

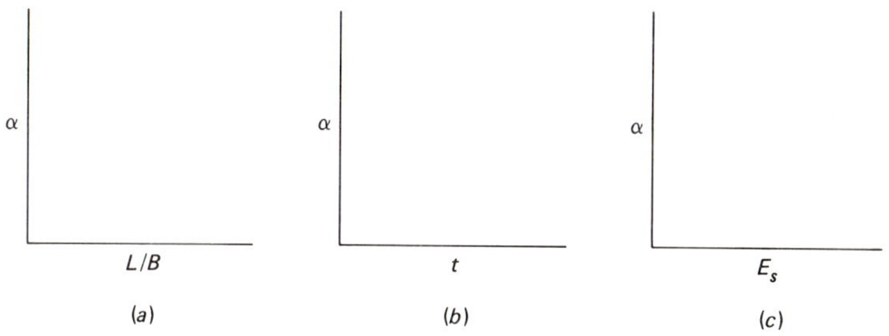

1.19 A 40 x 50 x 3.5 ft concrete mat foundation rests on soil having a modulus of 1500 lb/in². Determine if the foundation is rigid, somewhat flexible, or very flexible according to Eq. (1.19).

1.20 (*a*) For the same set of conditions as given in Prob. 19, except that L varies from 20 to 120 ft, plot the $\bar{\alpha}$ versus L/B response and label the appropriate areas.

(*b*) Plot $\bar{\alpha}$ versus t and label the areas when $L = 50$ ft and t varies from 1 to 7 ft

(*c*) Plot $\bar{\alpha}$ versus E_s and areas when $L = 50$ ft, $t = 3.5$ ft, and the modulus of the soil varies from 100 to 5000 lb/in².

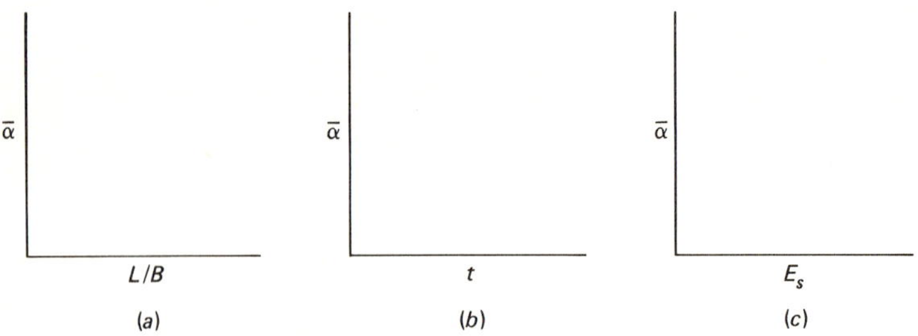

1.21 Calculate the immediate settlement of a 5 x 25 ft footing having a contact pressure of 2000 lb/ft^2 if the soil is

    (*a*) Soft clay

    (*b*) Silty sand

    (*c*) Dense sand

(Use average values of E and μ from Tables 1.7 and 1.8, respectively.)

1.22 If the footings discussed in Prob. 21 were 5 ft beneath the ground surface, what would the settlement be?

1.23 Calculate the primary consolidation settlement of a 20-ft-thick Boston blue clay layer under the conditions shown in Fig. P1.23*a* assuming that (*a*) a permeable layer exists beneath the clay and (*b*) an impermeable layer exists beneath the clay, and plot the results as shown in Fig. P1.23*b*.

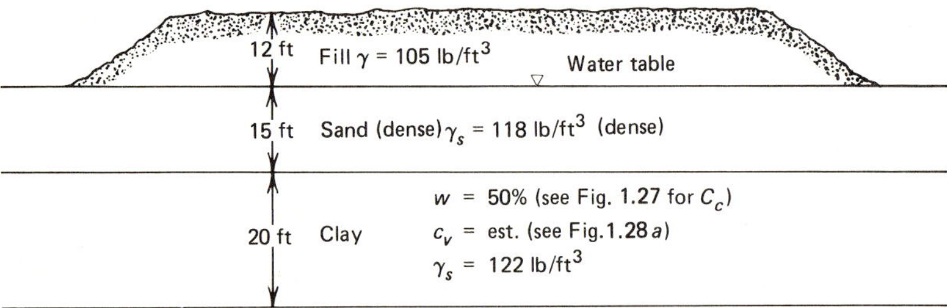

Figure P1.23*a*

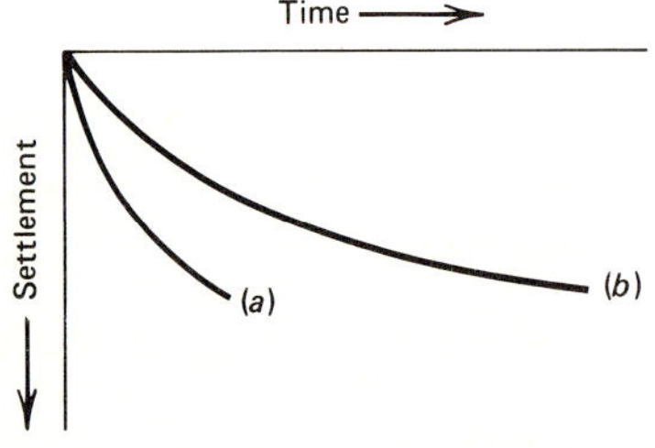

Figure P1.23*b*

1.24 For a building with a column spacing of 25 ft, and using data in Table 1.11 as criteria, what amount of differential settlement would result in

    (*a*) Structural cracking

    (*b*) Structural strength reduction

    (*c*) Impaired use of the structure

1.25 (*a*) For isolated footings on fine-grained soils, what is the limiting settlement according to Skempton and MacDonald (see Table 1.12) that will prevent structural damage?

    (*b*) If a mat foundation were used instead, what would be the maximum allowable settlement?

REFERENCES

1. Sowers, G. F., "Shallow Foundations," Chap. 6 in *Foundation Engineering,* G. A. Leonards (ed.), McGraw-Hill, New York, 1962, pp. 525–632.
2. Terzaghi, K., *Erdbaumechanik,* F. Deuticke, Leipzig and Wien, 1925.
3. Terzaghi, K., *Theoretical Soil Mechanics,* John Wiley and Sons, New York, 1943.
4. Bowles, J. E., *Foundation Design and Analysis,* McGraw-Hill, New York, 2d ed., 1977.
5. Sowers, G. B., and G. F. Sowers, *Introductory Soil Mechanics and Foundations,* Macmillan, New York, 1961.
6. Yong, R. N., and B. P. Warkentin, *Introduction to Soil Behavior,* Macmillan, New York, 1966.
7. Jumikis, A. R., *Thermal Soil Mechanics,* Rutgers, New Brunswick, NJ, 1966.
8. Tsytovich, N. A., in *The Mechanics of Frozen Ground,* G. K. Swinzow (ed.), Scripta, Washington, DC, 1975, pp. 55–58.
9. Andersland, O. B., and D. M. Anderson, *Geotechnical Engineering for Cold Regions,* McGraw-Hill, New York, 1978.
10. Thorson, B. M., and J. S. Braun, "Frost Heaves—A Major Dilemma for Ice Arenas," *Civil Eng., ASCE,* March 1975, pp. 62–64.
11. ———, *Scour at Bridge Waterways,* NCHRP Publication No. 5, Highway Research Board, National Academy of Engineering, Washington, DC, 1970.
12. Jones, D. E., and W. G. Holtz, "Expansive Soils—the Hidden Disaster," *Civil Eng., ASCE,* August 1973, pp. 49–51 (see also related articles in the August 1973, October 1974, July 1974, and October 1978 issues).
13. Chen, F. H., *Foundations on Expansive Soils,* Elsevier, New York, 1975.
14. Singh, A., and J. K. Mitchell, "General Stress-Strain-Time Function for Soils," *J. Soil Mech. Found. Eng. Div., ASCE,* Vol. 94, No. SM1, January 1968, pp. 21–46.
15. Mitchell, J. K., "Fundamentals of Soil Behavior," John Wiley and Sons, New York, 1976.
16. Hirst, R. J., and J. K. Mitchell, "Compositional and Environmental Influences on the Stress-Strain-Time Behavior of Soils," Report No. TE 68-4, Bureau of Reclamation, U.S. Dept. of the Interior, University of California, Berkeley, April 1968.
17. ———, in *Proc. First Int. Symp. on Land Subsidence,* Int. Assoc. of Hydrological Sciences, Tokyo, 1969.
18. ———, in *Proc. Second Int. Symp. on Land Subsidence,* Int. Assoc. of Hydrological Sciences, Anaheim, 1976.
19. ———, in *Proc. Evaluation and Prediction of Subsidence,* S. K. Saxena (ed.), ASCE/Eng. Foundation Conference, Pensacola Beach, FL, January 1978.
20. Scott, R. F., "Subsidence—A Review," in ibid., pp. 1–25.
21. Pottgens, J. J. E., "Ground Movements by Coal Mining in the Netherlands," in ibid., pp. 267–282.
22. Singh, M. M., "Experience with Subsidence due to Mining," in ibid., pp. 92–112.
23. Barden, L., A. McGown, and K. Collins, "The Collapse Mechanism in Partly Saturated Soil, *Eng. Geol.* (Amsterdam), Vol. 20, 1973, pp. 49–60.
24. Sultan, H. A., "Collapsing Soils, State-of-the-Art," in *Proc. Seventh Int. Conf. on Soil Mech. and Found. Eng.,* Vol. 5, Mexico City, Mexico, 1969, pp. 1–28.
25. Dudley, J. H., "Review of Collapsing Soils," *J. Soil Mech. Found. Eng. Div., ASCE,* Vol. 96, No. SM3, May 1970, pp. 925–947.
26. Clemence, S. P., and A. I. Inbarr, "Design Considerations for Collapsible Soils," *J. Geotech. Eng. Div., ASCE,* Vol. 107, No. GT3, March 1981, pp. 305–317.
27. Gibbs, H. J., and W. Y. Holland, *Petrographic and Engineering Properties of Loess,* Engineering Monograph No. 28, U.S. Bureau of Reclamation, Denver, 1960.
28. Knight, K., "The Origin and Occurrence of Collapsing Soils," in *Proc. Third Reg. Conf. for Africa on Soil Mech. and Found., Eng.,* Vol. 1, Salisbury, So. Rhodesia, 1963, pp. 127–130.
29. Lobdell, G. T., "Hydroconsolidation Potential of Palouse Loess," *J. Geotech. Eng. Div., ASCE,* Vol. 107, No. GT6, June 1981, pp. 733–742.
30. Knill, J. L., "Geology and Foundations," Chap. 2 in *Foundation Engineering in Difficult Ground,* F. G. Bell (ed.), Newnes-Butterworths, London and Boston, 1978, pp. 116–131.

31. Herak, M., and V. T. Springfield (eds.), *Karst,* Elsevier, New York, 1972.
32. Sowers, G. F., "Failures in Limestones in Humid Subtropics," *J. Geotech. Eng. Div., ASCE,* Vol. 101, No. GT8, August 1975, pp. 771–787.
33. Tschebatorioff, G. P., *Soil Mechanics, Foundations, and Earth Structures,* McGraw-Hill, New York, 1951.
34. Vesic, A. S., "Bearing Capacity of Deep Foundations in Sand," *Highway Res. Rec.,* Vol. 39, 1963, pp. 112–153.
35. Prandtl, L., "Über die Harte plastischer Korper," Nachuchten der Ges. Wiss., Göttingen, Math.-Phys. Klass, 1920, pp. 74–85.
36. Bazant, A., *Methods of Foundation Engineering,* Elsevier, New York, 1979.
37. Vesic, A. S., "Bearing Capacity of Shallow Foundations," Chap. 3 in *Foundation Engineering Handbook,* H. F. Winterkorn and H.-Y. Fang (eds.), Van Nostrand Reinhold, New York, 1975.
38. Craig, R. F., *Soil Mechanics,* 2d ed., Van Nostrand Reinhold, New York, 1978.
39. Meyerhof, G. G., "The Bearing Capacity of Foundations under Eccentric and Inclined Loads," in *Proc. Third Int. Conf. on Soil Mech. and Found. Eng.,* Vol. 1, Zurich, Switzerland, 1953, pp. 440–445.
40. Cronin, H. J., "Post-Tensioned Foundations as Economical Alternative," *Civil Eng., ASCE,* March 1980, pp. 62–65.
41. Hetenyi, M., *Beams on Elastic Foundations,* University of Michigan, Ann Arbor, 1946.
42. Vesic, A. S., "Bending of Beams Resting on Isotropic Elastic Solid," *J. Eng. Mech. Div., ASCE,* Vol. 87, No. EM2, April 1961, pp. 35–53.
43. Timoshenko, S., and S. Wionowsky-Krieger, *Theory of Plates and Shells,* 2d ed., McGraw-Hill, New York, 1959.
44. Zeevaert, L., *Foundation Engineering for Difficult Subsoil Conditions,* Van Nostrand Reinhold, New York, 1972.
45. Lee, I. K., "Bearing Capacity of Foundations with Particular Reference to the Melbourne Area," *Proc. Inst. Eng., Australia,* vol. 12, 1962, pp. 283–291.
46. Fox, E. N., "The Mean Elastic Settlement of a Uniformly Loaded Area at a Depth Below the Ground Surface," in *Proc. Second Int. Conf. on Soil Mech. and Found., Eng.,* Vol. 1, Rotterdam, 1948, pp. 129–132.
47. Teng, W. C., *Foundation Design,* Prentice-Hall, Englewood Cliffs, NJ, 1962.
48. Meyerhof, G. G., "Penetration Tests and Bearing Capacity of Cohesionless Soils," *J. Soil Mech. Found. Eng. Div., ASCE,* Vol. 82, No. SM1, 1956, pp. 1–19.
49. Mesri, G., and Rokhsar, A., "Theory of Consolidation for Clays," *J. Geotech. Eng. Div., ASCE,* Vol. 100, No. GT8, August 1974, pp. 889–904.
50. Taylor, D. W., *Fundamentals of Soil Mechanics,* John Wiley and Sons, New York, 1949.
51. Lambe, T. W., and R. V. Whitman, *Soil Mechanics,* John Wiley and Sons, New York, 1969.
52. Sowers, G. B., and G. F. Sowers, *Introductory Soil Mechanics and Foundations,* 3d ed., Macmillan, New York, 1970.
53. Skempton, A. W., and D. H. McDonald, "The Allowable Settlement of Buildings," *Proc. Inst. Civil Eng.,* Vol. 5, Pt. 3, 1956, pp. 727–784.
54. Grant, R., J. T. Christian, and E. H. Vanmarcke, "Differential Settlement of Buildings," *J. Geotech. Eng. Div., ASCE,* Vol. 100, No. GT9, September 1974, pp. 973–992.

DEEP FOUNDATIONS

2.1 OVERVIEW

In this chapter, piles, piers, and caissons are described, with an emphasis on their design and construction. Thus, both geotechnical and construction engineering aspects are combined. Complete coverage, however, is difficult because of the wealth of material written on the subject over the years. And such material is still expanding since new systems are emerging continually, e.g., offshore drilling platforms in the 1970s, which are a great challenge from both the design and construction viewpoints. So for this chapter, those procedures were selected for discussion which seem to be most relevant to the topic and to form the essentials upon which an extended study into the area could be conducted.

2.2 HISTORY, CONCEPTS, AND TYPES

The design and construction of deep foundations for the purpose of transferring the weight of surface loads down through unsuitable soils to underlying firm bearing strata is an exciting and fascinating aspect of foundation engineering. However, it is far from a new technology. Pile-supported structures are known to have existed in prehistoric times, and references to cedar timber piles in Babylon can be found in the Bible. Later, the Romans used piles to support bridge piers, while in England shortly thereafter, the original Winchester Cathedral had oak piles supporting it. In the Middle Ages, pile foundations supported a wide assortment of structures in Venice and Holland to the point where a Dutch nursery rhyme identified not only the existence of pile foundations, but also the possibility of their failure.

Piles during these many years were made from trees, whose branches were trimmed, and were driven down in a reversed position, i.e., with the small diameter at the bottom, until penetration of the soil was no longer possible. This condition, known as *refusal,* was a combined function of the soil stratigraphy and the limits of the driving mechanism. Chellis[1] notes that driving was probably by hand mauls, hand-operated machine mauls, ratchet winch rams, treadmill drivers, water wheel drivers, and gang-operated rams.

The Industrial Revolution, with its utilization of steam power, changed the situation dramatically. The first steam-activated pile hammer (the forerunner of today's single-acting hammers) was invented by Nasmyth in 1845[1] and ushered in modern pile driving. Parallel to these developments, structures larger than those previously built, e.g., bridge piers and large buildings, made their appearance and required pile capacities greater than those already available, thus leading to the development of piers and caissons. Piles, piers, and caissons together make up the general topic of deep foundations as we will discuss it in this chapter.

The general uses of deep foundations in today's construction world are indeed many. Some are shown in Fig. 2.1, where a variety of pile situations are illustrated, and in Fig. 2.2, where types of deep foundations other than piles are illustrated. While the design and construction of pile foundations will be emphasized in this chapter, the methods discussed are applicable to the design and construction of all types of deep foundations since the basics of support required from soil (or rock) below and adjacent to foundations are similar for any system.

To distinguish the subject of this chapter from that of Chap. 1, it should be noted that:

Deep foundations have a least width generally less than their depth beneath the ground surface.

The shear strength of the soil adjacent to the deep foundation can, and often does, contribute to its stability.

2.3 TYPES OF DEEP FOUNDATIONS

The various types of deep foundations are generally designated as *piles, piers,* and *caissons,* although this terminology is by no means clear-cut these days. No longer can one rely upon a "pile" being a long, slender structural unit driven into the soil since many are large in diameter (up to 72 in) and placed in excavated holes or vibrated in place, or upon a caisson being a prefabricated unit placed in the soil as excavation proceeds since many are now poured in place in an augered and/or cased hole. Thus care in labeling the precise type of deep foundation, its fabrication, and placement method is necessary to properly describe a given situation. With this caution, we will proceed to discuss piles, piers, and caissons along with the construction equipment and methods required for their installation.

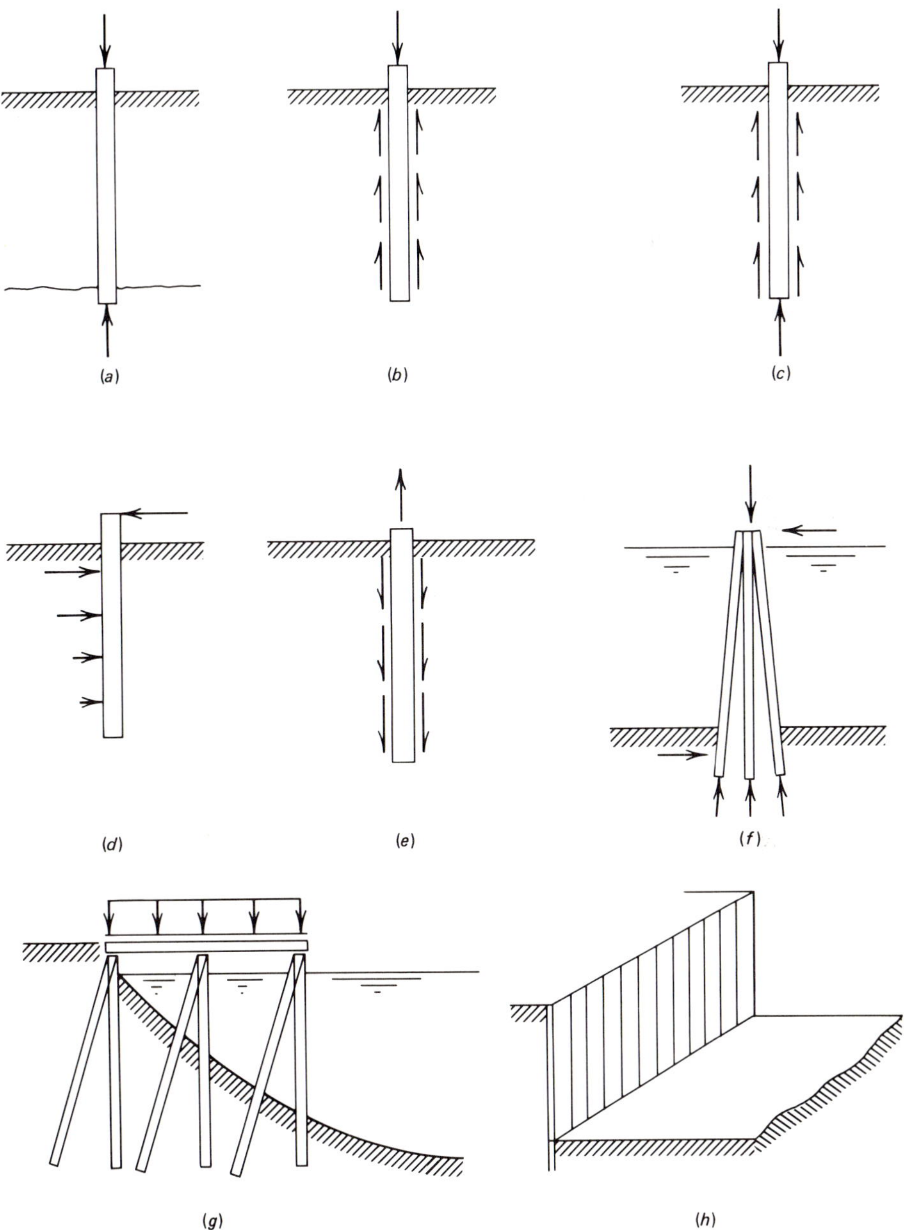

Figure 2.1 Some uses of pile foundations. (*a*) End-bearing pile; (*b*) friction pile; (*c*) combined end-bearing and friction; (*d*) lateral load support; (*e*) uplift support; (*f*) dolphins; (*g*) relieving platforms; (*h*) sheeting systems (full or partial).

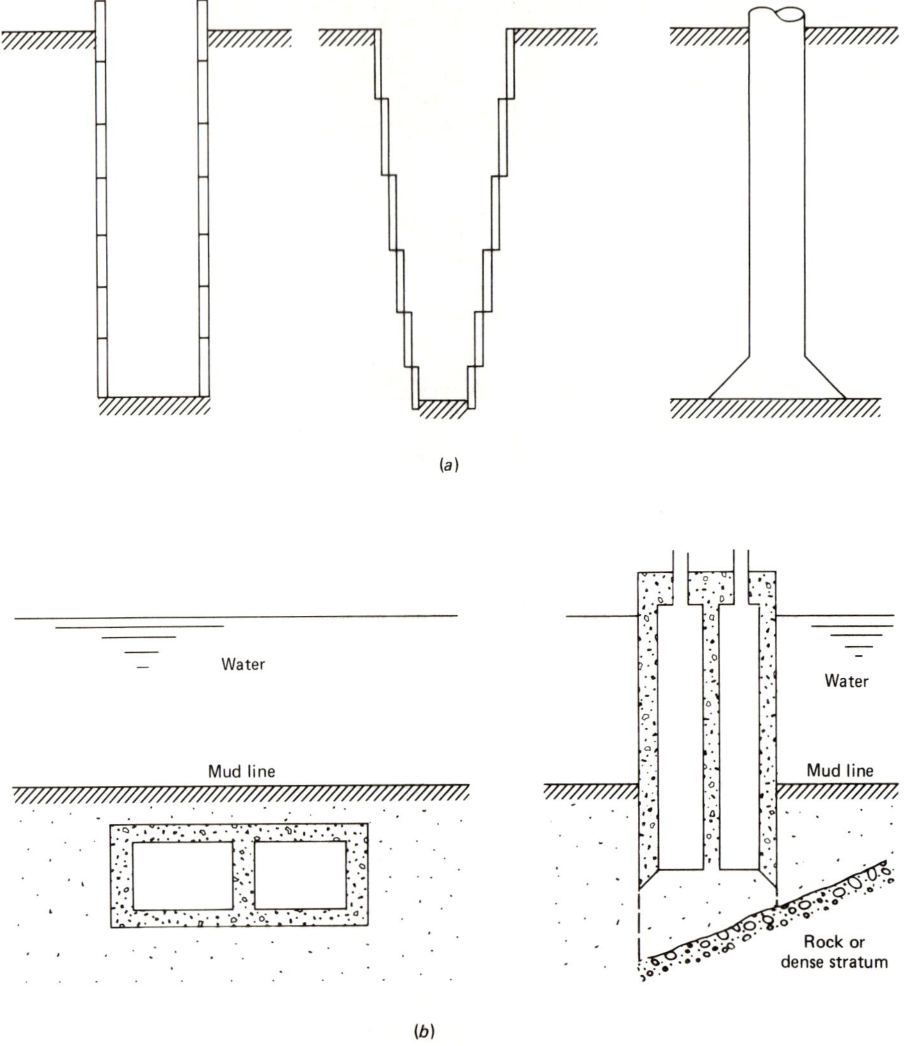

Figure 2.2 Deep foundation types other than piles. (*a*) Piers of various types; (*b*) caissons of various types.

2.3.1 Piles

In terms of historical perspective, as well as of modern day practice, a basic understanding of pile foundations is necessary to the foundation engineer and contractor. In this section, pile types, methods of installation, and the equipment needed to do so will be described. Regarding pile types, the basic construction material is a convenient way to categorize them. Using the NAVDOCKS[2] listing, the following types of bearing piles are currently available:

Timber
Steel
Precast concrete (including prestressed)
Cast-in-place concrete (thin shell driven with mandrel)
Cast-in-place concrete (shells driven without mandrel)
Cast-in-place concrete (shells withdrawn), also called pressure-injected footings
Concrete-filled steel pipe
Composite
Auger placed, pressure-injected concrete

Table 2.1[2] lists each of these pile types along with pertinent technical information which is valuable for gaining insight into their uses. It can easily be seen that of the many types, the cast-in-place concrete piles offer the most flexibility. It should be noted that many subtypes within the overall categories are patented and are best suited for particular situations.

The pile types described in Table 2.1 are usually installed by means of a *pile hammer*. Pile hammers are either *drop* (rarely used for high-capacity piles), *single-acting* (steam or air merely lifts the ram and is exhausted, the ram then falling by gravity alone), *double-acting* (similar to single-acting hammers except that the exhaust is cycled above the ram in a closed unit and used to push it down, thus aiding gravity), *differential-acting* (similar to double-acting), or *diesel* (a self-contained combustion chamber simultaneously lifts the ram and reacts on the pile, then the ram falls adding its own impulse). These various types are shown in Fig. 2.3 and listed in Table 2.2. A more complete list of current hammers, along with their technical specifications, is given in Appendix C.

With the increasing concern over noise pollution and its reduction through the enforcement of numerous federal, state, and local regulations, pile hammers are forbidden in some situations. In those instances, piles can be installed using a *vibratory driver*. Such units are often called vibratory "hammers," but placement is by vibration, not impact, so the term "hammer" is really a misnomer. The basic principle utilizes two counterrotating eccentric weights which have horizontal forces that cancel but vertical downward forces that are additive and are transmitted to the pile. The pile vibrations impart an acceleration to the adjacent soil which gives it a low frictional resistance so that the pile can then penetrate into the soil. A schematic diagram is shown in Fig. 2.4 and the various types of vibratory drivers are listed in Table 2.3. A more complete list, along with technical specifications, is given in Appendix D.

The basic construction unit which positions and places the pile along with its driving hammer is a suitable crawler or truck-mounted crane. Such cranes are adapted for pile driving by adding vertical pile "leaders" to the crane boom and a

Table 2.1 Design criteria for bearing piles†

	Pile type			
Criteria	Timber	Steel	Precast concrete (including prestressed)	Cast-in-place concrete (thin shell driven with mandrel)
Consider for length of:	30–60 ft	40–100 ft	40–50 ft for precast 60–100 ft for prestressed	100 ft
Applicable material specifica-tions:	TS-2P3	TS-P67	TS-P57	ACI Code 318—for concrete
Maximum stresses:	Measured at most critical point, 1200 lb/in^2 for Southern Pine and Douglas Fir. See U.S.D.A. Wood Handbook No. 72 for stress values of other species.	12,000 lb/in^2	For precast—15% of 28-day strength of concrete, but no more than 700 lb/in^2. For prestressed—20% of 28-day strength of concrete, but no more than 1000 lb/in^2 in excess of prestress.	25% of 28-day strength of concrete with 1000 lb/in^2 maximum, measured at midpoint of length in bearing stratum.
Consider for design loads of:	10–50 tons	40–120 tons		
Disadvantages:	Difficult to splice. Vulnerable to damage in hard driving. Vulnerable to decay unless treated, when piles are intermittently submerged.	Vulnerable to corrosion where exposed. BP section may be damaged or deflected by major obstructions.	Unless prestressed, vulnerable to handling. High initial cost. Considerable displacement. Prestressed difficult to splice.	Difficult to splice after concreting. Redriving not recommended. Thin shell vulnerable during driving. Considerable displacement.

Advantages:	Comparatively low initial cost. Permanently submerged piles are resistant to decay. Easy to handle.	Easy to splice. High capacity. Small displacement. Able to penetrate through light obstructions.	High load capacities. Corrosion resistance can be attained. Hard driving possible.	Initial economy. Tapered sections provide higher bearing resistance in granular stratum.
Remarks:	Best suited for friction pile in granular material.	Best suited for end-bearing on rock. Reduce allowable capacity for corrosive locations.	Cylinder piles in particular are suited for bending resistance.	Best suited for medium-load friction piles in granular materials.

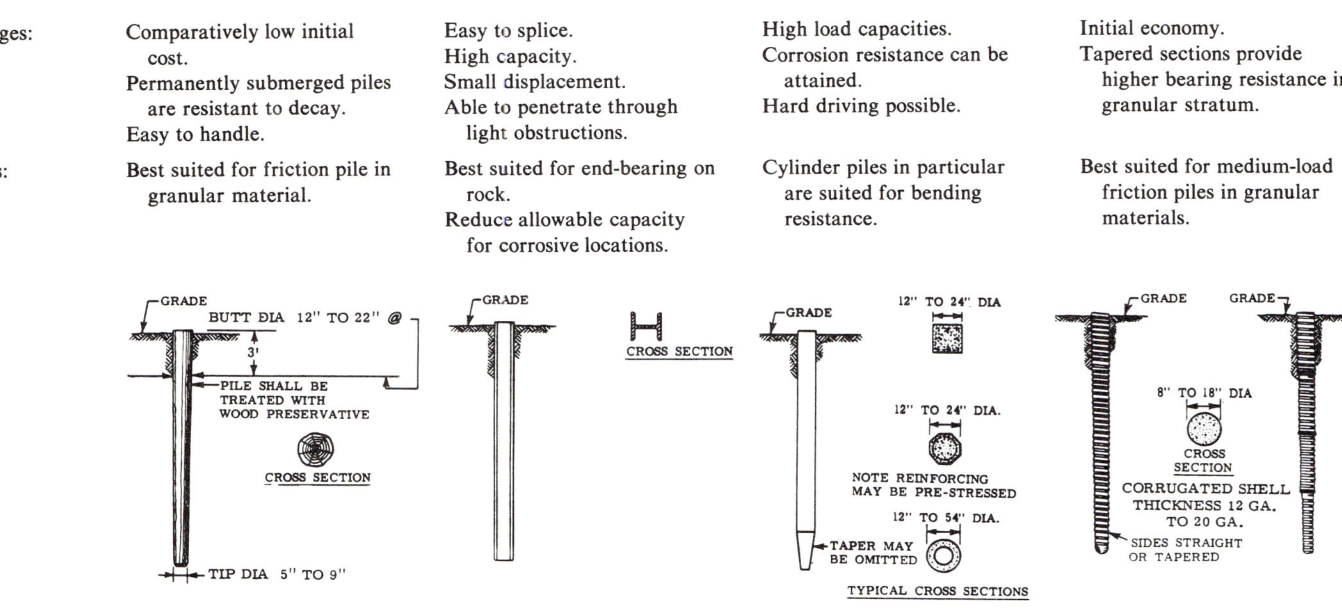

Table 2.1 Design criteria for bearing piles† (Continued)

Criteria	Pile type			
	Cast-in-place concrete piles (shells driven without mandrel)	Pressure-injected footings	Concrete filled steel pipe piles	Composite piles
Consider for length of:	30–80 ft	10–60 ft	40–120 ft	60–120 ft
Applicable material specifications:	ACI Code 318	TS-F16	ASTM A7—for core ASTM A252—for pipe ACI Code 318—for concrete	ACI Code 318—for concrete ASTM-36—for structural section ASTM A252—for steel pipe TS-P2—for timber
Maximum stresses:	25% of 28-day strength of concrete with maximum of 1000 lb/in^2 measured at midpoint of length in bearing stratum. 9000 lb/in^2 in shell.	25% of 28-day strength of concrete, with a maximum of 1000 lb/in^2. 9000 lb/in^2 for pipe shell if thickness greater than $\frac{1}{2}$ in.	9000 lb/in^2 for pipe shell. 25% of 28-day strength of concrete with a maximum of 1000 lb/in^2. 12,000 lb/in^2 on steel cores.	25% of 28-day strength of concrete with 1000 lb/in^2 maximum. 9000 lb/in^2 for structural and pipe sections. Same as timber piles for wood composite.
Consider for design loads of:	50–70 tons	60–120 tons	80–120 tons without cores 500–1500 tons with cores	30–80 tons
Disadvantages:	Hard to splice after concreting. Considerable displacement.	*Base* of footing cannot be made in clay. When clay layers must be penetrated to reach suitable material, special precautions are required for *shafts* if in groups.	High initial cost. Displacement for closed-end pipe.	Difficult to attain good joint between two materials.

Advantages:	Can be redriven. Shell not easily damaged.	Provides means of placing high-capacity footings on bearing stratum without necessity for excavation or dewatering. Required depths can be predicted accurately. High blow energy available for overcoming obstructions. Great uplift resistance if suitably reinforced.	Best control during installation. No displacement for open-end installation. Open-end pipe best against obstructions. High load capacities. Easy to splice.	Considerable length can be provided at comparatively low cost.
Remarks:	Best suited for friction piles of medium length.	Best suited for granular soils where bearing is achieved through compaction around base. Minimum spacing 4 ft 6 in on center. For further design requirements see Philadelphia Building Code 4-1710.	Provides high bending resistance where unsupported length is loaded laterally.	The weakest of any material used shall govern allowable stresses and capacity.

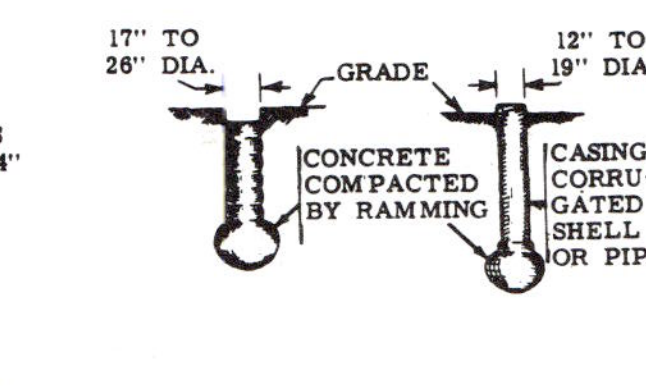

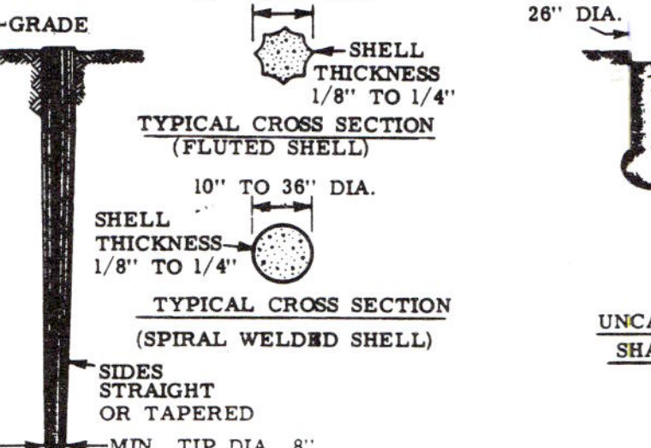

Table 2.1 Design criteria for bearing piles† (Continued)

Criteria	Pile type	
	Auger-placed, pressure-injected concrete piles	General notes
Consider for length of:	30–60 ft	1. Stresses given for steel piles are for noncorrosive locations. For corrosive locations, estimate possible reduction in steel cross section or provide protection from corrosion.
Applicable material specifications:	TS-2P69	
Maximum stresses:	25% of 28-day strength of concrete with maximum of 1000 lb/in².	2. Lengths and loads indicated are for feasibility guidance only. They generally represent current practice.
Consider for design load of:	35–70 tons	3. Design load capacity should be determined by soil mechanics principles limiting stresses in piles and type and function of structure.
Disadvantages:	More than average dependence on quality fieldwork Not suitable through peat or similar highly compressible material.	

Advantages:	Economy. Completely nondisplacement. No driving vibration to endanger adjacent structures. High skin friction. Good contact on rock for end bearing. Convenient for low-headroom underpinning work. Visual inspection of augered material. No splicing required. 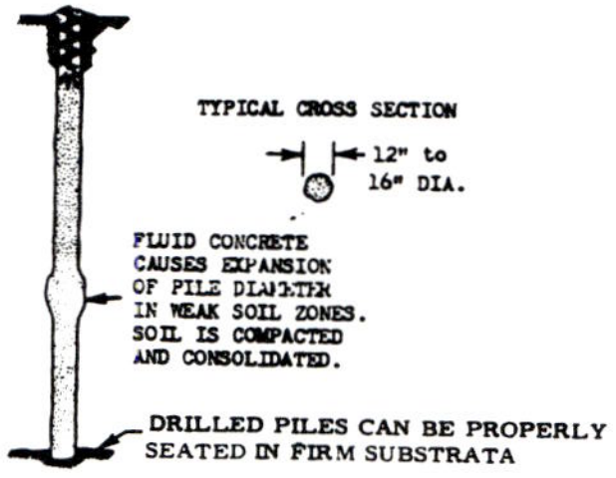
Remarks:	Process patented.

TYPICAL CROSS SECTION

12" to 16" DIA.

FLUID CONCRETE CAUSES EXPANSION OF PILE DIAMETER IN WEAK SOIL ZONES. SOIL IS COMPACTED AND CONSOLIDATED.

DRILLED PILES CAN BE PROPERLY SEATED IN FIRM SUBSTRATA

†After NAVDOCKS, Ref. 2.

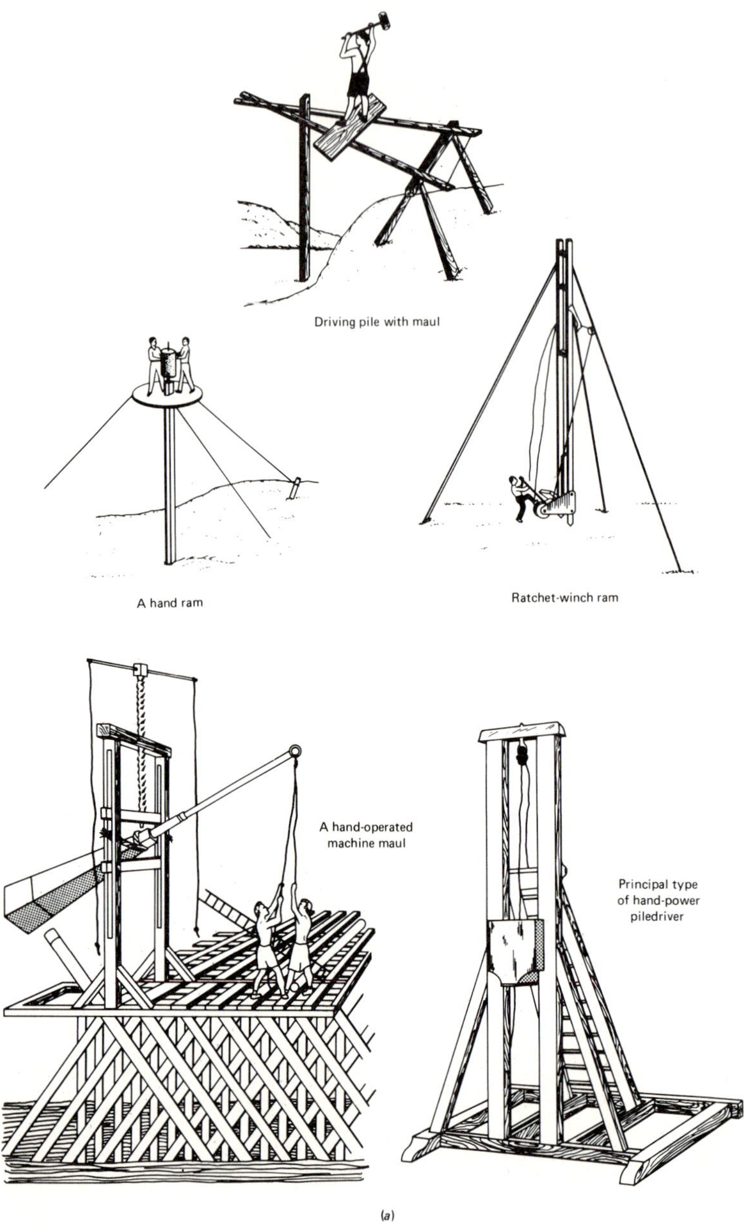

Figure 2.3 (*a*) Examples of early pile driving involving drop hammers; (*b*) example of single-acting steam-air hammer; (*c*) example of double-acting steam-air hammer; (*d*) example of differential-acting steam-air hammer; (*e*) examples of open-end single-acting and closed-end double-acting diesel hammers. (*After Compton, Ref. 5.*)

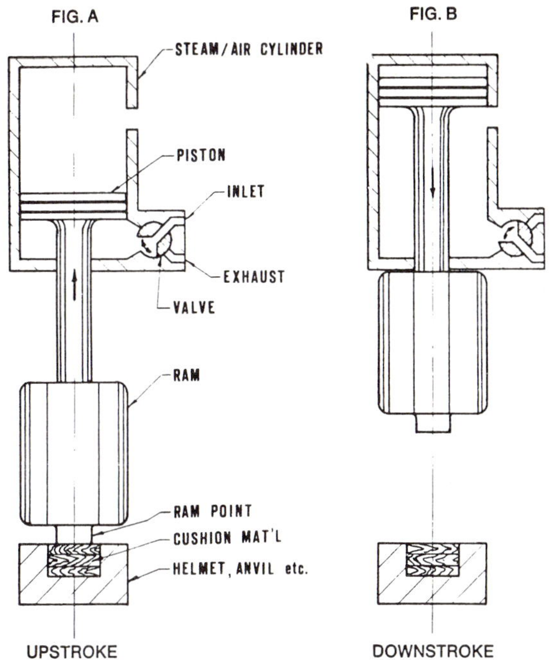

SINGLE ACTING
STEAM/AIR HAMMER

(*b*)

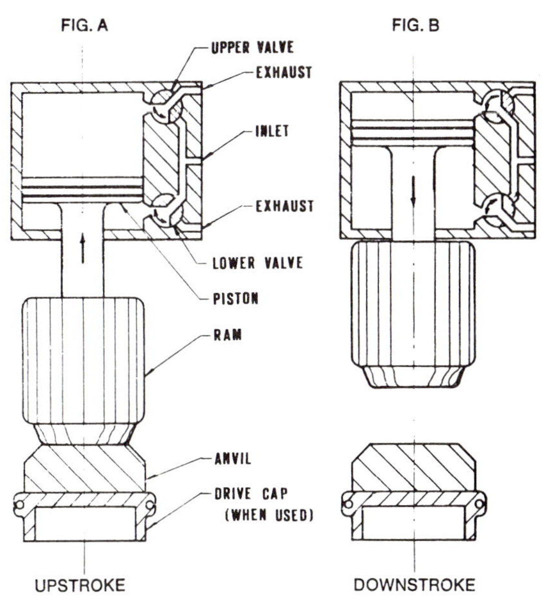

DOUBLE ACTING
STEAM/AIR HAMMER

(*c*)

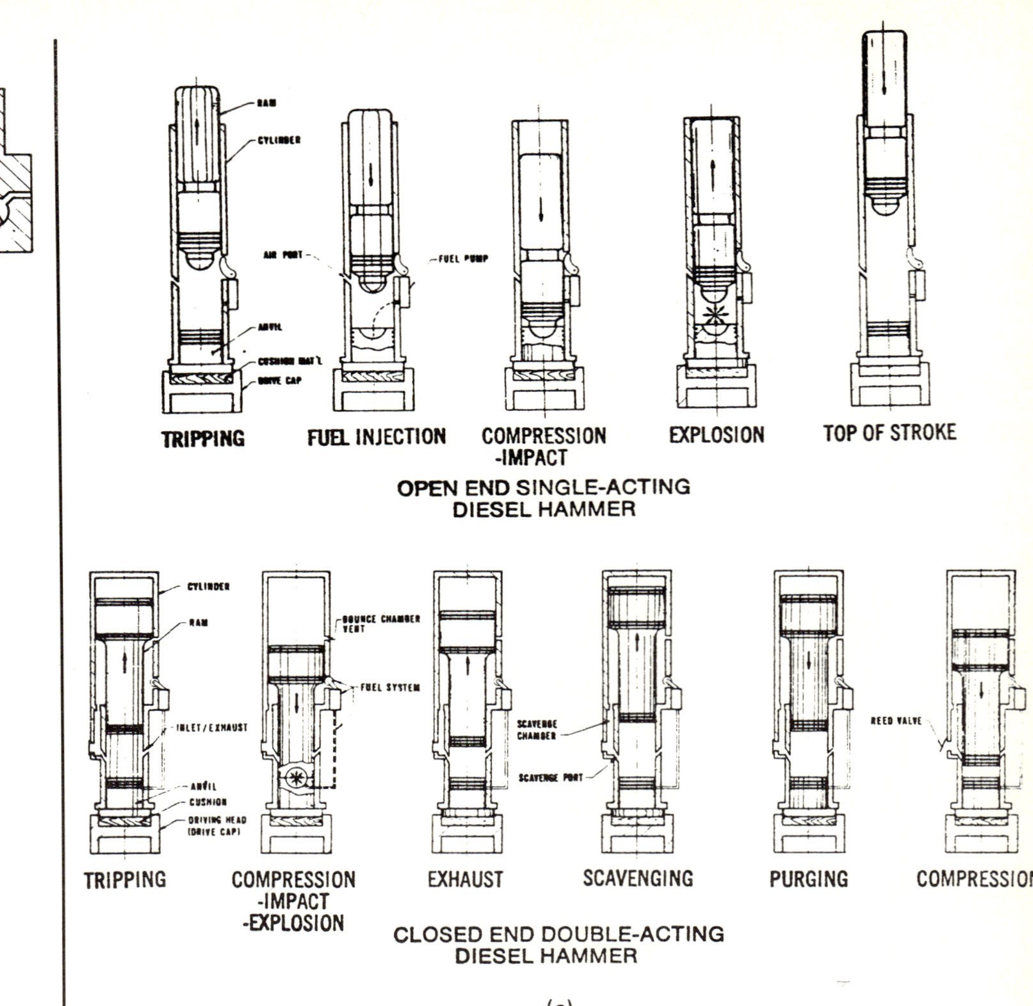

FIG. A
FIG. B
EXHAUST
VALVE
INLET
PISTON
RAM
CUSHION MAT'L
HELMET, ANVIL etc.
UPSTROKE
DOWNSTROKE
DIFFERENTIAL ACTING STEAM/AIR HAMMER
(d)
RAM
CYLINDER
AIR PORT
FUEL PUMP
ANVIL
CUSHION MAT'L
DRIVE CAP
TRIPPING
FUEL INJECTION
COMPRESSION -IMPACT
EXPLOSION
TOP OF STROKE
OPEN END SINGLE-ACTING DIESEL HAMMER
CYLINDER
RAM
BOUNCE CHAMBER VENT
FUEL SYSTEM
INLET/EXHAUST
SCAVENGE CHAMBER
SCAVENGE PORT
ANVIL
CUSHION
DRIVING HEAD (DRIVE CAP)
REED VALVE
TRIPPING
COMPRESSION -IMPACT -EXPLOSION
EXHAUST
SCAVENGING
PURGING
COMPRESSION
CLOSED END DOUBLE-ACTING DIESEL HAMMER
(e)

Table 2.2 Pile hammer comparison based on rated energy†

Rated energy, kip/ft	Make of hammer	Model no.	Type	Blows per min	Stroke at rated energy	Weight striking parts, kips	Total weight, kips
180.0	Vulcan	060	Single-acting	62	36	60.0	121.0
130.0	MKT‡	S-40	Single-acting	55	39	40.0	96.0
120.0	Vulcan	040	Single-acting	60	36	40.0	87.5
113.5	S-Vulcan§	400C	Differential	100	16.5	40.0	83.0
97.5	MKT	S-30	Single-acting	60	39	30.0	86.0
79.6	Kobe	K42	Diesel	52	98	9.2	22.0
60.0	Vulcan	020	Single-acting	60	36	20.0	39.0
60.0	MKT	S20	Single-acting	60	36	20.0	38.6
56.5	Kobe	K32	Diesel	52	98	7.0	15.4
50.2	S-Vulcan	200C	Differential	98	15.5	20.0	39.0
48.7	Vulcan	016	Single-acting	60	36	16.2	30.2
48.7	Raymond	0000	Single-acting	46	39	15.0	23.0
44.5	Kobe	K22	Diesel	52	98	4.8	10.6
42.0	Vulcan	014	Single-acting	60	36	14.0	27.5
40.6	Raymond	000	Single-acting	50	39	12.5	21.0
39.8	Delmag	D-22	Diesel	52	N/A	4.8	10.0
37.5	MKT	S14	Single-acting	60	32	14.0	31.6
36.0	S-Vulcan	140C	Differential	103	15.5	14.0	27.9
32.5	MKT	S10	Single-acting	55	39	10.0	22.2
32.5	Vulcan	010	Single-acting	50	39	10.0	18.7
32.5	Raymond	00	Single-acting	50	39	10.0	18.5
32.0	MKT	DE-40	Diesel	48	96	4.0	11.2
30.2	Vulcan	OR	Single-acting	50	39	9.3	16.7
26.3	Link-Belt	520	Diesel	82	43.2	5.0	12.5
26.0	MKT	C-8	Double-acting	81	20	8.0	18.7
26.0	Vulcan	08	Single-acting	50	39	8.0	16.7
26.0	MKT	S8	Single-acting	55	39	8.0	18.1
24.4	S-Vulcan	80C	Differential	111	16.2	8.0	17.8
24.4	Vulcan	8M	Differential	111	N/A	8.0	18.4
24.3	Vulcan	0	Single-acting	50	39	7.5	16.2
24.0	MKT	C-826	Double-acting	90	18	8.0	17.7
22.6	Delmag	D-12	Diesel	51	N/A	2.7	5.4
22.4	MKT	DE-30	Diesel	48	96	2.8	9.0
24.4	Kobe	K13	Diesel	52	98	2.8	6.4
19.8	Union	K13	Double-acting	110	24	3.0	14.5
19.8	MKT	11B3	Double-acting	95	19	5.0	14.5
19.5	Vulcan	06	Single-acting	60	36	6.5	11.2
19.2	S-Vulcan	65C	Differential	117	15.5	6.5	14.8
18.2	Link-Belt	440	Diesel	88	36.9	4.0	10.3
16.2	MKT	S5	Single-acting	60	39	5.0	12.3
16.0	MKT	DE-20	Diesel	48	96	2.0	6.3
16.0	MKT	C5	Compound	110	18	5.0	11.8
15.1	S-Vulcan	50C	Differential	120	15.5	5.0	11.7
15.1	Vulcan	5M	Differential	120	15.5	5.0	12.9

Table 2.2 Pile hammer comparison based on rated energy† (*Continued*)

Rated energy, kip/ft	Make of hammer	Model no.	Type	Blows per min	Stroke at rated energy	Weight striking parts, kips	Total weight, kips
15.0	Vulcan	1	Single-acting	60	36	5.0	10.1
15.0	Link-Belt	312	Diesel	100	30.9	3.8	10.3
13.1	MKT	10B3	Double-acting	105	19	3.0	10.6
12.7	Union	1	Double-acting	125	21	1.6	10.0
9.0	Delmag	D5	Diesel	51	N/A	1.1	2.4
9.0	MKT	C-3	Double-acting	130	16	3.0	8.5
9.0	MKT	S3	Single-acting	65	36	3.0	8.8
8.8	MKT	DE-10	Diesel	48	96	11.0	3.5
8.7	MKT	9B3	Double-acting	145	17	1.6	7.0
8.2	Union	1.5A	Double-acting	135	18	1.5	9.2
8.1	Link-Belt	180	Diesel	92	37.6	1.7	4.5
7.2	Vulcan	2	Single-acting	70	29.7	3.0	7.1
7.2	S-Vulcan	30C	Differential	133	12.5	3.0	7.0
7.2	Vulcan	3M	Differential	133	N/A	3.0	8.4
6.5	Link-Belt	105	Diesel	94	35.2	1.4	3.8
4.9	Vulcan	DGH900	Differential	238	10	0.9	5.0
3.6	Union	3	Double-acting	160	14	0.7	4.7
3.6	MKT	7	Double-acting	225	9.5	0.8	5.0
0.4	Union	6	Double-acting	340	7	0.1	0.9
0.4	Vulcan	DGH100A	Differential	303	6	0.1	0.8
0.4	MKT	3	Double-acting	400	5.7	0.06	0.7
0.3	Union	7A	Double-acting	400	6	0.08	0.5

†After Parola, Ref. 3.
‡McKiernan-Terry.
§Super-Vulcan.

horizontal "spotter" or "catwalk" for stabilizing the entire assembly; see Fig. 2.5. The leaders serve to guide the pile hammer and the pile for proper positioning and placement. Piles can be driven vertically or at an angle, i.e., at a "batter," by pushing the leads out or pulling them in (or rotating them if a "moonbeam" is available) with the spotter as need be.

With the weight of the pile, hammer, leads, spotter, and associated equipment on the front of the crane, a large counterbalancing force is required in the back for overall stability. This counterbalancing force is often provided by air compressors when using single- or double-acting hammers. But weights of any type can be used.

The entire pile-driving rig when fully assembled is relatively heavy, and when one considers that the soils being penetrated are often weak, settlement and instability beneath the crane are common. For this reason, large timber mats made from interconnected 12 × 12 in timbers are often used as working platforms.

The pile-driving business, playing a major part in heavy construction, has developed over the years many accessories and "tricks of the trade" to assist in proper placement of piles. Table 2.4 gives an overview of some of these procedures. But even with this backlog of information from past practice to work with, many unanticipated problems can occur during construction. Some helpful remedies for such field problems are given in Table 2.5. Both Tables 2.4 and 2.5 are from Ref. 2.

The Vibratory Driver/Extractor System

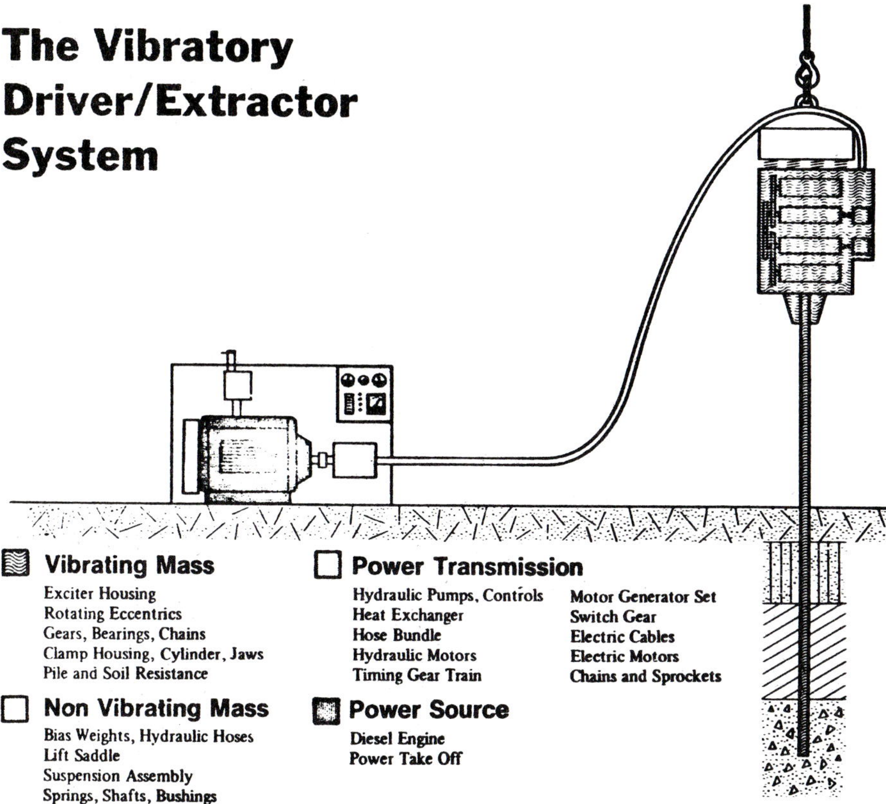

Figure 2.4 Schematic diagram of vibratory driver. *(After Compton, Ref. 5.)*

2.3.2 Piers

A *pier* (also called a "drilled pier," a "concrete pier," and sometimes even a "drilled caisson") is a type of deep foundation constructed by excavating a hole and filling it with concrete. In a sense, it is similar to a bored, cast-in-place, concrete pile but is usually higher in its load-carrying capacity, thus being larger in dimensions and spaced more widely apart from other piers than are piles. A distinction sometimes noted is that piers are larger than 3 ft (0.76 m) in diameter, while piles are smaller.[6] Regarding load-carrying capacity, drilled piers generally support axial loads in excess of 100 tons ($\simeq$1000 kN). Figure 2.6 illustrates typical types of drilled pier foundations.

The historic development of the modern high-capacity drilled pier began with hand-excavated piers supported by wood lagging that reached down to a competent load-bearing stratum. Complications during excavation came about when water-

Table 2.3 Vibratory pile drivers†

Make	Model	Total weight, kips	Available horsepower	Frequency range, Hz	Force,‡ kips Frequency, Hz
Foster	2-17	6.2	34	18–21	
(France)	2-35	9.1	70	14–19	62/19
	2-50	11.2	100	11–17	101/17
Menck	MVB22-30	4.8	50		48/
(Germany)	MVB65-30	2.0	7.5		14/
	MVB44-30	8.6	100		97/
Muller	MS-26	9.6	72		
(Germany)	MS-26D	16.1	145		
Uraga	VHD-1	8.4	40	16–20	43/20
(Japan)	VHD-2	11.9	80	16–20	86/20
	VHD-3	15.4	120	16–20	129/20
Bodine	B	22	1000	0–150	63/100–175/100
(U.S.A.)					
(Russia)	BT-5	2.9	37	42	48/42
	VPP-2	4.9	54	25	49/25
	100	4.0	37	13	44/13
	VP	11.0	80	6.7	35/7
	VP-4	25.9	208		198/

†After Davisson, Ref. 4.

‡Forces given are present maximums. These can usually be raised or lowered by changing weights in the oscillator.

bearing granular soils, or sensitive clay soils, were encountered. Different types of bracing systems were devised, e.g., concrete rings of the Chicago or Gow types and steel sets, but the work was very labor-intensive and hence slow and costly. The situation changed radically after World War II with the introduction of truck-mounted drilling rigs. Today, rigs capable of drilling piers greater than 100 ft in depth and up to 12 ft in diameter with 25-ft-diameter enlarged bases are in use. See Fig. 2.7 for a typical such rig and Fig. 2.8 for examples of the various types of drilling tools used to advance holes and create underreamed bases.

Advancement of the hole can be a problem with drilled piers in the following soil types:

Dry sands and gravels
Saturated sands and gravels
Soils with boulders or rock fragments
Sensitive clays

(See Appendix A for soil classification systems and Appendix B for miscellaneous soil details and properties.)

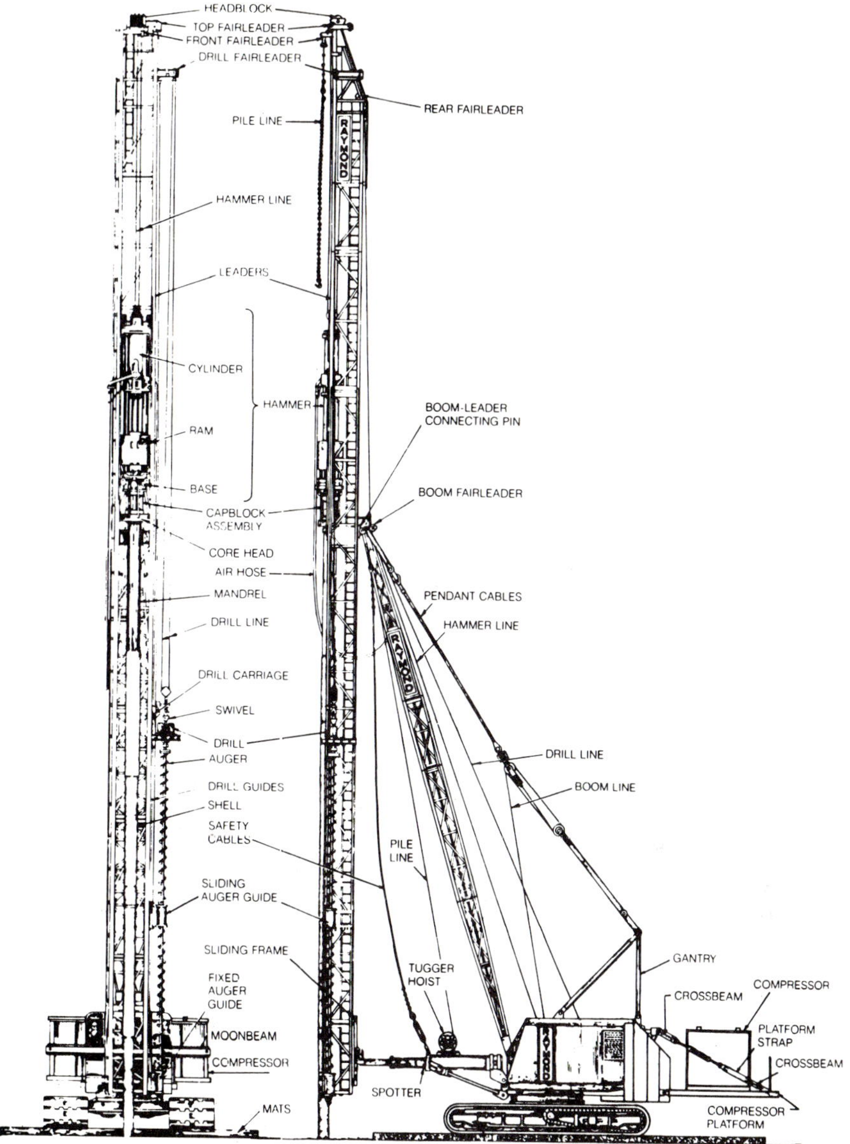

Figure 2.5 Example of a complete pile-driving rig. *(Courtesy Raymond International, Inc.)*

To adequately handle these conditions and to keep deformation of adjacent soils (and structures founded thereon) to a minimum, cased holes or holes filled with a bentonite slurry can be used. Casing the hole with steel pipe of adequate diameter is time-consuming and expensive but often necessary to complete it in a safe and stable condition prior to the placement of concrete. Because of the cost of the casing, it is often pulled as concreting proceeds. This is a difficult and tedious procedure and can

Table 2.4 Supplementary procedures and appurtenances used in pile driving†

Method	Equipment and procedure utilized	Applicability
Means of reducing driving resistance above bearing stratum: Temporary casing	Open-end pipe casing driven and cleaned out. May be pulled later.	a. To drive through minor obstructions. b. To minimize displacement. c. To prevent caving or squeezing of holes. d. To permit concreting of pile prior to excavation to subgrade of foundation.
Precoring	By continuous flight auger or churn drill, a hole is formed into which the pile is lowered. Pile is then driven to bearing below the cored hole.	a. For driving through thick stratum of stiff to hard clay. b. To avoid displacement and heave of surrounding soil. c. To avoid injury to timber and thin-shell pipes. d. To eliminate driving resistance in strata unsuitable for bearing.
Spudding	Heavy structural sections or closed-end pipes are alternately raised and dropped to form a hole into which pile is lowered. Pile is then driven to bearing below the spudded hole.	a. For driving past individual obstruction. b. To drive through strata of fill with large boulders or rock fragments.
Jetting	Water, air, or mixture of both forced through pipe at high pressures and velocity; jets are sometimes built into piles.	a. Used in practically all soils to reduce friction in strata unsuitable for bearing during driving of pile.
Means of increasing driving resistance in bearing stratum: Upside down piles	Tapered piles, specifically timber, driven with large butt downward.	a. For end-bearing timber piles where it is necessary to minimize penetration into bearing stratum. b. To avoid driving through to incompressible but unsuitable bearing material.

Table 2.4 Supplementary procedures and appurtenances used in pile driving†

Method	Equipment and procedure utilized	Applicability
Lagging	Short timber or steel sections connected by bolting or welding to timber or steel pipes.	a. To increase frictional resistance along sides of pile. b. Increase of end-bearing resistance when mounted near tip.
Means of overcoming obstructions:		
Shoes and reinforced tips	Metal reinforcing, such as bands and shoes for all types of piles.	a. To provide protection against damage of tip. b. To provide additional cutting power.
Explosives	Drill and blast ahead of pile tip.	a. To remove obstructions to open end piles under very severe conditions.
Pre-excavation	Hand or machine excavation.	a. Used for removal of obstruction close to ground surface.
Special equipment for advancing piles:		
Jacking	Hydraulic or mechanical screw jacks are used to advance pile. Pile is built up in short, convenient lengths.	a. Used instead of pile hammer where access is difficult. b. To eliminate vibrations.
Vibration	High amplitude vibrators.	a. Advantageous for driving in water-logged sands and gravel. b. Advantageous for driving sheetpiling.
Follower	Temporary filler section between hammer and pile top, preferably of same material as pile.	a. To drive pile top to elevation below reach of hammer or below water.

†After NAVDOCKS, Ref. 2.

easily result in cavitated or undersized pier shafts; see Ref. 7. The alternative procedure is to advance the hole through a bentonite slurry (approximately 1 part bentonite clay to 20 parts water) which keeps the walls of the excavated hole open until the design depth is reached. The concrete is then placed from the bottom of the hole upward by means of a tremie pipe. The slurry is displaced by the concrete, collected at the ground surface, cleaned of soil and displaced stones, and reused in the next hole. The action of the slurry in retaining stability of the side walls is a topic of active research.[8] See Sec. 3.4.3 for additional details on the action of the slurry itself.

Table 2.5 Treatment of field problems encountered during pile driving†

Description of problem	Procedures to be applied
Obstructions to driving:	
Old foundations, boulders, rubble fill, cemented lenses, and similar obstacles to driving.	Excavate or break up shallow obstructions, if practical. For deeper obstructions use spudding, jetting, temporary casings, or use drive shoes and reinforced tips where pile is strong enough to be driven through obstructions.
General problems:	
Vibration in driving: May compact loose granular materials causing settlement of existing structures near piles. Effect most pronounced in driving displacement piles.	Select pile type with minimum displacement, and/or precore or jet with temporary casing or substitute jacking for pile driving.
Damage to thin shells: Driven shells may have been crimped, buckled, torn, or be leaking at joints as the results of driving difficulties or presence of obstructions.	Each pile is inspected with light beam. If diameter at any location varies more than 15% from original diameter or if other damage to shell cannot be repaired, pile is abandoned and filled with sand and a replacement is driven. Concrete shall be placed in dry shell only.
Inappropriate use of pile-driving formula: Piles driven to a penetration determined solely by driving resistance may be bearing in a compressible stratum. This may occur in thick strata of silty fine sand, varved silts and clays, or medium stiff cohesive soils.	Unsuitable bearing strata should be determined by exploration program. Piles should not be permitted to stop in these strata, regardless of driving resistance. For bearing in stiff and brittle cohesive soils and in soft rock, load tests are particularly important.
Difficulties at pile tip:	
Fracturing of bearing materials: Fracturing of material immediately below tips of piles driven to required resistance as a result of driving adjacent piles. Brittle weathered rock, clay-shale, shale, siltstone, and sandstone are vulnerable materials. Swelling of stiff fissured clays or shales at pile tip may complicate this problem.	For piles bearing in these materials specify driving resistance test on selected piles after completion of driving adjacent piles. If damage to the bearing stratum is evidenced, require redriving until specified resistance is met.
Steeply sloping rock surface: Tips of high-capacity end-bearing piles may slide or move laterally on a steeply sloping surface of sound hard rock which has little or no overlying weathered material.	Provide special shoes or pointed tips or use open-end pipe pile socketed into sound rock.
Loss of ground: May occur during installation of open-end pipe piles. Materials vulnerable to piping, particularly fine sands or silts, may flow into pipe under the influence of an outside differential head, causing settlement in surrounding areas or loss of ground beneath tips of adjacent piles.	Avoid cleaning in advance of pile cutting edge, and/or retain sufficient material within pipe to prevent inflow of soil from below.

Table 2.5 Treatment of field problems encountered during pile driving†

Description of problem	Procedures to be applied
Movement of piles subsequent to driving:	
Heave: Completed piles rise vertically as the result of driving adjacent piles. Particularly common for displacement piles in soft clays and medium-compact granular soils. Heave becomes serious in soft clays when volume displaced by piles exceeds $2\frac{1}{2}\%$ of volume of soil enclosed within the limits of the pile foundation.	For piles of solid cross sections (timber, steel, precast concrete), survey top elevations during driving of adjacent piles to determine possible heave. For piles that have risen more than 0.005 ft: redrive to at least the former tip elevation, and beyond that as necessary to reach required driving resistance. Heave is minimized by driving temporary open-end casing, precoring or jetting so that total volume displaced by pile driving is less than 2 or 3% of total volume enclosed within limits of pile foundation.
Lateral movement of piles: Completed piles move horizontally as the result of driving adjacent piles.	Survey horizontal position of completed piles during the driving of adjacent piles. Movement is controlled by procedures used to minimize heave.

†After NAVDOCKS, Ref. 2.

Depending upon the particular situation and the nature of the subsurface soil or rock conditions, drilled piers can be classified according to their load-carrying capacity; see Table 2.6.[9] Additional detail about the design aspects of drilled pier foundations will be given in Sec. 2.4.

2.3.3 Caissons

A *caisson* (French for "box") is a hollow foundation structure which is prefabricated off-site and shipped to location, or built at the site and lowered into place as a single unit. Caissons are usually large and quite often serve as bridge pier foundations in marginal soils, such as swamps, marshes, etc., or are placed directly in rivers or bodies of open water. They have also been used as foundations for offshore drilling platforms and other large offshore installations. The structures they support are of a critical and very expensive nature so that every detail must be fully explored in both the design and construction stages (see Refs. 10 through 12 for further details).

Caissons can be grouped into different categories, and we will review each type here.

Open caissons These are structural units which are hollow at both top and bottom so that excavation and lowering can proceed directly from the ground or water surface. The unit is often positioned from within a cofferdam or temporary island. Excavation is usually done by clamshell or grab bucket and when the outer (cutting) edge

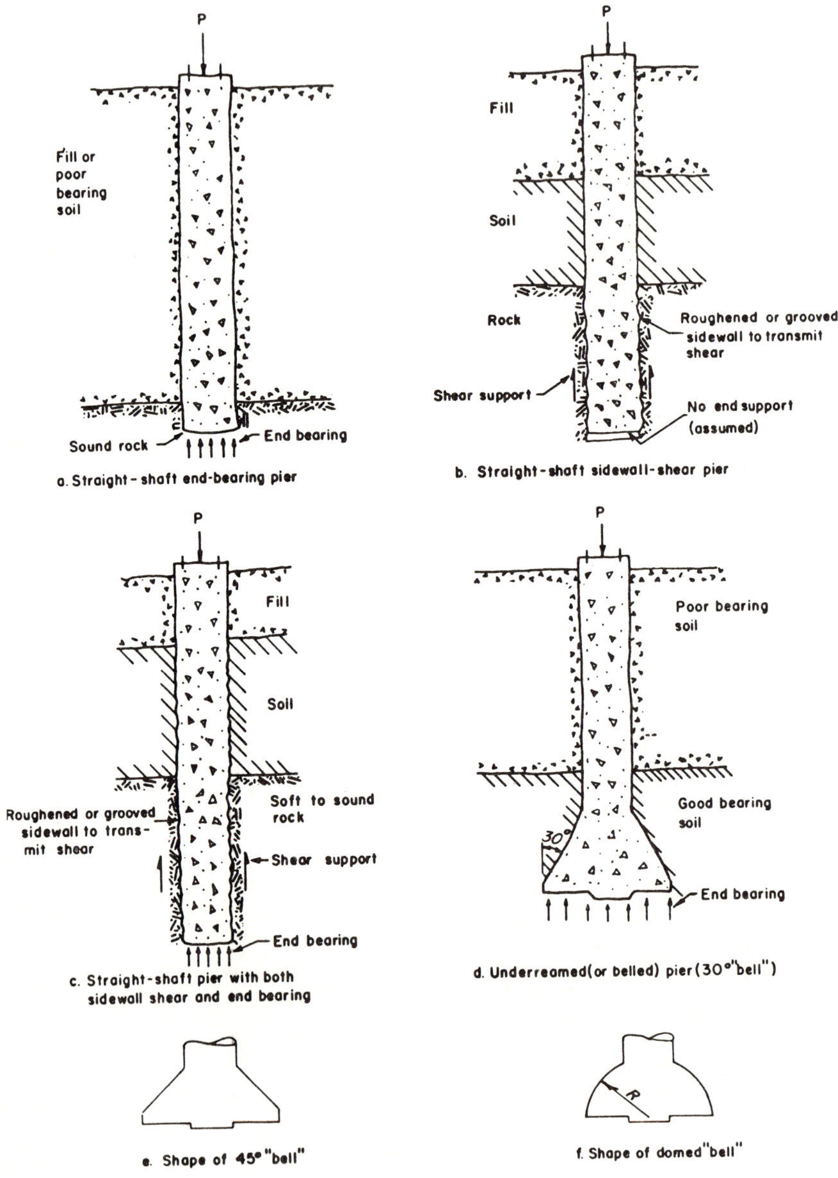

Figure 2.6 Typical types of drilled-pier deep foundations. *(After Woodward et al., Ref. 7.)*

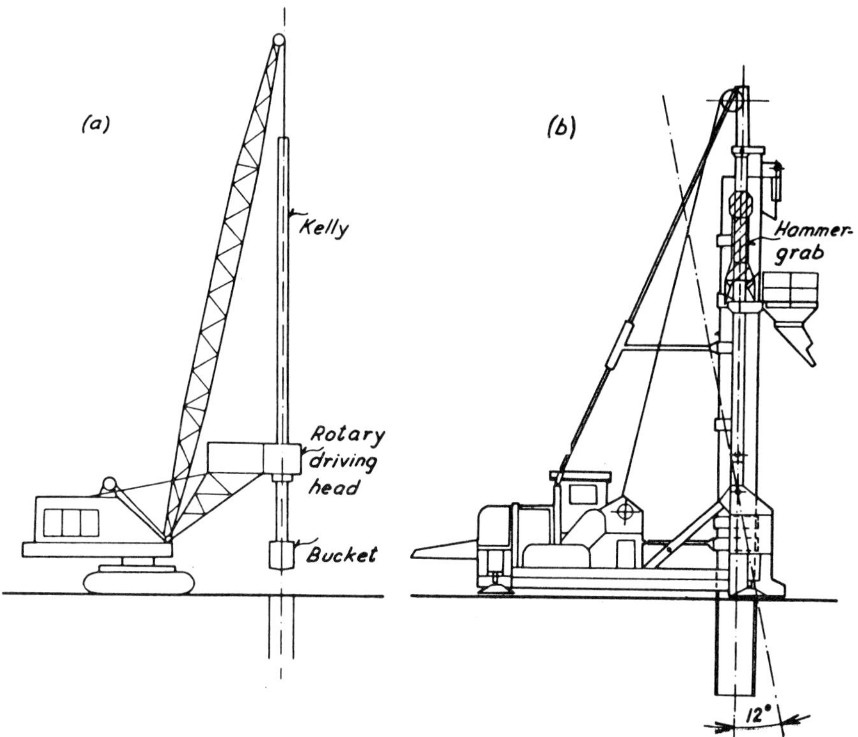

Figure 2.7 Typical drill rig for excavating pier foundations. *(After Woodward et al., Ref. 7.)*

of the caisson is undermined, the unit sinks by gravity. The size of these structures, which must support loads up to 100,000 tons ($\simeq$1,000,000 kN), requires them to be segmented into cells that can be circular, square, or rectangular.

Examples of open caissons are the pier foundations for the Verrazano Narrows Bridge in New York City (the largest suspension bridge in the world and the point from which the New York City Marathon starts each year); see Fig. 2.9.[13] The outer dimensions of each caisson are 38.3 × 68.5 × 48.2 m and internally they each consist of sixty-six 5.2-m circular cells through which the soil was excavated. To prevent high external hydrostatic pressures from causing instability of the bottom soil while the caissons were being sunk, wellpoints surrounding the area reduced the pressure.

As with most caissons, when the lowest elevation is reached a carefully constructed bottom seal of concrete is placed on the rock or dense soil upon which the structure will be founded. When underwater, as is usually the case, tremie concrete provides the seal and the use of divers is recommended to assure proper placement. Once the seal is made, concreting of the cells can proceed on a production basis.

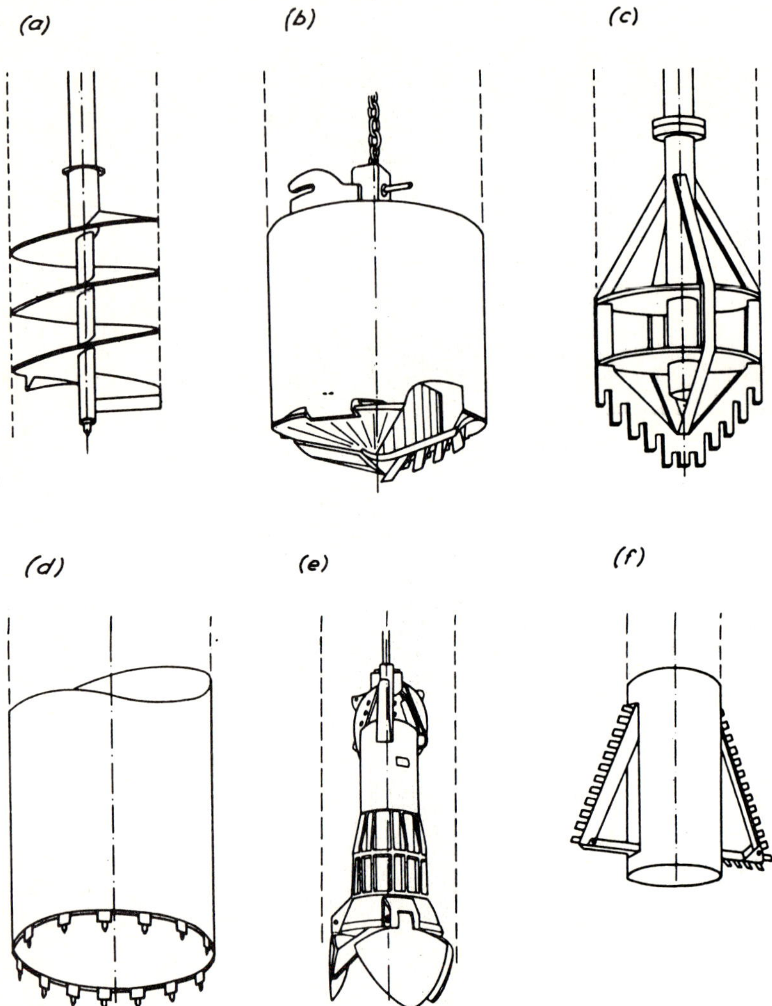

Figure 2.8 Various drilling tools used in construction of drilled piers. (*a*) Open-helix auger; (*b*) drilling bucket; (*c*) bit with teeth and reverse circulation; (*d*) core barrel; (*e*) hammer-grab; (*f*) underreaming bucket. *(After Bazant, Ref. 10.)*

Pneumatic caissons These are similar to open caissons but are closed at the top so that compressed air can be maintained within them during sinking. The pressure of the compressed air must be carefully controlled to equal the external pressure of the soil and water at the bottom of the caisson. This, of course, creates a tremendous work obstacle, for a separate airlock must be made to interface between the high internal caisson pressures and atmospheric pressure. For materials used in the excavation, such as lumber, concrete, etc., the compression or decompression process that takes place within the airlock can be very rapid, but for workers the process must be

Table 2.6 Load-carrying classification of drilled piers†

	Pier classification					
	Homogeneous soil		End-bearing in hard soil		End bearing in rock	
Design factor	Straight shaft	Enlarged base	Socketed into bearing stratum	Enlarged base	Bearing on competent rock	Socketed in competent rock
Approximate range of working load	10–150 tons	50–500 tons	50–250 tons	100–3000 tons	200–7000 tons	300–7000 tons
Usual limiting design criterion	Settlement	Settlement and bearing capacity	Bearing capacity	Bearing capacity	Shaft concrete	Bond between concrete and rock or shaft concrete
Major component of resistance at working load	Shaft adhesion	Shaft adhesion and end bearing	Shaft adhesion in bearing stratum	End bearing	End bearing	Shaft adhesion in rock
Major component of resistance at ultimate load	Shaft adhesion	End bearing	Shaft adhesion and end bearing	End bearing	End bearing	Shaft adhesion and end bearing
Usual method of obtaining design working load	Analytical consideration of shaft adhesion	Presumptive bearing stress or analysis of ultimate bearing capacity	Presumptive bearing stress supplemented by analysis of ultimate bearing capacity, when possible, or qualitative evaluation of bearing material		Presumptive bearing stress plus qualitative evaluation of rock quality	
Requirement for down-the-hole inspection	On selected piers	Every pier	Every pier	Every pier, probe, or core below selected piers	Every pier, probe, or core below selected piers	

†After D'Appolonia et al., Ref. 9.

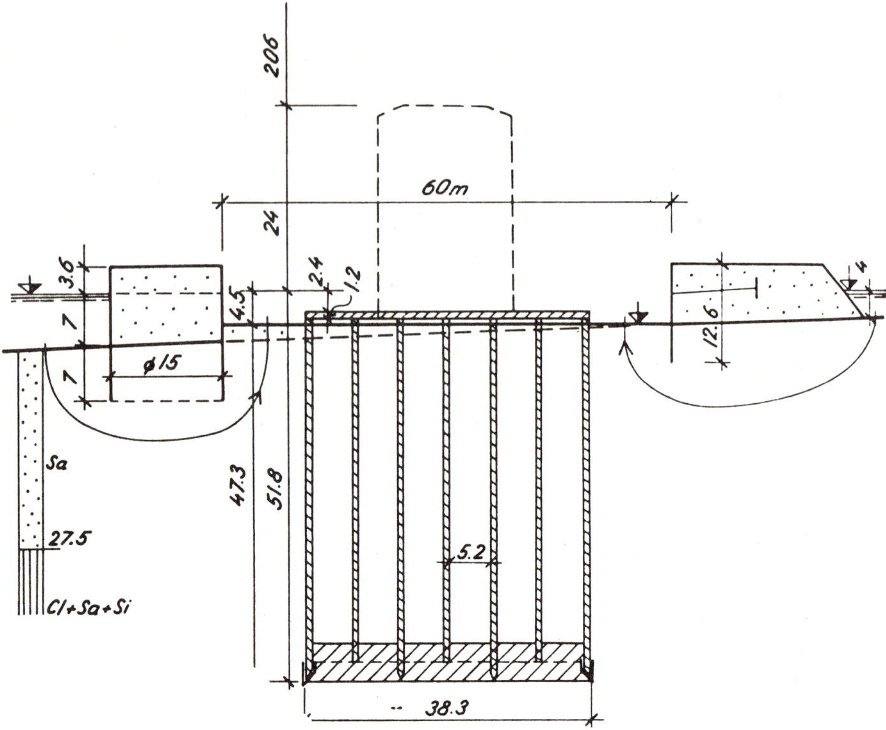

Figure 2.9 Schematic diagram of open caissons used for piers of Verrazano Narrows Bridge in New York City. *(After Hoffman, Ref. 13.)*

very gradual. (An interesting historical account of the effects of compressed air, and in particular of decompression which can cause caisson sickness or the "bends," is given by McCullough[14].) Thus two separate airlocks, one for materials and the other for workers, are customarily designed; see Fig. 2.10 for a number of possible design schemes. The necessity of using compressed air has the advantage of permitting a carefully controlled sinking but the disadvantage of introducing relatively high labor costs (at 40-lb/in^2 pressure, New York City codes only allow 2 h of work per 8-h shift) as compared with the use of open caissons or box caissons.

Box caissons These are distinct from open and pneumatic caissons in that they are structural units which are closed on the bottom (and sometimes at the top also), floated into the proper position, and lowered onto a prepared foundation at a river bottom. They can be used advantageously in relatively shallow tidal waters where suitable load-bearing soils are located at or within a few feet of the water bottom. They are often constructed on land, or in shipyards, and their site installation costs are the least of those for all the types of caissons. However, horizontal and vertical positioning is difficult and a ± 1.0 ft tolerance must be included in their design. A

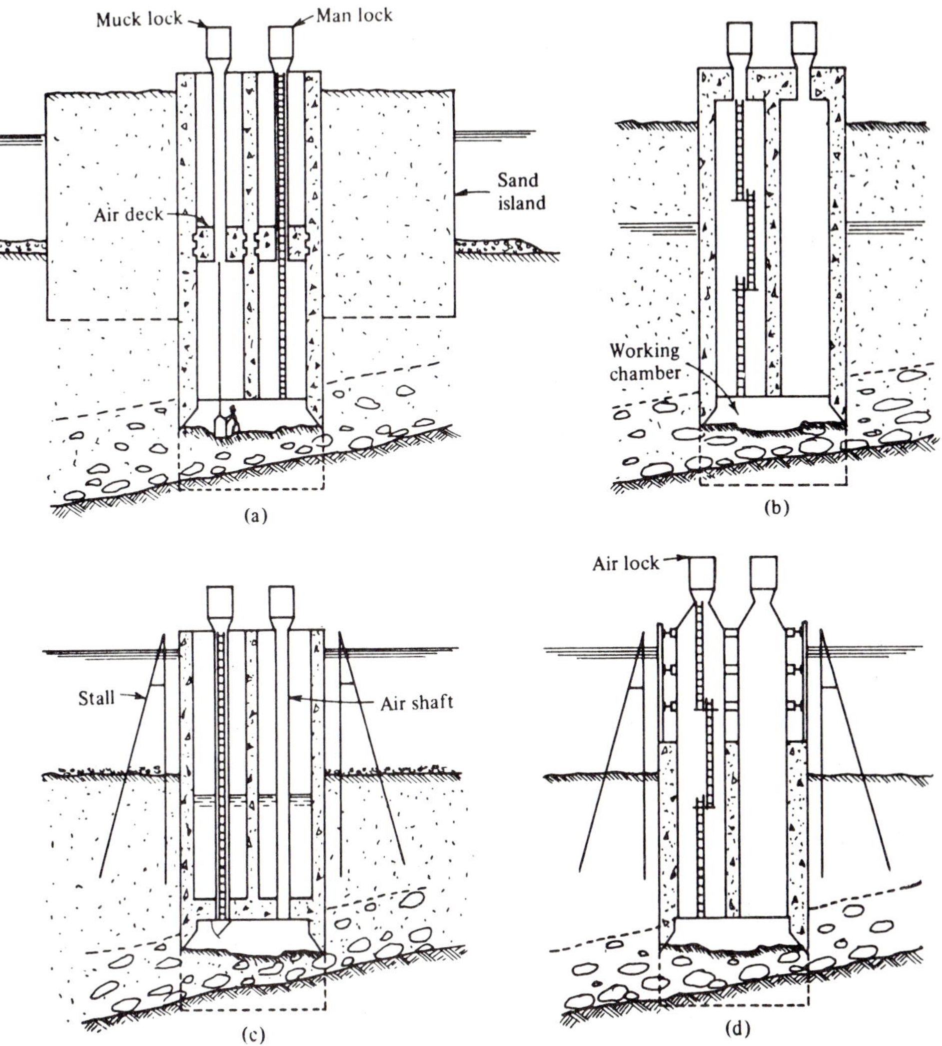

Figure 2.10 Various pneumatic caisson arrangements. (*a*) Temporary island; (*b*) land; (*c*) floating with ballasted thin walls; (*d*) floating with thick walls or tubes. (*After Swatek, Ref. 11.*)

schematic diagram of a typical construction sequence for one is shown in Fig. 2.11*a*, and a completed waterfront structure in Fig. 2.11*b*. The extension of the box caisson practice of constructing the structural unit in a drydock and towing it to the site where it is lowered onto a prepared bed has recently been used in the construction of river tunnels and subways. In such cases, tube sections with temporary bulkheads at each end are positioned in place, joined together underwater, properly ballasted, and pumped out. When fully connected, the bulkheads are removed and the tunnel's sections appear as a continuous unit. Excellent examples of this construction method

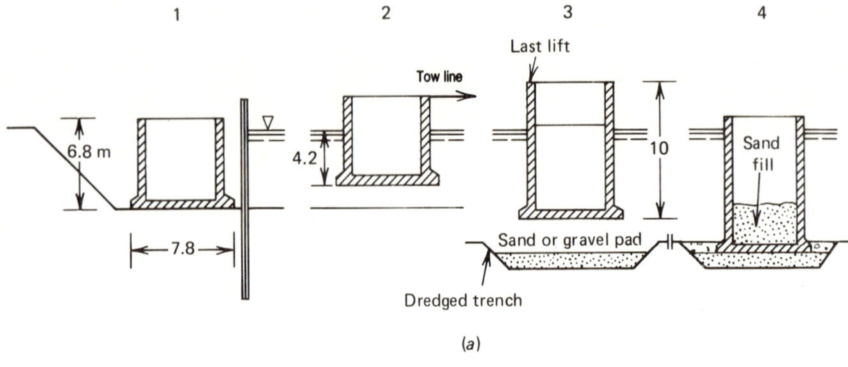

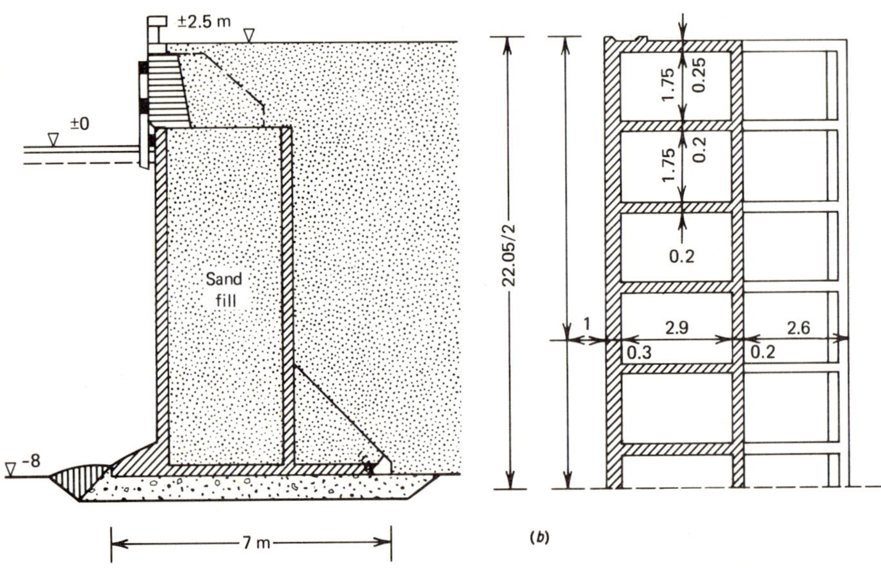

Figure 2.11 Box caisson arrangements. (*a*) Typical construction sequence: 1, concreting in cofferdam; 2, box caisson towing; 3, casting of last lift after arrival in position; 4, sand filling of box caisson sunk to bottom. (*b*) Completed waterfront structure. (*After Bazant, Ref. 10.*)

are the Oakland Bay crossing of the San Francisco subway system[15] and the underwater tunnels of the Chesapeake Bay Bridge and Tunnel crossing[16]; see Fig. 2.12 for a typical cross section of such a unit.

2.4 DESIGN OF DEEP FOUNDATIONS

The essential step in the construction of a deep foundation system lies in the site exploration and in the system's design. Already at these initial stages, a clear insight into the final product, including how it will probably be constructed and its projected

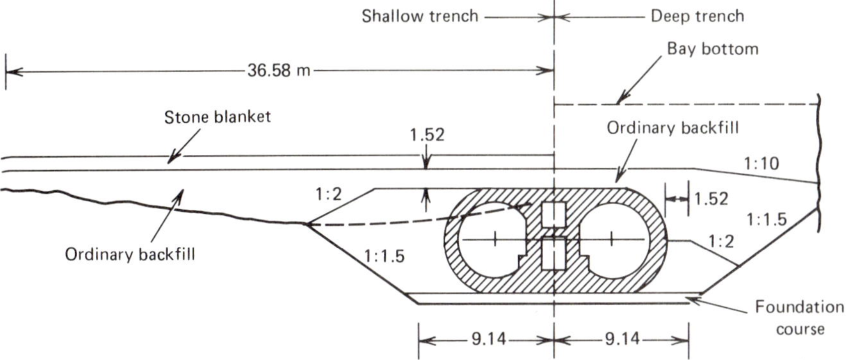

Figure 2.12 Box caisson underwater tube, San Francisco-Oakland Bay Crossing of BART. *(After Ref. 15.)*

performance, is absolutely necessary. It is equally important that the contractor have an insight into the design methodology, for if changed conditions come about during construction, the interface between contractor and designer will then be much smoother, to the benefit of both parties and ultimately to the benefit of the facility's owner and users. This section will discuss the types of loads generally encountered and the various design techniques used for deep foundation systems.

2.4.1 Loads

Most deep foundation systems for buildings, bridges, machines, etc., are designed on the basis of static load conditions consisting of dead loads, usually oriented vertically downward, and live loads. Of the two, the dead loads are usually far greater than the live loads, and so it is customary to design systems with the the vertical, downward, gravitational load as the primary design consideration. As in many other design situations, the magnitude and direction of the loads are often determined by a structural engineer, who then gives the information to the geotechnical engineer designing the system. Vertical loads of 10 to 300 tons on pile foundations, 50 to 7000 tons for pier foundations, and up to 100,000 tons or higher for caisson foundations are quite common.

Other possible orientations of static loads are lateral and uplift. In the lateral load condition, the deep foundation element acts in a cantilever mode (although this can be quite complicated depending on how the stiffness of the adjacent soil varies), and in the uplift load condition the deep foundation element acts as a tension member (versus acting as a compression member in the usual vertical, downward case). Obviously, loads at any orientation other than vertical or horizontal can be encountered. In such cases, the various components of the resultant force are statically determined and treated separately. Under proper design conditions, superposition of results can then usually be applied to determine the resultant force.

Dynamically applied loads (of any orientation) are significantly more difficult to design for than are static loads. Indeed, the area is being actively researched not only in regard to steady state vibrations (as in machine vibrations) but also in regard to the more complex random vibrations of wave forces against offshore platforms, of seismic forces, of wind forces against tall structures founded on deep foundations, etc.

Unusual circumstances can also cause other kinds of loads (both static and dynamic) of concern to designers to be applied to deep foundations. These include time-dependent settlement of fill soils around previously installed end-bearing piles, which creates downdrag or negative skin friction, and areal subsidence, which can cause distress to entire pile groups.

As with shallow foundations (see Sec. 1.2), one must be cognizant in regard to deep foundations of the entire geological and humanly engineered setting in which any system is to be placed before embarking on its design. Serious consideration must be given to unusual, and potentially dangerous, circumstances which may interfere with the system as finally designed and constructed. All too often that is not done and disputes over the completed project end up in a court of law where all the parties involved suddenly display excellent, but belated, insights into the problems encountered. In order to avoid this, an adequate soil exploration program should be undertaken *before* the problems arise. No deep foundation project should ever be initiated without previous soil borings at the site and the identification of the various strata to be found there.

2.4.2 Classical Dynamic Methods

Beginning as early as 1850,[17] the concept of forming an energy balance to predict pile capacity was suggested. The idea stems from the fact that a known energy input from the pile hammer (see Table 2.2 and Appendix A) must in some way be related to the soil resistance times the pile penetration. Thus the following relationships should, in some way, model the phenomenon:

$$W_H H = Q_0 S \qquad (2.1)$$

where W_H = hammer (ram) weight
H = fall of hammer (ram)
Q_0 = ultimate pile resistance
S = pile penetration (set)

Equation (2.1) is quite inappropriate, however, because it does not include a number of practical considerations, for example:

Energy loss due to friction
Energy loss due to pile bounce
Energy loss due to compression of hammer parts
Energy loss due to compression of pile
Uncertainties in soil response
Uncertainties in pore water response
Nonuniformity of pile resistance along shaft
Nonuniformity of pile resistance at tip

While this problem is recognized by most engineers and contractors working in the field, it has not stopped attempts at pursuing modifications to Eq. (2.1) which over time have added to it a greater amount of detail. At present there are over 200 such formulas available, of which a few of the more common examples will be given in this section.

A widely used (and abused) formula, originally developed for end-bearing piles driven through soft soil sediments in New Jersey, is the *Engineering News Record* (ENR) formula. It assumes the losses are a constant, C, which allows for the following equation for ultimate pile capacity:

$$Q_0 = \frac{W_H H}{S + C} \tag{2.2}$$

Q_0 is usually seen as an allowable pile load Q_A carrying a factor of safety (FS) of 6. So we arrive at Eq. (2.3), the final ENR formula:

$$Q_A = \frac{2W_H H}{S + C} \tag{2.3}$$

where Q_A = allowable pile load (based on FS = 6)
$\quad W_H$ = hammer (ram) weight
$\quad H$ = fall of hammer (ram), ft
$\quad S$ = pile penetration (set), inches per blow
$\quad C$ = constant, where C = 1.0 for drop hammers and 0.1 for steam/air hammers

An example of a more detailed dynamic formula is the Hiley formula[18] which, in its complete form, is given by Chellis[1,19]:

Formula for use with drop hammers and single-acting hammers:

$$R_u = \frac{e_f W_r h}{s + \frac{1}{2}(C_1 + C_2 + C_3)} \frac{W_r + e^2 W_p}{W_r + W_p} \tag{2.4}$$

Formula for use with double-acting and differential-acting hammers and diesel hammers:

$$R_u = \frac{12 e_f E_n}{s + \frac{1}{2}(C_1 + C_2 + C_3)} \frac{W_r + e^2 W_p}{W_r + W_p} \tag{2.5}$$

where R_u = ultimate carrying capacity of pile (considered as ultimate resistance to driving), lb, before applying any factor of safety, i.e., equal to Q_0 in Eq. (2.2).
$\quad W_r$ = weight of falling mass, lb (usually the ram, although the casing cylinder in many European hammers).
$\quad E_n$ = rated energy of hammer per blow, ft·lb, as published by manufacturers. Hammers should be run at maximum speeds listed, to obtain greatest efficiency.
$\quad h$ = length of free fall for drop hammers, and normal stroke of ram for single-acting hammers, in.
$\quad e_f$ = efficiency. The following percentages are suggested for use in computing bearing capacities (when computing fiber stresses in piles it is well to increase by 10 percent values below 100 percent):
100 percent for drop hammers released by trigger.
75 percent for drop hammers actuated by rope and friction winch (may decrease with

small drop, large drag, friction-band effect, or if operator keeps taut line, or may increase with large drop and small drag).

85 percent for McKiernan-Terry single-acting hammers.

75 percent for Warrington-Vulcan single-acting hammers.

85 percent for McKiernan-Terry, Industrial Brownhoist, National, and Union double-acting hammers.

75 percent for differential-acting hammers.

100 percent for diesel hammers. (Rated energies are intended by manufacturers to be available energies. However, engineers usually downgrade them because of difficulties and variable methods used in determinations. McKiernan-Terry diesels are least subject to reduction.)

80 percent for BSP semiautomatic single-acting hammers.

W_p = weight of pile, lb, including shoe, driving cap or anvil, follower, and weight of earth wedged in pile. For driving of pipe piles, Monotube shells, and steel mandrels, use steel weight only, and not weight of concrete which is poured later.

l = length of pile, in, measured from head to center of driving resistance. For end-bearing piles, use full length.

L = length of pile, ft, measured from head to center of driving resistance.

e = coefficient of restitution: varies from 0 for broomed or deteriorated wood-pile head or soft-cushion block material to 0.55 for steel hitting steel with no cushion. Use 0.40 for compact cushion over steel pile; 0.25 on wood piles or wood cap over concrete piles; other values proportionately.

s = final set, in/blow.

C_1 = temporary compression allowance for pile head and cap, in (see Table 2.7). Include additional value for followers.

C_2 = temporary compression of pile, in:

$$C_2 = \frac{R_u l}{AE} \tag{2.6}$$

(include additional value for followers).

C_3 = temporary compression allowance for quake of ground, in. Assume as 0.1 in for nominal value, with range from 0.2 in for resilient soils to 0 in for hardpan.

A = average of cross-sectional areas of pile at butt and center of resistance to driving, in^2. (Shell only if Monotube or pipe piles, shells and core for cast-in-place piles, or mandrel. For precast concrete piles, transform reinforcing to equivalent concrete area.)

E = modulus of elasticity for pile material being driven, lb/in^2.

p_1 = stress per square inch on driving cushion or on pile head:

$$p_1 = \frac{R_u}{\text{area of pile head}} \tag{2.7}$$

It must be clearly understood that classical dynamic formulas, such as the ENR and Hiley formulas just discussed, are approximate at best. The results of the Michigan pile test program[20] (see Table 2.8) show that the true factors of safety vary over a wide range for all pile formulas and pile load ranges. Thus these order-of-magnitude-only formulas *should never be used for pile design or pile capacity prediction.* Indeed, their only constructive use at this point in time is to give the field inspector and contractor an indication of pile movement near the end of driving. This is done by using the capacity of the pile (as predicted by other methods illustrated in the following sections) as a known value and back-calculating the pile set, which is given in units of inches per blow. The inverse of the value found for the pile set (i.e., the value for blows per inch) can be used in a somewhat meaningful manner.

Table 2.7 Temporary compression allowances C_1 for the pile head and cap†

Material to which blow is applied	Easy driving, p_1 = 500 lb/in^2 on cushion, or butt if no cushion, in	Medium driving, p_1 = 1000 lb/in^2 on head or cap, in	Hard driving, p_1 = 1500 lb/in^2 on head or cap, in	Very hard driving, p_1 = 2000 lb/in^2 on head or cap, in
Head of timber pile	0.05	0.10	0.15	0.20
3–4-in packing inside cap on head of precast concrete pile	0.05 + 0.07‡	0.10 + 0.15‡	0.15 + 0.22‡	0.20 + 0.30‡
$\frac{1}{2}$–1-in mat pad only on head of precast concrete pile	0.025	0.05	0.075	0.10
Steel-covered cap, containing wood packing, for steel piling or pipe	0.04	0.08	0.12	0.16
$\frac{3}{16}$-in red electrical fiber disk between two $\frac{3}{8}$-in steel plates, for use with severe driving on Monotube piling	0.02	0.04	0.06	0.08
Head of steel piling or pipe	0	0	0	0

†After Chellis, Ref. 1.

‡The first figure represents the compression of the cap and wood dolly or packing above the cap, whereas the second figure represents the compression of the wood packing between the cap and the pile head.

Example 2.1 For a pile designed for an allowable load of 40 tons driven by an air hammer with a rated energy of 15,000 ft·lb, what is the approximate terminal resistance of the pile using the ENR formula?

SOLUTION

$$Q_A = \frac{2W_H H}{S + C}$$

$$S = \frac{2W_H H}{Q_A} - C$$

$$= \frac{2(15,000)}{80,000} - 0.1$$

Table 2.8 Summary of range of factors of safety for equations used in the Michigan pile test program†

| | Upper and lower limits of FS $= Q_o/$ Q_d‡ Range of Q_o, kips | | |
Formula	0–200	200–400	400–700
Engineering News Record	1.1–2.4	0.9–2.1	1.2–2.7
Hiley	1.1–4.2	3.0–6.5	4.0–9.6
Pacific Coast Uniform Building Code	2.7–5.3	4.3–9.7	8.8–16.5
Redtenbacher	1.7–3.6	2.8–6.5	6.0–10.9
Eytelwein	1.0–2.4	1.0–3.8	2.2–4.1
Navy-McKay	0.8–3.0	0.2–2.5	0.2–3.0
Rankine	0.9–1.7	1.3–2.7	2.3–5.1
Canadian National Building Code	3.2–6.0	5.1–11.1	10.1–19.9
Modified *Engineering News Record*	1.7–4.4	1.6–5.2	2.7–5.3
Gates	1.8–3.0	2.5–4.6	3.8–7.3
Rabe	1.0–4.8	2.4–7.0	3.2–8.0

†After data given in Ref. 20.

‡Q_o = *ultimate* test load.

Q_d = *design* capacity, using the safety factor recommended for the equation (values range from 2 to 6, depending on the formula).

$$= 0.27 - 0.1$$

$$S = 0.17 \text{ inches per blow}$$

$$\frac{1}{S} \simeq 6 \text{ blows per inch}$$

The field inspection aspect of pile placement is critical in the construction of a safe and efficient deep foundation system. The procedure of counting blows per inch of pile penetration is important but, as just illustrated, can only be performed when using an impact hammer. In order to assess pile penetration when using a vibratory pile driver, Davisson[21] has derived the following relationship:

$$r_p = \frac{550H_p - fS_LQ_0}{Q_0 - W_H} \tag{2.8}$$

where r_p = final rate of pile penetration, ft/s

H_p = horsepower of energy delivered to the pile (see Table 2.3 and Appendix B)

f = frequency of driver (see Table 2.3 and Appendix B)

S_L = loss factor, ft/cycle (see Ref. 21)

Q_0 = ultimate pile load, lb

W_H = weight of vibratory driver

2.4.3 Static Methods

The manner of pile installation being mainly the contractor's decision, it appears that pile design should be more heavily weighted toward existing and modified soil conditions, as influenced by final pile type, than toward anything else. Thus we have what might be considered a rational pile design method, the so-called "static design," based on the pile-plus-soil system *after* installation. Conceptually, the technique considers the pile point and shaft resistances separately and sums the results as follows:

$$Q_0 = Q_p + Q_s \tag{2.9}$$

$$Q_0 = A_p p_0 + A_s s_0 \tag{2.10}$$

where Q_0 = ultimate pile load
$\quad Q_p$ = load carried by pile point
$\quad A_p$ = area of pile point
$\quad p_0$ = ultimate shear resistance of soil or rock beneath pile point
$\quad Q_s$ = load carried by pile shaft
$\quad A_s$ = surface area of pile shaft
$\quad s_0$ = ultimate shear resistance of soil adjacent to pile with pile material

In this formulation, A_p and A_s are functions of the pile itself, while p_0 and s_0 are related to the shear strength properties of the soil. More specifically, p_0 is related to the bearing capacity of the soil or rock beneath the pile point (as with shallow foundation theory) and s_0 is related to Mohr/Coulomb failure theory. The following is recommended for general pile design. For point resistance p_0:

$$p_0 = cN_c + qN_q + \tfrac{1}{2}\gamma BN_\gamma \tag{2.11}$$

where the third term is neglected to compensate for the weight of the pile itself and because of the usually small value of B, the pile width. Thus, depending on the type of soil the pile point is founded in:

$$p_0 = qN_q \quad \text{for granular soils} \tag{2.12}$$

$$p_0 = cN_c \quad \text{for cohesive soil (sometimes used as } 9c) \tag{2.13}$$

$$p_0 = cN_c + qN_q \quad \text{for } c - \phi \text{ soils} \tag{2.14}$$

where p_0 = unit point resistance
q = overburden pressure related to $\gamma'z$
γ' = effective unit weight of overburden soil
z = depth of deep foundation
c = cohesion
N_c, N_q = bearing capacity factors which are related to the soil's angle of shearing resistance, ϕ

The value of q should be modified by a factor α, where α is a function of depth and soil strength as shown in Table 2.9.

For the bearing capacity factors, N_c can be taken directly from Table 1.5, but some amount of controversy surrounds the value of N_q. Since the mode of failure of a deep foundation generally differs from that of a shallow foundation, values of N_q from Table 1.5 cannot be used. Vesic[23] has contributed significantly to the topic, and using his bearing capacity theory for deep foundations (which is both logical and conservative), one obtains the curve shown in Fig. 2.13. That curve should be used for determining N_q in pile design.

It is also to be noted that p_0, like most other soil properties, has been related to the standard penetration blow count N, as in Table 2.10. Those values, however, should only be taken as a general guide, or used only for preliminary design purposes.

Concerning the value of shear resistance between the soil adjacent to the pile and the pile material itself, the following procedure should be followed:

$$s_0 = c_a + \sigma_h \tan \delta \qquad (2.15)$$

$$s_0 = c_a + K\sigma_v \tan \delta \qquad (2.16)$$

$$s_0 = c_a + K\gamma'z \tan \delta \qquad (2.17)$$

where s_0 = unit skin resistance at the midpoint of the stratum in question
K = earth pressure coefficient (see Table 2.11)
γ' = effective unit weight of soil
c_a = adhesion, which is a function of c (see Table 2.12)
δ = pile-to-soil friction angle, which is a function of ϕ (see Table 2.12)
z = average depth of stratum in question

Table 2.9 Reduction values α for overburden calculations†

	ϕ				
D/B	26°	30°	34°	37°	40°
5	0.75	0.77	0.81	0.83	0.85
10	0.62	0.67	0.73	0.76	0.79
15	0.55	0.61	0.68	0.73	0.77
20	0.49	0.57	0.65	0.71	0.75
≥25	0.44	0.53	0.63	0.70	0.74

†After Berenzantzev et al., Ref. 22.

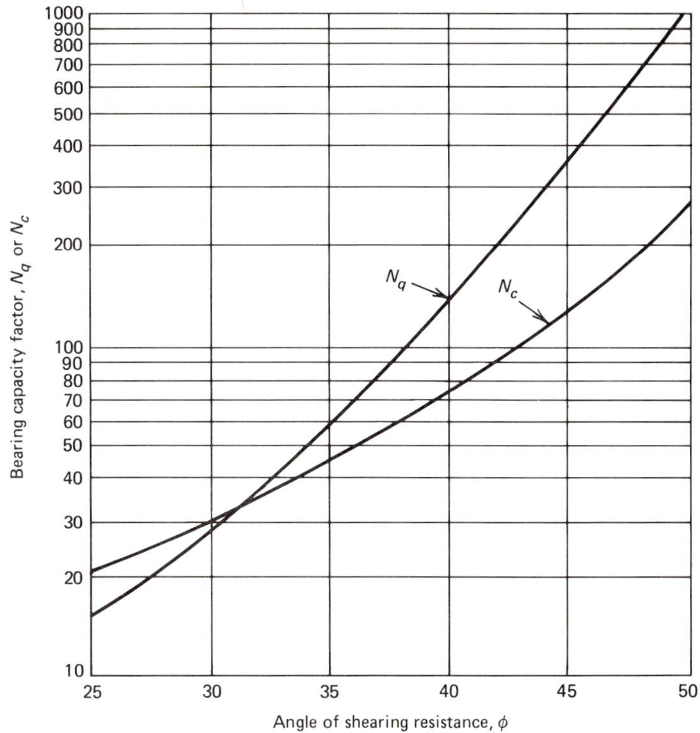

Figure 2.13 Bearing capacity factors N_c and N_q for deep foundations. Values of N_c are taken directly from Table 1.5. *(After Vesic, Ref. 23.)*

The calculated value of s_0, for the given situation under consideration, should be checked against known (or suggested) values of skin resistance, an example of which is given as Table 2.13.

Example 2.2 Determine the ultimate capacity of the following pile according to static design methods, and also the allowable capacity based on a factor of

Table 2.10 Relationships of point resistance values to standard penetration blow count N†

Soil type	$N < 15$	$N > 15$
Sand	$p_0 = 4N‡$	$p_0 = 60 + 2(N - 15)$
Silt	$p_0 = 2.5N$	$p_0 = 37.5 + 1.25(N - 15)$
Clay	$p_0 = 2N$	$p_0 = 30 + (N - 15)$

†After Meyerhof, Ref. 24.
‡Values of p_0 are in ton/ft^2.

Table 2.11 Earth pressure coefficients K for use in pile design[†]

Type of sand	Pile placement method	Value of K
Loose ($D_R < 50\%$)[‡]	Jetted	0.5–0.75
	Drilled	0.75–1.5
	Driven	2.0–3.0
Dense ($D_R > 85\%$)	Jetted	0.5–1.0
	Drilled	1.0–2.0
	Driven	3.0–5.0

[†]Modified from Sowers and Sowers, Ref. 25.

[‡]D_R = relative density; for intermediate density, interpolate between loose and dense values.

safety of 4.0. The pile is a driven 18-in-diameter cast-in-place concrete pile, as shown below.

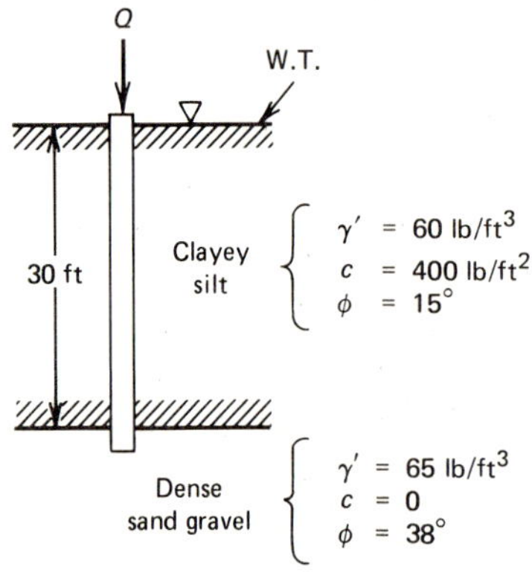

SOLUTION (*a*) Point resistance:

$$p_0 = qN_q$$

$$= (\alpha\gamma'z)N_q$$

$$= (0.72)(60)(30)(110)$$

$$= 142{,}000 \text{ lb/ft}^2 = 142 \text{ kips/ft}^2$$

(*b*) Average skin resistance:

$$S_0 = c_a + K\gamma'z \tan \delta$$

$$= (0.40)(400) + (1.0)(60)(15) \tan (0.50)(15)$$

$$= 160 + 900 \tan 7.5^0$$

$$= 280 \text{ lb/ft}^2 = 0.280 \text{ kips/ft}^2$$

Table 2.12 Proposed coefficients of skin friction between soils and construction materials†

$$f_\phi = \delta/\phi, f_c = \frac{c_a}{c}, f_{c,\max} = \frac{c_{a,\max}}{c_{\max}} \text{ ; without factor of safety}$$

Construction material			Sand		Cohesionless silt			Cohesive granular soil		Clay		
			$0.06 < D < 2.0$ mm		$0.002 < D < 0.06$			50% Clay + 50% sand		$D \leqq 0.06$ mm		
			Dry	Saturated	Dry	Saturated		Consistency index: 1.0–0.5		Consistency index: 1.0–0.73		
	Surface finish of construction material		Dense		Dense	Loose	Dense					
			f_ϕ	f_ϕ	f_ϕ	f_ϕ	f_ϕ	f_ϕ	f_c	f_ϕ	f_c	$f_{c,\max}$
Steel	Smooth	Polished	0.54	0.64	0.79	0.40	0.68	0.40	· · ·	0.50	0.25	0.50
	Rough	Rusted	0.76	0.80	0.95	0.48	0.75	0.65	0.35	0.50	0.50	0.80
Wood		Parallel to grain	0.76	0.85	0.92	0.55	0.87	0.80	0.20	0.60	0.4	0.85
		At right angles to grain	0.88	0.89	0.98	0.63	0.95	0.90	0.40	0.70	0.50	0.85
Concrete	Smooth	Made in iron form	0.76	0.80	0.92	0.50	0.87	0.84	0.42	0.68	0.40	1.00
	Grained	Made in wood form	0.88	0.88	0.98	0.62	0.96	0.90	0.58	0.80	0.50	1.00
	Rough	Made on adjusted ground	0.98	0.90	1.00	0.79	1.00	0.95	0.80	0.95	0.60	1.00

†After Potyondy, Ref. 26.

Table 2.13 Friction between piles and soils†

Material	Identification	Ordinary range of values per ft^2 of bounding area of pile for skin friction s_0, lb/ft^2	N number of blows per ft on 2-in OD sampling spoon from 140-lb weight falling 30 in
Fine-grained soils:			
Very soft clay	Fist penetrates several inches easily	125 ± 125	0–2
Mud	· · ·	250 ± 200	
Silt	· · ·	300 ± 200	
Soft clay	Thumb penetrates several inches easily	400 ± 200	2–4
Silty clay	· · ·	600 ± 200	
Sandy clay	· · ·	600 ± 200	
Medium clay	Thumb dents several inches—moderate effort	700 ± 200	4–8
Sandy silt	· · ·	800 ± 200	
Firm clay	Thumb dents several inches—great effort	900 ± 200	
Dense silty clay	· · ·	1200 ± 300	
Stiff clay	Thumb dents readily—great effort to penetrate	1500 ± 400	8–15
Very stiff clay	Thumbnail indents readily	3000 ± 1000	15–30
Hard clay	Thumbnail indents with difficulty	Over 4000	Over 50
Coarse-grained soils:			
Very loose sand and silt or clay	· · ·	100 ± 50	0–6
Medium sand and silt or clay	· · ·	500 ± 100	6–30
Dense sand and silt or clay	· · ·	700 ± 100	30–50
Very dense sand and silt or clay	· · ·	900 ± 100	Over 50
Sand:	· · ·		
Very loose	· · ·	300 ± 200	0–4
Loose	· · ·	500 ± 250	4–10
Medium	· · ·	800 ± 300	10–30
Dense	· · ·	1200 ± 400	30–50
Very dense	· · ·	1700 ± 500	Over 50
Sand and gravel	· · ·	2000 ± 1000	Over 50
Gravel	· · ·	2500 ± 1000	Over 50

†Modified from Chellis; Ref. 1.

(c) Calculate point load, skin load, total load, and allowable load.

$$Q_0 = Q_p + Q_S$$

$$= A_p p_0 + A_s s_0$$

$$= \frac{\pi 1.5^2}{4}\, 142 + (\pi 1.5)(30)(0.280)$$

$$= 250 + 40$$

$$= 290 \text{ kips}$$

(Note that the pile could be called a "point-bearing pile" since $250/290 = 86$ percent of its capacity is supported by the pile point. If the major proportion of the pile were supported by the pile shaft, that is, if $Q_s > Q_p$, then the pile would be a "skin friction" or "friction pile.")

The allowable pile capacity based on an FS of 4.0 is

$$Q_A = \frac{Q_0}{\text{FS}}$$

$$= \frac{290}{4.0}$$

$$= 72.5 \text{ kips}$$

Use 70 kips.

Miscellaneous comments on pile design by static methods

1. For deep piles, greater than 50 ft, the value of q in the point resistance equation becomes unreasonably large. Therefore a limit of approximately $20B$ is recommended as a maximum value of z.
2. There are numerous theories leading to different values of N_q; see Vesic.[23]
3. Because of the above two comments, unrealistically high values of p_0 can sometimes be calculated. Limiting values of p_0 have been recommended for granular soils as a function of relative density,[23] as shown in the table below:

D_R	Jetted or drilled, kips/ft^2	Driven, kips/ft^2
0	2.0	7.0
20	2.6	8.0
40	4.6	12.0
60	10.0	24.0
80	60.0	100.0
100	$\simeq 140.0$	$\simeq 1000.0$

4. The value of K for skin friction calculations varies widely between active and passive conditions, depending primarily upon the method of pile installation.

This, not being known during the design stage, must be estimated as it was in Example 2.2; see Table 2.11.

5. Since the theoretical distribution of s_0 is triangular, the value of z is the average, or midstratum, depth.

6. When the pile penetrates through different layers, each skin friction calculation must be made separately and the results added.

7. The values of c and ϕ are at the heart of the calculation. Without reasonably accurate estimates of them, the accuracy of the calculated pile capacity is quite suspect.

8. The philosophy of simultaneous mobilization of full point resistance and full skin friction seems reasonable when considering the estimate of ultimate pile capacity, that is, Q_0, but when a factor of safety is included to obtain the allowable pile capacity Q_A, some uncertainty regarding mobilization of skin friction is justified. This general topic, i.e., load transfer in deep foundations, is an area actively being researched.[27]

9. The design of drilled piers generally follows along the principles laid out in the above information. A notable exception, however, is drilled piers on rock as commonly seen in large buildings, factories, and bridges. Because of the expense of pile load tests, most urban areas have determined presumptive bearing capacities for their regions, as seen in Table 2.14. These values, generally conservative, usually call for widening the pier base so as to increase the point-carrying capacity. It should be noted that for a nonyielding rock foundation, skin friction should not be counted upon.

10. If the drilled pier does not have sufficient carrying capacity, it is common to socket it into the rock. Thus an adhesion between concrete and rock should be added to the point-carrying capacity. Table 2.15 gives some recommended values to be multiplied by the surface area involved.

11. In certain cases where the soil adjacent to an end-bearing pile settles with respect to the pile itself, skin friction forces will be set up that act downward on the pile, adding to the imposed load. This situation, called *negative skin friction* or *downdrag,* can be assessed by using a $-s_0$ value in Eq. (2.10). The resulting magnitude of $-Q_S$ can be enormous.

2.4.4 Exploration Design Methods

For every construction project where a failure might cause concern about human safety, property damage, economic loss, environmental damage, or even merely embarrassment to the parties involved, a soil exploration program should be undertaken before design begins. For deep foundations such a program could involve a simple method like taking disturbed auger borings and collecting jar samples for identification, or more complex ones like determining continuous in situ shear strength to refusal and then rock coring afterward. Table 2.16 gives a good overview of the methods currently available.[28] A single boring, however, is rarely—if ever— sufficient. Instead a set of borings is needed which enables one to trace individual

Table 2.14 Allowable bearing values for rock for drilled pier foundations from various city codes†

Code	Year	Massive crystalline bedrock, tons/ft²	Sound foliated rock, tons/ft²	Sound sedimentary rock, tons/ft²	Soft and broken rock excluding shale, tons/ft²	Shales, soft, tons/ft²	Broken shales, tons/ft²
Baltimore	1962	100	35	· · ·	10	· · ·	To be fixed by comissioner
BOCA‡	1970	100	40	25	10	4	1.5
Boston	1970	100	50	10	10	· · ·	To be fixed by building official
Chicago§,¶	1970	100	100	100			
Cleveland	Code 1951	· · ·	· · ·	25			
	Suppl. 1969	· · ·	· · ·				
Dallas	1968	· · ·	· · ·	Maximum value = 20% of ultimate crushing strength			
Detroit¶	1956	100	100	100	12	12	
Indiana	1967	· · ·	· · ·				
Kansas City	Code 1961	· · ·	· · ·	Not more than 20% of the ultimate crushing strength			
	Rev. 1969	· · ·	· · ·	Maximum value = 20% of ultimate crushing strength			
Los Angeles	1970	10	4	3	1	1	1
New York City¶	1970	60	60	60	8		
Ohio	1970	100	40	15	10	4	
Philadelphia	1969	50	20	15	10–15	8	
Pittsburgh¶	Code 1969	25	25	25	8	8	
	Ord. 1969						
Richmond	1968	100	40	25	10	4	1.5
St. Louis	Code 1960	100	40	25	10	1.5	1.5
	Ord. 1970						
San Francisco	1969	3–5	3–5	3–5			
Uniform Building							
Code	1970	Maximum value = 20% of ultimate crushing strength					

†After D'Appolonia et al., Ref. 9. The values in this table should not be used without first checking with the particular code for changes.

‡Building Officials Conference of America, Inc.

§Allows increase of 20 percent of bearing capacity for each foot of embedment (beyond initial required depth), but should not exceed twice given values.

¶Code uses term "Solid" or "Hard sound" rock; does not differentiate between kinds of rock.

strata from one location to another across the entire site and for a reasonable distance beyond. Such data mean that a cross section (sometimes called a "fence diagram") can be drawn, which is of tremendous value in foundation design.

Such a cross section can be relatively uncomplicated, as shown in Fig. 2.14, which was drawn from data supplied by three borings, or more complicated, as shown in Fig. 2.15, which was drawn from data supplied by eleven borings, or even

Table 2.15 Suggested values of adhesion between concrete and rock for drilled pier foundations†

Investigator	Measured values, lb/ft^2	Presumptive values allowed, lb/ft^2	Remarks
Schmidt (1956)	41,200		A "solid rock" formation
Moore (1964)	20,000 (Failure not reached)		Weathered sandstone and shale
ASCE and New York City	55,600 (Failure not reached)		Manhattan schist
Lovaas (1970)	184,300–288,000		Laboratory tests on 2-in-diameter sockets in limestone
New York City Building Code		28,800	Sound, hard rock (to be verified)
White (1967) and Coates (1967)		$0.05 f_c'$	$f_c' =$ Ultimate concrete strength

$$\begin{bmatrix} f_c, \text{lb/in}^2 & \text{Adhesion, lb/ft}^2 \\ 3000 & 21,600 \\ 4000 & 28,800 \\ 5000 & 36,000 \end{bmatrix}$$

†After D'Appolonia et al., Ref. 9.

very complicated, as shown in Fig. 2.16, which was drawn from data supplied by hundreds of borings. Each of these soil profiles were used to estimate deep foundation tip elevations.

In the first case, Fig. 2.14, the original intention was to install piles for a 19-story apartment building down to stratum 4, a dense, well-graded sand. It was subsequently decided to install pressure-injected footings into the upper portion of stratum 3. It should be noted that this system resulted in settlements of up to 3.4 in. The settlement was not a long-term concern, however, and settling ceased after the full dead load of the structure was completed.[29]

The second case, Fig. 2.15, illustrates the soil-rock profile along a river where a sheet pile bulkhead was to be constructed. Boring refusal was clearly illustrated by the blow counts, which increased rapidly approximately 20 ft beneath the ground surface. To be assured that it was solid rock at this depth, rock cores were taken periodically and the percent recovery determined. Steel sheeting was subsequently driven to refusal within ±1ft at the boring locations, and to ±3ft between borings as shown in the soil/rock profile of Fig. 2.15.

The third case, Fig. 2.16, illustrates the type of soil exploration program that was required before constructing a major bridge across a large river. The profile is along the center line of the approach span. It resulted eventually in steel pipe piles being driven into the dense sand and gravel layer underlying all of the fine-grained soil zones.

In each of these cases, pile design was heavily influenced by the soil profile as

Table 2.16 Subsurface exploration—exploratory boring methods†

Method	Procedure	Use	Limitations
Auger boring (ASTM D-1452)	Hand or power auger with removal of material at regular short intervals.	Identify changes in soil texture above water table. Locate groundwater.	Grinds soft particles—stopped by rock, etc.
Test boring (ASTM D-1586)	Drill hole, sample at intervals with 35-mm (1.4-in) ID, 50-mm (2-in) OD, split barrel sampler driven 0.45 m (18 in) in three 15-cm (6-in) intervals by 63.6-kg (140-lb) hammer falling 0.76 m (30 in). Below water, maintain hydrostatic balance with fluid.	Identify texture and structure; estimate density or consistency in soil or soft rock.	Gravel, hard seams.
Continuous core: soil (ASTM D-2113)	Force or drill tube into soil until resistance prevents further movement. Remove cuttings with air or water.	Identify soil texture and structure continuously in cohesive soils.	Gravel, hard seams, sands. Misleading squeeze in some clays.
Borehole camera, TV	View inside of bore hole.	Examine stratification in place.	Textural changes indistinct. Sometimes obscure below water table.

Table 2.16 Subsurface exploration—exploratory boring methods† (*Continued*)

Method	Procedure	Use	Limitations
Continuous core: rock (ASTM D-2113)	Rotate tube with diamond-studded bit to cut annular hole. Cuttings removed by circulating water. Core retained in tube by cylindrical wedge. Best with stationary inner tube to protect core.	Identify rock strata and structural defects continuously.	No data on soft seams, etc.
Dynamic sounding	Drive enlarged point on end of rod with weight falling fixed distance, in increments of 0.15 to 0.3 m (6 in to 1 ft).	Identify significant changes in density or consistency of materials.	Misleading in gravel, or cemented seams.
Static penetration (ASTM D-3441)	Force enlarged cone [Dutch cone: 3.6-mm (1.4-in) diameter, 60° angle] on end of rod into soil, measuring resistance of point at regular intervals. Other sizes can be adapted to local condition.	Identify subtle changes in density or consistency. Possibly identify soil by ratio of point load to skin friction.	Stopped by hard strata, misleading in gravel, hard seams.
Pits, trenches	Excavate pit or trench, by hand, large auger, and by excavator.	Visual examination of structure and stratification above water table.	Caving of walls, ground-water.

The following drilling methods are frequently used to advance the hole in test boring and core drilling; for quick but crude exploration, they are occasionally used independently.

Wash boring; rotary wet drilling; rotary air drilling	Chop with chisel bit or rotate toothed cutter. Cuttings washed to surface by circulating water or drilling mud through bit.	Identify coarser fraction from cuttings, hardness from drilling rate.	Misleading if appreciable fines present.
Churn or cable drilling	Pound and churn soil boulders and rock to slurry by dropping heavy chisel bit in wet hole. Bail water and cuttings at intervals.	Drill and identify broken rock, etc., from cuttings.	Strata difficult to define. Quick condition formed in sands.
Percussion drilling	Impact—drill with jack hammer; remove cuttings with compressed air.	Identify rock from cuttings, hardness from rate.	Plugged by wet soil.

†After Sowers, Ref. 28.

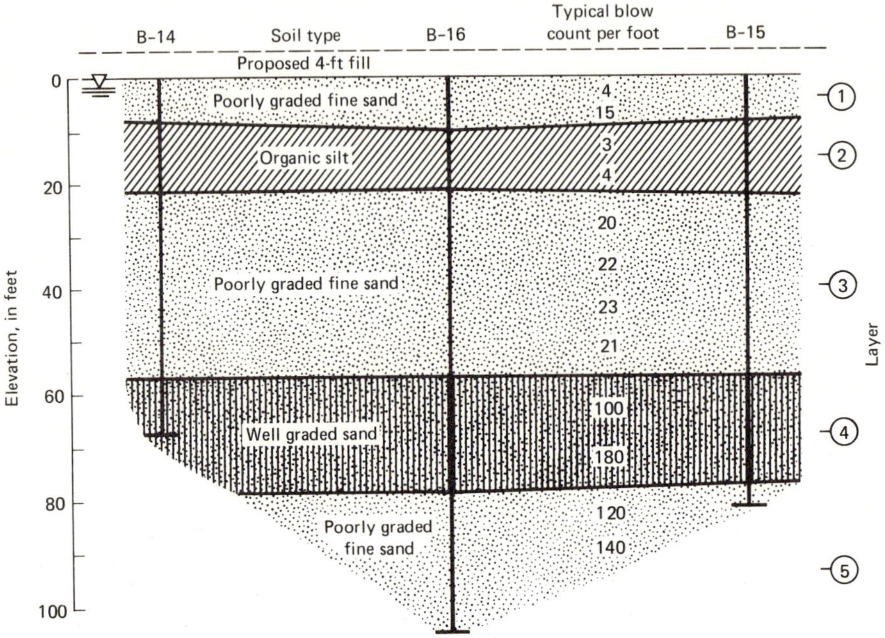

Figure 2.14 Soil profile and properties at site of high-rise concrete building. *(After Koerner and Partos, Ref. 29.)*

determined by exploratory soil borings and/or rock cores. When properly conducted, such tests can provide positive proof of the conditions at the boring location. A caution, however, is appropriate. The nature of the soil-rock strata *between* various boring locations can only be assumed. Such assumptions are regularly challenged in court when legal disputes involving deep foundation problems arise; yet making those assumptions is a technical necessity. If a valid concern exists regarding the possibility of strata irregularities between boring sites, it is obvious that additional borings must be taken. But there is a limit to the number of borings that can be taken, and engineering judgment must ultimately be relied upon in deciding on the final pile design. To emphasize that assumptions are involved, some engineers on their drawings will connect layers from boring to boring with a dashed line or with a dashed line containing question marks between the dashes; some engineers attach a disclaimer in regard to the assumptions the design plans and job specifications involve. It should also be pointed out that even though soil-rock profiles of necessity include assumptions, no *information* about the strata in question should ever be suppressed.

It should also be pointed out that a few techniques are available which can help to eliminate some of the uncertainty connected with assumptions about soil continuity between adjacent borings, i.e., nondestructive testing methods such as ground-probing radar or very low frequency electromagnetic probes. (These and other procedures will be covered in Sec. 9.4.)

In summary, the determination of pile tip elevations using soil-rock profiles, tempered with adequate information about the strengths of the strata in question, is a recommended procedure for many projects. The information on which that determination is based should be made accessible to all the parties involved in a project (owner, engineer, contractor, etc.), and those parties should also be informed of the intentions of and assumptions made by the design engineer.

2.4.5 Advanced Deep Foundation Design Methods

There are a number of advanced deep foundation design methods which we will discuss here since they seem to hold significant promise for the future. They are already being used to design most of the major deep foundation systems currently being undertaken, e.g., offshore drilling platforms, power plant foundations, etc., and should eventually find their way into simplified procedures for common use. The methods are known as *wave equation analysis,* the *finite-element method,* and the *in situ monitoring method.*

Wave equation analysis This method was originally developed by Smith[30,31] and subsequently extended by others, some of whom are employed by the Raymond Pile Company. Relatively complete reviews of its history have been made by Hirsch et al.[32] and Bowles.[33]

The solution is basically an adaptation of the finite-difference method, and as such it is a numerical technique, requiring a computer. Computer codes are available for general solutions (see Bowles[33]). In the analysis the pile being analyzed is represented as segmented into a number of discrete weights, with soil resistance acting upon each of them, the weights being connected by springs as shown in Fig. 2.17. The hammer ram and pile cap are also represented by weights, while the cap block and cushion block are drawn as spring elements. Beneath the last weight is a force representing the point resistance. The solution to the problem thus results in this value, that is, Q_p, and the individual side friction resistance forces, that is, Q_s, the sum of Q_p and the Q_s's being equal to the pile capacity Q_0 at various times during the driving process. Smith assumed the soil along the shaft to behave as shown in Fig. 2.18, where *OABC* represents loading and *DEFG* represents unloading due to the dynamic action of the pile hammer. At the pile point only compression can occur, so the path is *OABCF*. These characteristics are defined by the terms Q and R_u, which are the soil quake (maximum elastic deformation) and ultimate soil resistance (plastic behavior of soil), respectively. A separate response can be generated for each spring. In the following equations, developed by Smith, $K'(m) = R_u(m)/Q(m)$, where $K'(m)$ is the elastic spring constant for external spring m.

$$D(m, t) = D(m, t - 1) + 12 \, \Delta t \, V(m, t - 1) \qquad (2.18)$$

$$C(m, t) = D(m, t) - D(m + 1, t) \qquad (2.19)$$

$$F(m, t) = C(m, t)K(m) \qquad (2.20)$$

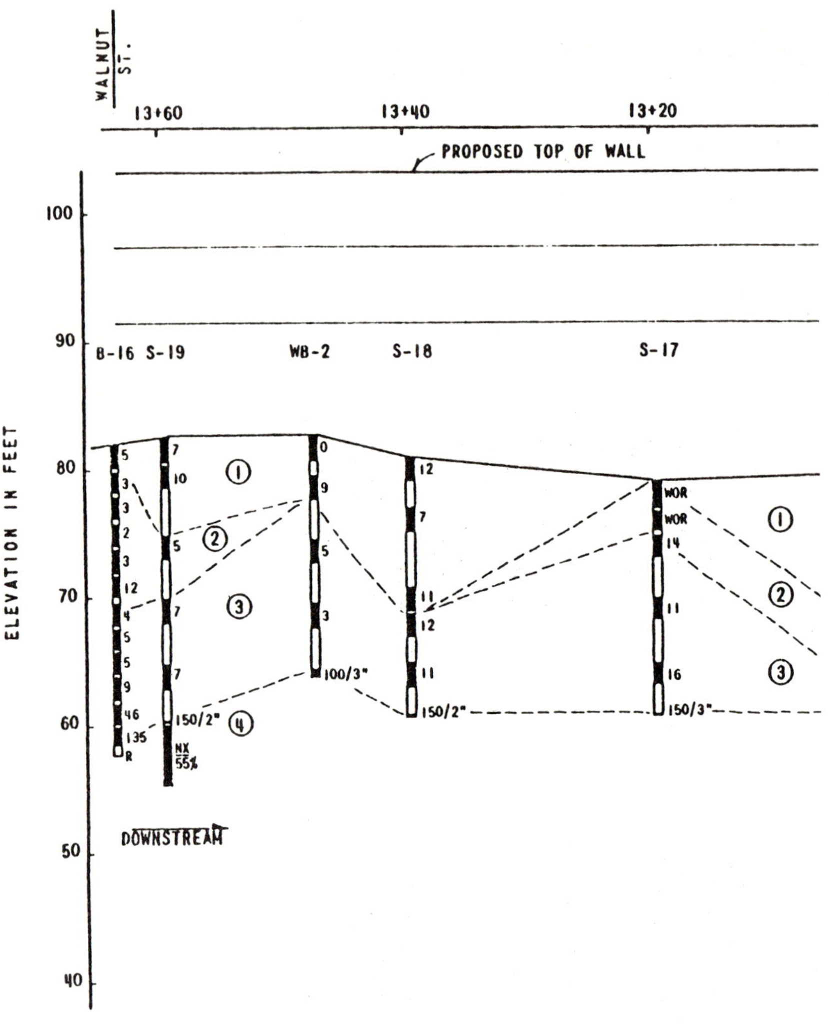

Figure 2.15 Soil profile and properties at site of river bulkhead. *(Compliments of City of Philadelp Woodward-Clyde Consultants and Delta Group, Inc.)*

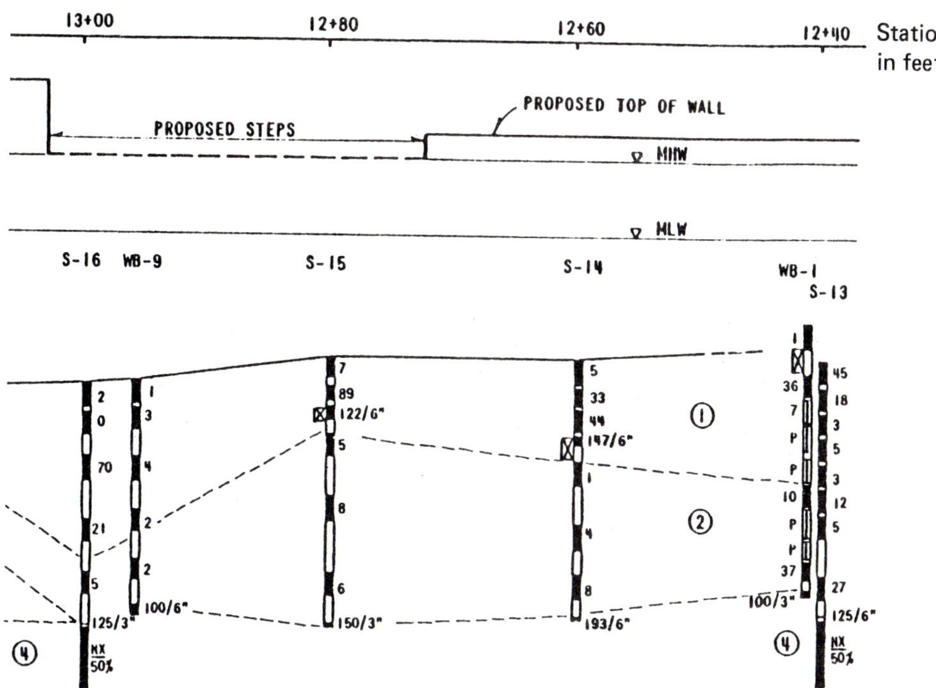

<u>LEGEND:</u>

  ① FILL - BOULDERS, CONCRETE, TIMBER, CINDERS, BRICK, ETC. IN MATRIX OF BLACK SOFT SILT AND CLAY AND LOOSE SAND

  ② CLAY AND SILT - VERY SOFT TO FIRM BROWN, GRAY AND DARK GRAY FINE SANDY SILT, CLAYEY SILT, AND SILTY CLAY WITH TRACES OF VEGETATIVE MATERIAL

  ③ SAND - VERY LOOSE TO MEDIUM DENSE GRAY TO BLACK COARSE TO FINE SAND WITH TRACES OF FINE GRAVEL, SILT, AND CLAY

  ④ DECOMPOSED TO INTACT ROCK - VERY DENSE, PARTIALLY FRIABLE WEATHERED TO STAINED INTACT BROWN TO GRAY MICA SCHIST.

NOTE: B - DRILLED IN 1968

     S - DRILLED IN 1980

     WB - DRILLED FOR THIS INVESTIGATION

     ⊠ - TIMBER DRILLED THROUGH BY BORING

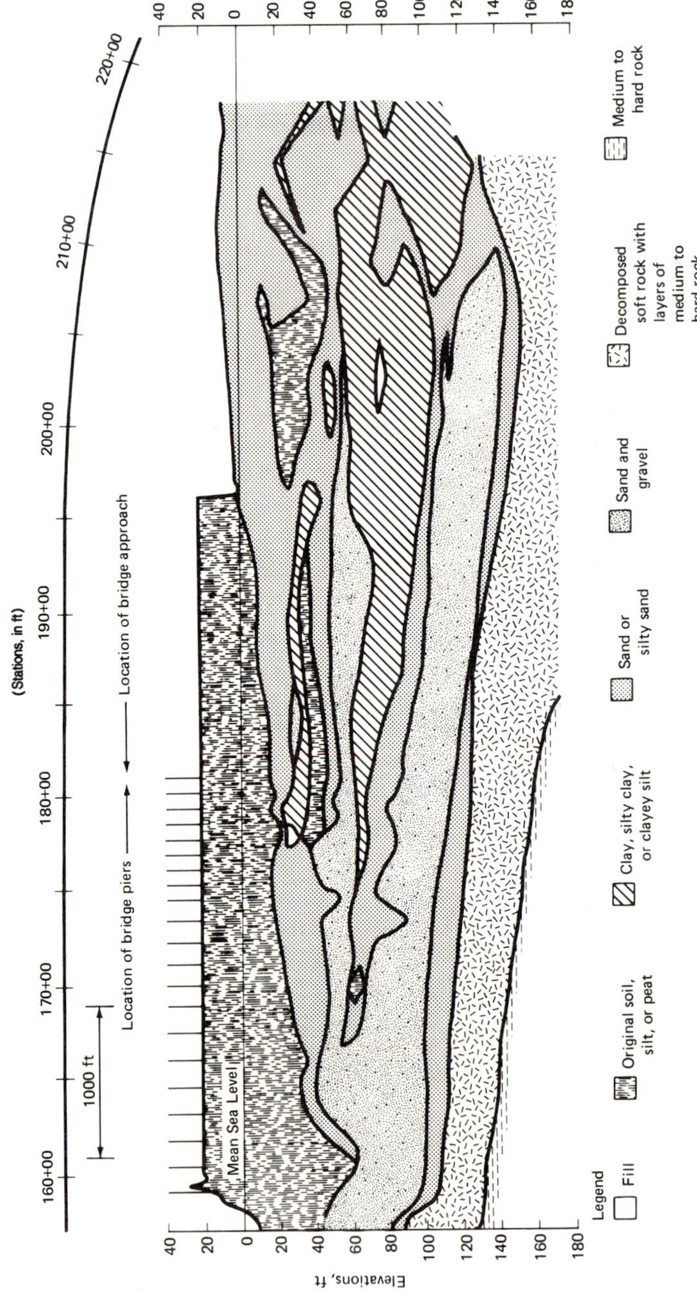

Figure 2.16 Soil profile at approach to a major river bridge crossing. *(Compliments of Delaware River Port Authority.)*

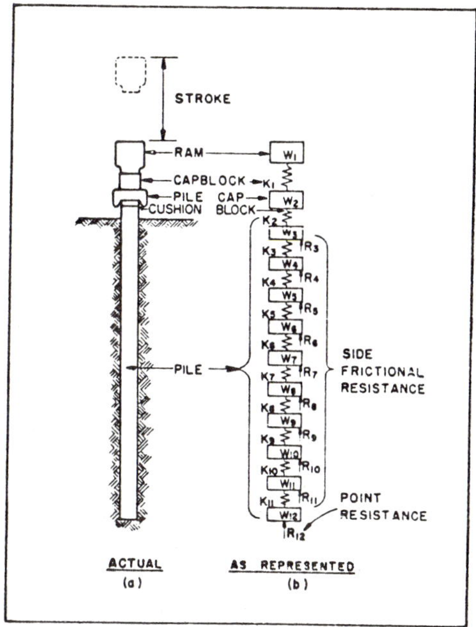

Figure 2.17 Method of representing pile for purpose of analysis. *(After Smith, Refs. 30 and 31.)*

$$R(m, t) = [D(m, t) - D'(m, t)]K'(m)[1$$
$$+ J(m)V(m, t - 1)] \tag{2.21}$$

$$V(m, t) = V(m, t - 1) + [F(m - 1, t)$$
$$- F(m, t) - R(m, t)]\frac{g\,\Delta t}{W(m)} \tag{2.22}$$

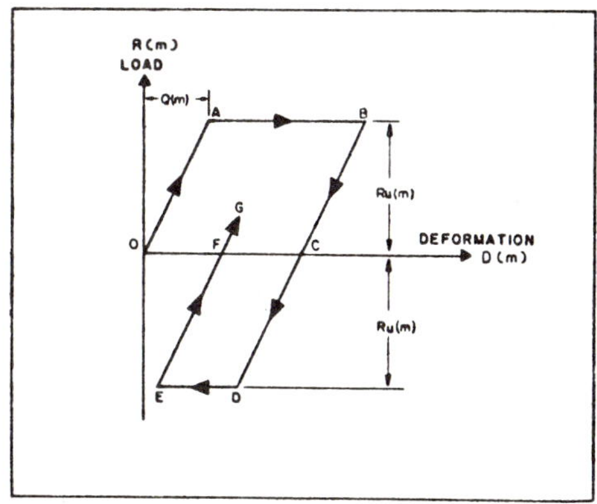

Figure 2.18 Load-deformation characteristics assumed for soil spring *m*. *(After Smith, Refs. 30 and 31.)*

where ( ) = functional designation

$\qquad m$ = element number

$\qquad t$ = number of time interval

$\qquad \Delta t$ = size of time interval, s

$\qquad C(m, t)$ = compression of internal spring m in time interval t, in

$\qquad D(m, t)$ = displacement of element m in time interval t, in

$\qquad D'(m, t)$ = plastic displacement of external soil spring m in time interval t, in

$\qquad F(m, t)$ = force in internal spring m in time interval t, lb

$\qquad g$ = acceleration due to gravity, ft/s^2

$\qquad J(m)$ = damping constant of soil at element m, s/ft

$\qquad K(m)$ = spring constant associated with internal spring m, lb/in

$\qquad K'(m)$ = spring constant associated with external soil spring m, lb/in

$\qquad R(m, t)$ = force exerted by external spring m on element m in time interval t, lb

$\qquad V(m, t)$ = velocity of element m in time interval t, ft/s

$\qquad W(m)$ = weight of element m, lb

The computations proceed as follows:

1. The initial velocity of the ram is determined from the properties of the pile hammer. Other time-dependent quantities are initialized at zero or to satisfy static equilibrium conditions.
2. Displacements $D(m, 1)$ are calculated by Eq. (2.18). It is to be noted that $V(1, 0)$ is the initial velocity of the ram.
3. Compressions $C(m, 1)$ are calculated by Eq. (2.19).
4. Internal spring forces $F(m, 1)$ are calculated by Eq. (2.20).
5. External spring forces $R(m, 1)$ are calculated by Eq. (2.21).
6. Velocities $V(m, 1)$ are calculated by Eq. (2.22).
7. The cycle is repeated for successive time intervals.

Figures 2.19 and 2.20 illustrate the types of results that can be generated by this kind of pile-soil modeling. Figure 2.19, after Raamot,[35] quantifies the driving stresses in the pile at various times after impact as well as the force in the pile tip and the predicted pile set. The use of the technique has been compared to actual pile load tests for a number of piles (three steel pipes, six step tapers, two uniform tapers, and one Monotube) with lengths ranging from 31 to 80 ft, driven through a clayey silt into saturated medium-to-dense granular soils with the results shown in Fig. 2.20.[34] Here it can be seen that all the results predicted by wave equation analysis are within ± 20 percent of the pile load test results.

Finite-element method (FEM) This method is a numerical technique used to obtain an approximate solution for a wide range of problems and boundary conditions. Initiated in the aerospace industry, it rapidly moved to the mechanics area, first structural engineering and then all other disciplines, geotechnical engineering included.[36]

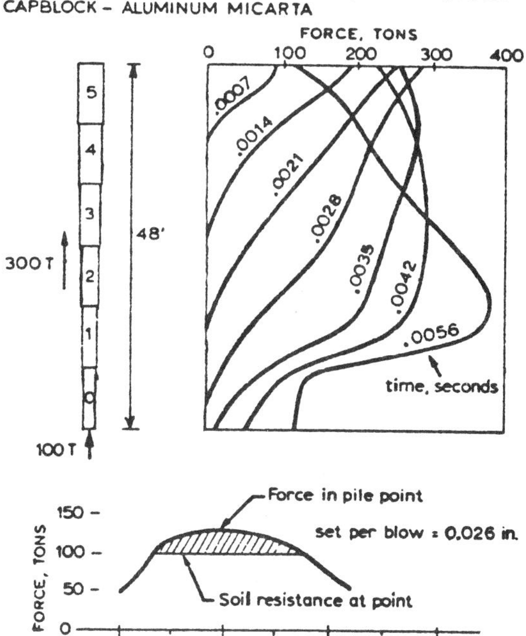

Figure 2.19 Typical results of wave equation analysis. *(After Raamot, Ref. 35.)*

The technique is reasonably well defined and the following steps are required to obtain a solution:

Discretizing the continuum
Selecting the interpolation functions
Determining the element properties
Assembling the system equations
Solving the equations to determine the unknown
Calculating other related information

Such procedures have been applied to deep foundations schemes, and references by Ellison et al.,[37] Desai et al.,[38] and Amir and Sokolov[39] will serve to show the wide range of their use and the elegance of the FEM method.

Ellison et al.,[37] in evaluating bored piles in clay soil, used quadrilateral pile elements and a FEM grid as shown in Fig. 2.21. Using a piecewise linear stress-strain curve, principal stresses and deformations were plotted at various load levels. The results are shown in Fig. 2.22 for three load levels and in Fig. 2.23 for displacements at various elevations under the maximum load of 260 tons. These values were compared with actual test pile results and Table 2.17 shows what an excellent correlation there was between predicted and observed values.

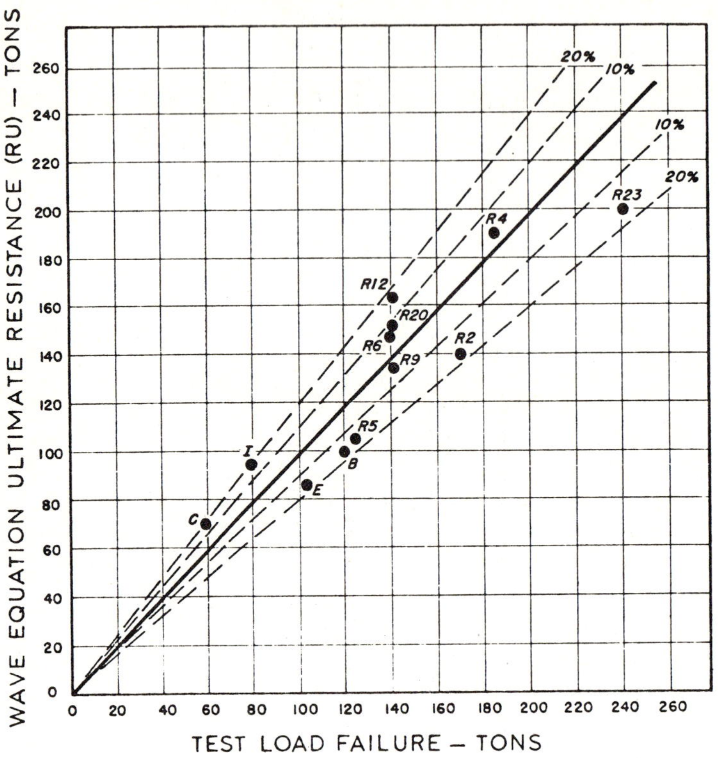

Figure 2.20 Comparison of ultimate pile loads from load tests versus predicted value by wave equation analysis. *(After Mosley, Ref. 34.)*

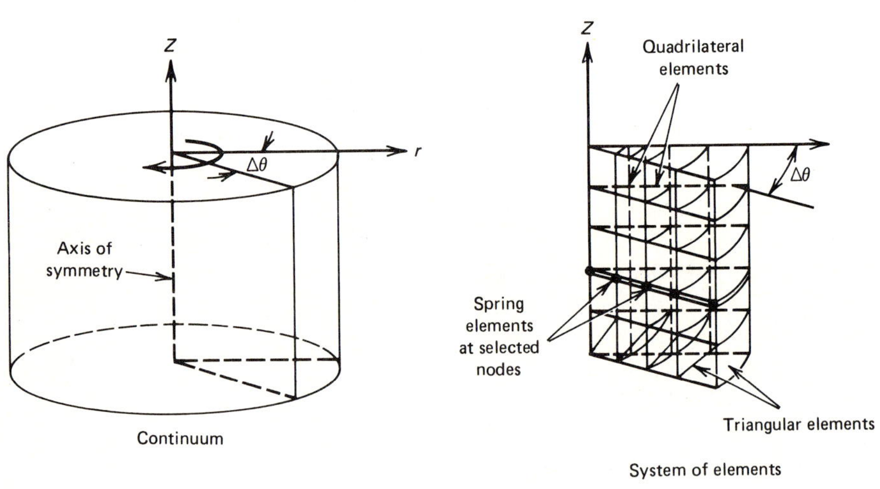

Figure 2.21 Elements and model for buried pile in clay soil. (*a*) Arrangement of elements for axis symmetry continuum. (*b*) Finite-element model. (*c*) Detail at pile tip for finite-element model. *(After Ellison et al., Ref. 37.)*

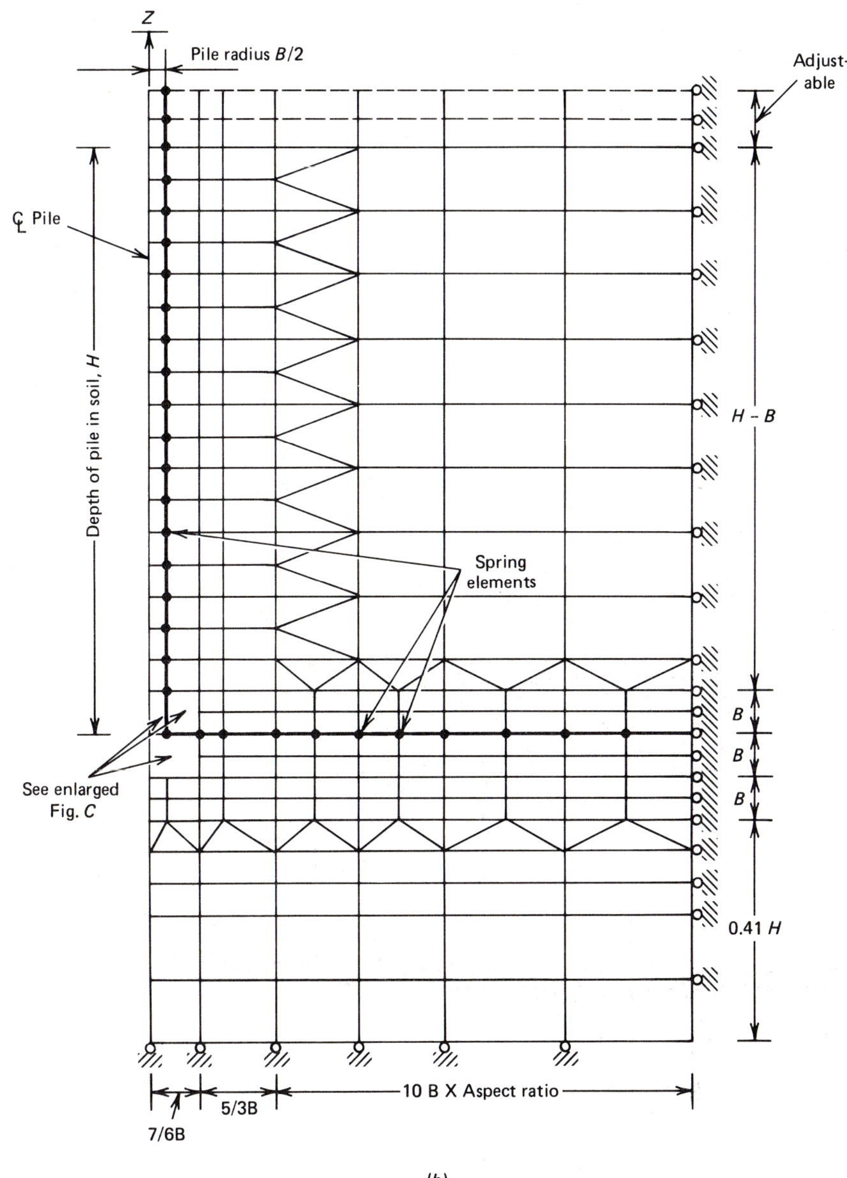

(b)

Figure 2.21 (*Continued*)

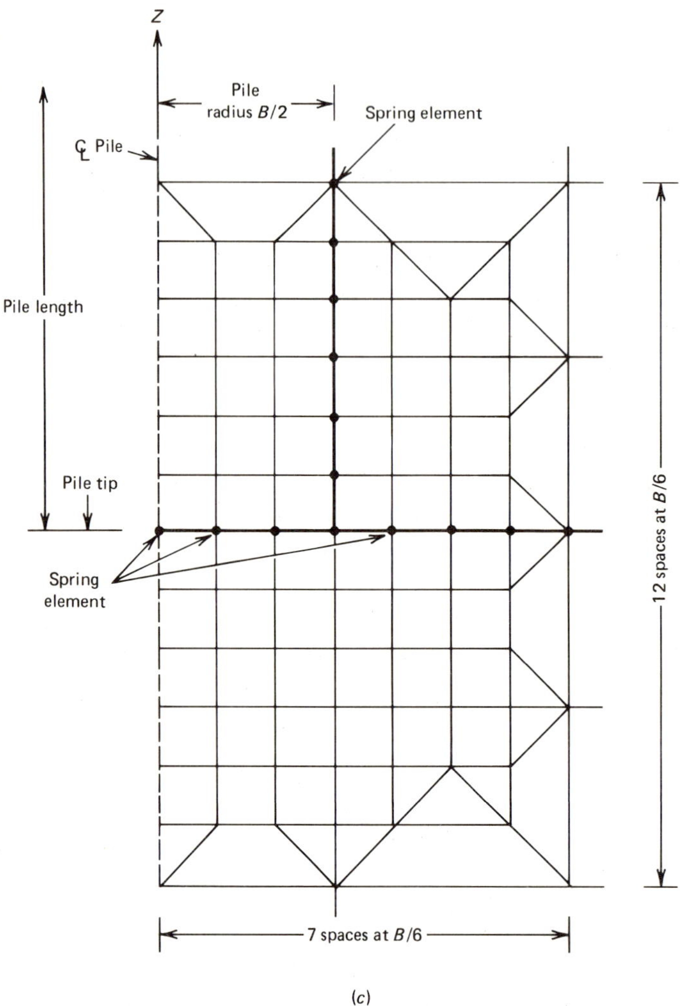

Figure 2.21 (*Continued*)

Other areas where FEM methods have been used in connection with deep foundations include piles in expansive clays (see Fig. 2.24) and the modeling of an entire river lock and its foundation (see Fig. 2.25). Indeed, the basic understanding of deep foundation–soil-or-rock interaction has progressed rapidly via the finite-element method. Furthermore, the method's potential has not yet been fully realized.

In situ monitoring Utilization of wave dynamics principles for field monitoring while pile driving has proved to result in quite accurate pile load predictions. The

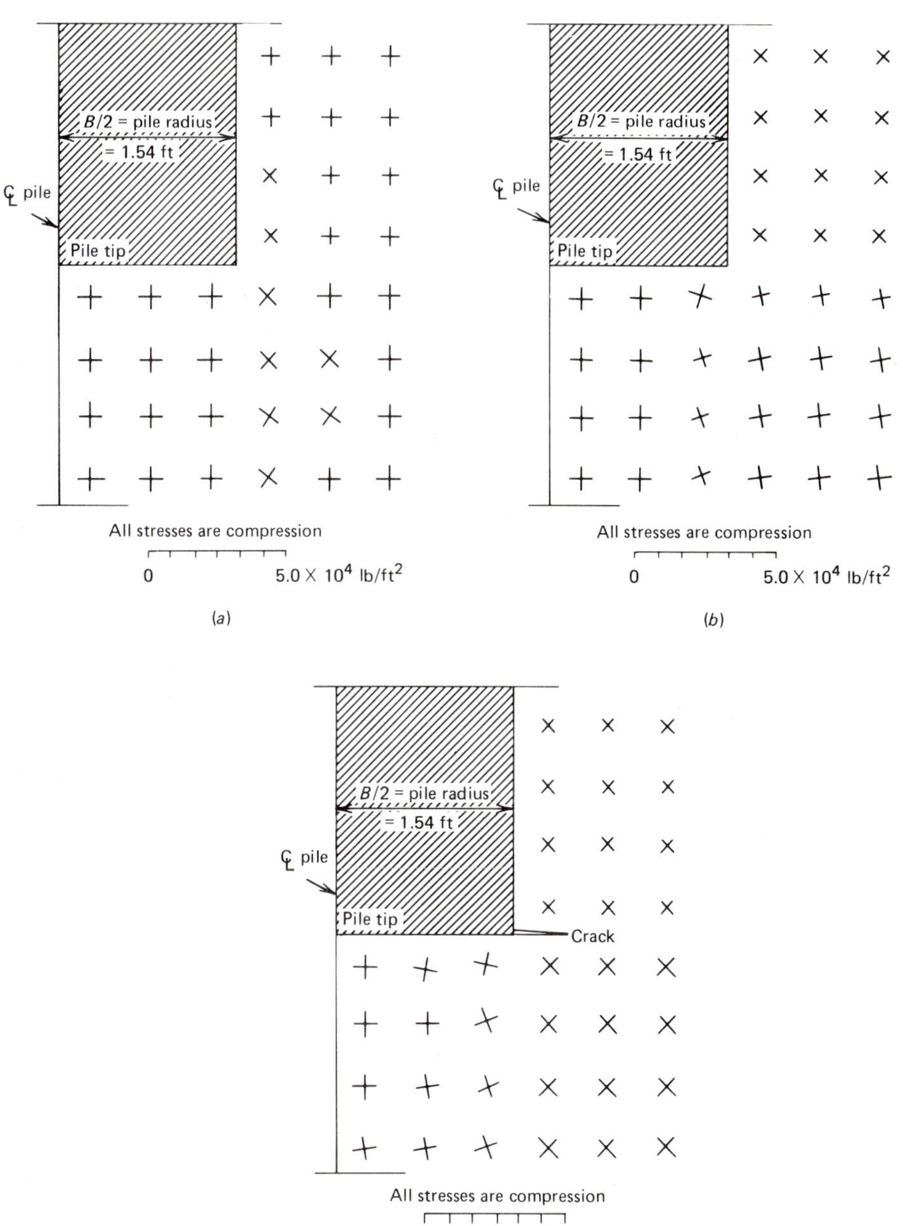

Figure 2.22 Principal stresses in clay at the tip of pile. (*a*) Gravity load. (*b*) Applied load of 290 tons. (*c*) Applied load of 360 tons. (*After Ellison et al., Ref. 37.*)

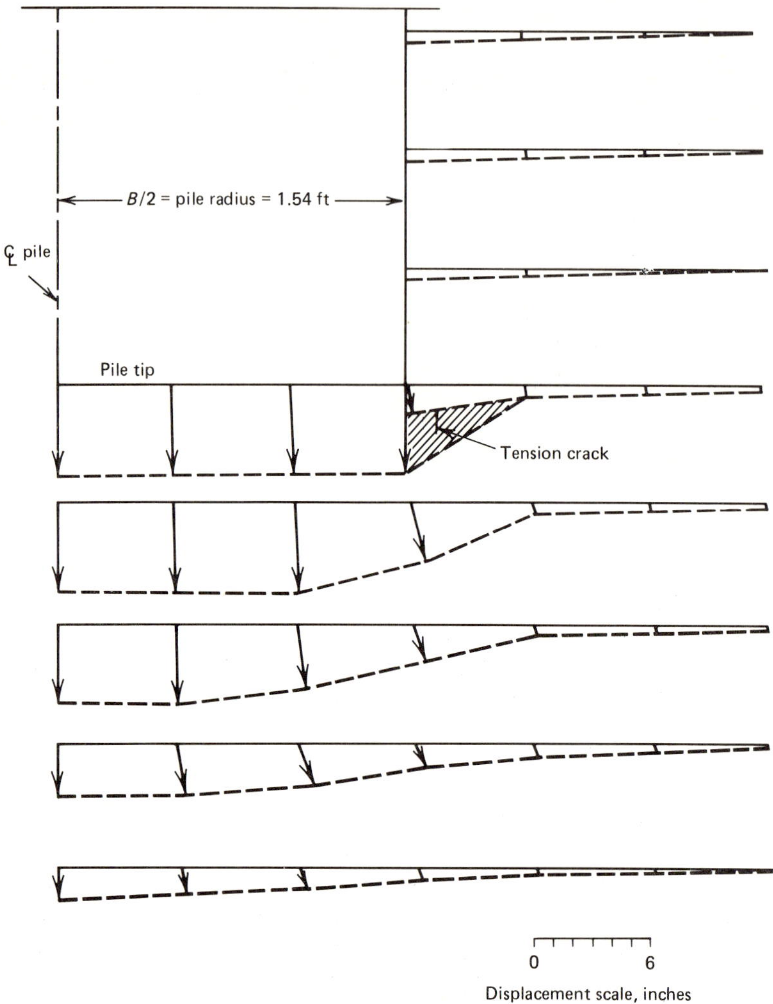

Figure 2.23 Final displacement of clay at tip of pile N for applied load of 360 tons. *(After Ellison et al., Ref. 37.)*

method, developed by Goble, Rausche, and others at Case Western University,[40–42] monitors force and acceleration at the top of the pile during impact driving. Using a lumped mass system, similar to that shown in Fig. 2.17, a predictor-corrector numerical integration is performed using the known values of acceleration as boundary conditions and adjusting the soil resistance properties until the computed output force equals the measured input force. Results of this method are compared with static load test values in Fig. 2.26, and they show an excellent correlation with test values. Statistical analysis of predicted values by this method gives a correlation coef-

Table 2.17 Summary of observed and predicted ultimate loads for test piles G, H, and K†

Test pile	Ultimate tip load, tons			Ultimate skin friction load, tons			Ultimate total load, tons		
	Observed	Predicted‡	Difference, as a percentage	Observed	Predicted‡	Difference, as a percentage	Observed	Predicted‡	Difference, as a percentage
G	50	59	+18.0	100	102	+2.0	150	163	+8.7
H	77	65	−15.6	143	150	+4.9	220	215	−2.3
K	77	69	−10.4	198	205	+3.8	275	274	−0.3

†After Ellison et al., Ref. 37.
‡Calculated load at a pile butt movement of 5 in.

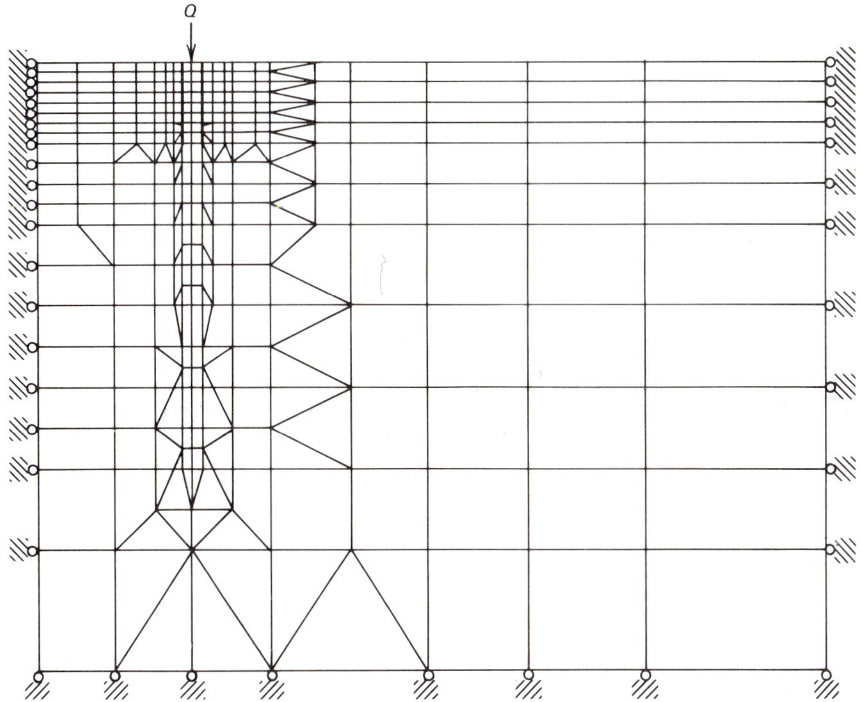

Figure 2.24 FEM mesh for deep foundation in expansive clay. *(After Amir and Sokolov, Ref. 39.)*

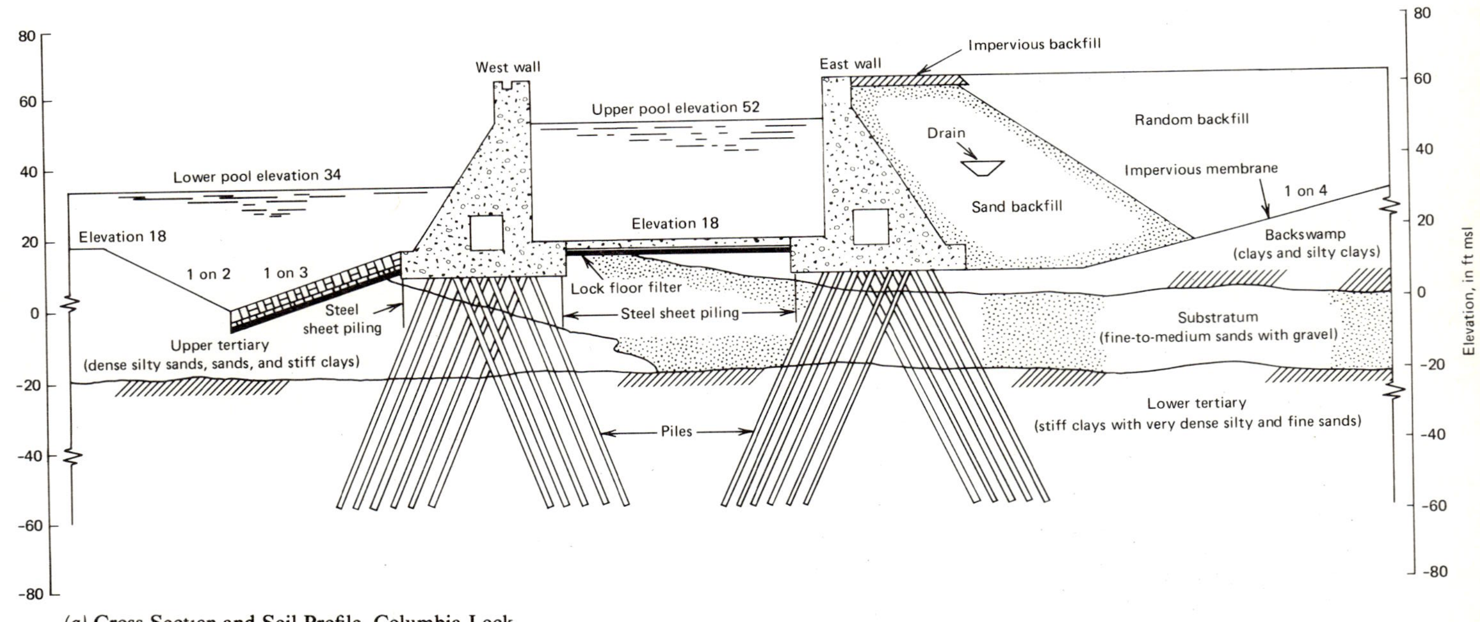

(a) Cross Section and Soil Profile, Columbia Lock

Figure 2.25 Actual section and FEM mesh for lock and deep foundation system. (*After Desai et al., Ref. 38.*)

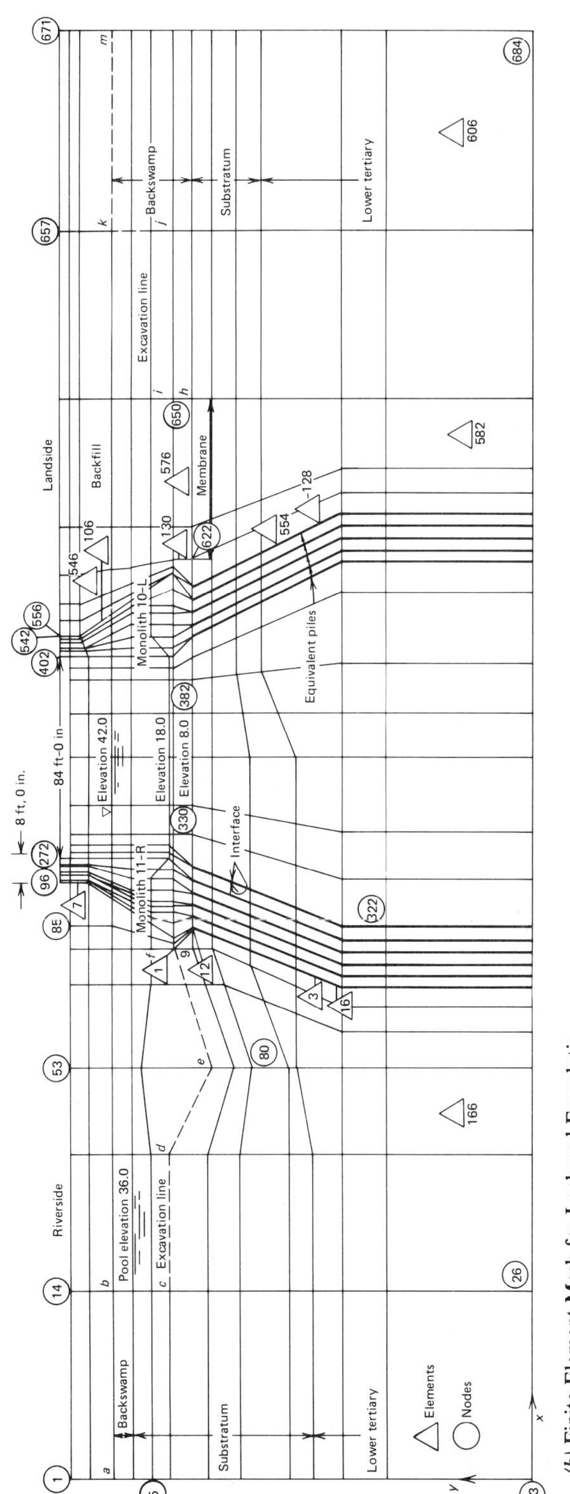

(b) Finite Element Mesh for Lock and Foundations

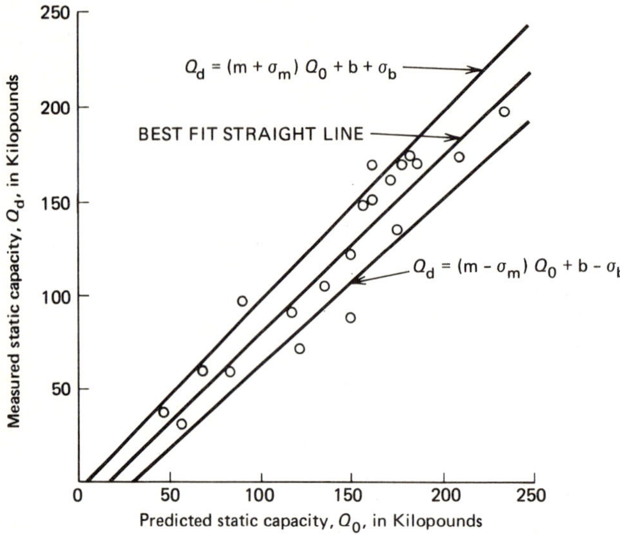

Figure 2.26 Results from statistical analysis for 20 predictions. *(After Rausche et al., Ref. 41.)*

ficient of 0.94 to test results, in comparison with a correlation coefficient of 0.29 for data found using the ENR formula and a correlation coefficient of 0.72 for data found using the Hiley formula. (Recall that the latter two were the classical dynamic formulas given in Sec. 2.4.2.)

2.4.6 Pile Load Tests

If the "proof of the pudding were in the eating," then the results of a pile load test should provide the answer to the actual load-carrying capacity of a given pile and set of soil conditions. But while pile load tests are indeed informative, they fall short of providing all the answers, for a number of reasons:

1. Pile load tests are generally not taken to failure. They are usually taken to twice design load and judged on the basis of deformation.
2. The interpretation of the load deflection curve generated during a pile load test taken to failure is generally not straightforward.
3. When the pile load test is taken to failure and properly interpreted, the meaning of the results versus the behavior of the pile group (which is what we are mainly interested in) is not clear.

Since such pile load tests are common, however, they will be described here.

Many forms of the test exist, but most contractors seem to favor the use of a hydraulic jack which mobilizes the load against a reaction load or frame as shown in Fig. 2.27. The load-versus-deflection curve from the test can be generated in a

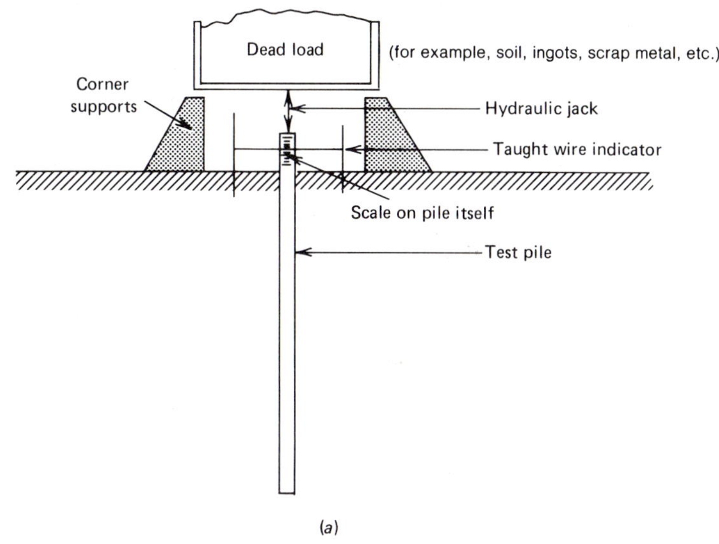

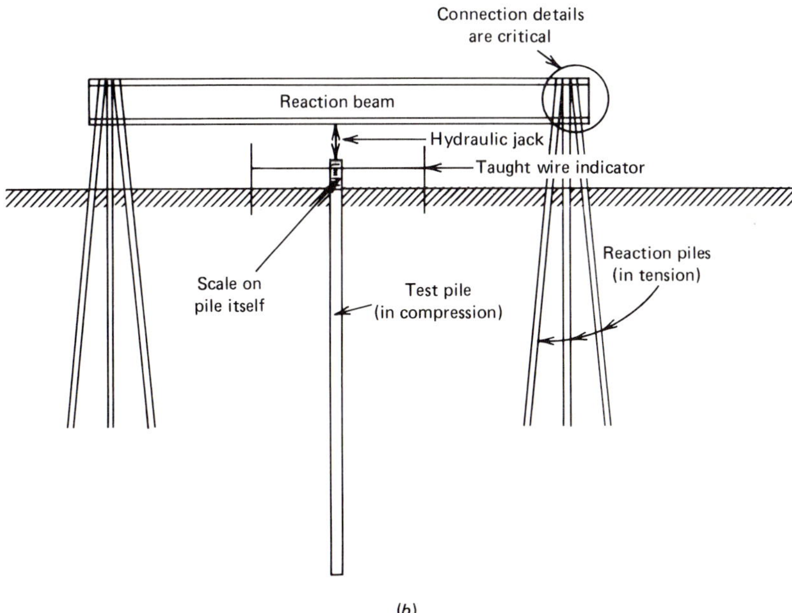

Figure 2.27 Schematic diagrams of typical pile load test configurations. Typical configurations for pile load test (*a*) up to (*b*) beyond 50 tons.

number of ways and most metropolitan areas have codes prescribing details of the test. Typical of such specifications is the Building Officials Council of America procedure. The pile load test curves shown in Fig. 2.28 were generated using this procedure, as explained in the following paragraphs:

Allowable Pile Load by Load Test—At least one test pile in each area of uniform conditions shall be loaded by a method that will maintain constant load under increasing settlement. The test load shall be twice the proposed load value of the pile; it shall be applied in seven increments equal to 50 percent 75, 100, 125, 150, 175 and 200 percent of the proposed working load. Readings of settlements and rebounds shall be recorded to 0.001 ft. for each increment or decrement of load.

After the proposed working load has been reached and for each increment of the load above that, the test load shall remain in place until there is not more than 0.001 ft. settlement in a two-hour period. The total test load shall remain in place until settlement does not exceed 0.002 ft. in 48 hours. The total load shall be removed in decrements not exceeding one-fourth of the total load at intervals of not less than one hour. The rebound shall be recorded after each decrement is removed, and the final rebound shall be recorded 24 hours after the entire test load has been removed. The maximum allowable pile load shall not exceed one-half that which causes a net settlement of not more than $\frac{1}{2}$ in. or a gross settlement of 1 in., taking into account the elastic shortening of the pile. (There is an opportunity here for exercise of engineering judgement to assess the possible damage and practical economy of permitting loads that might cause greater settlement.)

Note: For greater information on foundation conditions and total pile capacity at least some piles should be designated for test to failure. Additional cost will be small and knowledge gained for the owner and the profession will be great.

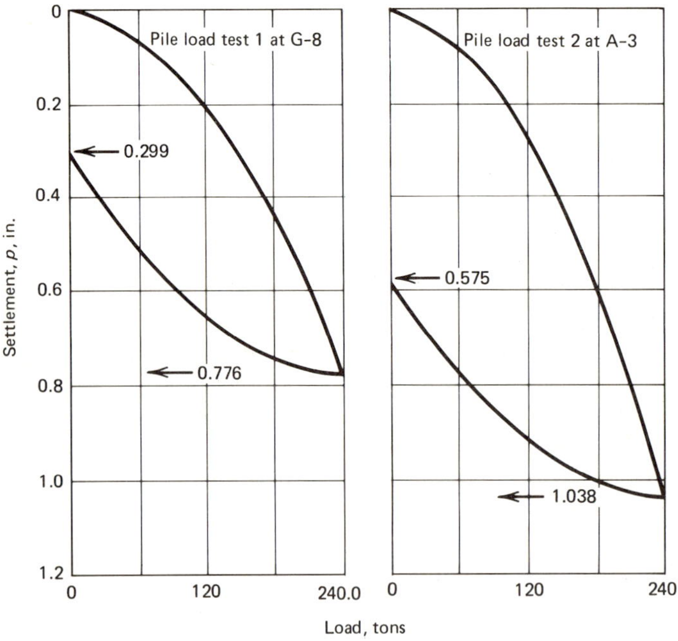

Figure 2.28 Pile load test response curves. *(After Koerner and Partos, Ref. 29.)*

Example 2.3 Determine if the pile load test illustrated in Fig. 2.28 at A-3 is acceptable for a 16-in-diameter concrete pile 25 ft long with a design load of 120 tons.

SOLUTION (*a*) Determine the elastic shortening at design load:

$$\Delta H_e = \frac{PL}{AE}$$

$$= \frac{(120)(2000)(25)(12)}{\dfrac{\pi(16)^2}{4} \, 4{,}000{,}000}$$

$$= 0.087 \text{ in}$$

(*b*) Check gross settlement:

$$\Delta H_G \le (1.000 + \Delta H_e)$$

$$\le 1.087 \text{ in}$$

where $\Delta H_g = 1.038$ in < 1.087 in
    (Answer is OK.)
    (*c*) Check net settlement:

$$\Delta H_N \le (0.500 + \Delta H_e)$$

$$\le 0.587 \text{ in}$$

where $\Delta H_N = 0.575$ in < 0.587 in
    (Answer is OK.)
    Use 120-ton design load.

Regarding how to determine the value of ultimate load of a pile from the resulting load-versus-deflection curve, a number of approaches are possible. They are as follows:

Van der Veen's method[43]: Van der Veen proposes an empirical relation between the load and settlement. The settlement versus the ratio of load acting on pile to assumed ultimate load is plotted on semi-log paper. If the plot is a straight line, the choice of ultimate load is considered to be correct. But a number of researchers, including Buisson,[44] disagree with this conclusion. Buisson asserts that the ratio of load and ultimate load, when plotted against settlement under the load, will not be a straight line on semi-log paper, since the latter part of such a curve approaches an inclined tangent when plotting is carried out with respect to ordinary coordinates.

Whitaker's method[45]: According to Whitaker, the ultimate load capacity can be determined from a test in which the maximum load required for a constant rate of penetration is observed. This would undoubtedly yield an unambiguous result.

In clays, and occasionally other soils, e.g., peats, the ultimate load is clearly defined. But in sands, the total strain and the rates of strain needed in testing to reach the ultimate load would require very sophisticated equipment and applying such large loads would be extremely difficult. Such a method is suitable for model tests but impractical for full-scale testing. It fails to give any indication of the load settlement relation and for clays offers little advantage over simple load tests.

ASCE pile load test method[46]: The load is indicated in this method by the intersection of lines tangent to the flatter part and to the steeper part of the load-versus-deflection curve (factor of safety = 1.5 to 2).

Los Angeles Building Code[19]: The load beyond which, by inspection, there is a disproportionate rate of settlement (factor of safety = 2).

Pacific Coast Uniform Building Code[19]: The load at which the total settlement is not more than 0.01 in/ton of test load (factor of safety = 2).

RABE method[19]: The load at which the settlement begins to exceed 0.03 in/ton of additional load (factor of safety = 2).

Terzaghi's method[19]: The load at which the settlement is one-tenth of the pile diameter (factor of safety = 2.5).

Indian practice[47]: The load at $\frac{1}{2}$ in of settlement (factor of safety = 1.5).

The situation is obviously not clear and research into the area is ongoing and indeed warranted.

2.4.7 Pile Group Capacity

Since one rarely installs a single pile to support an entire structure, the question always arises of what load a group of piles will support. Still further, can the maximum capacity of the group be inferred from the maximum capacity of a single pile, i.e., from the results of a pile load test? It should be mentioned that by a *pile group* we mean more than one pile at pile spacing to diameter ratios of 2.5 to 10. It should also be noted that many references on this topic are available, e.g., see Ref. 48.

The earliest attempts to determine group pile capacities came out of geometric relationships based on surface area, which resulted in a pile group efficiency E. This value, being less than 1, was multiplied by the sum of the capacities of the individual piles to give the group capacity. But such formulas are fundamentally deficient in that they do not take into account the nature of the soil in which the piles are founded, nor are they applicable for point bearing piles. They are not recommended for use.

Other formulas taking a fractional portion of the total load as influenced by the number of neighboring piles suffer from the same drawbacks as efficiency formulas.

It is best, for design purposes, to use a rational approach by considering the nature of the soil after the pile group has been placed. For pile groups founded in granular soils, the relative density of the soil is increased by the driving action. Possible exceptions to this are when soil of nearly 100 percent relative density is displaced or when piles are placed by jetting or augering down to the pile tip. But since both conditions are unusual, simply adding the sum of the individual pile loads to determine the group capacity is recommended for piles founded in granular soils. (This is in essence using a group efficiency factor of 1.)

For pile groups founded in cohesive or $c - \phi$ type soils, a group failure analysis by block action is recommended. As can be seen in Fig. 2.29, one must calculate the bearing capacity beneath the pile group and the shear strength adjacent to the group to arrive at the pile group load capacity:

$$Q_g = BLp_0 + (2B + 2L)D_f s_0 \tag{2.23}$$

where Q_g = pile group capacity
$\quad B$ = group width
$\quad L$ = group length
$\quad D_f$ = group depth
$\quad p_0$ = point resistance of soil beneath group
$\quad s_0$ = average skin friction adjacent to the group

There are numerous modifications to the above equation,[25,48] most of which are empirical, that attempt to relate its results to field performance results. Since field results are usually settlement-controlled, and quite often monitor the structure itself rather than a specific pile group load test, the subject of single-pile and pile group settlement should be addressed, as we do next.

2.4.8 Single-Pile Settlement

While it is generally recognized that piles do indeed settle, the recognition is a bit clandestine since any pile was placed in the first place to prevent the settlement that a shallow foundation would have resulted in. Thus, calculations for settlement of piles have not received a great deal of attention. A few comments on the subject, however, are in order here.

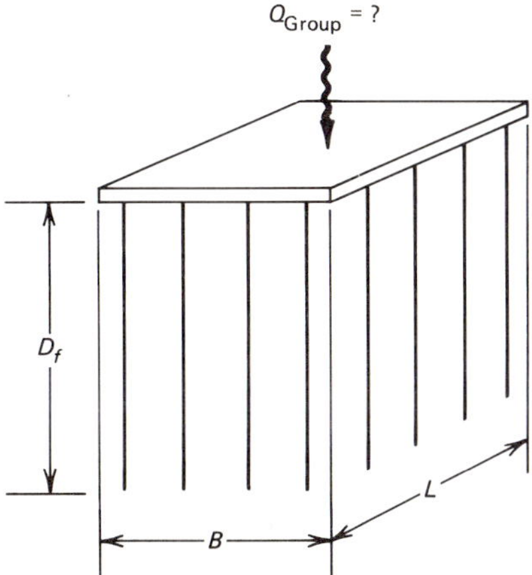

Figure 2.29 Procedure to determine pile group capacity assuming a block failure mode.

Philosophically, if the deep foundation is properly designed, with a reasonable factor of safety, settlements should be in the elastic range. This is particularly the case for end-bearing piles, where the applicability of the theory of elasticity seems to be a reasonable assumption. However, rather than using the conventional Boussinesq theory, it seems more appropriate to use Mindlin's theory, which utilizes a point load within an elastic solid. Poulos and Davis have thoroughly covered the topic in book form,[49] and Fig. 2.30, based on their work, shows the influence factors for use in the following equation:

$$\Delta H = \frac{P}{LE} I_\rho \tag{2.24}$$

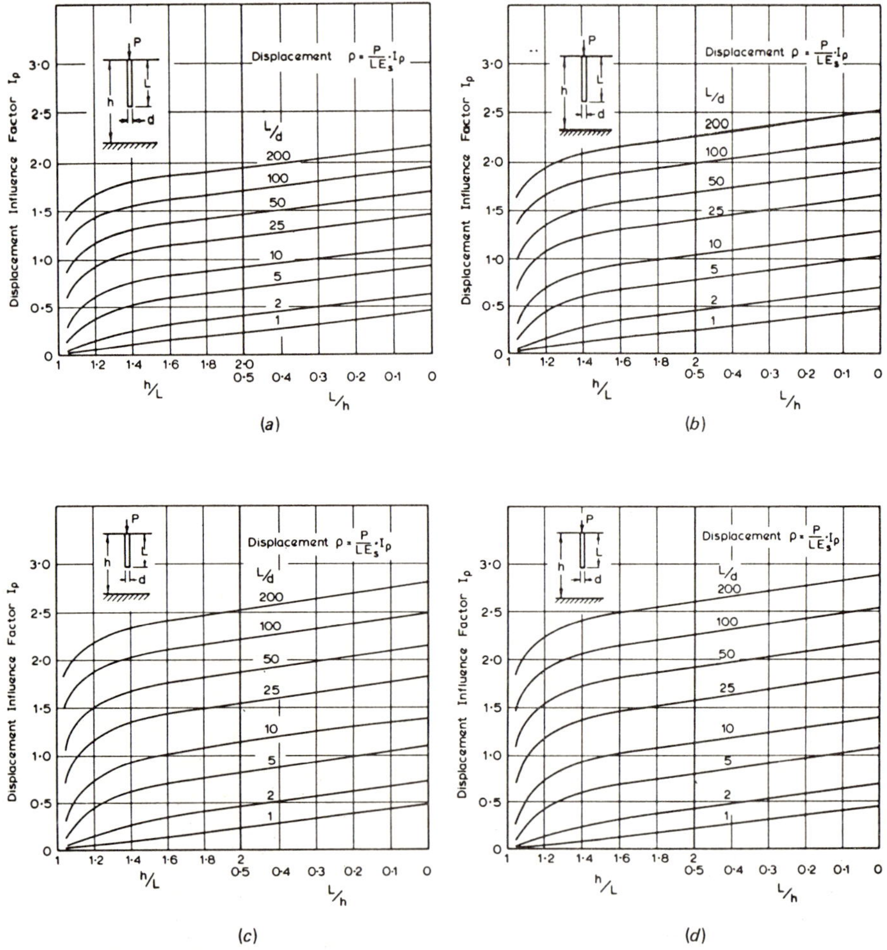

Figure 2.30 Influence factors for vertical pile settlement as a function of Poisson's ratio where (a) v_s = 0, (b) v_s = 0.2, (c) v_s = 0.4 and (d) v_s = 0.5. *(After Poulos and Davis, Ref. 49.)*

where ΔH = pile settlement
P = applied surface load
L = pile length
E = modulus of elasticity of soil at pile tip
I_ρ = influence factor taken from Fig. 2.30
ν_s = Poisson's ratio of soil at pile tip (needed to use Fig. 2.30)

Note that the value from the above equation should be added to the elastic deformation of the pile itself ($\Delta H = PL/AE$) and, depending on circumstances, to a contribution of load transmitted along the pile shaft; see Vesic.[27]

Example 2.4 For a 120-ton pile, 16 in in diameter and 30 ft deep founded in a dense clayey gravel with $E = 10,000$ lb/in^2 and $\nu_s = 0.2$, what is the estimated settlement in addition to elastic pile shortening?

SOLUTION Using $L/h = 30/\infty = 0$, and $L/d = 30/1.33 = 22.5$, from the curves shown in Fig. 2.30 for $\nu_s = 0.2$, one obtains $I_\rho = 1.5$. Then

$$\Delta H = \frac{P}{LE}\, I_\rho$$

$$= \frac{120(2000)(1.5)}{30(10,000)(12)}$$

$$= 0.10 \text{ in}$$

That is, ΔH is very small, which is typical of properly designed end-bearing piles.

The case of friction piles, however, can be very different. Here the modulus of elasticity is much lower than that given in Example 2.4 and, even more important, the phenomenon may be completely different than elasticity theory would predict. In the case of saturated, fine-grained soils, consolidation would occur and actual settlements would exceed, by far, those predicted on the basis of elasticity theory. In such instances, standard consolidation theory should be used, at least to determine an approximate value of settlement.

2.4.9 Pile Group Settlement

The increase in vertical settlement of a pile due to the installation of an adjacent pile is covered by Poulos and Davis[49] for both end-bearing pile groups and friction pile groups. Curves are presented that result in an interaction factor which, when applied to the settlement of a single pile, can be used to predict the group behavior.

A different approach, however, is a completely empirical one in which a growing body of case histories has resulted in the development of a relationship of group settlement to pile load test data for a single pile. Meyerhof[50] extended earlier work to develop that relationship:

$$\frac{\Delta H_G}{\Delta H_1} = \frac{s(5 - s/3)}{(1 + 1/r)^2} \tag{2.25}$$

where ΔH_G = pile group settlement
   ΔH_1 = settlement of a single pile at design load in a pile load test
   s = ratio of pile spacing to pile diameter
   r = number of rows in pile group

In this formula, r is made to represent an equivalent square group, e.g., a pile group consisting of three rows with six piles per row, $r = \sqrt{18} = 4.2$. Using this technique, Koerner and Partos[29] estimated the average group settlement of a large foundation to be 2.2 in while the actual measured settlement was 2.5 in.

2.4.10 Lateral Movement

As in determining various types of pile settlements, in determining lateral movement the elastic solutions discussed by Poulos and Davis[49] should be calculated before more elaborate methods are undertaken. For horizontal displacement of a single pile under horizontal load and moment, the elastic solution is:

$$\Delta H = I_{\rho H}\left(\frac{H}{E_s L}\right) + I_{\rho M}\left(\frac{M}{E_s L^2}\right) \tag{2.26}$$

where ΔH = horizontal displacement
   E_s = modulus of elasticity of soil
   L = length of pile
   H = horizontal force
   M = moment at top of pile
   $I_{\rho H}$ = influence factor taken from Fig. 2.31
   $I_{\rho M}$ = influence factor taken from Fig. 2.31
   K_R = pile flexibility = $E_p I_p / E_s L^4$
   E_p = modulus of elasticity of pile material
   I_p = moment of inertia of pile

The rotation θ at the top of a free-head pile is given by the following:

$$\theta = I_{\theta H}\left(\frac{H}{E_s L^2}\right) + I_{\theta M}\left(\frac{M}{E_s L^3}\right) \tag{2.27}$$

where $I_{\theta M}$ = influence factor taken from Fig. 2.31
   $I_{\theta H} = I_{\rho M}$

For a fixed-head pile, the rotation is zero and the horizontal displacement is found from the following equation:

$$\Delta H = I_{\rho F}\left(\frac{H}{E_s L}\right) \tag{2.28}$$

where $I_{\rho F}$ = influence factor taken from Fig. 2.31

Note that for all the solutions shown in Fig. 2.31, Poisson's ratio is 0.5. Variations of this value have relatively little effect on the calculated displacement and rotation values.

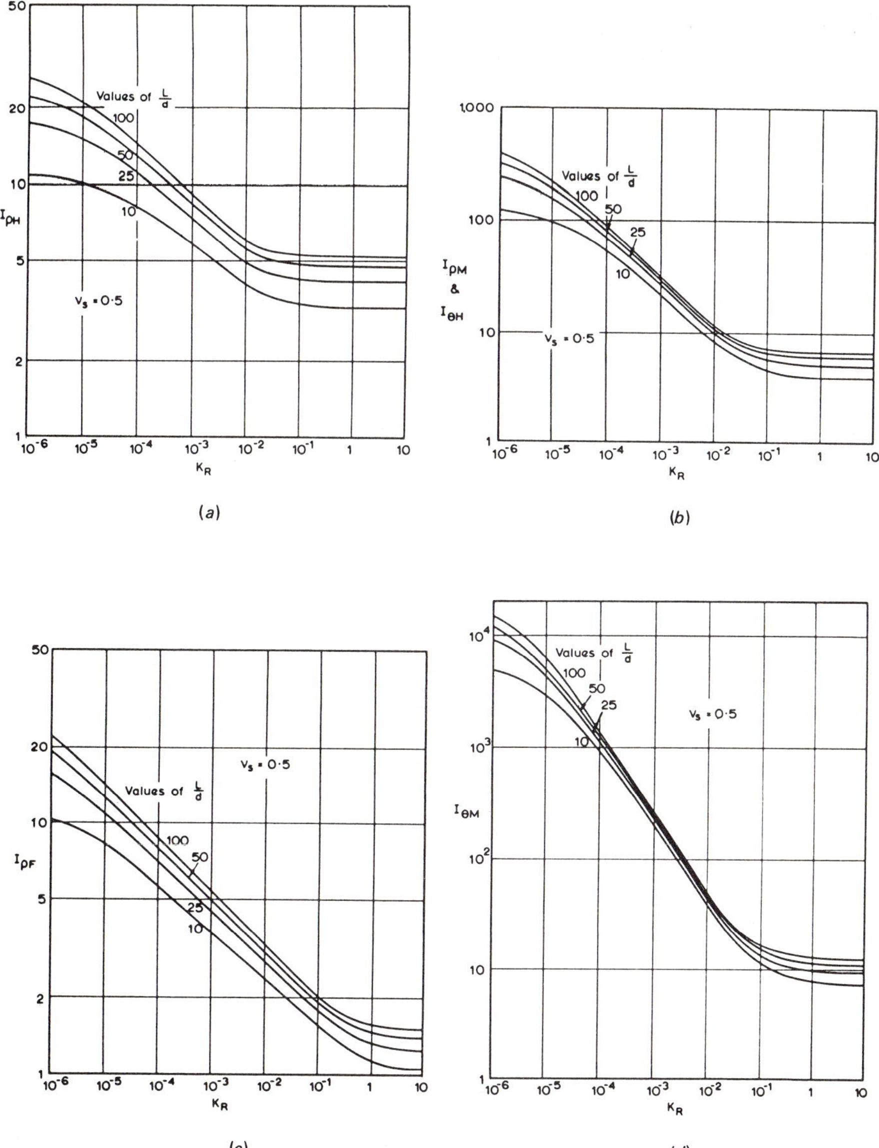

Figure 2.31 Influence factors for horizontal pile movements and rotations. *(After Poulos and Davis, Ref. 49.)* Influence factor (*a*) $I_{\rho H}$ for free-head pile, (*b*) $I_{\theta H}$ and $I_{\rho M}$ for free-head pile, (*c*) $I_{\rho F}$ for fixed head pile, (*d*) $I_{\theta M}$ for free-head pile.

Poulos and Davis treat the horizontal displacements of pile groups in a similar manner to how they treat vertical movement of a single pile. Reference 49 should be consulted for details.

Other approaches to laterally loaded pile design are based on beam methods, elastic foundation theory, and, more recently, on finite-element methods; see Sogge.[51]

PROBLEMS

2.1 In regard to the driving of pile foundations, three classes of hammers were reviewed in Chap. 2: drop, steam or air, and vibratory. Rate these according to frequency and amplitude of impact and also according to likelihood of vibrations beyond the normal work area.

Method of pile placement	Typical frequency, Hz	Typical amplitude, in	Likelihood of transmitting vibrations
Drop			
Steam or air			
Vibratory			

2.2. (*a*) While driving a pile through saturated soil, a contractor stops for a short time and then starts driving it again. The resistance of the soil to driving increases dramatically. Why?

(*b*) This increase in driving resistance proves to be a function of the time interval that the driving is stopped. In order for maximum resistance to be offered, what typical times are required for sands? For silts? For clays?

(*c*) Should the practice of stopping pile driving then starting it again be considered "cheating" on the part of the pile driving contractor? Similarly, is it "dumb" on the part of the pile driving inspector not to notice it?

2.3. Table 2.3 lists several vibratory pile drivers and shows that the frequency varies from negligible to 150 Hz. Of what significance is this number?

2.4. In regard to timber piles, Table 2.1 stated that they are "vulnerable to damage in hard driving." What happens to the pile tip? What happens to the pile butt? How can both of these situations be prevented?

2.5. (*a*) In steel piling, what is the difference between an I beam, a wide flange beam, and a bearing pile?

(*b*) What are the geometric properties of a 14 BP 117? Of a 12 BP 74? Of an 8 BP 36?

(*c*) In the driving of steel bearing piles through soils containing large pieces of gravel or boulders, or in driving such piles through trash dumps and landfills, the web of the pile sometimes fails. This splitting allows the two flanges to crumple or curl up and return toward the surface of the ground. How can this obviously disastrous situation be prevented from occurring?

2.6. Table 2.1 described precast concrete piles as "unless prestressed, vulnerable to handling." What are the most vulnerable times in handling such piles? Why does prestressing help those situations?

2.7. Cast-in-place concrete piles (shells driven with mandrel) are usually placed in a thin sacrificial steel shell, as illustrated in Table 2.1. The fresh concrete is often dropped from the top of the shell and allowed to free-fall to the bottom. What is the problem with this practice? Comment also on the situation where a cage of steel reinforcement bars is first placed in the shell and then the concrete is placed.

2.8. Pressure-injected footings, which usually have the casing withdrawn during installation, exhibit "more than average dependence on quality of field work." Why is this the case?

2.9. For all of the pile types shown in Table 2.1, the tip is as small or smaller than the butt, with the exception of the pressure-injected footings. On a conceptual basis, why is this type of pile an advantage?

2.10. Do you attach any importance to differences between the weight of the pile hammer and the weight of the pile itself? If so, why? (Present your answer in terms of W_h/W_p ratio, where W_h is the weight of the hammer and W_p is the weight of the pile.)

2.11. (*a*) Timber, concrete, and closed-end pipe piles are sometimes referred to as "displacement piles" while bearing piles and open-ended pipe piles are not. What are the implications of a displacement pile?

(*b*) When driving a large number of closely spaced displacement piles (say, for the foundation of an oil storage tank), how should the contractor proceed: center outward, outside to center, or progressively from one side to the other? Why?

(*c*) If the above job is paid for on a lump sum basis, which way would the contractor like to do the job? If it is paid for on a unit price basis ($/lin ft of pile), which way would the contractor prefer?

2.12. Drilled piers, of the types shown in Figs. 2.6 through 2.8, rely on end bearing for most or all of their strength. How does one inspect the soil or rock at this depth before placement of concrete? What kinds of tests do you recommend for assessing the strength of that soil or rock, which will eventually be supporting the pier and superstructure loads placed upon it?

2.13. A belled out, or underreamed, pedestal at the bottom of a drilled pier is highly desirable for end-bearing strength. It is usually formed with a bucket, as shown in Fig. 2.8*f*. In what types of soils is this type of bell difficult to achieve? Why?

2.14. When working under compressed air, as divers dealing with the pneumatic caissons described in Sec. 2.3.3 do, a common sickness (which can cause permanent disability or death) is known as "the bends." What is this sickness physiologically and how does it attack the body's systems?

2.15. In an airlock system, as shown in Fig. 2.10, what would happen if the internal air pressure dropped below the external soil and/or water pressure? What would happen if it rose above that external pressure?

2.16. Using the ENR formula, calculate the allowable bearing capacity of a pile being driven to a resistance of 10 blows per inch using the following hammers:

    (*a*) Conmaco Model 100
    (*b*) Kobe diesel KC45
    (*c*) MKT 11 B 3
    (*d*) Raymond "0"
    (*e*) Vulcan 08

2.17. For an ultimate pile load of 300 kips and a factor of safety of 5.0, what would be the pile set (in units of blows per inch) predicted by the ENR formula using the following pile hammers?

    (*a*) Conmaco model 50
    (*b*) Kobe diesel KC 25
    (*c*) MKT 9 B 3
    (*d*) Raymond differential 15 M(a)
    (*e*) Vulcan 1

2.18. Using the Hiley formula for predicting pile capacity, what are the ultimate load and the allowable load (based on a factor of safety of 3.5) of a pile driven by the hammers listed below. The pile is a 14 BP 117 steel pile, 80 ft long, driven as an end-bearing pile, using a cushion to a final set of 0.10 inches per blow.

    (*a*) Conmaco model 115 (e_f = 85 percent)
    (*b*) Vulcan 06

2.19. Resolve Prob. 18 with all the same conditions except that now the pile being driven is a pre-stressed, precast concrete pile of square cross section 14 × 14 in.

2.20. Using the static (rational) pile design method discussed in Sec. 2.4.3, determine Q_s, Q_p, Q_0, and

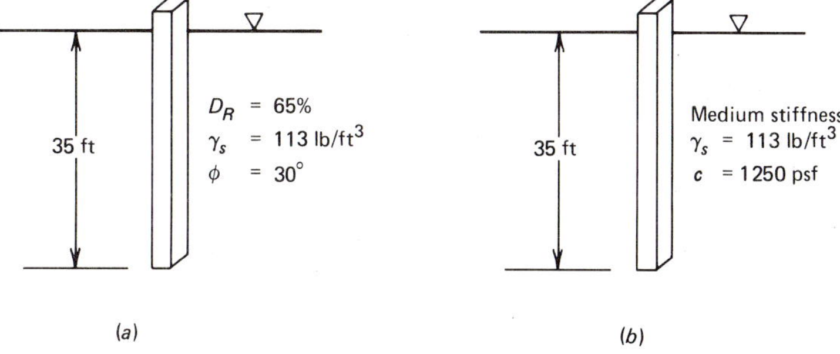

Figure P2.20 (*a*) Sand soil; (*b*) clay soil.

Q_{allow} for the situations illustrated in Fig. P2.20a and b. Use FS = 4.0 for Q_{allow}. Both piles are 14 $\times$ 14 in prestressed, precast concrete made in metal forms in a precasting yard.

2.21. Resolve Prob. 20 using all the same information except that now both piles are in sand, one with no tip (Fig. P2.21a), the other with an enlarged precast tip of 3-ft diameter (Fig. P2.21b).

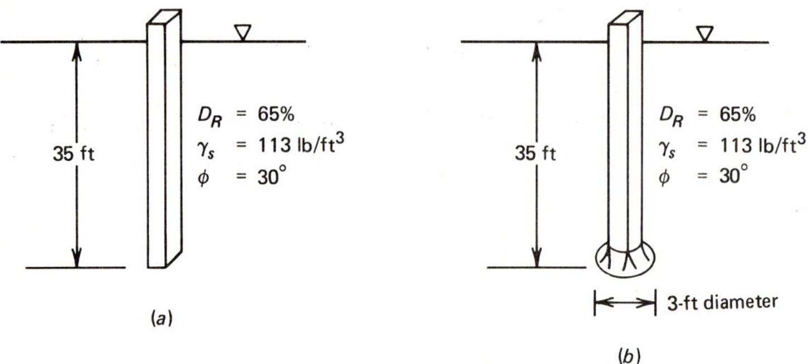

Figure P2.21 (a) Straight pile, no tip; (b) pile with enlarged tip.

2.22. Determine the ultimate bearing capacity of the following pile types, all driven as shown in Fig. P2.22, using the static design method:

(a) Timber—butt 16 in in diameter; tip 8 in in diameter; length 40 ft

(b) Steel—12 BP 74; length 75 ft

(c) Prestressed concrete or cast-in-place—18 in in diameter; length 75 ft

(d) Cast-in-place concrete—butt 16 in in diameter; tip 8 in in diameter; length 40 ft

(e) Concrete-filled pipe pile—18 in in diameter; length 130 ft

(f) Composite pile—16-in-diameter concrete upper section, 40 ft long; 16-in-diameter butt, 8-in-diameter timber lower section, 40 ft long

The soil profile is as follows:

Layer	Depth, ft	Soil	γ', lb/ft^3	c, lb/ft^2	ϕ, deg	N, blows/ft	D_R, %
1	0–12	Rubble	90	0	0		
2	12–35	Silty clay	58	600	0	6	
3	35–45	Silty sand	64	100	28	27	80
4	45–70	Soft clay	52	400	0	4	
5	70–100	Sandy clay	60	1000	10	29	
6	100–130	Sand and gravel	66	0	33	75	98
7	>130	Rock (sound)	$\cdots$		$\cdots$	60	

Comment on the adequacy of each pile type. Comment on potential settlement of each pile type.

2.23. (a) For the soil profile shown in Fig. 2.14, calculate the ultimate capacity of a cast-in-place concrete pile (shell withdrawn) with a pedestal, i.e., a pressure-injected footing, to a tip elevation of 25 ft below ground. The pile has a 16-in-diameter shaft and a 36-in-diameter pedestal.

(b) Redo the problem on the basis of driving a 14 BP 117 bearing pile to an elevation of 65 ft below ground.

(c) Considering that a number of these piles will support an 18-story concrete building, which alternative would you select and why?

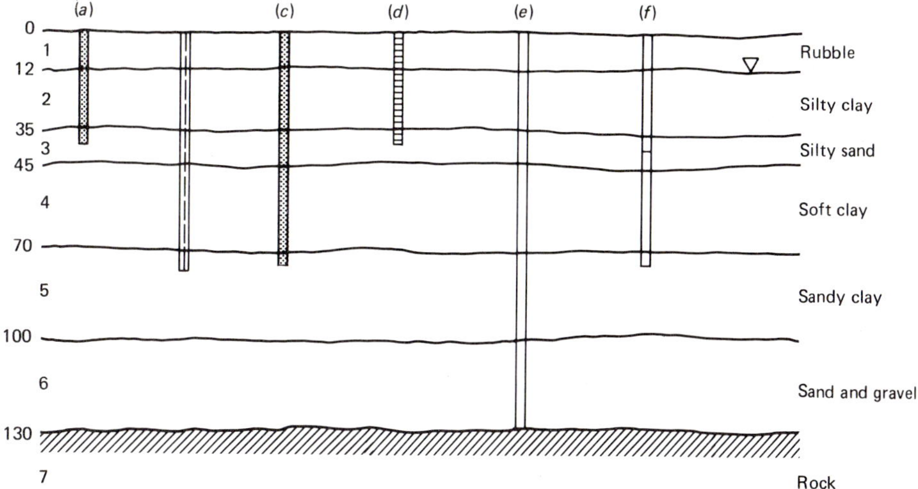

Figure P2.22

2.24. (*a*) In driving PZ 27 steel sheet piling for a bulkhead structure into the soil whose profile is shown in Fig. 2.15 with a Foster model 1700 vibrating pile driver, at what elevation do you expect refusal between the borings S-18 and S-16?

(*b*) If a hardened driving shoe were placed on the tip of the individual sheets, what would the anticipated driving elevation be?

2.25. (*a*) A major bridge is to be placed over the soil whose profile is shown in Fig. 2.16. In the vicinity of station 185 + 00, how deep would you expect 18-in closed-end pipe piles to be driven?

(*b*) Considering that the upper soil stratum is recently placed dredged river silt, is there a chance that negative skin friction (i.e., downdrag) might occur?

(*c*) If so, what remedial measure(s) would you suggest?

2.26. For the pile load test data shown in Fig. 2.28 as "pile load test No. 2 at A-3," what is the ultimate pile load according to:

   (*a*) Van der Veen's method
   (*b*) Whitaker's method
   (*c*) ASCE pile load test method
   (*d*) Los Angeles Building Code
   (*e*) Pacific Coast Uniform Building Code
   (*f*) RABE method
   (*g*) Terzaghi's method (pile diameter was 16 in)
   (*h*) Indian practice

2.27. (*a*) For a 12-pile group (4 × 3 configuration) consisting of 50-ft-long 18-in-diameter concrete piles in the soil shown in Fig. P2.27 and attached at the top by a heavily reinforced thick footing, what is the ultimate group capacity and the allowable group capacity based on FS = 4.5?

Soil properties
0 to 45 ft below surface, sandy silt:
   $s_0 = 800$ lb/ft^2
45 to 50 ft below surface, medium sand:
   $s_0 = 1500$ lb/ft^2
   $p_0 = 5$ ton/ft^2

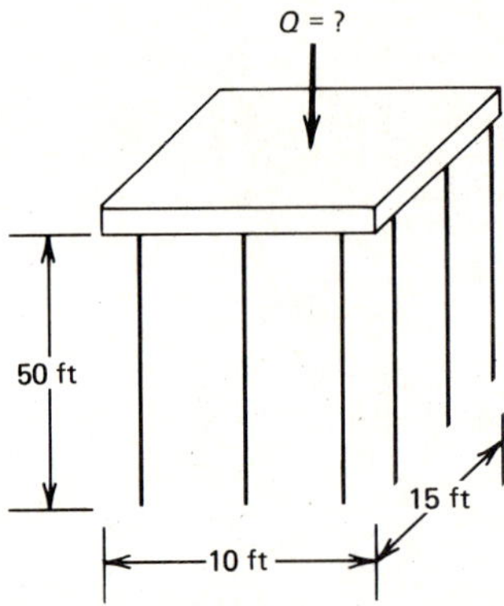

Figure P2.27

(*b*) If each of the individual piles could sustain 170 tons (calculated by the static design method), what would be the efficiency of the group?

(*c*) If the soil profile were completely sand of medium relative density, what would be the ultimate group capacity and the allowable group capacity based on a factor of safety of 4.5?

2.28. If a pile load test on a single pile in the group described in Prob. 27 resulted in a settlement of 0.242 in at design load, what would be the anticipated group settlement according to Eq. (2.25)?

2.29. For the soil profile shown in Fig. P2.29, determine the influence of the elastic constants on the anticipated single-pile settlement. (Vary the modulus from 500 to 10,000 lb/in^2 holding Poisson's ratio at 0.4; then hold the modulus at 3000 lb/in^2 and vary Poisson's ratio from 0 to 0.5.)

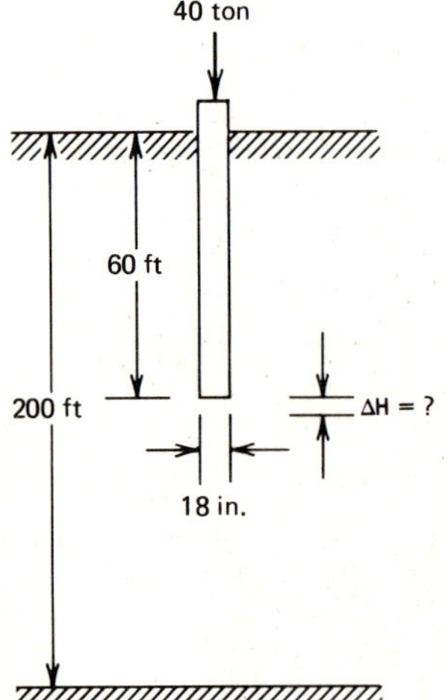

Figure P2.29

2.30. What is the influence of the modulus of elasticity on the horizontal movement of a pile in the situation illustrated in Fig. P2.30? The pile is 14-in square concrete.

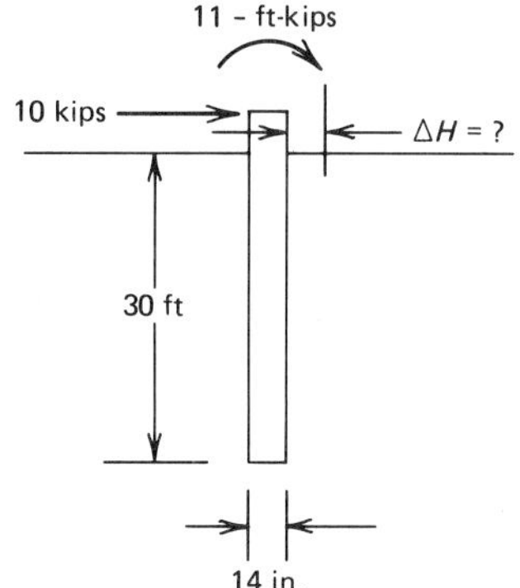

Figure P2.30

REFERENCES

1. Chellis, R. D., "Pile Foundations," Chap. 7 in *Foundation Engineering*, G. A. Leonards (ed.), McGraw-Hill, New York, 1962, pp. 633–768.
2. *Design Manual: Soil Mechanics, Foundations, and Earth Structures*, NAVDOCKS DM-7, Dept. of the Navy, Bureau of Yards and Docks, Washington, D.C., 1962.
3. Parola, L. F., "Mechanics of Impact Pile Driving," Ph.D dissertation, University of Illinois, Urbana, Ill., 1970.
4. Davisson, M. T., "Pile Hammers, Pile Driving and Driving Formulas," ASCE Lecture Series, American Society of Civil Engineers, New York, May 1966.
5. Compton, G. R., Jr., "Selecting Pile Installation Equipment," *Assoc. Pile and Fitting Corp. Pile-talk Seminar*, San Francisco, 1977, Assoc. Pile and Fitting Corp., Paramus, N. J.
6. Schousboe, I., "Suggested Design and Construction Procedures for Pier Foundations," *J. Amer. Concr. Inst.*, Vol. 69, August 1972, pp. 461–480.
7. Woodward, R. J., W. S. Gardner, and D. M. Greer, *Drilled Pier Foundations*, McGraw-Hill, New York, 1972.
8. Gill, S. A., "Applications of Slurry Walls in Civil Engineering," *J. Constr. Div., ASCE*, Vol. 106, No. CO2, June 1980, pp. 153-167.
9. D'Appolonia, E., R. D. Ellison, and D. J. DiApplonia, "Drilled Piers," Chap. 20 in *Foundation Engineering Handbook*, H. F. Winterkorn and H.-Y. Fang (eds.), Van Nostrand Reinhold, New York, 1975, pp. 601–615.
10. Bazant, A., *Methods of Foundation Engineering*, Elsevier, Amsterdam, 1979.
11. Swatek, E. P., Jr., "Pneumatic Caissons," Chap. 21 in *Foundation Engineering Handbook*, H. F. Winterkorn and H.-Y. Fang (eds.), Van Nostrand Reinhold, New York, 1975, pp. 616–625.
12. Tomlinson, M. J., *Foundation Design and Construction*, 2d ed., Wiley-Interscience, New York, 1969.
13. Hoffman, J. F., "Wellpoints Dewater Sites for Narrows Bridge Piers," *Civil Eng., ASCE*, Vol. 30, October 1960, pp. 71–73.
14. McCullough, D. G., *The Great Bridge*, Simon and Schuster, New York, 1972.
15. ———, "Placing the Underwater Tubes for BART," *Civil Eng., ASCE*, Vol. 37, September 1967, pp. 33–35.
16. "A New Era Opens—Chesapeake Bay Bridge Tunnel," Chesapeake Bay Bridge - Tunnel Authority, Cape Charles, Virginia, April 15, 1964.
17. Terzaghi, K., and R. B. Peck, *Soil Mechanics in Engineering Practice*, 2d ed., John Wiley and Sons, New York, 1968.

18. Hiley, A., "Pile Driving Calculations with Notes on Driving Forces and Ground Resistance," *The Structural Engineer,* Vol. 8, July–August 1930.

19. Chellis, R. D., *Pile Foundations,* McGraw-Hill, New York, 1961.

20. Carried out by the Michigan State Highway Commission in 1965. See Bowles, J. E., *Foundation Design and Analysis,* 2d ed., McGraw-Hill, New York, 1977.

21. Davisson, M. T., "BRD Vibratory Driving Formula," *Found. Facts,* Vol. 6, No. 1, 1970, pp. 9–11.

22. Berezantzev, V. G., V. S. Khristoforov, and V. N. Golubkov, "Load Bearing Capacity and Deformation of Piled Foundations," in *Proc. Fifth Int. Conf. on Soil Mech. and Found. Eng.,* Vol. 2, Paris, 1961, pp. 11–15.

23. Vesic, A. S., "Ultimate Loads and Settlements of Deep Foundations in Sand," Lecture 6 in *Bearing Capacity and Settlement of Foundations,* A. S. Vesic (ed.), Duke University, Durham, NC, 1967, pp. 53–68.

24. Meyerhof, G. G., "Compaction of Sands and Bearing Capacity of Piles," *J. Soil Mech. Found. Eng. Div.,* ASCE, Vol. 85, No. SM6, December 1959, pp. 1–29.

25. Sowers, G. B., and G. F. Sowers, *Introductory Soil Mechanics and Foundations,* 3d ed., Macmillan, New York, 1970.

26. Potyondy, J. G., "Skin Friction between Various Soils and Construction Materials," *Geotechnique,* Vol. 11, 1961, pp. 339–353.

27. Vesic, A. S., "Design of Pile Foundations," National Cooperative Highway Research Publication No. 42, Transportation Research Board, Washington, D.C., 1977.

29. Koerner, R. M., and A. Partos, "Settlement of a Building on Pile Foundation in Sand," *J. Geotech. Eng. Div., ASCE,* Vol. 100, No. GT3, March 1974, pp. 265–278.

29. Koerner, R. M. and A. Partos, "Settlement of a Building on Pile Foundation in Sand," *J. Geotech. Eng. Div., ASCE,* Vol. 100, No. GT3, March 1974, pp. 265–278.

30. Smith, E. A. L., "Impact and Longitudinal Wave Transmission," *Trans. ASME,* Vol. 77, 1955, pp. 963–973.

31. ———, "Pile-Driving Analysis by the Wave Equation," *J. Soil Mech. Found. Eng. Div., ASCE,* August 1960, pp. 35–61.

32. Hirsch, T. J., L. L. Lowery, H. M. Coyle, and C. H. Samson, Jr., "Pile Driving Analysis by One Dimensional Wave Theory: State of the Art," *Highway Res. Rec.,* Pub. 333, 1970, Highway Research Board, Washington, D.C., pp. 33–53.

33. Bowles, E., *Foundation Analysis and Design,* McGraw-Hill, New York, 1968.

34. Mosley, E. T., "Wave Equation Analysis," *Found. Facts,* Vol. 3, No. 2, 1967, pp. 15–17.

35. Raamot, T., "Analysis of Pile Driving by Wave Equation," *Found. Facts,* Vol. 3, No. 1, 1967, pp. 10–12.

36. Desai, C. S. (ed.), *Proc. Symp. on Application of Finite Element Methods in Geotech. Eng.,* U.S. Army Corps of Engineers, Vicksburg, MS, 1972.

37. Ellison, R. D., E. D'Appolonia, and G. R. Thiers, "Load-Deformation Mechanism for Buried Piles," *J. Soil Mech. Found. Eng. Div., ASCE,* Vol. 97, No. SM4, April 1971, pp. 661–680.

38. Desai, C. S., L. D. Johnson, and C. M. Hargett, "Analysis of Pile Supported Gravity Lock," *J. Geotech. Eng. Div., ASCE,* Vol. 100, No. GT9, September 1974, pp. 1009–1030.

39. Amir, J. M., and M. Sokolov, "Finite Element Analysis of Piles on Expansive Media," *J. Geotech. Eng. Div., ASCE,* Vol. 102, No. GT7, July 1976, pp. 701–720.

40. Goble, G. G., and F. Rausche, "Pile Load Test by Impact Driving," Highway Research Record Pub. No. 333, Highway Research Board, Washington, DC, 1970, pp 102–130.

41. Rausche, F., F. Moses, and G. G. Goble, "Soil Resistance Predictions from Pile Dynamics," *J. Soil Mech. Found. Eng. Div., ASCE,* Vol. 98, No. SM9, September 1972, pp. 917–937.

42. Goble, G. G., F. Rausche, and G. E. Likins, Jr., "The Analysis of Pile Driving, A State-of-the-Art," in *Proc. Int. Seminar on Appl. of Stress Wave Theory on Piles,* Stockholm, Sweden, June 4–5, 1980, 1981, pp. 86–98.

43. Van der Veen, C., "The Bearing Capacity of a Pile," *Proc. Third Int. Conf. on Soil Mech. and Found. Eng,* Vol. II, Zurich, 1953, pp. 84–90.

44. Buisson, M., "Piles and Pile Foundations, Settle of Pile Foundations," in *Proc. Third Int. Conf. on Soil Mech. and Found. Eng.,* Vol. III, Zurich, 1953, p. 176.

45. Whitaker, T.W., and R.W. Cooke, "A New Approach to Pile Testing," *in Proc. Fifth Int. Conf. on Soil Mech. and Found. Eng.* Vol. II, Paris, 1961, pp. 171–176.
46. Mansur, C. I., and Focht, J. A., "Pile Loading Tests," ASCE Trans., Vol. 121, 1956, pp. 555–587.
47. Mohan, D., G. S. Jain, and V. Kumar, *Inst. Eng. J. India,* Vol. 3, 1960, pp. 32–46.
48. Moorehouse, D. C., and J. V. Sheehan, "Predicting Safe Capacity of Pile Groups," *Civil Eng., ASCE,* October 1968, pp. 44–48.
49. Poulos, H. G., and E. H. Davis, *Elastic Solutions for Soil and Rock Mechanics,* John Wiley and Sons, New York, 1974.
50. Meyerhof, G. G., "Compaction of Sand and Bearing Capacity of Piles," *J. Soil Mech. Found. Eng. Div., ASCE,* Vol. 85, No. SM6, December 1959, pp. 1–29.
51. Sogge, R. L., "Laterally Loaded Pile Design," *J. Geotech. Eng. Div., ASCE,* Vol. 107, No. GT9, September 1981, pp. 1179–1199.

THREE

SOIL STABILITY

3.1 BACKGROUND, CONCEPTS, AND METHODS

In construction and foundation engineering the realization that unbraced or even braced soil masses may become unstable is all too easily arrived at after a *soil stability* failure occurs and the damage becomes apparent. There have been many infamous situations that underline the importance of proper design, construction, and maintenance methods in achieving soil stability, and perhaps the 1889 failure of the earth dam upstream from Johnstown, PA, will serve to illustrate the significance of the problem (see Fig. 3.1 for an artist's sketch of that failure[1]). However, earth dam engineering, which certainly is related to the topics to be discussed in this chapter, is well beyond the scope of this book (see Refs. 2 and 3 for an in depth treatment of the subject). We will limit ourselves here to a consideration of unbraced embankment slope stability and a variety of braced excavation methods.

The need for braced as opposed to unbraced excavations comes about by virtue of the particular locales where projects are situated. Slopes adjacent to a proposed foundation will always be laid back (no bracing needed) if:

Sufficient room adjacent to the excavation is available
There is no particular need to get very close to the excavation
The slopes are stable at a reasonably steep angle, e.g., the water table is beneath the
  foundation elevation

In urban areas, these conditions are usually not met, and bracing is then required to create steep slopes or, as is usually the case, vertical ones.

Figure 3.1 Artist's conception of failure of earth dam contributing to Johnstown flood of 1889.

The evolution of the different procedures currently used for unbraced and braced excavations is not particularly clear but a few comments about their development seem appropriate here.

An analysis of the behavior of unbraced slopes took major steps forward with the introduction of the theory of circular arc failure, the method of slices, and the concept of a factor of safety, all of which were proposed by the Swedish Geotechnical Institute in the early 1900s. Design procedures have since evolved using effective stress parameters and incorporating the use of pore water pressures, random slip surfaces, and, most recently, probability theory.

Conversely, the practices associated with braced slopes have developed over the years less through theory than by trial and error in actual practice. Bracing undoubtedly began as *skeleton sheeting*, where vertical boards are placed at intervals in a trench and are braced either from one side of the trench to the other or by means of braces sloped downward into the bottom of the excavation. Where this procedure was inadequate because of the type of soil being contained or because of the pressures on the braces that developed, *continuous sheeting* was required. As deeper excavations requiring bracing were undertaken, structural members with heavier cross sections were developed: first concrete, then steel. Today, various hybrids of steel and timber braces together with cast-in-place concrete walls are commonly used, as well as a wide variety of bracing systems, the most recent innovation in this area being the procedure called *soil nailing*. Each type of bracing method will be reviewed in the sections that follow.

3.2 UNBRACED EXCAVATIONS

For excavations made in soil where the total depth is less than 10 to 15 ft (3 to 5 m), the slope can be laid back on a constant angle in accordance with conventional geotechnical engineering design practice[2,3] or estimated from empirical values developed over the years; see Table 3.1 for granular soils[4] and Table 3.2 for cohesive soils.[5] In general, the longer the excavation remains open, the more conservative one should be. This is particularly the case for cohesive soils where their shear strength is known to be time-dependent. It is also good practice to protect all soil slopes from rainfall runoff and erosion by either a geotextile or a geomembrane (see Chap. 7), whichever the particular circumstances call for.

For excavations made in soil where the total depth is greater than 10 to 15 ft (3 to 5 m), the slope angle should be determined on the basis of a stability analysis for the particular situation. A number of design charts are currently available for this purpose, of which Taylor's[6] seem reasonable to use, at least for a first approximation.

For normally consolidated and underconsolidated fine-grained soils where the shear strength can be approximated by its undrained strength a stability number concept (which makes use of stability curves) can be used.[6] Here, the factor of safety is:

$$FS = \frac{c}{N_s \gamma H} \tag{3.1}$$

where FS = factor of safety

c = undrained strength

γ = total unit weight

H = vertical height of slope

N_s = stability number taken from Fig. 3.2

Stability curves illustrate a number of problem geometries which are of practical interest. It should be added that they can also be used to determine the critical slope angle β or the critical height H, by using a factor of safety of 1 in Eq. (3.1).

The extension of this stability number concept to soils containing both cohesion

Table 3.1 Slope angles for shallow unbraced excavations in granular soil†

Soil type	Temporary slope		Permanent slope	
	β, deg	tan β	β, deg	tan β
Gravel with boulders	53	1:0.75		
Sand gravel or angular sand	39	1:1.25	34	1:1.5
Rounded coarse sand	34	1:1.5	30	1:1.75
Rounded fine sand	30	1:1.75	27	1:2
Sand with water emerging on slope	22–16	1:2.5–3.5		

†After Bazant, Ref. 4.

Table 3.2 Slope angles for shallow unbraced excavations in cohesive soils[†]

Soil type	Plasticity index PI[‡]	Depth of excavation, m	β, deg	$\tan \beta$
Clayey silt	≤ 10	0–3	39	1:1.25
		3–6	32	1:1.60
		6–9	30	1:1.75
Silty clay	10–20	0–3	39	1:1.25
		3–6	39	1:1.25
		6–9	35	1:1.40
Plastic clay	≥ 20	0–3	39	1:1.25
		3–6	39	1:1.25
		6–9	39	1:1.25

[†]After Ref. 5.

[‡]PI = liquid limit minus plastic limit.

and friction has also been made by Taylor[6] and is presented in Fig. 3.3. The same relationship, Eq. (3.1), is used for the calculations.

It should be noted that beyond these approximation methods lie methods of considerable sophistication in the form of various types of stability analyses, beginning with total stress analysis using the method of slices and going all the way up to probabilistic methods which assume nonuniform slip surfaces. Although these are beyond the scope of this text, they are covered in Refs. 2 and 3.

Depending on the exposure time of the slope in question, its protection against scour must be assured. Erosion protection methods run the gamut from seed and mulch, to geotextiles, to paving with a number of materials such as articulated blocks, asphalt, concrete, etc. However, once a slope is in danger of failing, or has failed, remedial actions are required. Table 3.3 gives a number of recommendations in this regard.[7]

3.3 BRACED EXCAVATIONS

3.3.1 Overview

Where conditions for an unbraced excavation are not possible, some form of lateral support, i.e., bracing, must be constructed to retain the soil. The various systems described in this section are invariably more time-consuming and expensive to construct than to merely lay back the slope of the adjacent soils as described in Sec. 3.2. Bracing systems to be described here are the following:

Skeleton sheeting
Soldier beams and lagging
Solid sheeting (usually steel)

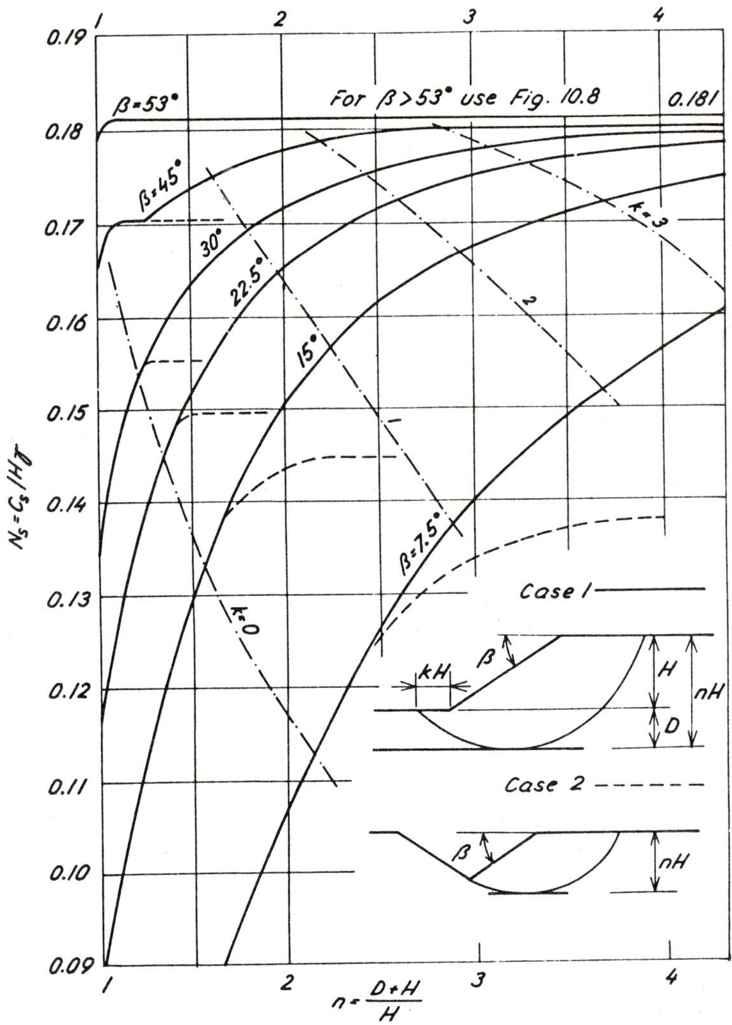

Figure 3.2 Stability curves for soils whose strengths can be approximated by undrained conditions. *(After Taylor, Ref. 6.)*

They are necessary to protect the workers within the excavation from slope failures and "cave-ins," but also to protect property and equipment in the immediate vicinity. While contractors have long recognized the safety problems involved in excavations, it was the Occupational Safety and Health Administration (OSHA) regulations for excavation, trenching, and shoring[8] that set the tone for a number of subsequent studies.[9–11] The argument as to who should be the designer of temporary bracing systems (the contractor or owners' consultant) is still unsettled (see, for example, the discussion and conclusion of Ref. 9). But whoever it is, the designer should be a

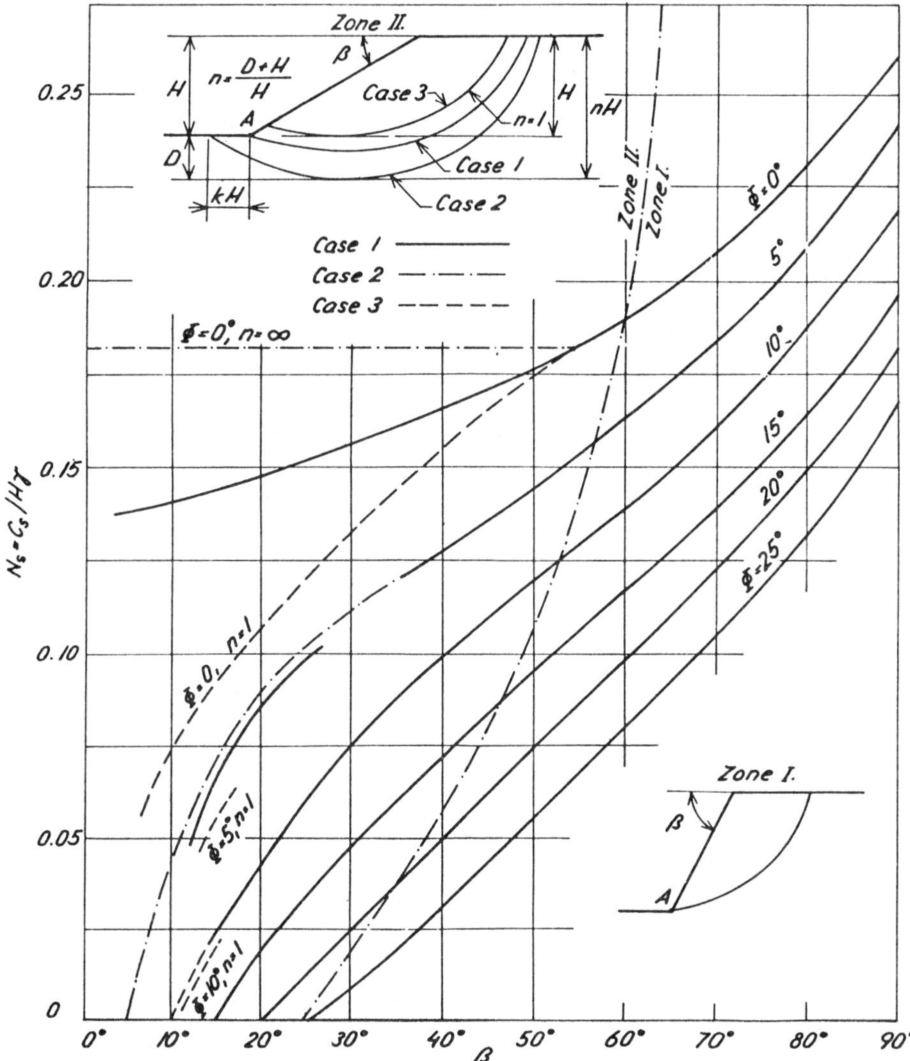

Figure 3.3 Stability curves for soils whose strengths come about from cohesion and friction. *(After Taylor, Ref. 6.)*

competent and registered engineer who has the following information:

Requirements of the structure being built (depth, size, time constraints)
Type of soil
Soil properties (or reasonable estimates thereof)
Location of groundwater

Table 3.3 Methods of slope stabilization†

Grading of slope

Excavation at top of slope. Appropriate for deep masses of cohesive soil where rotational failure is likely. Less effective in creep movements or shallow transitional failures.

General flattening of slope. Applicable for deep masses of cohesive soil where rotational failure is likely and for relatively small slide areas where toe has been oversteepened.

Benching of slope. Appropriate on steep slopes where flattening is difficult and surface sloughing occurs. Assists in erosion control and in catching debris of small slides. Benches should be sloped to collect runoff and convey it off the slide.

Complete removal of unstable mass. Feasible for relatively small slide masses or shallow creep movements.

Earth fills at toe of slope. Placed as counterweight or stabilizing berm at locations where upward movement of slide mass may occur. Careful study is required so that addition of fill does not increase driving force.

Rock or gravel fills at toe of slope. Serves as counterweight, also controls toe erosion and seepage exit gradients, preventing backward erosion.

Surface drainage

Open ditch surrounding slide area. Useful in a variety of situations. Ditching within slide mass should be used with caution. Discharge should be directed away from potentially unstable areas.

Surface treatment by seeding, paving, or drainage blanket. Applied in critical locations for erosion control and to collect, control, or redirect surface seepage, preventing it from infiltrating unstable mass.

Regrading of surface to improve runoff. Useful for slide masses with sag ponds, depressions, or troughs which trap surface runoff. Frequently coupled with drainage ditches or trenches.

Sealing joint planes, cracks, and fissures. Appropriate for slides in stiff brittle materials where single large cracks or jointed zones can be sealed at the surface with mastic or clay or cement slurry.

Subsurface drainage

Drainage by drilled horizontal or sloping drain holes. Frequently the most economical method of drainage. Appropriate for large and deep slide mass, particularly where pervious water-bearing strata can be intercepted by drain holes before reaching failure zone. Thorough investigation of groundwater conditions is a prerequisite.

Diversion of water by deep drainage trenches. Appropriate where shallow or perched groundwater flow influences stability directly or delivers large quantity of water to slide mass. Usually limited to conditions where flow can be intercepted at depths of less than 10 to 15 ft.

Interception of seepage by tunnels. Useful for draining large slide masses and where structures of great value are threatened. Usually should be accompanied by drilled drainage holes extending outward from tunnel.

Vertical drain wells. Joined with horizontal drain holes or may be pumped directly to intercept strata or lenses of permeable material. Utilized in special cases to lower perched water table by draining into underlying pervious stratum. Accelerates consolidation and gain in strength of normally consolidated strata.

Retaining structures

Rock and earth fill buttresses at toe of slope. To be effective, volume of fill should be between one-quarter and one-half of volume of total slide mass and should extend between 5 to 10 ft below failure zone to provide adequate shear key.

Cribs or gravity retaining wall. Appropriate to prevent undercutting of toe or for resisting small slides, not suitable for large unstable masses.

Pile walls or caissons at toe of slope. Increases resistance on failure surface, must be driven to sub-

Table 3.3 Methods of slope stabilization† (*Continued*)

Retaining structures

stantial depth below failure zone. Applicable only for slides of limited dimensions and frequently misapplied for slides of large extent. Thorough analysis should precede use of pile wall.

Barriers at toe of slope anchored by tie-back to firm support. Useful in control of surface sloughing and small slides. Thorough analysis should precede use in large unstable masses.

Increase of shear strength

Injection and cementation. Appropriate for soils of moderate permeability. Silicates, cement, asphalt emulsion are utilized.

Freezing. Used as a temporary stabilization method during construction. Process is slow and costly.

Electro-osmosis. Applicable to soft, fine-grained materials. Costs may be high.

Compaction. Conventional procedure for embankment construction. Also utilized in removal and replacement of portion of unstable mass.

Miscellaneous

Rock bolting. Appropriate for steep slopes of weathered rock, fractured hard rock, or stiff soils where shallow sloughing or failure occurs on joint system.

Riprap protection. Prevents toe erosion and undercutting of slope at water's edge.

Blasting at toe of slope. Temporary expedient which interrupts slip surface and facilitates drainage of seeping water in lower portion of slide.

†After NAVDOCKS, Ref. 7.

Contractor's approximate installation procedure and plans (excavation, staging area, etc.)

It should also be borne in mind by the designer that each bracing system is composed of simple structural members which act both individually and collectively as a complete system.

3.3.2 Skeleton Sheeting

Skeleton sheeting is the least expensive temporary bracing system and can only be used in soils which possess some amount of cohesion. This includes soils in the GC, SC, CL, ML, CH, and MH categories of the Unified Soil Classification system (see Appendix A). The general configuration of skeleton sheeting is shown in Fig. 3.4, where it can be seen that the soil must support itself between uprights when such bracing is used. Obviously the lower the cohesion of the soil, the closer the uprights must be. Uprights are customarily 2 × 10 in or 3 × 12 in construction-grade lumber at spacings of 10 ft (in firm clays of 2000-lb/ft² cohesion or greater) to 3 ft (in softer clays of 300- to 500-lb/ft² cohesion). Weaker clays or granular soils with no cohesion require the uprights to be placed so close together that they might just as well be classified as continuous sheeting (see Sec. 3.3.3).

There is no standard design procedure for skeleton sheeting, but it should be noted that problems sometimes occur at the bottom of the upright, which "kicks out"

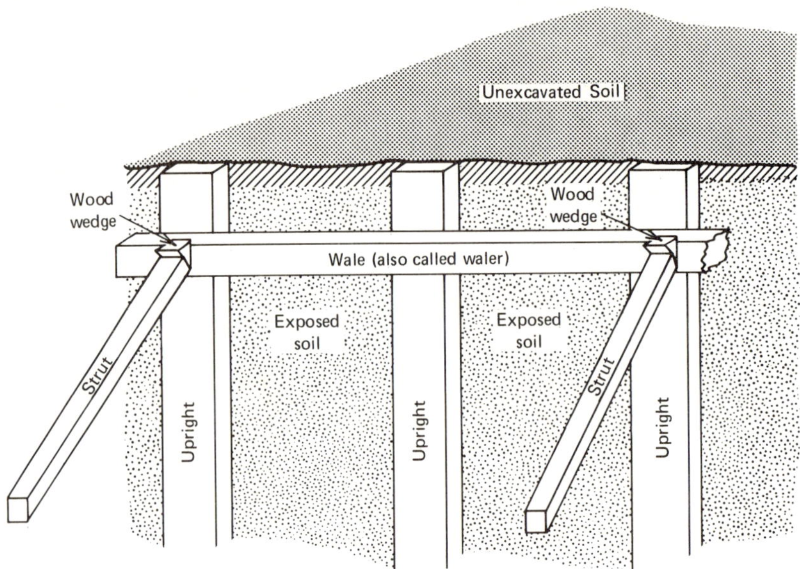

Figure 3.4 Basic components for skeleton sheeting.

into the trench because of high soil pressures and lack of embedment. This lack of embedment comes about from the typical construction practice of placing the uprights as excavation proceeds, unlike other sheeting methods where the vertical members are driven before excavation begins and usually to a depth greater than the lowest excavation elevation. Thus a waler and a set of struts is sometimes placed at the bottom of the trench below subgrade elevation (sometimes called a "mud sill"), or stakes can be driven against the uprights to give some additional support.

3.3.3 Design Elements of Continuous Sheeting

Continuous sheeting of any type will mobilize the full lateral pressure of the soil, a subject which has been studied considerably over the years. Critical in any of the standard formulations of such bracing is the evaluation of K_a, the coefficient of active earth pressure. This value, which is a function of the angle ϕ of shearing resistance of the soil, can be obtained from Rankine theory in the simplest case:

$$K_a = \tan^2\left(45 - \frac{\phi}{2}\right) \tag{3.2}$$

A sophisticated theory based on curved slip surfaces has been developed by Caquot and Kerisel,[12] and is given in Appendix B.

It must also be recognized that the support systems under consideration are quite flexible in comparison with the permanent wall systems for which most theories have been developed. And with this flexibility comes about a very different earth pressure diagram than the linearly increasing one customarily associated with rigid, permanent walls. A set of recommended pressure distributions, following those developed by Peck[13] and by Goldberg et al.,[14] are given in Fig. 3.5.

For strut design, the appropriate pressure distribution is assumed to be subdivided between each strut. This is in essence equivalent to assuming that a plastic hinge exists at each strut, and it makes a statically indeterminate problem into one that is statically determinate. Example 3.1 illustrates this design technique.

Example 3.1 Strut design for temporary sheeting

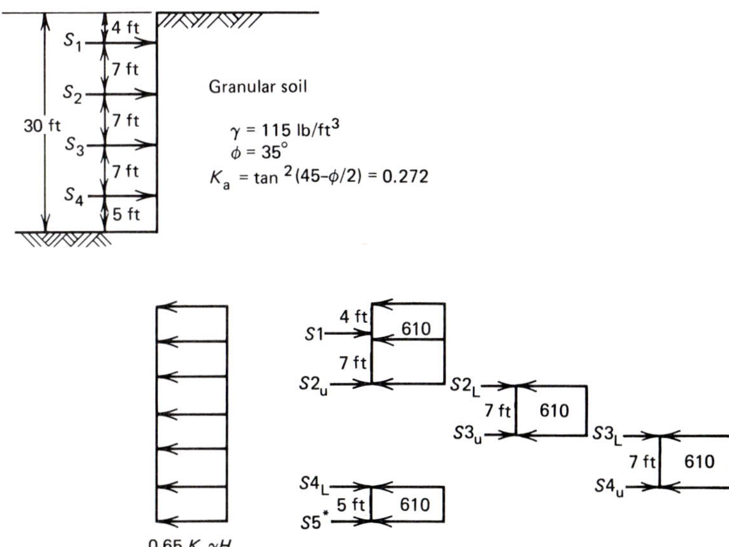

Statics are applied to each figure for loads; that is:

$$\Sigma M_{s2U} = 0; \quad 610(11)(5.5) = S_1(7) \quad \therefore S_1 = 5272 \text{ lb/ft}$$

$$\Sigma F_H = 0; \quad 610(11) = 5272 + S_{2U} \quad S_{2U} = 1438 \text{ lb/ft}$$

$$\Sigma F_H = 0; \quad 610(3.5) = S_{2L} \quad S_{2L} = 2135 \text{ lb/ft}$$

$$\therefore S_2 = 3573 \text{ lb/ft}$$

$$\Sigma F_H = 0; \quad 610(3.5) = S_{3U} \quad S_{3U} = 2135 \text{ lb/ft}$$

$$\Sigma F_H = 0; \quad 610(3.5) = S_{3L} \quad S_{3L} = 2135 \text{ lb/ft}$$

$$\therefore S_3 = 4270 \text{ lb/ft}$$

$$\Sigma F_H = 0; \quad 610(3.5) = S_{4U} \quad S_{4U} = 2135 \text{ lb/ft}$$

$$\Sigma F_H = 0; \quad 610(2.5) = S_{4L} \quad S_{4L} = 1525 \text{ lb/ft}$$

$$\therefore S_4 = 3660 \text{ lb/ft}$$

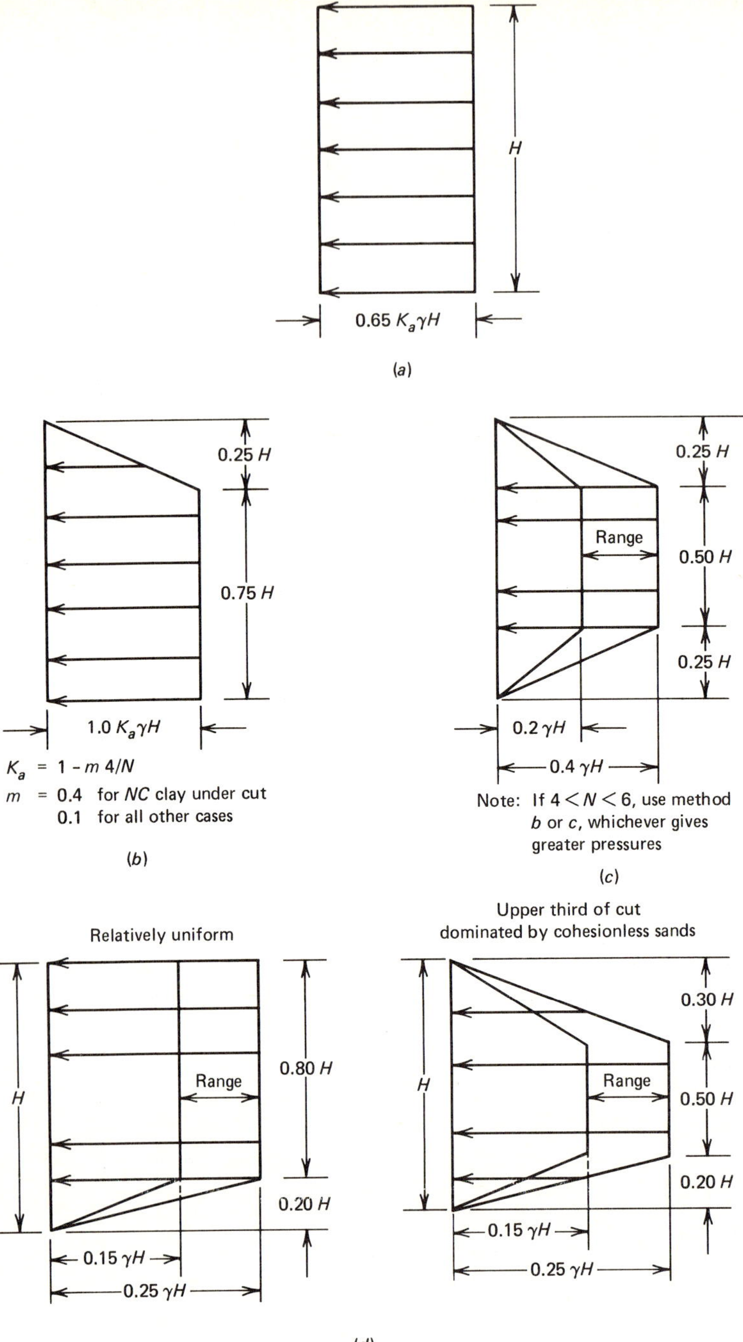

Figure 3.5 Pressure distribution behind temporary (and flexible) bracing systems. (*a*) Sands [use $K_a = \tan^2 (45 - \phi/2)$]. (*b*) Medium to soft clays. $N > 6$ where $N = \gamma H/c$. (*c*) Stiff clays. $N < 4$, where $N = \gamma H/c$. (*d*) Dense cohesive sands; very stiff sandy clays. (*After Peck, Ref. 13, and Goldberg et al., Ref. 14.*)

Therefore, for struts at 10 ft center to center:

$$S_1 = 53 \text{ kips}$$

$$S_2 = 36 \text{ kips}$$

$$S_3 = 43 \text{ kips}$$

$$S_4 = 37 \text{ kips}$$

These values are not particularly well-balanced; and if there are no constraints, the entire bracing system should be moved up about 1.5 ft.

3.3.4 Soldier Beams and Lagging

An extremely popular method of temporary wall construction is the use of soldier beams and lagging (see Figs. 3.6 and 3.7). The procedure is to drive H piles prior to any excavation at the location where the face of the wall is desired. The piles are driven with the flanges parallel to the excavation area. When they have been driven down to the desired elevation (usually 5 to 10 ft beneath the proposed excavation level when in soil, or socketed when in rock), excavation begins in stages. Depending upon the cohesion of the soil at the site, first-stage excavation can be as deep as the location of the uppermost strut. At this point timber lagging, cut to fit between the webs of adjacent soldier piles and placed in back of the front flanges of the piles is set one piece on top of the other with only a small spacer between. The lagging is usually 3×12 in lumber. (An alternative method, called *contact sheeting,* involves using random lengths of wooden lagging placed on the front face of the soldier pile, which are held onto the flange by means of J bolts through the lagging and hooked onto the back side of the soldier pile flange. A washer and nut are then used to draw the lagging against the pile face.) Once the lagging is set down to the first strut level, the horizontal waler is placed against the piles and the struts are placed at predesigned intervals. If horizontal struts are to be used, as in utility trenches, the excavation and lagging operation must proceed down both sides of the trench simultaneously. If sloping struts down to the final elevation are to be used, as in excavations of building foundations, the struts are sloped downward to the final grade. Thus only a berm of soil is left adjacent to the partially unexcavated wall. (Tiebacks eliminate the need for this relatively complicated procedure; see Sec. 3.5.) At this point the upper section of the wall is braced and excavation proceeds to the next strut level, with the process continuing there in the manner just described. Figure 3.6 shows some photographs of typical jobs and Fig. 3.7 some construction sequence diagrams.

Soldier beams and lagging are widely used because contractors are familiar with the system and because they have found it to perform relatively well. However, the following comments should be kept in mind when using it:

The soldier piles cannot be allowed to deviate when being driven so that they move from their intended vertical alignment *into* the excavation area. This often hap-

Figure 3.6 Photographs of various projects using soldier beams and lagging to temporarily retain adjacent soil and nearby structures. (Soil tiebacks are used for support in these cases.) *(Compliments of Schnabel Foundation Company.)*

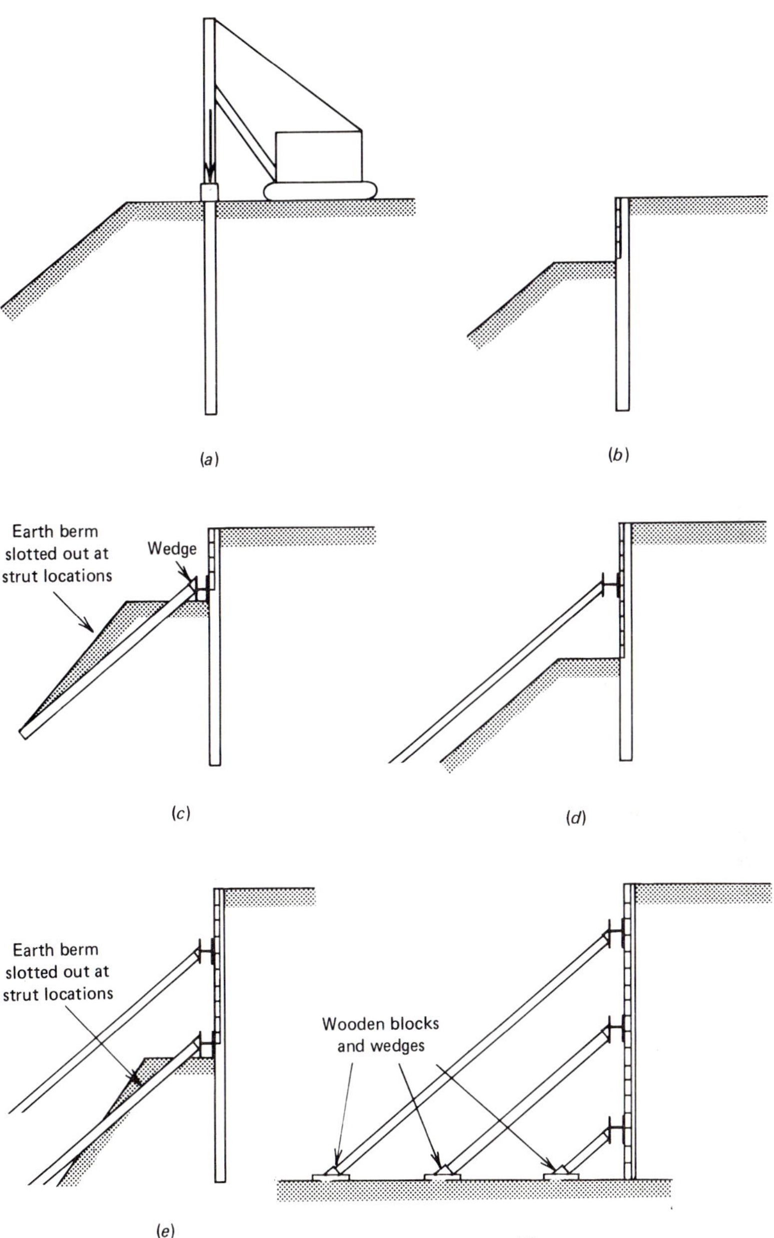

Figure 3.7 Construction sequence for installing of soldier beams and lagging. (Struts and rakers placed into the excavation are used for support.) (*a*) Driving H-pile soldier beams. (*b*) First-stage excavation and installation of lagging. (*c*) Installation of uppermost waler and struts. (*d*) Second-stage excavation and installation of lagging. (*e*) Installation of second waler and struts. (*f*) Process repeated until completion.

pens when dense, granular soils are encountered. In such cases, expensive pre-drilling may be necessary.

The system is not watertight, but use of hay behind the lagging helps avoid minor seepage problems.

A gravel-filled drainage ditch at the toe of the excavation will also aid in controlling seepage that comes through the lagging.

If the seepage flow is too great to be handled by sump pumping, diversion of water by dewatering methods may be necessary (see Sec. 5.2).

The past use of struts extending down into the excavation, as seen in Fig. 3.7, creates so many construction problems, not only while they are being installed, but during excavation, wall formwork, splicing of reinforcement, patching in the hole in the wall, release during backfilling behind the wall, etc., that their future use must be questioned. Certainly with the advent of soil tiebacks (see Fig. 3.6), struts into the excavation seem antiquated.

A major consideration in regard to any temporary retaining system is how the soil behind the wall will move horizontally and vertically.

Moreover, this is obviously of concern in regard to other nearby structures founded on the soil as well. A interesting study was conducted by Lambe et al., who first challenged the current way of predicting these movements[15] (as well as the way of predicting strut loads and wall stress), then presented actual measurements of an installation in Boston[16] to support that challenge (see Fig. 3.8). Reviewing the two papers by Lambe et al., it can be seen that the current state of the art regarding prediction of soil movement is far from perfect, but with adequate factors of safety structures and retaining systems can be made safe and better than order-of-magnitude movements can be predicted.

When movements of soil are estimated to be excessive, a stiffer wall system is required, e.g., a tangent pile or a slurry wall, both of which will be considered later.

As excavation proceeds, the ever increasing difference between soil levels on each side of the sheeting can cause a base failure by upheaving within the excavation. This situation is particularly prone to occur in clay soils, as described by Peck et al.[17] In their book, those authors equate the bearing capacity equation [recall Eq. (1.6)] to the equivalent surcharge load as follows:

$$q_0 = cN_c = \gamma H$$

$$N_c = \frac{\gamma H}{c}$$

where γ = unit weight of soil behind wall
    H = difference in height between excavation depth and soil behind wall
    c = undrained shear strength of soil below excavation level
    N_c = bearing capacity factor, where $N_c < 6$ (base movements are small), $6 \leq N_c \leq 8$ (base movements are measurable), or $N_c > 8$ (base movements are intolerably large)

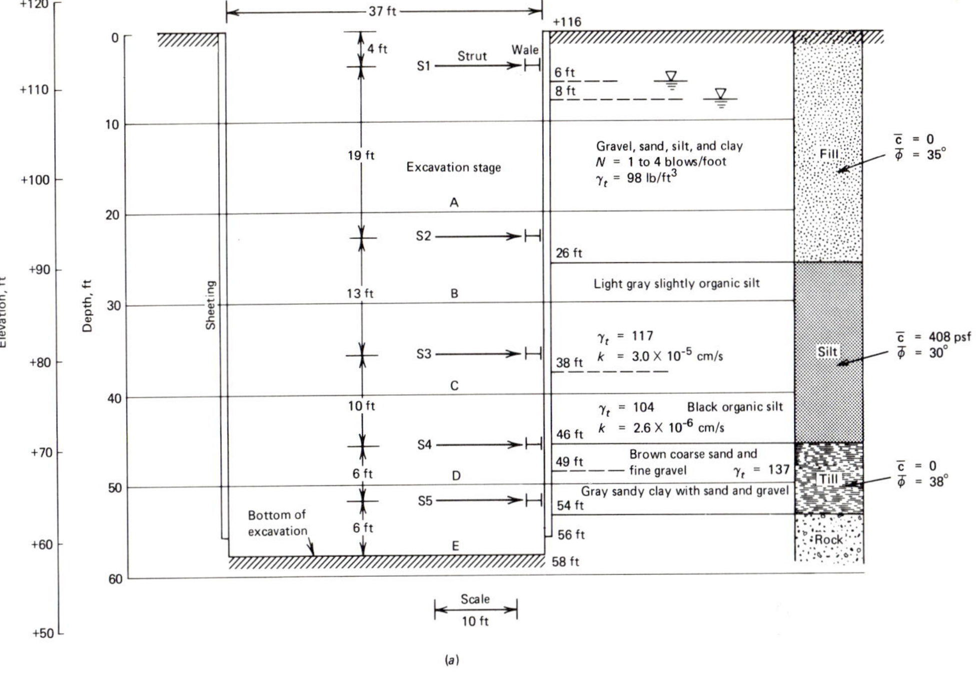

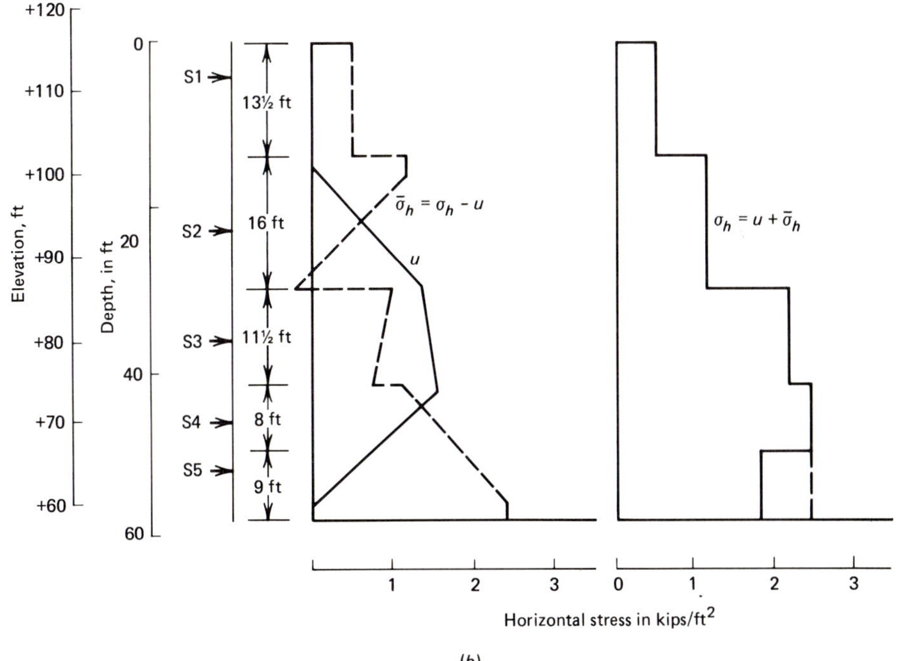

Figure 3.8 Details of soldier beam and lagging installation in Boston, MA, with measured values for stresses, strut loads, and movements. (*a*) Cross section at station 76 + 30. (*b*) Apparent horizontal stresses on wall. *(After Ref. 16.)*

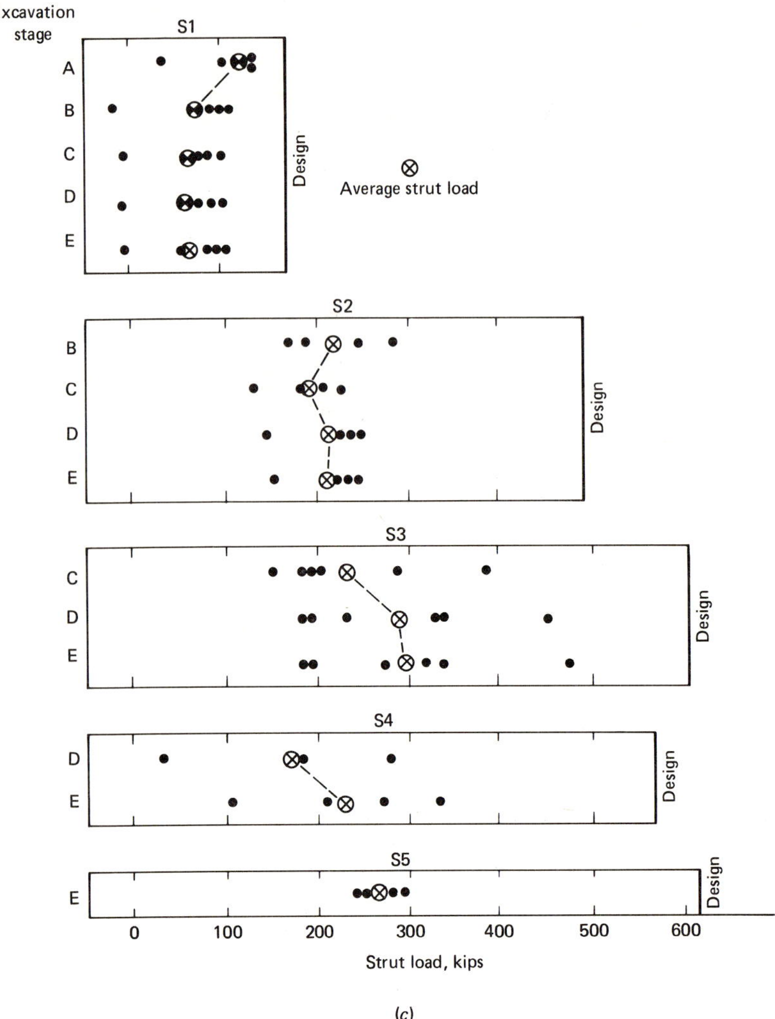

(c)

Figure 3.8 (c) Strut loads.

Under the last conditions, where $N_c > 8$, simple sheeting such as that described in this section should not be used. Continuous steel sheeting driven considerably beneath the base of the excavation is required (see Sec. 3.3.5).

Upon completion of the final wall or pile installation, the soldier piles are removed using a pile extractor. During extraction some of the upper pieces of lagging will come up to the ground surface, where they can be recovered. However, soil compaction of the backfill between the temporary and the permanent walls is necessary to prevent future ground subsidence. Thus some of the timber lagging

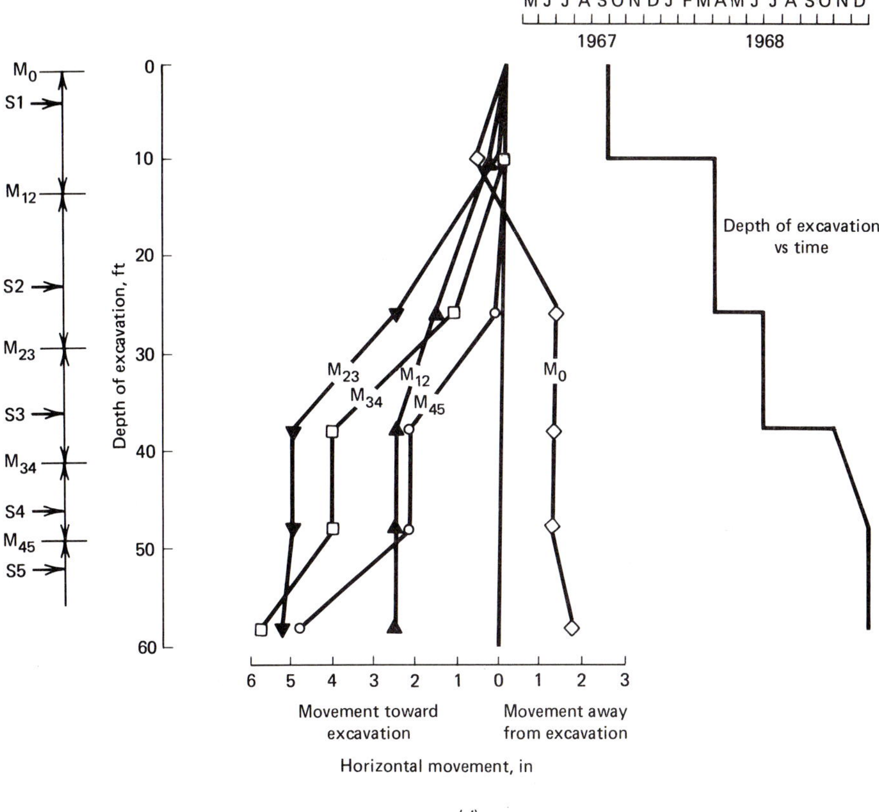

(*d*)

Figure 3.8 (*d*) Movement of east wall.

may be buried. As far as deterioration of the remaining lagging in the backfill is concerned, it probably creates more harm than good to try to retrieve all the pieces. They should be left in place.

3.3.5 Continuous Sheeting

In terms of construction practice, continuous sheeting resembles skeleton sheeting where the uprights are placed side by side, and in terms of design, it resembles soldier beams and lagging. See Sec. 3.3.3 for design details. The materials used can be wood, concrete, or steel. The wood and concrete sheets are sometimes tongue-and-grooved but are generally uniform, rectangular sections. The steel sheets are interlocking and are manufactured by many steel companies. Typical sections are shown in Fig. 3.9, and sketches and photographs of several completed wall sections are shown in Fig. 3.10.

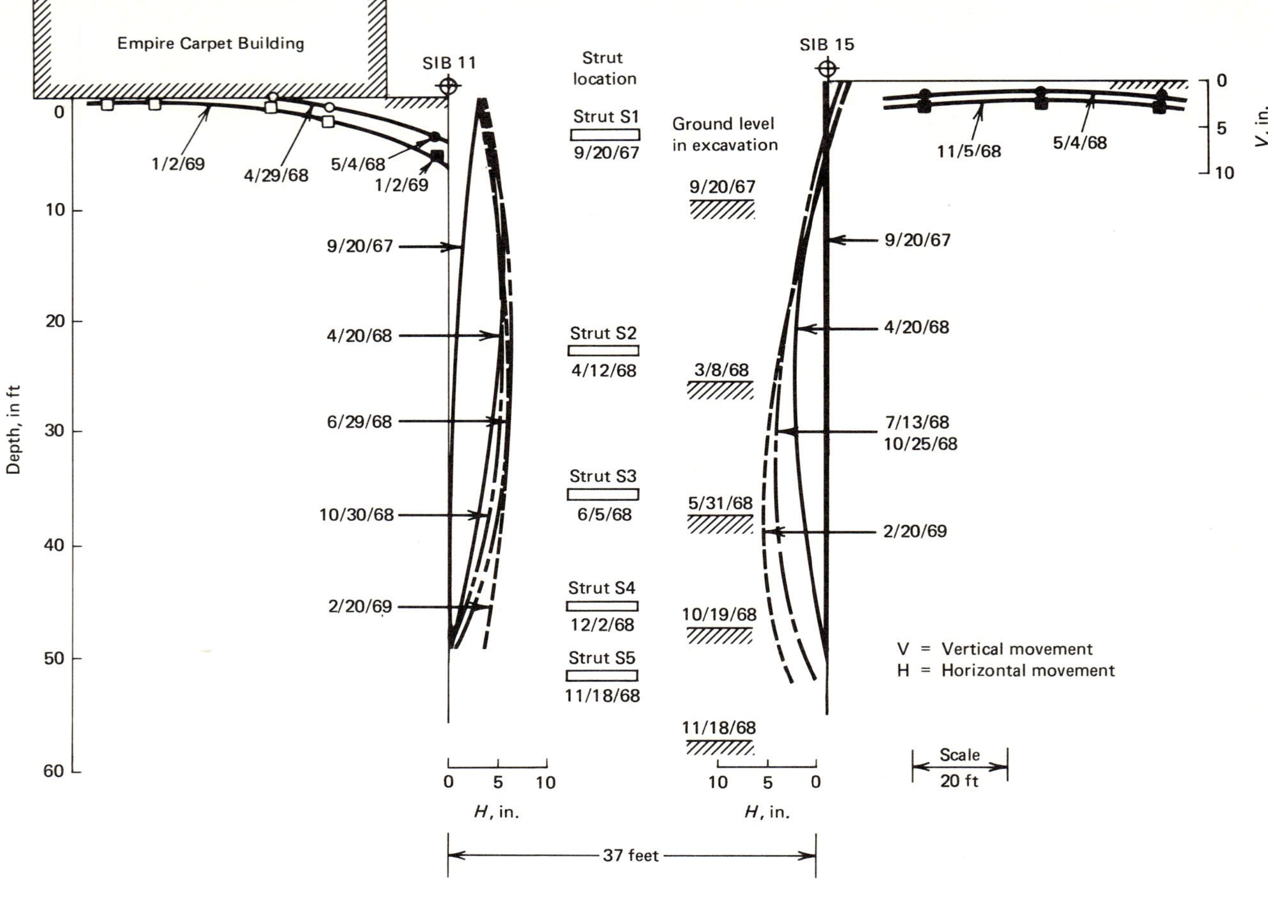

Figure 3.8 (*e*) Movements near excavation.

158

Profile	Interlock	Designation	District Rolled	Driving Distance per Pile	Weight		Web Thickness	Section Modulus		Area	Moment of Inertia	
					Per Foot	Per Square Foot of Wall		Per Pile	Per Foot of Wall	Per Pile	Per Pile	Per Foot of Wall
				In.	Lbs.	Lbs.	In.	In.³	In.³	In.²	In.⁴	In.⁴
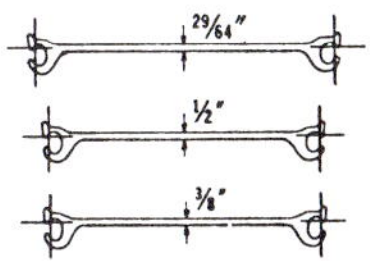	Interlock with Each Other	PSX 32	H.	16½	44.0	32.0	29/64	3.3	2.4	12.94	5.1	3.7
		PS 32*	H.S.	15	40.0	32.0	½	2.4	1.9	11.77	3.6	2.9
		PS 28	H.S.	15	35.0	28.0	⅜	2.4	1.9	10.30	3.5	2.8
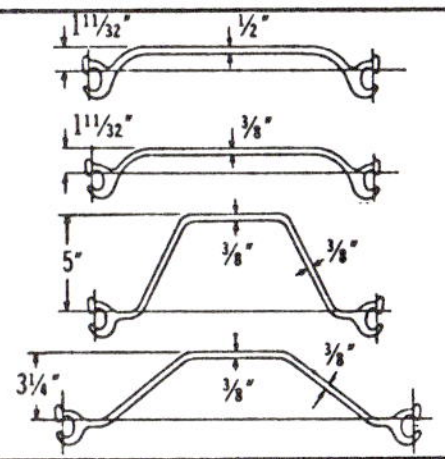	Interlock with Each Other	PSA 28*	H.	16	37.3	28.0	½	3.3	2.5	10.98	6.0	4.5
		PSA 23	H.S.	16	30.7	23.0	⅜	3.2	2.4	8.99	5.5	4.1
		PDA 27	H.S.	16	36.0	27.0	⅜	14.3	10.7	10.59	53.0	39.8
		PMA 22	H.S.	19⅝	36.0	22.0	⅜	8.8	5.4	10.59	22.4	13.7
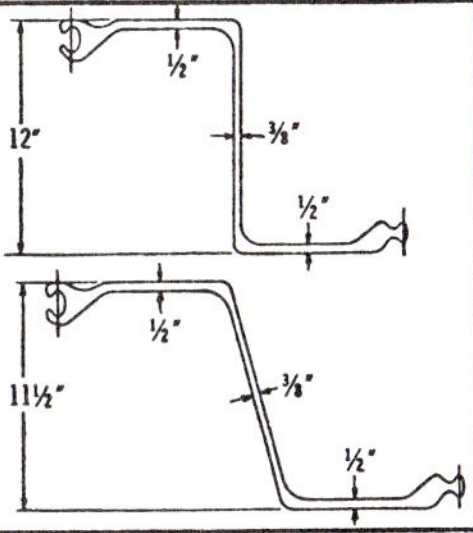	Interlock with Each Other and with PSA 23 or PSA 28	PZ 38	H.	18	57.0	38.0	⅜	70.2	46.8	16.77	421.2	280.8
		PZ 32	H.	21	56.0	32.0	⅜	67.0	38.3	16.47	385.7	220.4
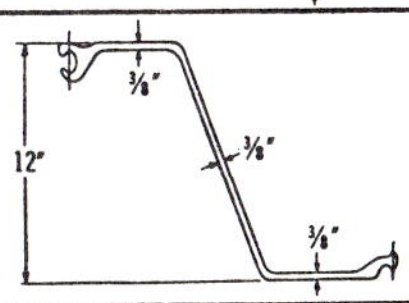	Interlocks with Itself and with PSA 23 or PSA 28	PZ 27	H.	18	40.5	27.0	⅜	45.3	30.2	11.91	276.3	184.2

*Sections PS 32 and PSA 28 are infrequently rolled and we do not advise their use in a design unless an adequate tonnage can be ordered at one time to assure a minimum rolling.

Complete data regarding these sections will be found in a separate publication entitled "USS Steel Sheet Piling."

H—Homestead, Pa.
(Eastern Steel Division)

S—South Chicago
(Central Steel Division)

Suggested Allowable Design Stresses—Sheet Piling

Steel Brand or Grade	Minimum Yield Point, psi	Allowable Design Stress, psi*
ASTM A328	38,500	25,000
ASTM A572 GR 50 (USS EX-TEN 50)	50,000	32,000
ASTM A690 (USS MARINER GRADE)	50,000	32,000

*Based on 65% of minimum yield point. Some increase for temporary overstresses generally permissible.

Figure 3.9 Types of interlocking U.S.S. steel sheeting (also called *sheet piling*).

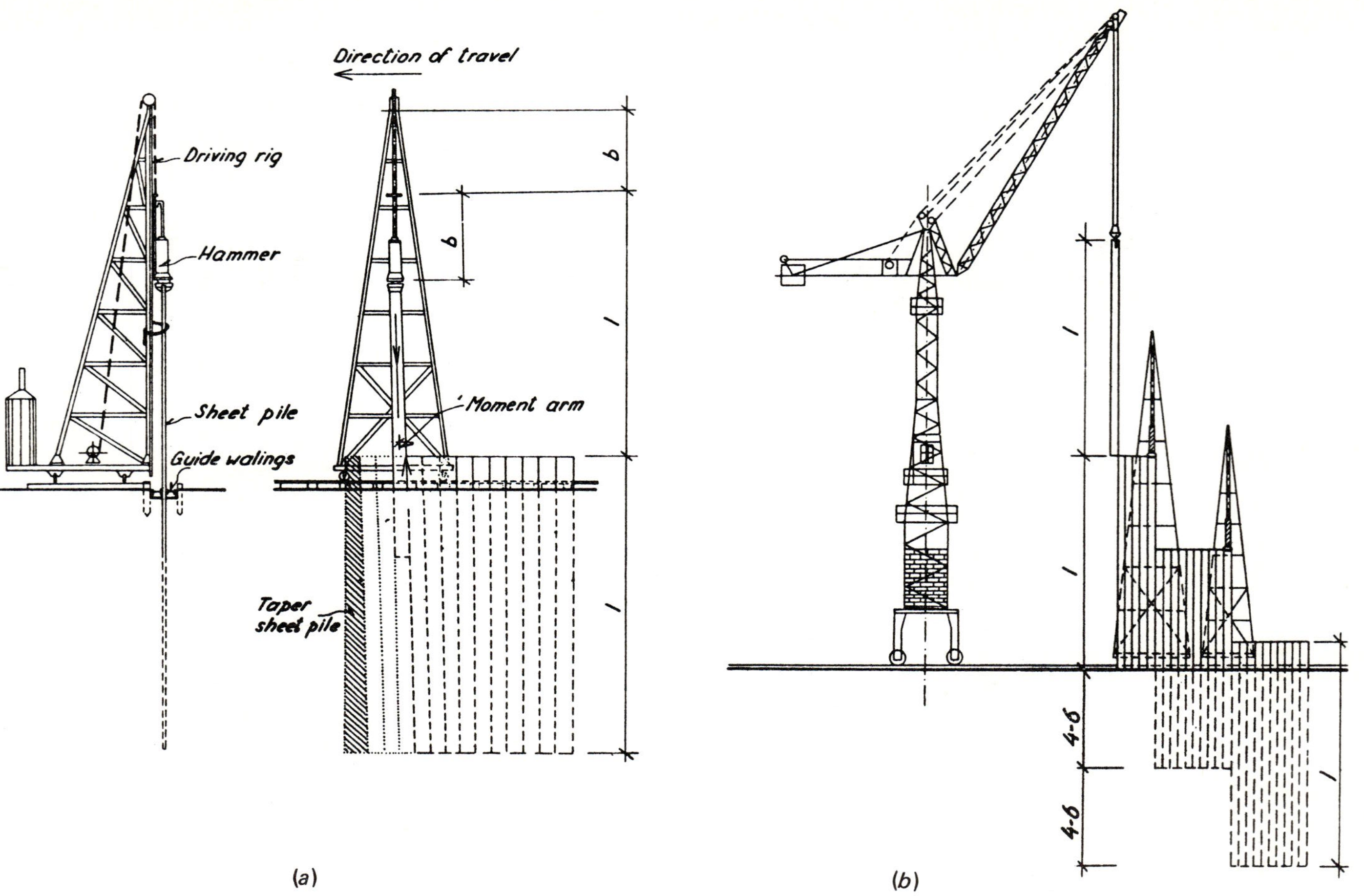

Figure 3.10 Methods of driving interlocking steel sheet piling using (*a*) full depth and (*b*) stage method. *(After Bazant, Ref. 4.)* (*c*) Construction of interlocking steel sheet piling walls (where work is being done from a barge floating in the water).

(c)

The installation of the steel sheets is difficult because they must be treaded into one another and driving usually proceeds in stages. Thus the driving of each new sheet must be started when its neighbor is only about one-third driven. A worker is required to sit on the partly driven sheet to guide the new one in place—a precarious job to say the least. Driving these sheets in high wind is not recommended and even the mere threat of wind sometimes means elaborate falsework is required.

Splitting of the web during driving is not uncommon, particularly when obstructions or dense granular soils are being penetrated. Specially fabricated, case-hardened driving shoes are available for those conditions.

There is a certain amount of dimensional tolerance in the interlocks, which permits 5 to 10° turning from one sheet to the next. An entire circle can be formed in this manner. There are special sections, for 90° turns, however, as well as specialty items such as T's, Y's, and other shapes. The sheets are not watertight because of the approximately $\frac{1}{4}$-in tolerance in each of the joints.

Waler and strut details are similar to those described for soldier beams and lagging in Sec. 3.3.4. If used for temporary support, the sheets are removed when the support is no longer needed, but they are often used for permanent support as well. Those latter cases usually involve waterfront construction for bulkheads, wharves, harbors, marinas, etc. The topic will be covered in Sec. 3.4 on permanent construction walls.

3.4 PERMANENT CONSTRUCTION WALLS

In Sec. 3.3 we focused on braced excavations where the sheeting is of a temporary nature and removed after the work is completed. In many situations, however, the sheeting must be left in place—either because of its required service life, resistance against deformation of retained soils, and/or the disturbance that its removal would cause. Three different situations of that sort will be described in this section:

Sheet pile bulkheads
Continuous concrete piles
Diaphragm (slurry) walls

3.4.1 Sheet Pile Bulkheads

The differences between continuous sheeting of a temporary nature, as described in Sec. 3.3.5, and permanent bulkhead walls are relatively subtle, yet great enough to warrant a separate discussion of permanent walls. The conditions resulting in the need for permanent walls are as follows:

The soils requiring permanent bulkhead walls are usually soft, recently transported river silts and clays adjacent to rivers or loose, granular soils found near oceans.
The anchorage used to restrain the upper part of the wall is often at a considerable distance from it.
Hydrostatic pressures are generally present, and particularly noticeable at times of receding tidal flow.
The lifetime of the wall and the elements that can affect the length of its life.

Typical installations are shown schematically in Fig. 3.11, the major difference between each having to do with the amount of surcharge load which is to be carried behind the wall. For small harbors and marinas, as shown in Fig. 3.11a, where loads are generally light, backfilling with free-draining soils can proceed directly to the elevation of the pile cap. Asphalt paving is often placed directly up to the wall. Concrete paving should not be used because of the low bearing capacity of the backfill soils and the high likelihood of soil loss through the sheeting interlocks. Geotextiles should be placed against the inside of the sheeting to prevent this type of internal erosion. For heavy surcharge loads, which would transmit large lateral loads to the sheeting, relieving platforms are constructed as shown in Fig. 3.11b. These platforms are completely independent of the sheet pile wall and must be supported on a pile structure of their own.

The design of the walls has evolved over a long period of time, beginning with direct structural analysis and including later modifications to provide for flexibility, up to current design concepts based on probable deflected shape. The free-earth support (FES) method was long used for design (see Fig. 3.12), where Rankine earth pressures were used for P_a and P_p:

$$P_a = \tfrac{1}{2}\gamma(H + D)^2 K_a \tag{3.3}$$

$$P_p = \tfrac{1}{2}\gamma D^2 K_p \tag{3.4}$$

$$K_a = \tan^2\left(45 - \frac{\phi}{2}\right) \tag{3.5}$$

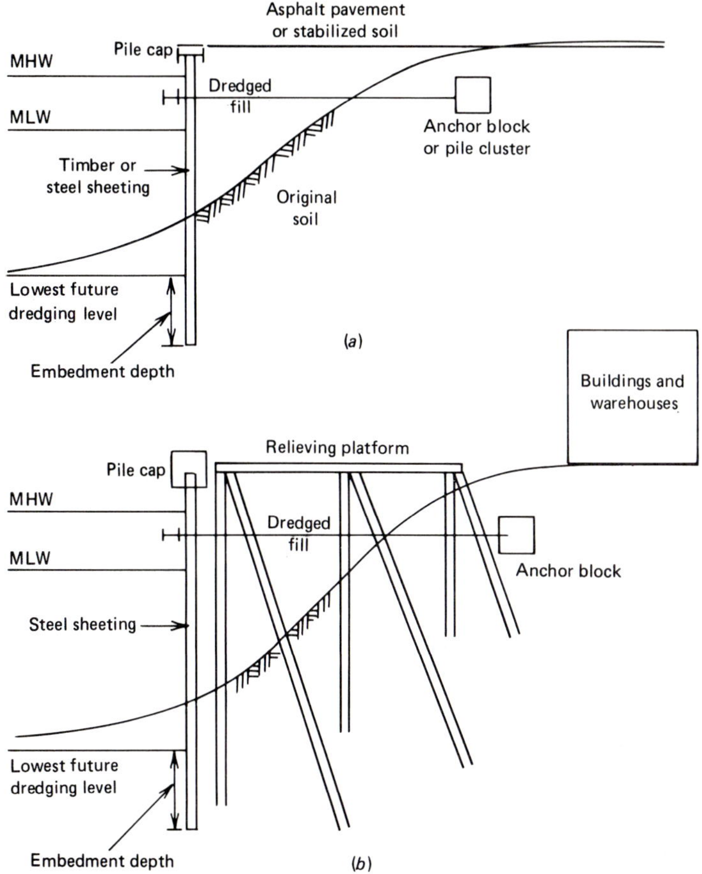

Figure 3.11 Typical details of permanent bulkheads for (*a*) small harbor or marina, and (*b*) major industrial area.

$$K_p = \tan^2\left(45 + \frac{\phi}{2}\right) \tag{3.6}$$

Since the embedment depth is usually unknown, the analytic procedure in this method is to take moments about the anchor force to solve for D (the resulting equation is a cubic) and then take force summation in the horizontal direction for the anchor force (A). All units are in force per unit length of wall. Knowing P_a, P_p, and A allows one to draw shear and moment diagrams in the conventional manner to obtain the maximum moment. Using the standard equation for maximum stress

$$f = \frac{M_{max}c}{I} \tag{3.7}$$

$$f = \frac{M_{max}}{S} \tag{3.8}$$

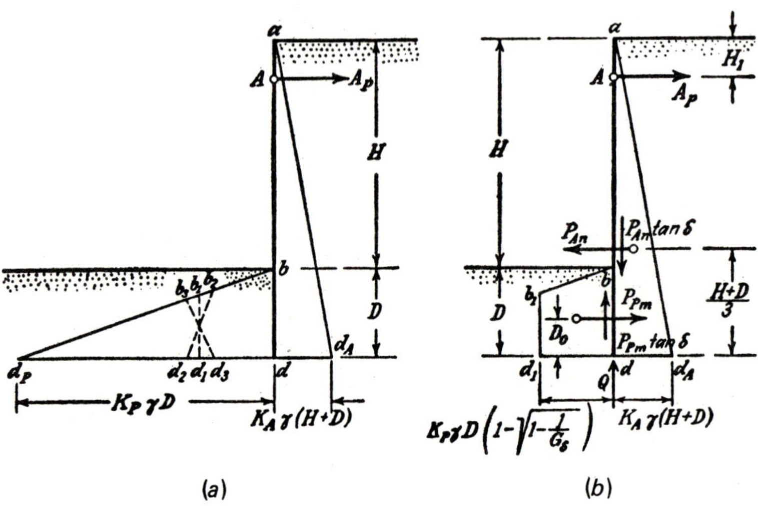

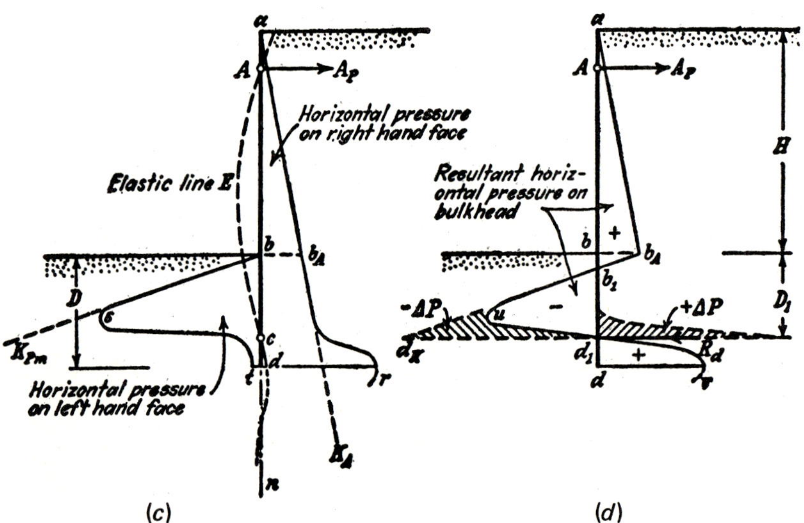

Figure 3.12 Assumptions and design details for bulkheads using free-earth support (upper) and fixed-earth support (lower) methods. (*a*) Various assumptions regarding distributions of earth pressure on buried part of bulkhead with free-earth support; (*b*) customary assumptions concerning the forces which act on such a bulkhead. (*c*) Actual and (*d*) assumed distributions of horizontal pressures on the two sides of a bulkhead with fixed-earth support. (*After Terzaghi, Ref. 18.*)

one can solve for the required section modulus $S = M_{max}/f$ to select the appropriate sheet section modulus (see Fig. 3.9 for typical values).

This procedure, however, does not take into account the fact that the sheets are relatively flexible (certainly with respect to the soil and water pressures being exerted on them) and will deform substantially. For this reason there are a number of modifications which have been suggested to lower M_{max} and A. Rowe[19] simply uses a modification whereby λ_M and λ_A are applied directly to the FES values (see Fig. 3.13):

$$M_{design} = \lambda_M M_{FES} \tag{3.9}$$

$$A_{design} = \lambda_A A_{FES} \tag{3.10}$$

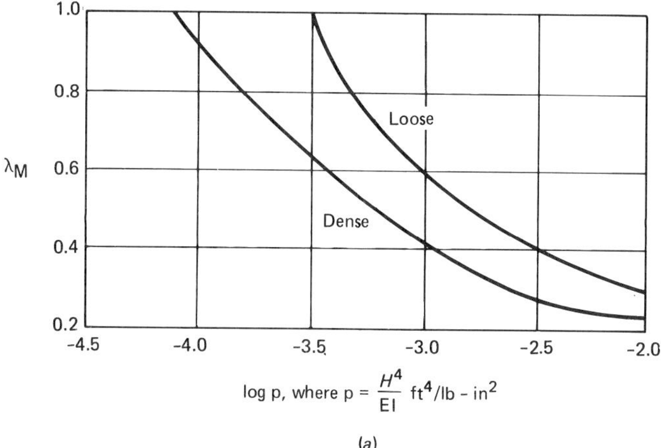

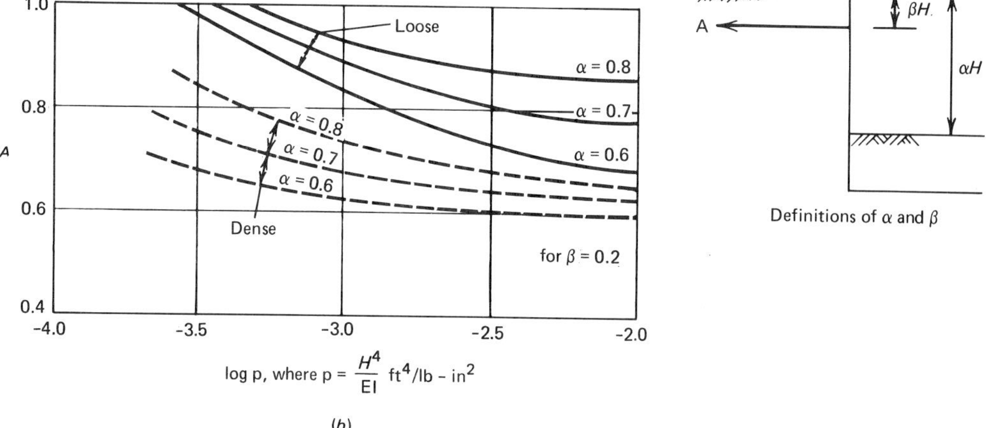

Figure 3.13 Maximum moment and anchor reduction factors to account for bulkhead flexibility. Reduction factor for (a) maximum bending moment and (b) anchor force. (*After Rowe, Ref. 19.*)

The calculated embedment depth is usually increased by a healthy factor of safety (1.5 to 2.0) to avoid toe failure during future dredging. The anchorage block or pile group is designed according to standard procedures, the anchor ties being spaced at 7- to 15-ft centers, and the calculations are based on a per unit length figure. It also must be remembered that the anchor block must be placed sufficiently behind the wall so that active and passive earth pressure zones do not overlap. See Example 3.2 for design details.

Regarding the expected lifetime of the structure, corrosion is by far the greatest enemy. Conventional corrosion rates of 0.005 in/year per year are simply too low under certain conditions, e.g., when the structure is located in salt water or brackish water. A study by the National Bureau of Standards[20] has shown that corrosion rates for different materials vary widely (up to 0.012 in/year), particularly in certain zones (see Table 3.4). The corrosion rate is greatly reduced by the use of pile-coating materials, as seen in Table 3.4. Caution, however, should be exercised in handling these coated piles so that they are not damaged during transport or installation.

In regard to the construction of bulkheads, the procedure followed is similar to that for installing continuous sheeting, except that the work is usually carried out from a barge floating on water. This greatly decreases the productivity and, hence, considerably increases costs.

The backfill soil is usually dredged from in front of the wall and placed behind it, and good compaction control is difficult, if not impossible, to achieve. Hence, the structure often suffers from large settlements during the first few years of its life. The anchorage also presents problems. These arise not so much from its construction, which is carried out on firm ground, but from the stressing of the cables or bars used

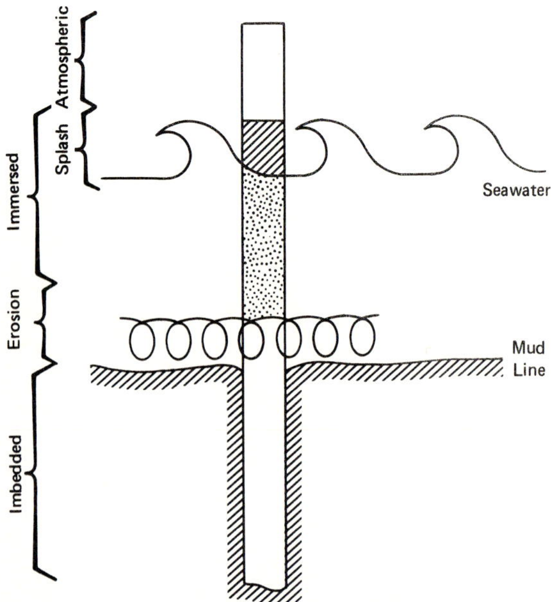

Zones encountered by an offshore structure for (Table 3.4)

Table 3.4 Corrosion rates of H piles based on flange thickness measurements†

System	Coating description	Overall average corrosion rate, mils/year‡	Average corrosion rate within zone, mils/year‡			
			Imbedded zone, 0–15 ft§	Erosion zone, 15–21 ft§	Immersed zone, 15–29 ft§	Atmospheric zone, 29–35 ft§
29A	Polyester glass flake	<0.10	<0.10	<0.10	<0.10	<0.10
17A	Phenotic mastic	0.14	0.11	0.11	0.15	0.21
4A	Coal tar epoxy	0.53	0.17	0.21	0.27	2.1
7A	Coal tar epoxy plus armor	0.55	0.13	0.07	0.07	2.7
6A	Coal tar epoxy	0.80	0.27	0.72	0.46	2.9
12A	Polyvinylidene chloride	2.4	0.81	4.9	3.6	3.5
30A	Bare carbon steel	4.9	0.9	8.9	6.7	10.5
1A	Bare carbon steel	6.0	1.8	9.7	7.9	12.2

†After Escalante and Iverson, Ref. 20.
‡1 mpy = 0.001 in/year = 25.4 μm/year
§Distance from bottom of pile.

to mobilize the anchor force between the sheeting and the anchorage. The stressing must be done simultaneously with backfilling so that the backfill soil does not buckle the sheeting outward nor does the anchor force pull it inward. For cases where the anchorage is developed by soil tiebacks or rock anchors, the proof-testing of the grouted bar or tendon can be difficult since there is no stable section for the proof loads and lockoff load to be mobilized against. Thus, either timber reaction frames are required to hold the sheeting before backfilling or the stressing must be done after backfilling.

For sheeting which is on firm rock where insufficient penetration is available to prevent a rotation of the toe about the anchorage cable, steel pins drilled into the rock in front of the toe have been used; see Fig. 3.14.

Relieving platforms that are constructed after the bulkhead is completed can be troublesome in that soil displacements and pore water pressures can severely distort the completed work. Relieving platform piles driven through anchor cables or the toe of the bulkhead sheeting are obviously disastrous.

Example 3.2 Bulkhead design

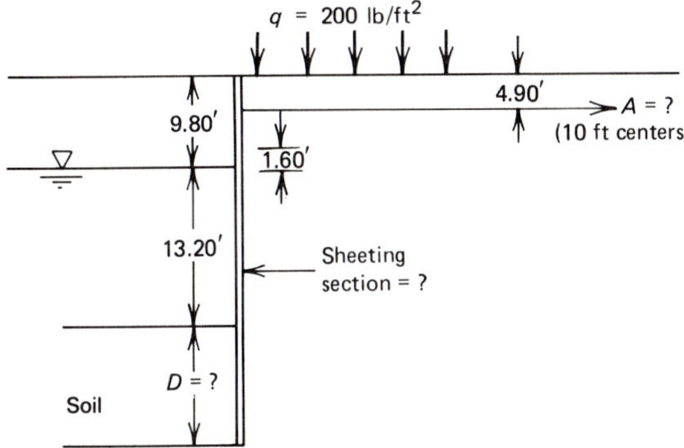

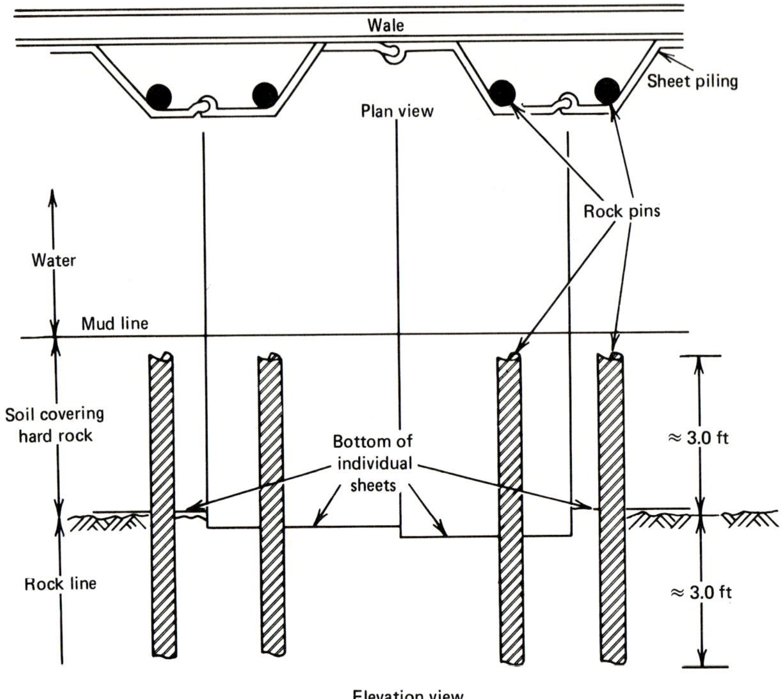

Figure 3.14 Use of steel pins socketed in bedrock to provide for toe resistance at bottom of steel sheet piling. *(After Ref. 21.)*

Soil properties

Location	γ, lb/ft^3	c, lb/ft^2	ϕ, °	δ, °
Above water table	113	0	30	20
Below water table	59	0	30	20
Below dredge line	69	0	35	23.3

where
$$K_a \tan \delta = \tan^2 (45 - \phi/2) \tan \delta$$
$$= 0.300 \qquad \text{for } \phi = 30°$$
$$= 0.247 \qquad \text{for } \phi = 35°$$
$$K_p = \tan^2 (45 + \phi/2)$$
$$= 3.70$$

SOLUTION

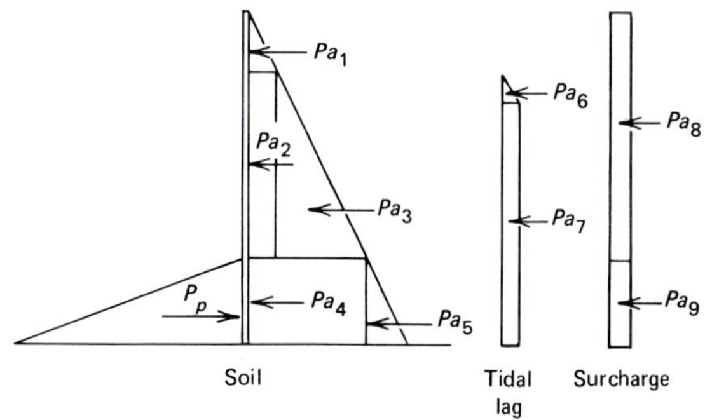

$$\Sigma M_A = 0;$$

$$P_p \times \text{arm} = \Sigma P_a \times \text{arms};$$

$$\tfrac{1}{2}69D^2(3.70)(18.1 + \tfrac{2}{3}D) = \tfrac{1}{2}(113)(9.80)^2(0.300)(1.64)$$

$$+ 113(9.80)(13.20)(0.300)(11.50)$$

$$+\tfrac{1}{2}(59)(13.20)^2(0.300)(13.70) + [113(9.80)$$

$$+ 59(13.20)] D(0.247)(18.1 + D/2) +\tfrac{1}{2}(69)(D^2)(0.247)(18.1 + \tfrac{2}{3}D)$$

$$+\tfrac{1}{2}(62.4)(1.6)^2(1.0)(5.1) + (62.4)(1.6)(13.2 + D)(1.0)[\tfrac{1}{2}(12.4 + D)$$

$$+ 5.70] + 200(23)(0.300)(6.6) + 200D(0.247)(18.1 + D/2)$$

which reduces to

$$79.6D^3 + 1806D^2 - 11952D - 100,228 = 0$$

and by T & E

$$D = 9.0 \text{ ft}$$

$$\Sigma F_H = 0; \quad A + P_p = \Sigma P_a$$

$$A + 10,380 = 16,530$$

$$A = 6150 \text{ lb/ft}$$

$$= 61,500 \text{ lb per cable}$$

$$S_{\text{reqd}} = \frac{M_{\max}}{f}$$

$$= \frac{57.6(12)}{32}$$

$$= 21.6 \text{ in}^3$$

Use PZ 27 $S_{\text{act}} = 30.2 \text{ in}^3$

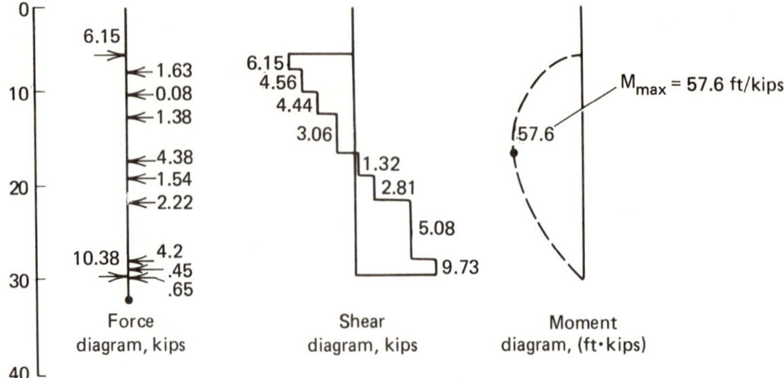

Check flexibility for possible reduction in S_{reqd} and A.

$$\rho = \frac{H^4}{EI}$$

$$= \frac{32^4}{(30 \times 10^6)(184.2)}$$

$$= 1.89 \times 10^{-4}$$

$$\log \rho = -3.72$$

$$\therefore \lambda_M = 0.95 \qquad M_{max} = 57.6(0.95)$$

$$= 54.7 \text{ ft/kips} \qquad \text{Use PZ27}$$

$$\lambda_A = 0.92 \qquad A = 6.15(0.92)$$

$$= 5.66 \text{ kips/ft}$$

$$= 57 \text{ kips/cable}$$

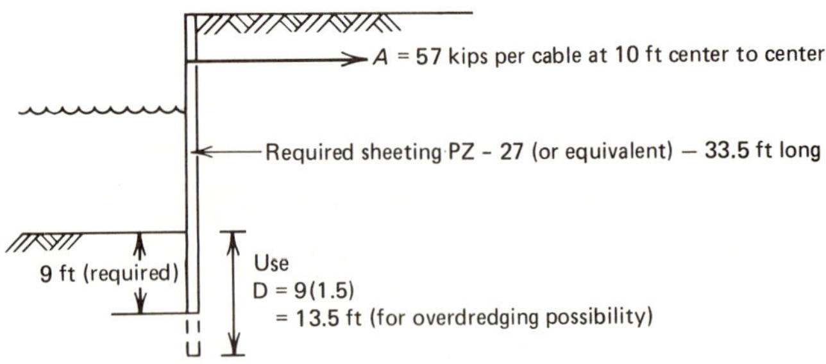

3.4.2 Continuous Concrete Piles

Continuous concrete piles placed directly adjacent to one another result in a rigid retaining wall, a method quite often used to provide lateral restraint to buildings instead of the method of direct underpinning. When so used, it could be called "lateral underpinning," and we could equally well discuss it in Sec. 3.6. The method, however, is still a wall and its design is based on the same principles as other retaining systems treated in this present section.

Shown in Fig. 3.15 is a photograph of a completed wall made from adjacent piles (a *tangent pile wall*) or from staggered piles (a *secant pile wall*). They are well adapted to placement immediately adjacent to existing structures since they are usually placed by continuous-flight hollow-stem augers or augers excavating within a slurry-supported hole. Both methods are essentially noise- and vibration-free. Concrete is placed through the auger or by means of a tremie pipe beneath the slurry to make a cast-in-place concrete pile. For additional stiffness, a reinforcement cage can be placed in the fresh concrete before it sets, or an H pile can be placed in it for even greater stiffness. The slurry method is more time-consuming and expensive, but necessary when granular soils with no cohesion are present.

When constructing a *tangent* pile wall, alternate piles are placed and allowed to cure. The intermediate ones are then added, and the auger should actually remove a small arc of concrete from those piles in place so that the retained soil will not pass between piles and some interlocking will also be gained. The *secant* pile system requires a bit more room since alternate piles are placed at distances of somewhat less than two diameters apart. The intermediate piles are then located as closely as possible to the original line, but the center of that line is slightly offset.

Figure 3.15 Photograph of continuous concrete piles used as lateral underpinning. *(Compliments of Intrusion Prepakt, Inc.)*

Restraint of these pile-constructed walls is achieved by suitable embedment at the bottom of the excavation and the use of struts or tiebacks located where necessary along the exposed lengths of the piles. It is customary to put a continuous pile cap across the top of the piles.

To design continuous concrete pile walls adequately (particularly when they are intended to prevent lateral movement as in underpinning), the design analysis must take into account the coefficient of earth pressure at rest, K_0. Thus, the pressures on them will be greater than the pressures on the walls described in previous sections of this chapter. For cohesionless soils, $K_0 = 1 - \sin \phi$, while for cohesive soils a zero lateral strain test must be made (see Bishop and Henkel[22]). Use of K_0 in the design is further warranted when using soil or rock tiebacks which further mobilize pressures above the minimum possible value of active earth pressure. The pressure distribution diagrams shown in Fig. 3.5a or b with K_0 replacing K_a are generally used since construction methods vary widely, depending on the conditions encountered and the particular contractor doing the work.

Since these walls usually have a stationary structure immediately adjacent to them, there are also other stresses on them that must be taken into account. This is done using conventional elasticity theory,[4] which for a concentrated load is

$$\sigma_x = \frac{3Q}{2\pi} \frac{x^2 z}{r^5} \tag{3.11}$$

and for a line load is

$$\sigma_x = \frac{2P}{\pi} \frac{x^2 z}{r^4} \tag{3.12}$$

where σ_x = horizontal stress increase due to load Q or P
     Q = concentrated load at ground surface
     P = continuous line load at ground surface
     x = shortest horizontal distance from load to wall
     r = radial distance from load to wall
     z = vertical distance from ground to point in question

Numerous charts and graphs are available to solve problems of this type; see Poulos and Davis.[23] Also, see Nendza[24] for a survey of pressures mobilized on walls by adjacent structures.

3.4.3 Diaphragm (Slurry) Walls

The use of diaphragm, or slurry-constructed, walls has increased dramatically over the past 20 years. These walls are constructed via a slurry-supported trench which is usually backfilled with tremie-placed concrete or precast concrete panels. The slurry, a mixture of bentonite clay (montmorillonite-type) and water in proportions of approximately 1:20, performs the critical function of supporting the narrow, vertically sided trench until the permanent wall is ready for forming or placing.

The uses of such walls are varied and include the following[25]:

Cutoffs for controlling seepage
Vertical construction shafts
Foundation pit walls
Bridge foundations
Basement walls
Lateral soil containment (in lieu of underpinning)
Cut and cover tunneling
Bulkheads and dock walls
Slope stabilization
Box culvert walls
Load-bearing elements
Vibration barriers

The construction of the walls usually proceeds as follows:

1. A concrete guide about 4 in wider than the proposed wall is formed at the ground surface.
2. Excavation of the wall begins by removing soil with a backhoe, clamshell, hydraulic grab bucket, or dragline bucket.
3. As the initial soil is removed, a bentonite slurry is placed in the partially excavated trench.
4. Excavation now proceeds through the slurry (which has the consistency of pea soup), more slurry being added as needed to maintain its level near ground surface. The slurry level must always be 2–4 ft higher than the static water table elevation for proper trench stability.
5. When the desired excavation depth has been reached panel end joints are placed (these are usually steel pipe piles or wide flange beams); see Fig. 3.16.
6. If required, a prefabricated reinforcement cage is then lowered into the slurry-filled trench (for some unknown reason, bonding of the tremied concrete with the reinforcement steel is never considered in this work).
7. Concrete is placed by the tremie method, i.e., slurry is displaced from the bottom of the trench upward by the concrete, until the panel is complete. Note that the slurry must be collected as it is displaced, stored in suitable containers, and cleaned of solid particles for reuse in constructing the next panel. This is a messy operation and, particularly when in urban areas, requires good "housekeeping" practices on the part of the contractor. It should be noted that during cold weather the spilled slurry will freeze when temperatures are low enough, creating hazardous conditions.
8. The panel joint members are removed when the concrete has cured, and the entire process is repeated for adjacent panels, and so on until the entire wall has been formed.
9. An alternative method to the use of cast-in-place tremie concrete just described is the use of precast panels which are placed directly in the slurry in a completed state.

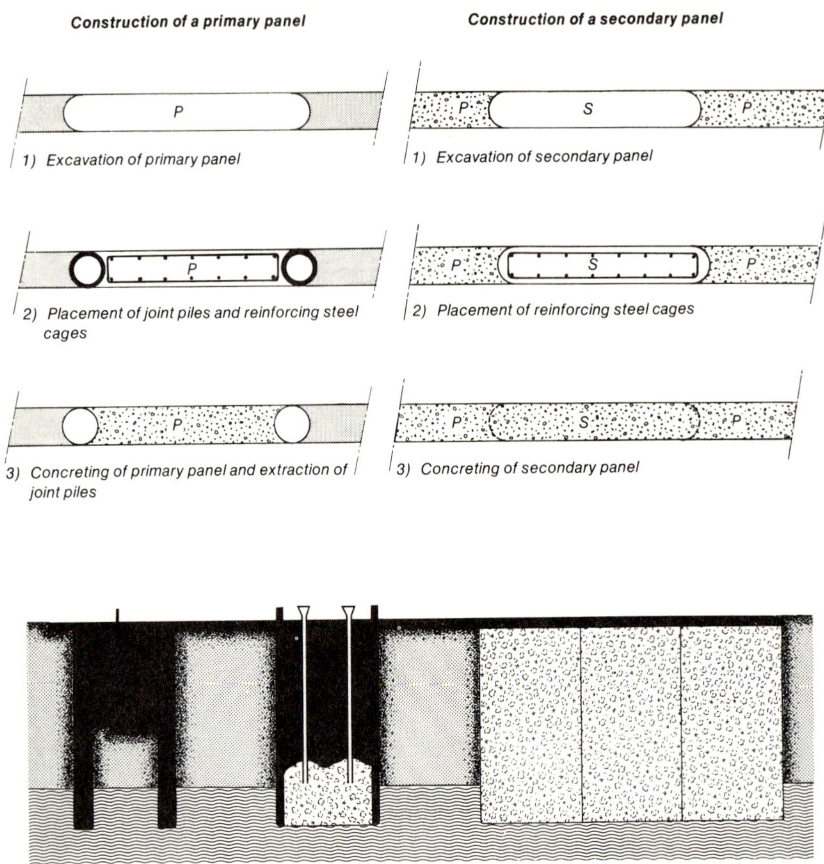

Figure 3.16 Stage construction of diaphragm (slurry-) constructed concrete wall. *(Compliments of Soletanche and Radio, Inc.)*

10. Upon completion of the wall, excavation proceeds and, when depths are great and bracing is required, it is installed in the conventional manner described previously for other wall systems. Quite often soil tiebacks or rock anchors are used for bracing systems.

There are many alternatives to the system as just described. Some of these are the following:

A reverse-circulation excavation procedure can be used which excavates horizontally using a chopping bit and brings the excavated soil (and some of the slurry) to the surface; see Fig. 3.17. The construction sequence for each panel follows as previously described.[25]

A "vibrating beam" method has also been devised whereby a specially adapted wide

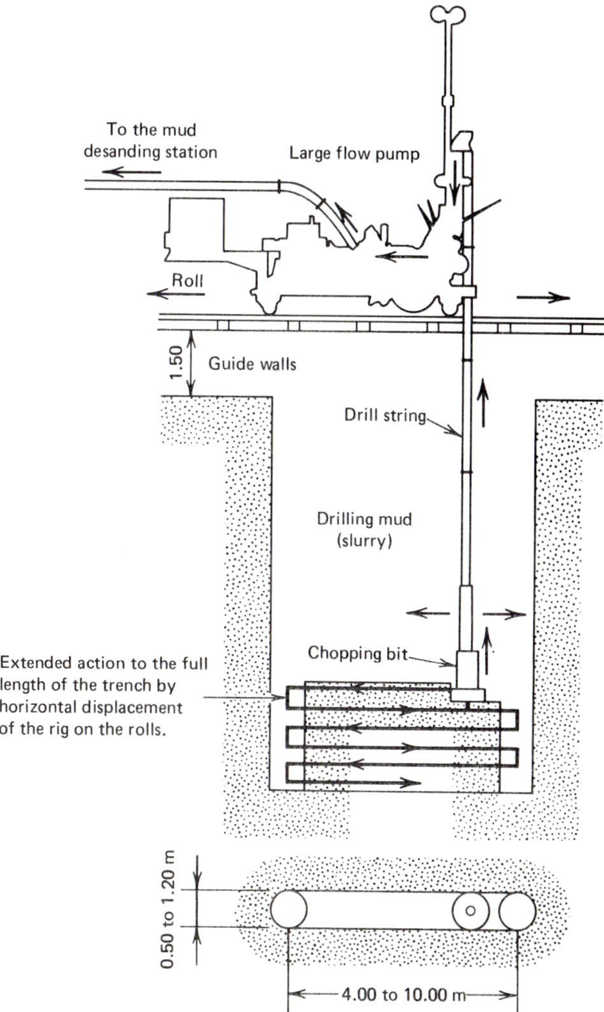

Figure 3.17 Reverse-circulation excavation of slurry-supported trench. *(After Nash, Ref. 26.)*

flange beam is vibrated into the soil and grout pipes attached to the beam deliver the backfill material to the desired area. This technique results in the narrowest trench width of any of the techniques known, thereby requiring a minimum amount of backfill material.

Slurry-excavated trenches can be backfilled by many materials other than concrete, depending upon the particular application. For example, as a horizontal seepage cutoff wall, a blended backfill material of soil and bentonite (SB) has been used and has been shown to be quite effective in reducing permeability; see Fig. 3.18. Also, specially blended bituminous backfill materials have been used with spe-

Table 3.5 Bentonite slurry properties and recommended specifications†

Property	Definition	Current test method	Recommended specifications
Concentration	kg of bentonite per 100 kg of water	$\cdots$	$>4\frac{1}{2}\%$
Density	Mass of given volume of slurry	Mud balance (e.g., by baroid)	>1.034 g/cm^3 <1.25 g/cm^3
Plastic viscosity, apparent viscosity, yield stress	For a slurry (behaving as a Bingham body) under shearing conditions: Shear stress $= T + V_p S$ (where $T =$ yield stress, $V_p =$ plastic viscosity, $S =$ shear rate); apparent viscosity $=$ shear stress/shear rate and is dependent upon shear rate for a Bingham body	Fann viscometer	<20 cP
Marsh cone viscosity	Time for fixed volume of slurry to drain from a standard cone	Standard Marsh cone, as used by drilling companies	30–90 s

10-min gel strength	Shear strength attained by the slurry after quiescent period of 10 min (slurry violently sheared before starting)	Fann viscometer, falling tube shearometer (*Note:* These two measurements give answers which commonly differ by up to a factor of 2)	>50 dyn/cm^2 <200 dyn/cm^2
pH	Logarithm of the reciprocal of the hydrogen ion concentration	pH meter, pH papers can give unreliable results	<11.7
Sand content	Percentage of sand greater than 200 mesh in suspension	API sand content test (basically 200-mesh screen)	>1% <25% <4% when concreting
Fluid loss	Volume of fluid lost in set time from fixed volume of slurry when filtered at set pressure through standard filter medium	Standard fluid loss apparatus as used by drilling companies (600 cm^3 of mud, 100 lb/in^2, 30 min, filter paper)	
Filter cake thickness	Thickness of filter cake built up under standard conditions	Measure filter cake built up in fluid loss test	

†After Hutchinson et al., Ref. 29.

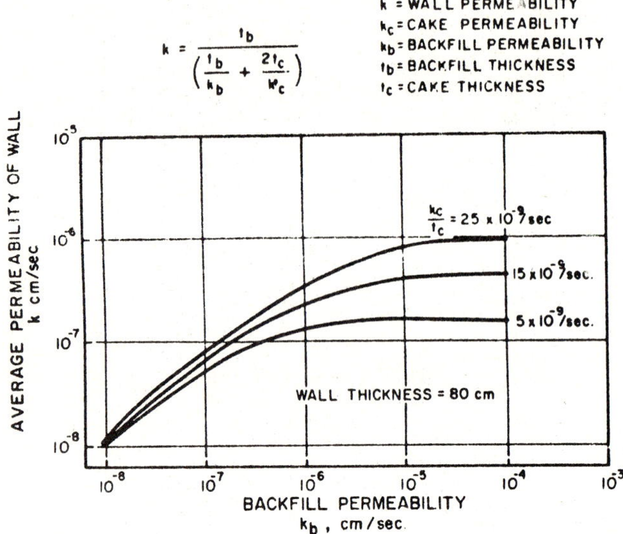

$$k = \frac{t_b}{\left(\dfrac{t_b}{k_b} + \dfrac{2t_c}{k_c}\right)}$$

Figure 3.18 Theoretical relationship between wall permeability and permeability of filter cake and backfill. *(After D'Appolonia, Ref. 27.)*

cific resistance to certain types of landfill leachates in hazardous materials areas[28]; see Chap. 8. Plastic sheets, for example, polyvinyl chloride, polyethylene, hypalon, etc. (properly called *geomembranes*) have also been placed in the slurry-supported trench for horizontal seepage control (see Chap. 8).

Concerning the nature and functioning of the slurry in supporting the vertically sided walls, much has been written. Recommended slurry properties are given in Table 3.5, and a review of current practice in the United States and typical specifications, carried out by Millet and Perez,[30] is available. Exactly how the slurry works, in situ, is still an area of active research. The probable phenomena involved are the following[25]:

Hydrostatic pressure of the slurry (probably accounts for 65 to 80 percent of the support)
Slurry gel strength
Resistance of the filter cake formed against the trench walls
Electrical potential between the slurry and the soil
Dynamic gradient of the slurry flowing into the soil
Rigidity of the slurry by the bentonite particles
Rigidity of the slurry by suspended cuttings
Action of the permeating slurry into the soil
And there may be others, as yet unknown.

3.5 ANCHORS AND TIEBACKS

The modern way to stabilize rock and soil masses is to utilize the inherent qualities of the materials in question by internally tying the mass together so that individual elements cannot fail. This concept is a radical departure from the older concept of building a sufficiently strong support system to resist movement of any, or all, of the material in question. To illustrate the point, note the incredibly complex bracing system shown in Fig. 3.19 versus the excavation-clear area shown in Fig. 3.20, which uses a tieback system for the stability of its wall system.

This section will discuss the design and construction of such systems, as illustrated in Figs. 3.21 and 3.22.

3.5.1 Rock Anchors

Instinctively, one would expect that a steel beam, rod, or wire cable, adequately sealed and locked into a rock mass, could support a measurable tensile load. And in

Figure 3.19 Temporary bracing and support system for deep building excavation in New York City, illustrating complexity and congestion that can occur in difficult foundation projects. *(Courtesy of Spencer, White and Prentice, Inc., New York, NY.)*

Figure 3.20 Tieback support system for deep building excavation in Philadelphia, PA.

fact there are many applications of this type, some of which are illustrated in Fig. 3.21. The method of construction is generally as follows:

Holes are drilled, usually by means of a drill rig, air track, or other reasonably stable piece of coring equipment, to the desired diameter and the desired depth.

The anchor itself is inserted into the hole. Numerous anchor types have been used, e.g., structural steel shapes, smooth and deformed rods, steel cables, etc.

A grout (usually cement) is pumped into the space between the hole and the anchoring member at a predetermined grouting pressure and allowed to properly cure.

A reaction and locking system is installed on the anchor at the ground surface. This can vary from a single washer and tapered lockoff nut to elaborate (and sometimes patented) systems.

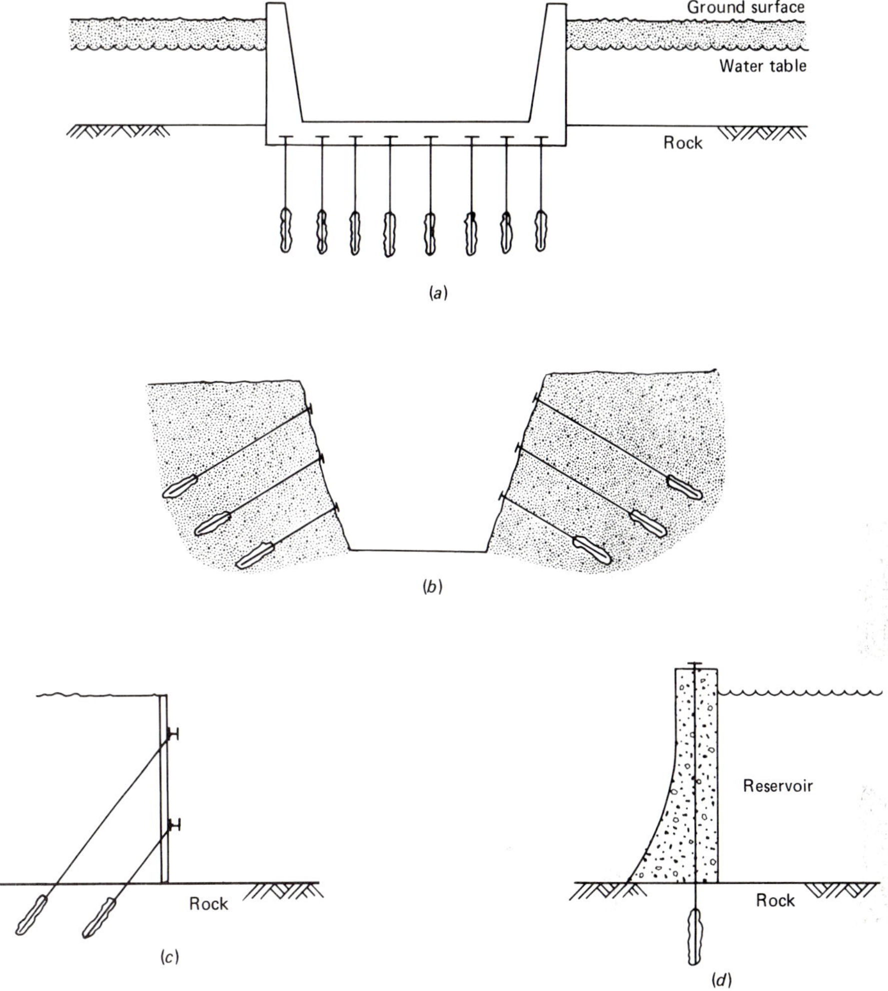

Figure 3.21 Various types of situations utilizing rock anchors. (*a*) Resisting hydrostatic uplift in areas of high water level or water table. (*b*) Bolting of rock slopes in open cuts (temporary or permanent). (*c*) Temporary anchorage. (*d*) Permanent anchorage.

A tensile stress is applied to the anchor by a hydraulic jack to a value somewhat greater than the anchor's design load. This value generally varies from 1.25 to 2.0 times the design load.

With the load applied to the anchor, the lockoff system is engaged and the applied tension from the jack is released.

A rock anchor system is designed mostly empirically. Different grouts, grout pressures, and rock systems result in different unit bond resistances, which, when multiplied by the surface area of the anchor in question, give the theoretical anchor

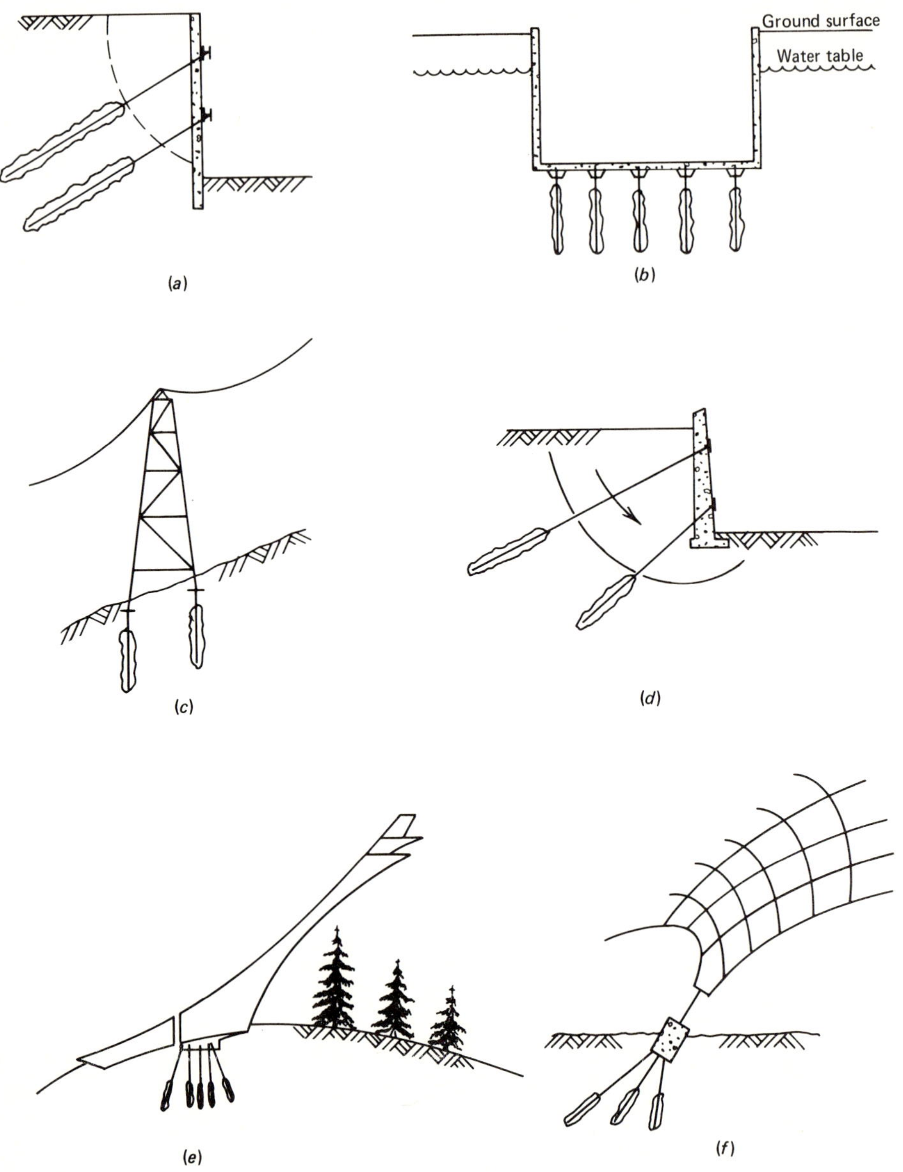

Figure 3.22 Various types of structure utilizing soil anchors. (*a*) Support temporary walls. (*b*) Resist hydrostatic uplift. (*c*) Resist uplift forces. (*d*) Resist unstable soil movement. (*e*) Ski-jump anchorage. (*f*) Anchorage for tent structures.

loads. Examples of approximate resistance values for different kinds of rock are given in Table 3.6. Obviously, the calculated load for an anchor cannot exceed the yield stress of the anchor steel itself, and often this is the limiting factor in the design. Field tests should always be conducted on selected anchors in different rock strata as well as for different construction methods.

There are, however, many unresolved questions regarding rock anchors, among which are the following:

How load is transferred along the length of the bonded anchor is largely unknown, but the transfer may be concentrated at the front portion of the loaded zone.

The loss of anchor prestress due to creep of grout, rock, anchor, and/or lockoff system can be a problem.

The group interaction of a set of rock anchors versus the behavior of a single anchor is largely unknown—recall the discussion on pile group behavior versus a single-pile behavior in Sec. 2.4.7.

How to design anchors at excavation corners or for nonuniform surfaces is not wholly clear.

The degree of corrosion, actually stress corrosion, anchors undergo in permanent installations is largely unknown.

3.5.2 Soil Tiebacks

The use of grouted anchors embedded in soil as opposed to their use in rock only was first introduced in Germany in 1958 by Bauer,[32] and the practice has grown rapidly from that time. Areas of application are shown diagrammatically in Fig. 3.22, and the overlap with applications for rock as illustrated in Fig. 3.21 is obvious. The basic difference here is that the intrinsic strength of soil, rather than the strength of rock, is being mobilized.

Common types of soil tiebacks are shown in Fig. 3.23. Each will be discussed briefly since the methods used to construct them differ considerably; see Sigourney[33] for further details.

Figure 3.23a shows the details of a soil anchor made from a structural steel section driven into the soil. It is driven as a pile and is generally used under existing

Table 3.6 Rock-grout bond stresses for rock tiebacks†

Rock type	Bond stresses between grout and rock, lb/in²
Sandstone	120–250
Soft shales	30–120
Slates and hard shales	120–200
Soft limestone	150–220
Hard limestone	300–400
Granite and basalt	250–800

†After Littlejohn and Bruce, Ref. 31.

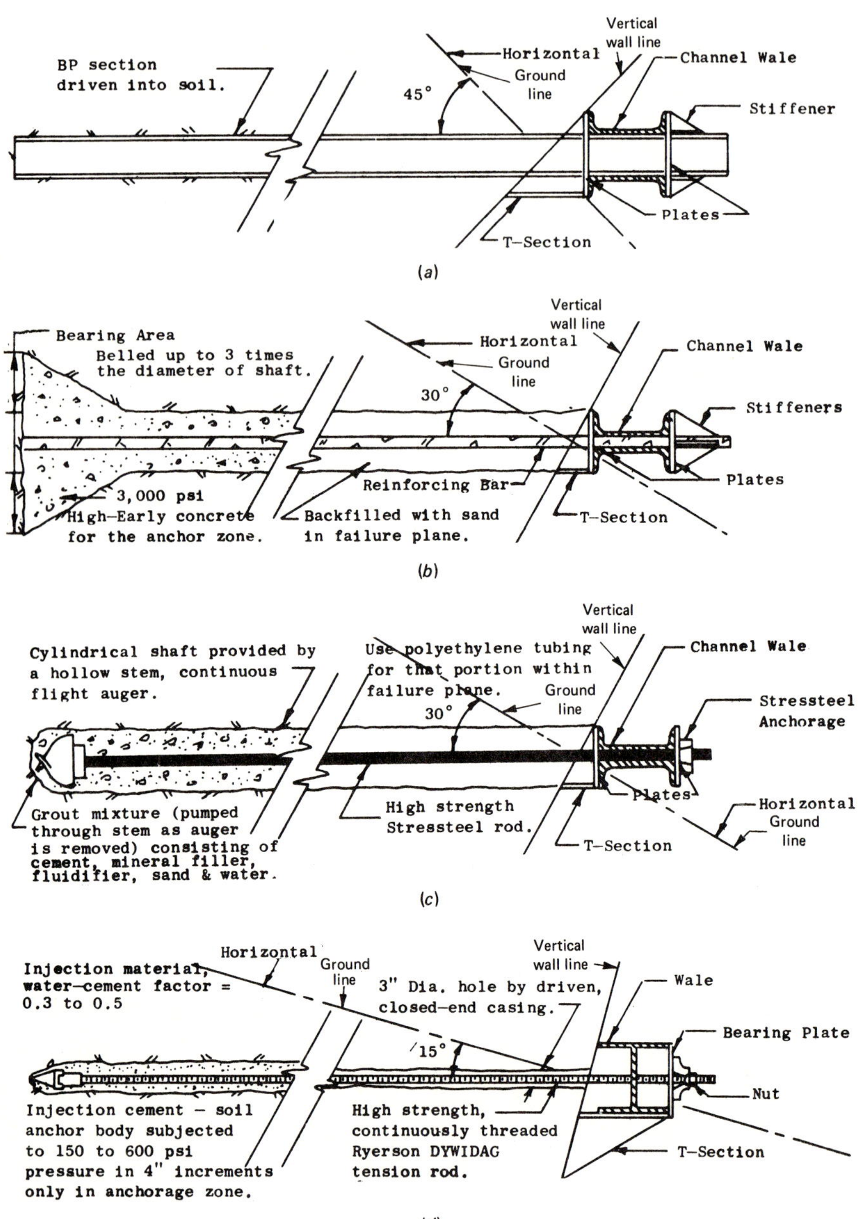

Figure 3.23 Various types of soil anchors. (*a*) Structural section. (*b*) Drilled pier type. (*c*) Pressure-grouted, uncased hole. (*d*) Pressure-grouted, cased hole. (*After Sigourney, Ref. 33.*)

structures which are very sensitive to settlement. It can be used where the presence of boulders and cobbles make other methods difficult to employ since to use them would require excavation of the boulders and pebbles. Testing is difficult, however, and the method is generally more expensive than others.

Figure 3.23*b* shows a soil tieback consisting of a drilled pier with a bell at the end. It is a tension member, but in function should be compared to the pier described in Sec. 2.3.2. It is installed by a conventional caisson rig, as shown in Fig. 3.24*a,* in cohesive soils (which have a measurable standup capability) using a continuous flight auger. A belling bucket is used to obtain an expanded base, whose size can be as much as three times the shaft diameter. Anchor tendons are reinforcing bars, high-strength steel tendons, or wire rope cables. The space between the soil and the anchor is filled with high-strength concrete, which is allowed to set before testing. Capacities of 120 kips in sandy clays of $N = 15$ to 30 blows per foot have been developed.

Figure 3.23*c* shows a pressure-grouted anchor which is usually installed using a continuous-flight, hollow-stemmed auger; see Fig. 3.24*b.* Upon reaching the desired depth, the anchor, usually a high-strength steel rod, is inserted into the center of the auger. The auger head is knocked off and the rod held in place as the auger itself is removed. During withdrawal, high-strength grout is placed which, when cured, allows for testing of the anchor. The method is useful in dry or saturated granular soils and, significantly, no casing is required during installation. A design capacity of up to 160 kips per anchor is common.

Figure 3.23*d* shows a pressure-grouted anchor of the Bauer type which requires a casing for proper ground stability. The anchor is placed after driving or drilling a closed-end casing to the required depth (see Fig. 3.24*c*). The anchor itself (reinforcing rod or high-strength rod) is inserted and is used to knock off the drive point at the end of the casing. High-strength grout is then placed around the anchor as the casing is withdrawn. After being properly cured, the anchor is tested. Capacities of 100 kips can be developed in granular soils, with somewhat lower values in cohesive soils primarily because of soil creep.

A popular type of high-strength bar which is used in soil tieback construction of the type just described is the Dywidag bar, manufactured by Dyckerhoff and Wedmann, Inc., and distributed worldwide. The steel conforms to ASTM A-322 and ASTM A-29 specifications. The rods have rolled deformations along their entire lengths, which serve to resist pullout and act as threads permitting anchorage or coupling hardware to be screwed onto them at any location. Table 3.7 gives some technical information about these bars.

When using various types of anchors, the manufacturer's values should be consulted. For example, 1-in-diameter steel bars can withstand an ultimate stress of 150 kips/in², an ultimate load of 128 kips, and a yield load of 109 kips, while 1.37-in-diameter steel bars can withstand an ultimate stress of 150 kips/in², an ultimate load of 234 kips, and a yield load of 199 kips. Seven-strand wire rope cable, 0.5 in in diameter, can withstand 270 kips/in² in ultimate stress, 41 kips in ultimate load, and 35 kips in yield stress. Often several cables are used to form a single anchor tieback.

Obviously, the capacity of a soil tieback depends upon the type of anchor system used, the soil it is anchored into, the type and pressure of the grout, etc., all of which

(a)

(b)

(c)

Table 3.7 Properties of Dywidag bars used as high-strength steel anchors in soil tiebacks†

Diameter, in	Area, in^2	Weight, lb/ft	Ultimate strength, kips/in^2	Prestressing force, kips			
				100%	80%	70%	60%
1.00	0.85	2.96	150	127.8	102.2	89.5	76.7
1.00	0.85	2.96	160	136.3	109.1	95.4	81.8
1.25	1.25	4.40	150	187.5	150.0	131.0	112.5
1.25	1.25	4.40	160	200.0	160.0	140.0	120.0
1.375	1.56	5.31	150	234.0	187.2	163.8	140.4

†Taken from a brochure by the Schnabel Foundation Company.

means the design of tiebacks must proceed largely empirically. However, some theoretical relationships have been established.[34] For belled tiebacks of the type shown in Fig. 3.23b, Eq. (3.13) can be used:

$$P = \pi(D^2 - d^2)N_c c \tag{3.13}$$

where P = ultimate tieback capacity
    D = bell diameter
    d = shaft diameter
    N_c = bearing capacity factor (approximately = 9; see Sec. 2.4.3)
    c = undrained shear strength of soil

For grouted tiebacks in granular soils of the type shown in Fig. 3.23c and d, Eq. (3.14), which is directly related to deep foundation theory as discussed in Sec. 2.4.3, can be used:

$$P = \pi dL(c_a + \sigma \tan \delta) \tag{3.14}$$

where P = ultimate tieback capacity
    d = tieback shaft diameter
    L = tieback length (active zone only)
    c_a = adhesion of grout to soil
    δ = friction angle between grout and soil
    σ = normal stress in shear plane (related to, but somewhat less than, the grout pressure)

However, calculations using Eq. (3.14) will almost always indicate tieback capacity that is greater than the structural strength of the anchor bar or cable being used. In fact, this has been described in the literature dealing with problems more often encountered in cohesive soils.

Figure 3.24 Types of construction equipment used to install soil-rock tiebacks. (a) Caisson rig drilling a belled tieback. (b) Hollow-stemmed auger installing a tieback. (c) Air track drilling rock for a tieback. *(Compliments of Schnabel Foundation Company.)*

For grouted tiebacks in cohesive soils, Eq. (3.15), along with data from Table 3.8, can be used:

$$P = \pi dLS \tag{3.15}$$

where P = ultimate tieback capacity

d = average diameter grout bulb (which is difficult to estimate)

L = anchor length behind potential shear plane

S = soil to grout bond stress

The approach represented by Eqs. (3.13) through (3.15) is undoubtedly simplified and research, both theoretical and field-oriented, is currently going on which should result in a better-defined methodology.

Because of the relatively high degree of uncertainty in the design of soil tiebacks, each installed tieback must be tested at least a nominal amount over its design load. Since each tieback must be stressed anyway, this is not an unreasonable demand. Figure 3.25*a* shows a typical tieback stressing operation. A tieback testing specification recommended by the Schnabel Foundation Company, see Weatherby,[34] is as follows:

1. Each tieback must be tested. The Contractor shall indicate on his plans the length of time the grout must cure prior to testing. A minimum of 5% of the tiebacks shall be tested in accordance with the performance test described herein. All other tiebacks shall be tested in accordance with the proof test described herein.

2. The performance test is performed by incrementally loading and unloading the tieback in accordance with the following schedule. At each increment, the movement of the end of the tendon shall be measured with a dial gage reading to .001 inches, and the jack load shall be maintained until the rate of movement is clearly approaching zero. The increments of load shall be:

0 Tons	.25 Design Load	.75 Design Load
2 Tons	.50 Design Load	.50 Design Load
.25 Design Load	.75 Design Load	.25 Design Load
2 Tons	1.00 Design Load	2 Tons
.25 Design Load	.75 Design Load	.25 Design Load
.50 Design Load	.50 Design Load	.50 Design Load
.25 Design Load	.25 Design Load	.75 Design Load
2 Tons	2 Tons	1.00 Design Load
.25 Design Load	.25 Design Load	1.20 Design Load
.50 Design Load	.50 Design Load	1.33 Design Load
.75 Design Load	.75 Design Load	1.20 Design Load
.50 Design Load	1.00 Design Load	1.00 Design Load
.25 Design Load	1.20 Design Load	.75 Design Load
2 Tons	1.00 Design Load	(Lock-off Load)

[*Note:* When conducted in this manner, the resulting data typically appear as shown in Fig. 3.25*b*.]

3. The proof test is performed by incrementally loading the tieback in accordance with the following schedule. The anchor movement shall be measured as in the performance test, and the movement criterion is the same as that in the performance test. The increments of load shall be:

0 Tons

2 Tons

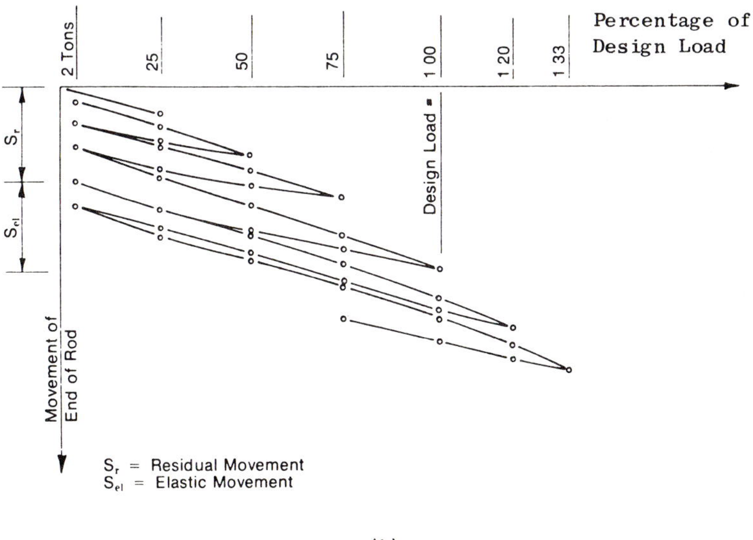

Figure 3.25 Tieback-stressing jack and related equipment and typical resulting data from performance test. (*a*) Tieback-stressing equipment. (*b*) Typical performance test data of a stressed tieback. (*Compliments of Schnabel Foundation Company.*)

Table 3.8 Soil-to-grout bond stress for grouted tiebacks in cohesive soils†

Soil type	Standard penetration resistances, blows/ft	Bond stress between grout and soil, kips/ft^2
Soft clay	2–4	0.50–0.75
Silty clay	3–6	0.50–1.00
Sandy clay	3–6	0.75–1.00
Medium clay	4–8	0.75–1.25
Firm clay	6–12	1.00–1.50
Stiff clay	8–15	1.00–2.00
Very stiff clay	15–30	1.5–2.5
Hard clay	Over 30	1.5–4.0

†After Weatherby , Ref. 34.

> .25 Design Load
> .50 Design Load
> .75 Design Load
> 1.00 Design Load
> 1.20 Design Load
> 1.00 Design Load
> .75 Design Load (Lock-off Load)

4. If a tieback fails to successfully carry the test load, the Contractor shall submit to the Engineer for review the method he intends to use to provide the required tieback capacity. This work shall be done at no additional cost to the owner.

It is important to realize that tiebacks extend for a considerably large distance behind the wall they are supporting. In many cases, they extend beneath city streets or under adjacent properties. In such cases, property easements must be obtained before construction is commenced.

3.5.3 Soil Nailing

Perhaps related to the recent practice in tunneling of using shotcrete and roof bolts directly behind a newly excavated face (called the *new Austrian tunneling method,* it does not allow for tunnel convergence and keeps stresses on a lining system to a minimum) is the concept of *soil nailing.*[35] Soil nailing involves a number of reinforcing bars being grouted or driven into the soil as it is exposed (thereby reinforcing it), the face of which is then supported by a wire-mesh-reinforced shotcrete facing. Shotcrete is a low water content sand and cement mixture (often with additives) which is sprayed onto a surface at high pressures in a manner similar to gunite. The various components used in the system are shown in Fig. 3.26. The construction procedure is as follows:

Excavation proceeds from the ground surface to the elevation of the first row of tiebacks.

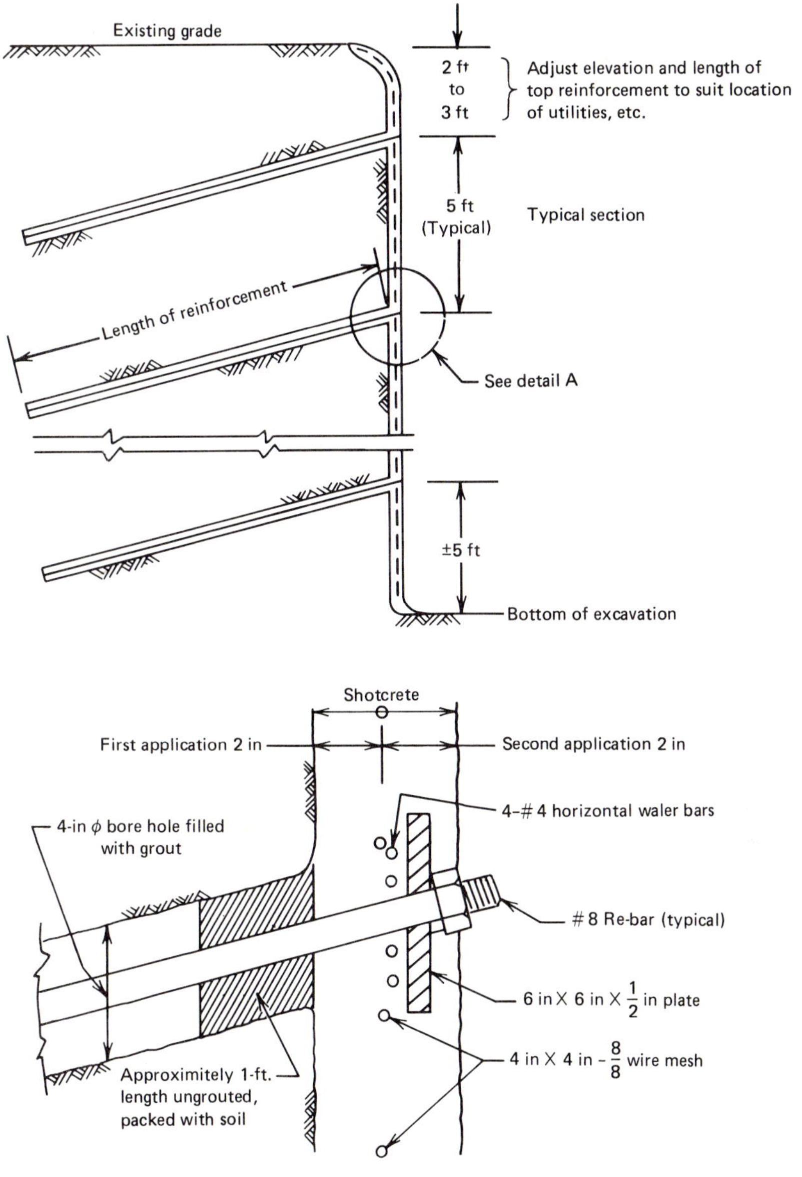

Figure 3.26 Various components of a slope protected and stabilized by soil nailing. *(After Shen et al., Refs. 35 and 36.)*

The tiebacks (usually No. 8 reinforcement bars of 60-kips/in^2 minimum yield stress) are placed in drilled holes and grouted in place or driven into the soil to be supported—note that they are not stressed.
A 2-in-thick layer of shotcrete is applied over the exposed soil surface.
Wire mesh is placed before complete cure occurs.
At tieback locations four No. 4 reinforcement bars are placed with a 6 × 6 × 0.5-in steel plate washer and lockoff nut being applied at each tieback.
A second 2-in shotcrete layer is placed over the entire area.
Excavation proceeds to the next level of tiebacks, approximately 5 ft deeper, and the process is repeated.
The last row of tiebacks is located about 5 ft from the bottom of the excavation.

The technique of soil nailing is quite new and no standardized design is available. In fact, there are numerous questions regarding the actual support mechanisms involved. Shear, tension, and bending of the tiebacks are all probably involved in the support of the retained soil. A design method needs to be developed for spacing of the tiebacks and bar stiffness effects need to be investigated further (pipes have been driven with good success). The technique, however, does offer a procedure for rapid construction and as such promises to be significantly less expensive than other ground support systems.

3.5.4 Reinforced Earth

The idea of mobilizing friction and/or cohesion to "grip" an anchor or tieback, thereby restraining a facing skin or element and thus forming a wall, has been incorporated into an accepted technology called *reinforced earth*. Here thin metal strips, or strips of wire mesh, which are connected to a wall facing element, are extended back into the backfill zone. When a layer of these strips is placed at one elevation, they are backfilled with granular soil and the entire process is repeated to the next, and higher, elevation. The construction sequence is shown in Fig. 3.27. Typical configurations have vertical separations between the reinforcing ties of from 1 to 3 ft and horizontal spacings of from 2 to 5 ft. When metal strips are used, they are usually $\frac{3}{16}$ in thick by 2 to 5 in wide. The welded wire mesh alternative uses wire of about $\frac{3}{8}$-in diameter in grids of 6 × 24 in. Both strips and wire mesh must be galvanized to protect them against corrosion. For particularly corrosive environments, materials to be used in reinforcement, connections, and facing elements must be critically evaluated, keeping in mind that in such environments those materials even when galvanized are often not very long-lasting vis-à-vis the anticipated lifetime of the structure.

Backfill soils should be free-draining granular sands and gravels (so as to avoid

Figure 3.27 Construction sequence for forming a reinforced earth wall. (*a*) Place footing and prepare subgrade. (*b*) Place first row of facing elements using temporary supports if necessary. (*c*) Connect reinforcing tie strips or mesh to facing elements. (*d*) Backfill to next level of reinforcing ties. (*e*) Repeat process for second level. (*f*) Continue until wall is complete. (*g*) Isometric of construction using wire mesh reinforcement. (*h*) Isometric of completed wall using steel strap reinforcement.

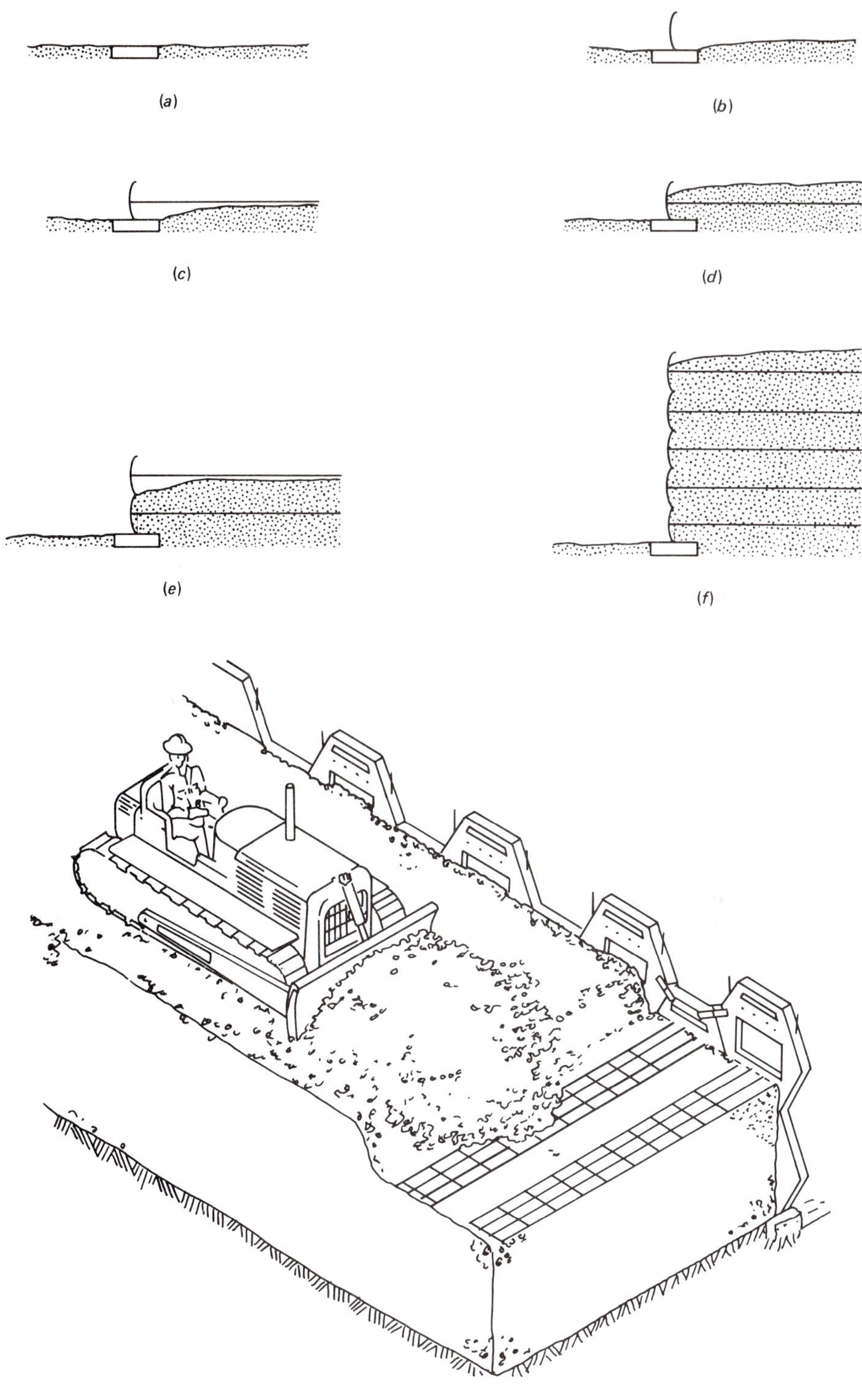

(a)

(b)

(c)

(d)

(e)

(f)

(g)

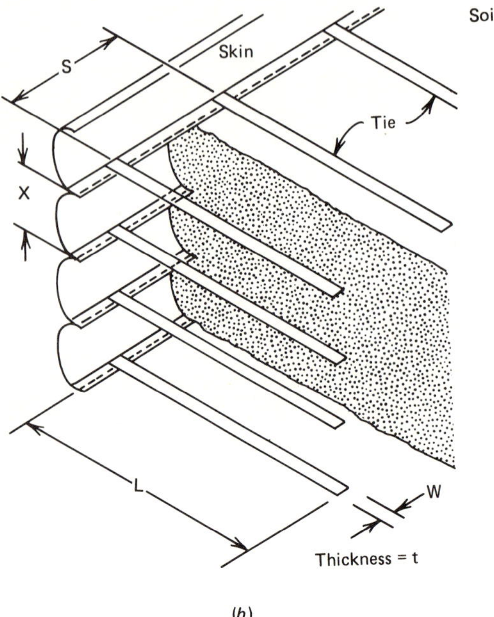

(h)

Figure 3.27 (*Continued*)

hydrostatic pressures), although 5 to 10 percent of fines passing the No. 200 sieve are helpful for good compaction. Compaction should be at 90 to 95 percent of maximum standard Proctor compaction and extra care must be exercised around the connections and immediately adjacent to the wall facing panels.

The designs for this kind of wall have been reasonably well established and are described in references by Lee et al.[37] and Al-Hussaini and Perry,[38] and in the commercial literature of specialty companies involved in this area.[39,40] Such designs are usually based on the assumption that active Rankine pressures are mobilized, which allows for the calculations of tie spacing and length. The relevant equations[37] are for geometry [Eqs. (3.16)] and for length [Eq. (3.17)]:

$$T_i = [\tfrac{1}{2}\gamma(ix)^2 K_a] - \Sigma T_{i-1} \tag{3.16a}$$

$$A_{\text{reqd}} = \frac{T_{\text{max}}}{\sigma_s} \tag{3.16b}$$

$$\text{FS}(T_i) = \gamma(ix)\tan\phi_u(2wL_i) \tag{3.17}$$

where T_i = tensile force in each strip
γ = unit weight of backfill soil
K_a = active Rankine earth pressure = $\tan^2(45 - \phi/2)$
ϕ = angle of shearing resistance of backfill soil
ϕ_u = angle of shearing resistance of soil to tie
i = numbering index from top of wall downward
x = vertical spacing of strips (= H/n if equally spaced)

H = wall height

n = number of vertical strips

σ_s = working stress in steel strip ($\simeq$20,000 lb/in^2, which includes the factor of safety)

A_{reqd} = width times thickness of strip ($w \times t$)

FS = factor of safety against tie pullout

w = width of tie

L_i = active length of tie behind Rankine failure plane (this plane makes an angle of $(45 - \phi/2)$ with the vertical at the base of the wall)

The entire process is quite different when using wire mesh as reinforcement. Instead of a friction phenomenon between solid tie and soil, there must be a bearing capacity type of phenomenon (similar to that shown in Fig. 1.18 except on a much smaller scale and oriented horizontally instead of vertically). The design equations[40] for pullout and breaking are the following:

$$\text{FS} = \frac{p}{T_{\text{max}}} db \left(\frac{L_b}{a} + 1 \right) \tag{3.18}$$

$$f_{\text{allow}} = \frac{T_{\text{max}}}{a_s} \tag{3.19}$$

where FS = factor of safety against pullout

f_{allow} = maximum allowable steel stress

p = maximum bearing resistance developed against crossbars (empirical value)

d = diameter of bars in mesh

b = width of mesh

L_b = bearing length of mesh

a = spacing of crossbars

a_s = cross-sectional area of longitudinal bars

Compared with conventional walls, such as reinforced concrete or metal bin crib walls, reinforced earth walls, as just described, are significantly less expensive. Figure 3.28 shows a cost comparison between different types of walls, where expense is expressed as a function of wall height. Building reinforced earth walls does not require large construction equipment, or skilled labor, but it does require a good amount of unskilled labor. And in situations where that can be used to advantage, the costs are even more appealing.

3.6 UNDERPINNING

By *underpinning* in the construction business, one means the building of a new foundation beneath an existing structure to prevent a stability failure or to eliminate excessive settlement. The main reasons for underpinning a structure are because it has an inadequate foundation to begin with or because of construction activity adja-

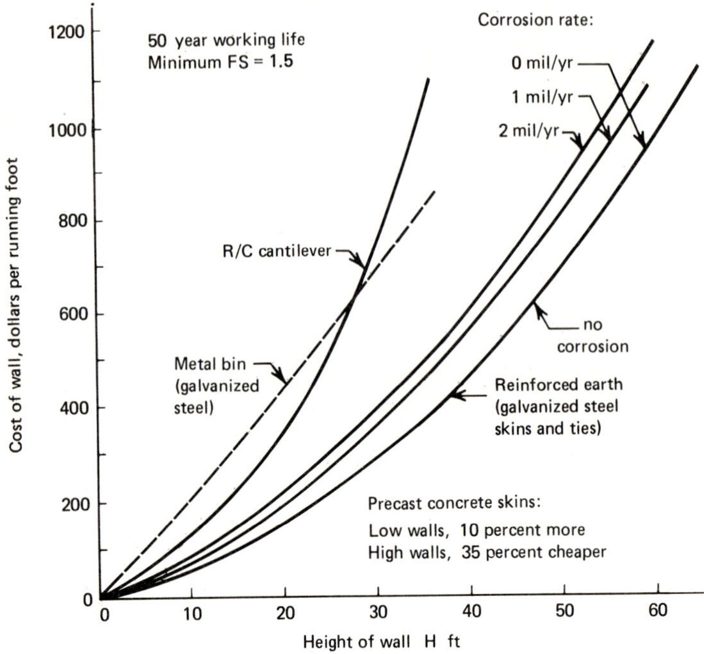

Figure 3.28 Comparative cost of different types of retaining walls illustrating cost-effectiveness of reinforced earth walls. *(After Lee et al., Ref. 37.)*

cent to it that might harm its stability. Theoretically there are two ways a structure can be underpinned:

1. Direct support methods, which include:
   Load transfer to adjacent soil
   Continuous pit method
   Intermittent pit method
   Grouting (penetration or compaction)
   Freezing
2. Lateral confinement methods, which include:
   Continuous cast-in-place augered piles
   Slurry-constructed walls
   Sheet pile walls

Regarding the load transfer of a wall or column load from its existing foundation to adjacent areas, the concept of a needle beam has been widely used; see Fig. 3.29. The work must be performed systematically and with great care and, when properly done, can result in fantastic accomplishments. For example, the nineteenth century Reading Railroad Train Shed in Philadelphia, PA, is a trihinged arch truss structure which is the largest of its type in the world (251-ft roof span) and is listed as a

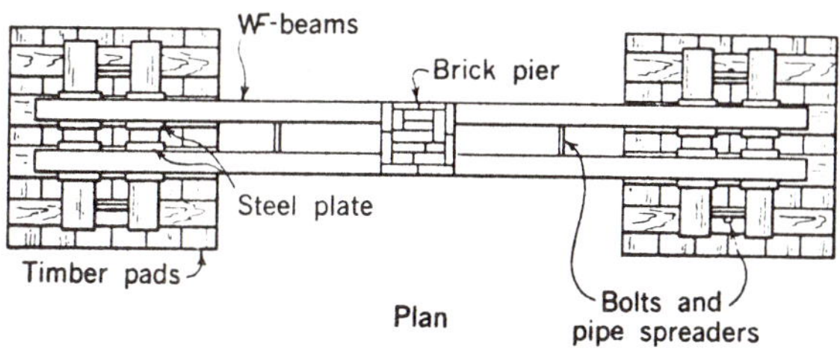

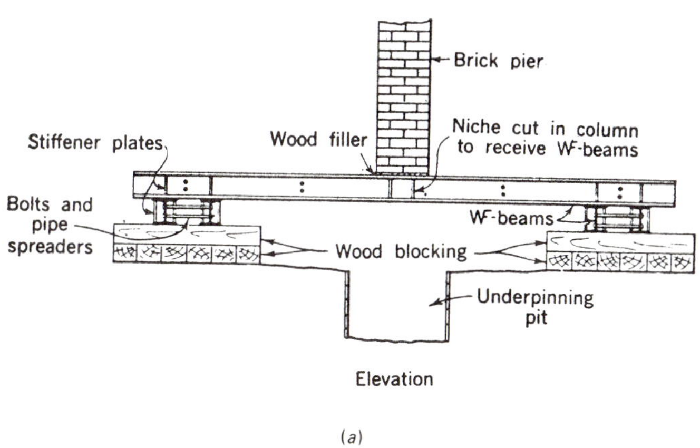

(a)

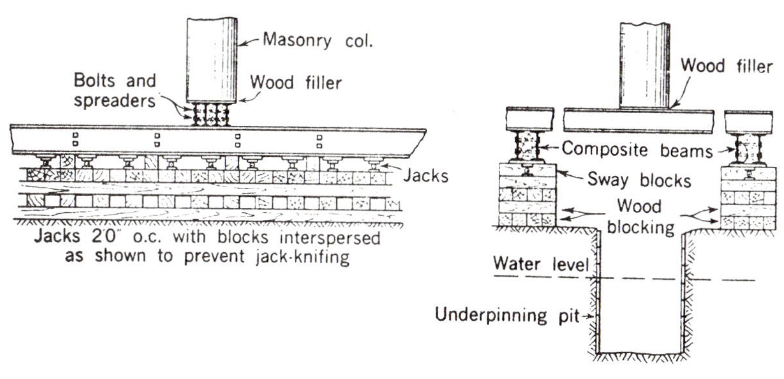

(b)

Figure 3.29 Use of needle beams on masonry or steel columns to transfer load to adjacent soil. (*a*) Needle beam method for masonry piers. (*b*) Needle beam method for masonry columns on soft ground with adjustable jack system. (*c*) Needle beam method for steel columns. (*After White, Ref. 41.*)

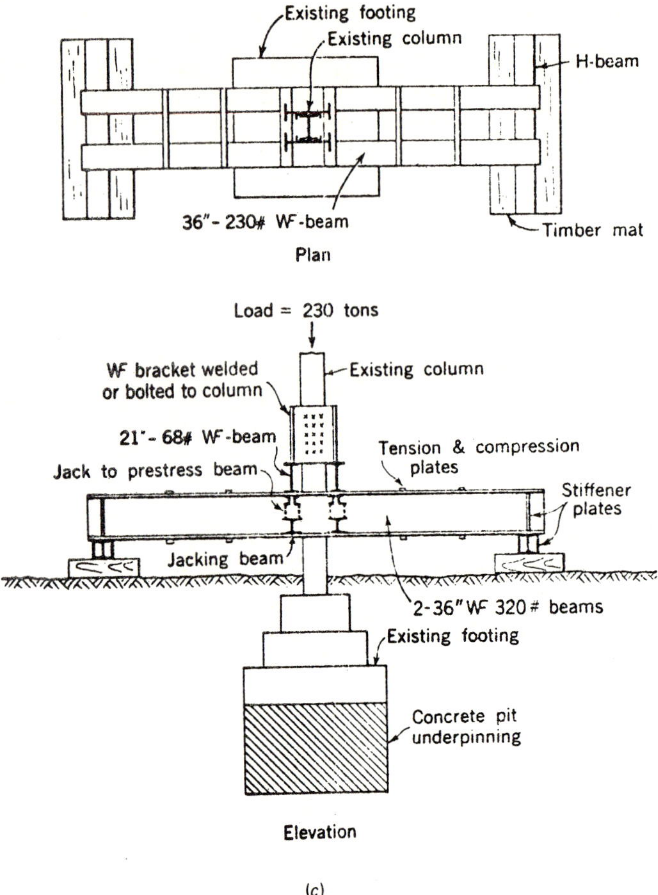

Figure 3.29 (*Continued*)

historic landmark. It stands in direct alignment with a commuter railroad tunnel which passes 50 ft beneath street level. The 55 steel columns, carrying from 374 to 661 kips each, are supported by a series of 75 caissons to subgrade beneath the new tunnel and by 15 heavily reinforced concrete grade beams. Eleven of them tie into the interior steel columns and four of them support the outside masonry piers. The interior connections are made via steel plates welded to the existing columns and the exterior piers are carried by steel needle beams spanning across adjacent grade beams. Figure 3.30 shows some of the scope of this project and the required detail.

The continuous pit underpinning method for continuous wall footings calls for hand excavation of small pits (3 × 3 ft to 6 × 6 ft) directly beneath the foundation in question down to the firm bearing soil, or rock, below. As with all underpinning methods it is slow, tedious, and demanding work which is very labor-intensive. Figure 3.31 shows the general technique of approaching the foundation from the side and, when its base is met, of moving directly beneath it down to the desired elevation.

Figure 3.30 Photographs of underpinning of Reading Railroad Train Shed, Philadelphia, PA, using load transfer method. *(Compliments of Philadelphia Railway Consultants Inc., Philadelphia, PA.)*

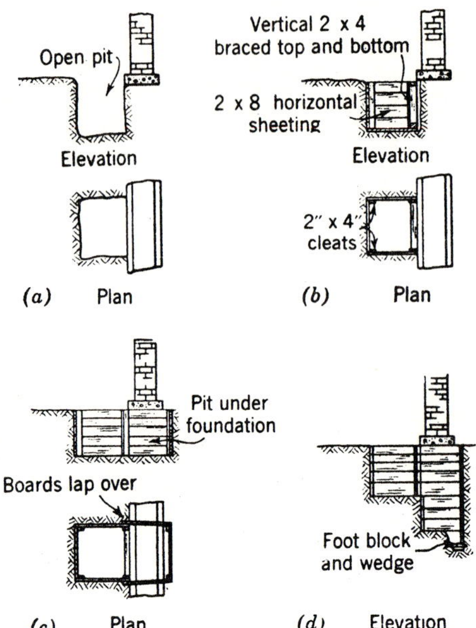

Figure 3.31 Sinking underpinning pit by horizontal sheeting. (*a*) Digging approach pit; (*b*) approach pit sheeted; (*c*) pit sheeted directly under foundation; (*d*) pit being sunk. (*After White, Ref. 42.*)

The pit is wood-framed throughout. Concreting proceeds to within 3 or 4 in of the existing footing and nonshrink cement "dry pack" is forced into the remaining space. The pits are completed on a staggered basis until the entire wall is underpinned. See Figs. 3.32 and 3.33.

For lighter loaded-wall foundations, the requirement of continuous pits is sometimes relaxed and intermittent pits are allowed; see Figs. 3.32 and 3.33. In such instances, the span must be carried by concrete or steel lintels. Possible options are reinforced with rods, beams, or arched configurations; see Fig. 3.34. Beneath these lintels, the space between the concrete pits must be filled with suitable sheeting.

In some situations, the soils beneath foundations of existing structures can be stabilized in situ by grouting (injection of cement, chemicals, or compaction grouting) or freezing (requires ground moisture and is only a temporary expedient). Both of these topics will be treated fully in subsequent chapters.

A totally different concept than directly supporting foundations is to provide only lateral confinement for the soils they are founded upon. This method requires a continuous wall to be placed around the site to achieve the depth needed for the project. It is critical, however, that the wall be sufficiently rigid so that local wall deformations do not cause lateral movements of the contained soils, which would result in vertical settlement of the foundation. Walls which meet this demand are continuous cast-in-place concrete walls (recall Sec. 3.4.2), slurry-constructed walls

---

Figure 3.33 Photographs of various types of direct pit underpinning methods. (See also page 202.) *Compliments of Schnabel Foundation Company.*

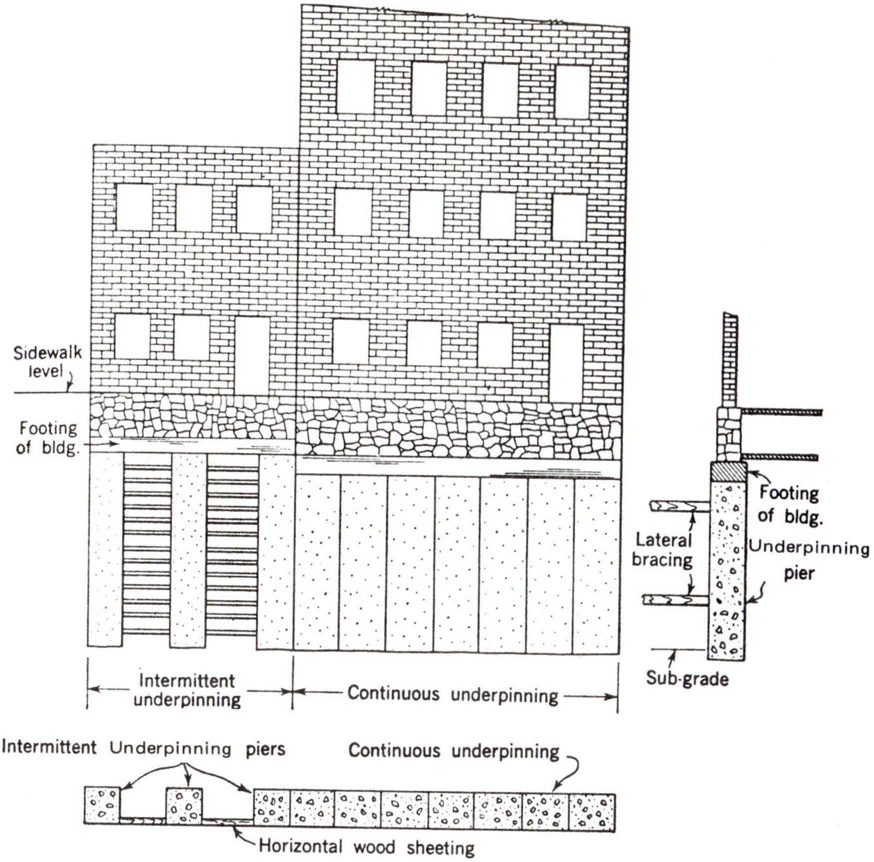

Figure 3.32 Sketch of continuous and intermittent pit underpinning methods. *(After White, Ref. 41.)*

Figure 3.33 (*Continued*)

(recall Sec. 3.4.3), and, sometimes, heavy "Z-section" sheet pile walls (recall Sec. 3.3.5). All of these are usually constructed with soil tiebacks or rock anchors (see Sec. 3.5.2, 3.5.1) and are instrumented for horizontal and vertical movements. Instrumentation of this type of situation will be discussed in Chap. 9.

PROBLEMS

3.1. When excavation at a construction site is carried out and sufficient room is available to lay the slopes back on a stable angle, the use of a geotextile or a geomembrane is recommended for control of rainfall runoff and for erosion protection. How do each of these materials fulfill these tasks?

3.2. (*a*) Determine the factor of safety for a slope of unbraced, fine-grained soil whose undrained shear strength is approximately 1200 lb/ft². The vertical height is 18 ft and the slope angle is 30°. The total unit weight of the soil is 122 lb/ft³ and the lower boundary is at the bottom of the excavation, that is, $D = 0$ ft in this case.

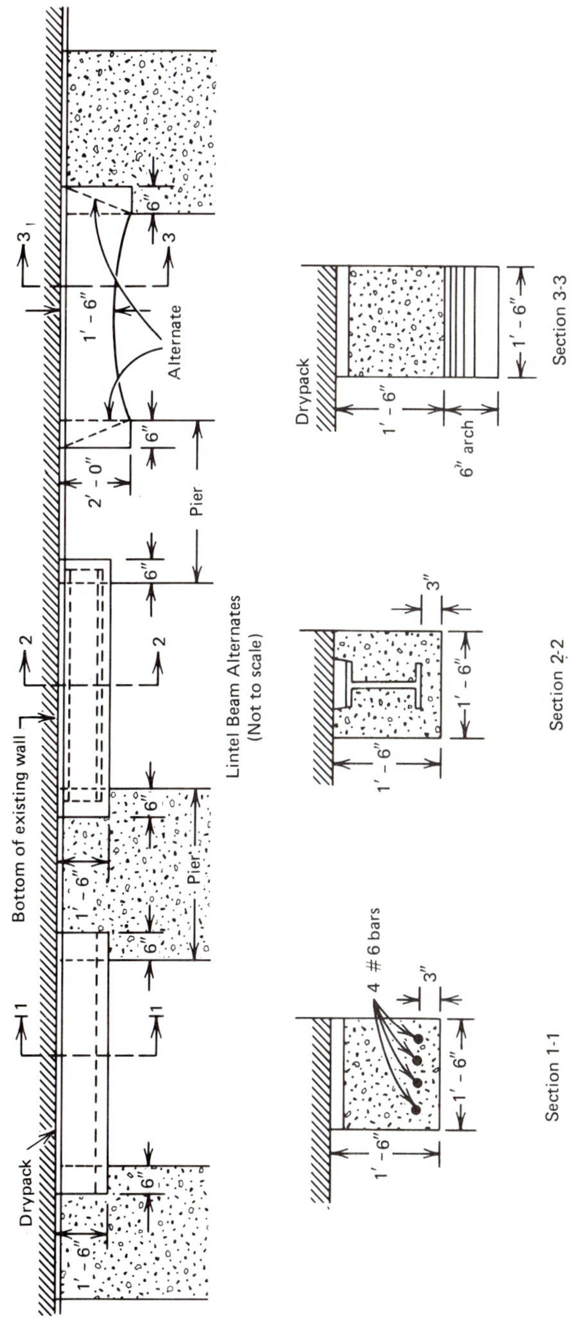

Figure 3.34 Types of lintels between underpinning pits. (*After White, Ref. 42.*)

203

(*b*) Resolve the problem for the case where the firm layer is 9 ft beneath the bottom of the excavation, that is, where $D = 9$ ft, so $n = 1.5$.

(*c*) Repeat part *a* for different slope angles and draw the response curve for slope angle (x axis) versus factor of safety (y axis). Comment on that curve.

3.3. (*a*) Determine the maximum depth to which an unbraced cut can be made at a 40° slope angle if the soil properties are $\phi = 10°$, $c = 500$ lb/ft^2, and $\gamma = 112$ lb/ft^3, and the factor of safety is 1.4.

(*b*) Determine the maximum slope angle to which an unbraced slope can be made for a 13-ft-deep excavation in a uniform soil whose properties are $c = 250$ lb/ft^2, $\phi = 8°$, and $\gamma = 108$ lb/ft^3, and where the factor of safety is 1.25.

3.4. The following three methods are used to predict the stability of unbraced slopes:

Empirical, as in Tables 3.1 and 3.2
Stability charts based on total stress analyses, as in Figs. 3.2 and 3.3
Complete analysis using undisturbed soil data, effective stress analyses, and a computer search for the
    critical slip circle

Which would you recommend in the following situations and why?

(*a*) For a sewer contractor constructing a slope who needs to lay pipe 5 ft deep in an open area, free of permanent buildings.

(*b*) For a sewer contractor constructing a slope who needs to lay pipe in a congested area, or at a depth greater than 10 ft.

(*c*) For a geotechnical consultant who is in the initial stages of planning to excavate an area for a bridge pier.

(*d*) For a geotechnical consultant designing a flood control dam which will only encounter hydrostatic pressures once every 20 years.

(*e*) For a municipal engineer working for a water company that is considering building an earth dam to impound water. The reservoir will be lined with a geomembrane (i.e., a plastic or rubber pond liner).

3.5. The Occupational, Safety, and Health Act (OSHA) requires that any excavation over 4 ft deep be temporarily supported using a sheeting system. Does skeleton sheeting of the type shown in Fig. 3.4 satisfy this requirement? If so, under what situations would the "letter of the law" be observed, but not the law's intent?

3.6. Most braced excavations of the types described in Sec. 3.3 are of a temporary nature. When the job is finished the contractor removes the sheeting or abandons it in place.

(*a*) Under those circumstances, who should design the sheeting and its bracing system, the contractor or the consulting engineer? Why?

(*b*) Where should the liability for a failure be placed in light of your answer to part *a*?

(*c*) Quite often the temporary sheeting and bracing are not paid for as separate items but are included in the unit price of the concrete or some other related item. What do you think of this practice?

(*Hint:* See Ref. 9 and the discussions about the paper that followed in the same journal.)

3.7. Temporary sheeting systems are obviously related to lateral earth pressure theory, which was developed in regard to permanent construction walls. However, there are some basic differences. Answer the following:

(*a*) The earth pressure against permanent walls is usually based on active conditions, yet temporary bracing (which is flexible and yields under pressure) can give significantly higher values. Why?

(*b*) Some of the pressure distributions for temporary bracing shown in Fig. 3.5 actually decrease with depth. This is in sharp contrast to what happens with permanent walls, where the distributions for both soil and water pressures increase uniformly with depth. Why is this the case?

(*c*) Can one expect the pressure behind a temporary wall to change with time? (Answer for both granular and cohesive types of soil.) If it does change, does it increase or decrease?

3.8. (*a*) For excavation in soil with the profile shown in Fig. P3.8, calculate the strut loads that can be anticipated if soldier beams and lagging are used for temporary sheeting.

(*b*) Would 6 × 6 in timber struts be adequate for those loads? Struts are at 10-ft centers.

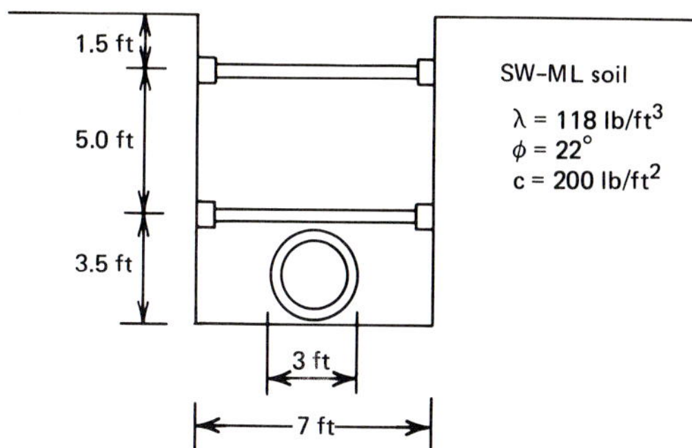

Figure P3.8

3.9. Using the configuration and layout of the soldier beam and lagging project shown in Fig. 3.8*a*, calculate the cross-trench strut loads S1 through S5. They are at 12-ft centers. Water must be kept at elevation 100, or higher, at all times. Compare your results to the average strut loads shown in Fig. 3.8*c*. (*Note:* Reference 16 contains background information on this very significant case history.)

3.10. (*a*) Design a continuous steel sheetpile wall for a small marina in an ocean bay using data given in Fig. P3.10 and following the steps listed here:

Determine embedment depth (then use FS = 1.5 for depth to be actually used)
Determine anchor force
Determine maximum moment
Modify for flexibility if warranted

(*b*) Considering how corrosive this environment can be, what is the anticipated lifetime of the sheeting?

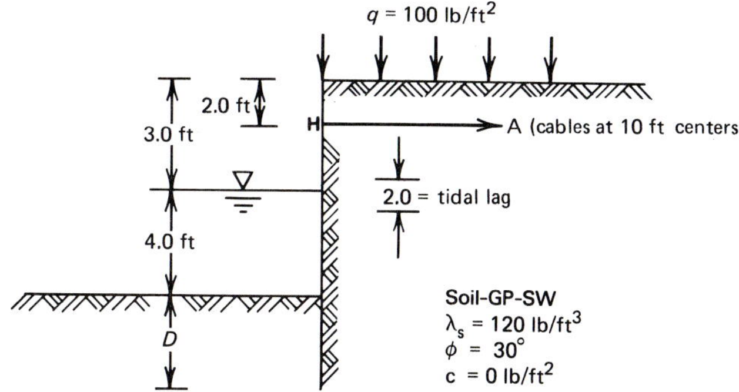

Figure P3.10

3.11. Sketch some possible configurations of how the anchor force in Prob. 3.10 could be resisted.

3.12. A major problem with bulkheads, docks, and related structures in freezing climates is ice jacking.
  (*a*) What is ice jacking and what is its result?
  (*b*) How can it be avoided?
  (*c*) If it cannot be avoided, at least how can it be controlled?

3.13. Compare and contrast the properties listed in the table below in relation to the types of retaining walls noted.

Property being compared or rated	Soldier beams and lagging	Continuous steel sheeting	Concrete secant piles
Time required for installation			
Likelihood of noise and vibration during installation			
Likelihood of horizontal movement between braces			
Likelihood of vertical movement behind wall			
Seepage control capability			
Possibility of reclaiming materials			

3.14. By what means are steel sheet piling and bearing piles reclaimed for later use on the same or future jobs?

3.15. Regarding the design and construction of a relieving platform as shown in Fig. 3.11*b:*
    (*a*) What is the basic design methodology?
    (*b*) How is such a platform connected to the sheeting or the pile cap?
    (*c*) Why are the outer piles always vertical or battered toward the land?
    (*d*) What happens if the dredged fill beneath the surface of the relieving platform settles?

3.16. Calculate the additional stress (as a function of depth) that a temporary construction wall, like that shown in Fig. P3.16, must take as the result of a crane being sustained in the zone above it.

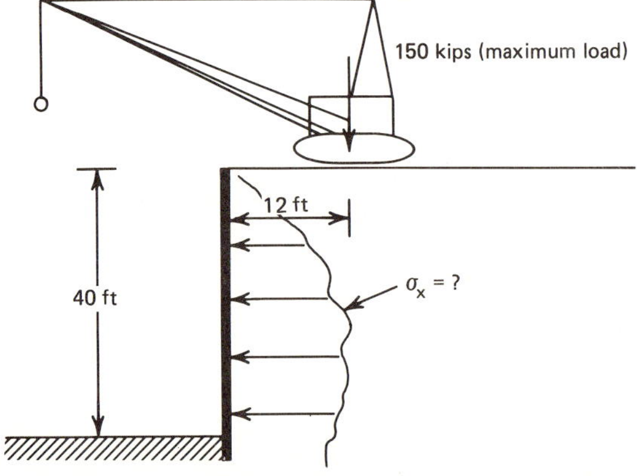

Figure P3.16

3.17. Regarding slurry-constructed concrete retaining walls as shown in Figs. 3.16 and 3.17:
    (*a*) How is an effective bottom seal made in rock while excavating through the slurry?
    (*b*) How is the concrete placed in the slurry-sustained trench from the bottom up?
    (*c*) Does bonding occur between the concrete and a reinforcement cage that has been immersed in the slurry and is coated with the slurry's clay soil?

3.18. For rock and soil anchors as illustrated in Figs. 3.21 through 3.23;
    (*a*) Why is there no measurable expense incurred by the contractor in testing each anchor to 1.25 times its design load?
    (*b*) Why is it often not possible to test to 2.0 times an anchor's design load?
    (*c*) How can prestress loss be measured?
    (*d*) How can anchor corrosion be prevented?

3.19 (*a*) Calculate the tieback capacity of a high-strength steel bar (Dywidag type) of 20-ft length (active zone) and 9-in diameter (grouted zone) in a SW-ML soil of 110 lb/ft³ having $\phi = 37°$ and $c = 550$ lb/ft² (see Fig. P3.19). (Use a friction factor of 0.8 and an adhesion factor of 0.6.)

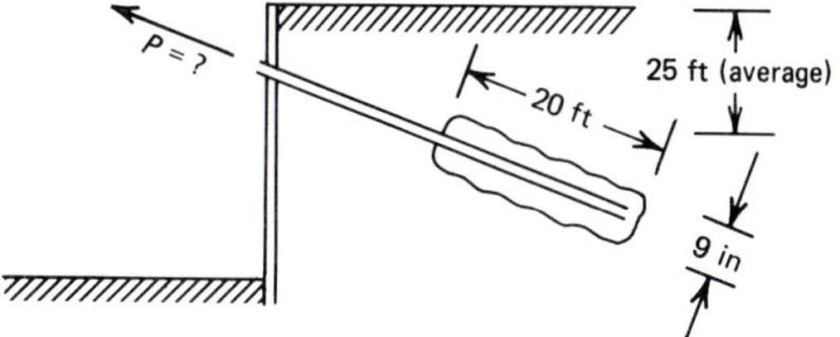

Figure P3.19

(*b*) What type of bar according to Table 3.7 would be satisfactory for this situation?

3.20. (*a*) Calculate the tieback capacity of the same steel bar in a stiff clay of average blow count of 12 blows per foot.

(*b*) What type of bar according to Table 3.7 would be satisfactory for this situation?

3.21. In regard to reinforced earth walls as shown in the sketches of Fig. 3.27, answer the following:

(*a*) What type of sustained load (creep) deformations would you anticipate?

(*b*) How would such walls behave under a dynamic or blast loading?

(*c*) Why are the walls not designed for hydrostatic loading?

3.22. For the reinforced earth wall shown in Fig. P3.22, which has steel strip reinforcement at 5-ft centers horizontally, a 20,000-lb/in² working stress, and a 1.75 factor of safety against pullout, design the size of the ties in terms of their width and thickness, and their lengths.

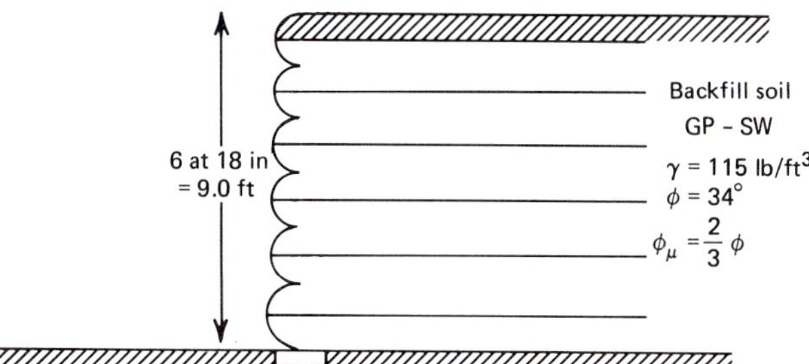

Figure P3.22

3.23. In regard to doing pit underpinning work as shown in Figs. 3.31 through 3.33, answer the following:

(*a*) Why is the work often done on a 24-hour basis even when the laborers are working at overtime rates?

(*b*) How do you determine the maximum size of the pit that can be safely excavated?

(*c*) When do you decide to use intermittent pit underpinning methods (with lintels) versus the more expensive continuous pit underpinning?

(*d*) Should tiebacks or struts be used to satisfy pit underpinning lateral support requirements?

3.24. In supporting an existing masonry building with a foundation 20 ft above the grade of a proposed tunnel, the underpinning can be done by direct support (as shown in Fig. 3.31) or by lateral confinement (as shown in Fig. 3.15). What are the major considerations that enter into deciding which method to use?

REFERENCES

1. McCullough, D. G., *The Johnstown Flood,* Simon and Schuster, New York, 1968.
2. Sherard, J. L., R. J. Woodward, S. F. Gizienski, and W. A. Clevenger, *Earth and Earth-Rock Dams,* John Wiley and Sons, New York, 1967.
3. Hirschfeld, R. C., and S. J. Poulos (eds.), *Embankment-Dam Engineering,* Casagrande Volume, John Wiley and Sons, New York, 1973.
4. Bazant, Z., *Methods of Foundation Engineering,* Elsevier, Amsterdam, 1979.
5. ————, "Empfehlungen für den Bau und die Sicherung von Boschungen," *Die Bautechnik,* Vol. 39, 1962, pp. 404–415.
6. Taylor, D. W., *Fundamentals of Soil Mechanics,* John Wiley and Sons, New York, 1948.
7. *Design Manual: Soil Mechanics, Foundations, and Earth Structures,* NAVDOCKS DM-7, Dept. of the Navy, Bureau of Yards and Docks, Washington, DC, 1982.
8. OSHA, "Safety and Health Regulations for Construction, Subpart P, Excavation, Trenching and Shoring," *Federal Register,* Vol. 39, No. 122, June 24, 1974.
9. Thompson, L. J., and R. J. Tanenbaum, "Responsibility for Trenching and Excavation Design," *J. Geotech. Eng. Div., ASCE,* Vol. 103, No. GT4, April 1977, pp. 327–338. (See also subsequent discussion about this paper in the same journal, and original authors' conclusions.)
10. ————, and ————, "Survey of Construction Related Trench Cave-Ins," *J. Constr. Div., ASCE,* Vol. 103, No. CO3, September 1977, pp. 501–512.
11. Yokel, F. Y., *Recommended Technical Provisions for Construction Practice in Shoring and Sloping of Trenches and Excavations,* NBS Building Series No. 127, U.S. Dept. of Commerce, Washington, DC, 1980.
12. Caquot, A., and J. Kerisel, *Traite de Mechanique des Sols,* 2d ed., Ganthier-Villars, Paris, 1967.
13. Peck, R. B., "Deep Excavations in Soft Ground," in *Proc. Int. Conf. on Soil Mech. and Found. Eng.,* State-of-the-Art Volume, Mexico City, Mexico, 1969, pp. 225–250.
14. Goldberg, D. T., W. E. Joworski, and M. D. Gordon, *Lateral Support Systems and Underpinning,* Report No. FHWA-RD-75-129, Vol. 2, Federal Highway Administration, U.S. Dept. of Transportation, Washington, DC, April 1976.
15. Golder, H. Q., J. P. Gould, W. T. Lambe, G. P. Tschebotaroff, and S. D. Wilson, "Predicted Performance of Braced Excavation," *J. Soil Mech. Found. Eng. Div., ASCE,* Vol. 96, No. SM3, May 1970, pp. 801–816.
16. Lambe, T. W., L. A. Wolfshill, and I. H. Wong, "Measured Performance of Braced Excavation," in ibid. pp. 817–836.
17. Peck, R. B., W. E. Hanson, and Thornburn, R. H. *Foundation Engineering,* 2d ed., John Wiley and Sons, New York, 1974.
18. Terzaghi, K., *Theoretical Soil Mechanics,* John Wiley and Sons, New York, 1943.
19. Rowe, P. W., "A Theoretical and Experimental Analysis of Sheet Pile Walls," *Proc. Inst. Civil Eng., London,* January 1955, pp. 32–86.
20. Escalante, E. M., and R. J. Iverson, "Corrosion of Steel Piles," *Mater. Performance,* Vol. 17, No. 10, October 1978, pp. 9–15.
21. ————, *Construction of Bulkhead for Schuylkill River Park,* Fairmount Park Commission, City of Philadelphia, 1980–1982.
22. Bishop, A. W., and D. J. Henkel, *The Measurement of Soil Properties in the Triaxial Test,* 2d ed., E. Arnold, London, 1962.
23. Poulos, H. G., and E. H. Davis, *Elastic Solutions for Soil and Rock Mechanics,* John Wiley and Sons, New York, 1974.
24. Nendza, H., "Sicherung tiefer Baugruben neben Bauwerken," *Tiefbau,* Vol. 15, 1973, pp. 698–702.
25. Gill, S. A., "Applications of Slurry Walls in Civil Engineering," *J. Constr. Div., ASCE,* Vol. 106, No. CO2, June 1980, pp. 155–167.
26. Nash, K. L., "Diaphragm Wall Construction Techniques," *J. Constr. Div., ASCE,* Vol. 100, No. CO4, December 1974, pp. 605–620.

27. D'Appolonia, D. J., "Soil—Bentonite Slurry Trench Cutoffs," *J. Geotech. Eng. Div., ASCE,* Vol. 106, No. GT4, April 1980, pp. 399–417.

28. Product literature on "Aspermix" slurry systems, Thatcher Engineering Corp., 7100 Industrial Ave., Gary, IN 46406.

29. Hutchinson, M. T., G. P. Daw, P. G. Shatton, and A. N. James, "The Properties of Bentonite Slurries Used in Diaphragm Walling and Their Control," in *Diaphragm Walls and Anchorages,* Institute of Civil Engineers., London, 1975.

30. Millet, R. A., and J.-Y. Perez, "Current USA Practice: Slurry Wall Specifications," *J. Geotech. Eng. Div., ASCE,* Vol. 107, No. GT8, August 1981, pp. 1041–1056.

31. Littlejohn, G. S., and D. A. Bruce, "Rock Anchors; State-of-the-Art; Part I—Design," *Ground Eng.* Vol. 8, No. 3, May 1975.

32. Bauer, K., "Injektionszuganker in nichtbindigen Boden," *Bau und Bauindustrie,* Vol. 16, 1960, pp. 520–522.

33. Sigourney, J. W., "Tieback Installations by Schnabel Foundation Company," in *Proc. Conf. in Des. and Constr. of Earth and Rock Tiebacks,* George Washington University, Washington, DC, February 1971, pp. 1–21.

34. Weatherby, D. E., *Tiebacks,* Report No. FHWA/RD-82/047, U.S. Department of Transportation, Federal Highway Administration, Washington, DC, July, 1982, 232 pps.

35. Shen, C. K., S. Bang, and L. R. Herrman, "Ground Movement Analysis of Earth Support System," *J. Geotech. Eng. Div., ASCE,* Vol. 107, No GT12, December 1981, pp. 1609–1624.

36. Shen, C. K., S. Bang, K. M. Romstad, L. Kulchin, and J. S. DeNatale, "Field Measurements of an Earth Support System," *J. Geotech. Eng. Div., ASCE,* Vol. 107, No. GT12, December 1981, pp. 1625–1642.

37. Lee, K. L., B. D. Adams, and J.-M. J. Vagneron, "Reinforced Earth Retaining Walls," *J. Soil Mech. Found. Eng. Div., ASCE,* Vol. 99, No. SM10, October 1973, pp. 745–764.

38. Al-Hussaini, M., and E. B. Perry, "Field Experiment of Reinforced Earth Wall," *J. Geotech. Eng. Div., ASCE,* Vol. 104, No GT3, March 1978, pp. 307–322.

39. Product literature, Reinforced Earth Company, Washington, DC 20037.

40. Product literature, VSL Corporation, Los Gatos, CA 95030.

41. White, E. E., "Underpinning," Chap. 9 in *Foundation Engineering,* G. A. Leonards (ed.), McGraw-Hill, New York, 1962, pp. 826–893.

42. ———, "Underpinning," Chap. 22 in *Foundation Engineering Handbook,* H. F. Winterkorn and H.-Y. Fang (eds.), Van Nostrand Reinhold, New York, 1975, pp. 626–648.

FOUR

IN SITU DENSIFICATION METHODS IN GRANULAR SOILS

4.1 INTRODUCTION

Without question, the single most important feature of granular soils is their *relative density*. Originally defined by Burmister[1] in terms of void ratios, relative density is usually used in its equivalent form of dry unit weights. The two equivalent relationships are expressed in the following equations:

$$D_R \equiv \frac{e_{max} - e}{e_{max} - e_{min}} \tag{4.1}$$

$$D_R = \frac{\gamma - \gamma_{min}}{\gamma_{max} - \gamma_{min}} \frac{\gamma_{max}}{\gamma} \tag{4.2}$$

where D_R = relative density (usually given as a percent and thus varies from 0 to 100 percent)

e_{max} = maximum void ratio (soil in its loosest state)

e_{min} = minimum void ratio (soil in its densest state)

e = in situ void ratio (soil in its natural state)

γ_{min} = minimum dry unit weight (soil in its loosest state)

γ_{max} = maximum dry unit weight (soil in its densest state)

γ = in situ dry unit weight (soil in its natural state)

e_{max} and/or γ_{min} in the above equations are found by placing the soil in its loosest possible state by carefully pouring it into a stationary container; see the ASTM test D2049-69 for Limiting Density States of Granular Soils. At the other end of the density spectrum, the values e_{min} and/or γ_{max} are found by placing the soil in its densest possible state. This is accomplished by vibrating the soil in layers in a confined container with a surcharge load on the upper surface. The test is also covered in ASTM D2049-69. The other terms in Eqs. (4.1) and (4.2) refer to the soil in its natural or in situ condition. To determine these data, field tests are required to obtain weight and volume information along with the soil's water content. The following ASTM tests are applicable:

ASTM D1556–64—*Density of Soil in Place by the Sand Cone Method*
ASTM D2167–66—*Density of Soil in Place by the Rubber-Balloon Method*
ASTM D2922–71—*Density of Soil in Place by Nuclear Method*
ASTM D2937–71—*Density of Soil in Place by the Drive-Cylinder Method*
ASTM D2216–71—*Laboratory Determination of Moisture Content of Soil*
ASTM D3017–72—*Moisture Content of Soil in Place by Nuclear Method*

In addition to actually testing for the properties desired, several empirical relationships are available. These, following Koerner,[2] relate the soil's limiting density states (or unit weights) to particle characteristics, as shown in Eqs. (4.3) and (4.4):

$$\gamma_{max} = 104 + \Delta\gamma_1 + \Delta\gamma_2 + \Delta\gamma_3 \qquad (4.3)$$

where γ_{max} = maximum dry unit weight, lb/ft³
$\quad \Delta\gamma_1$ = correction for particle size
$\qquad = +5$ lb/ft³ if $d_{10} > 2.0$ mm (gravel size)
$\qquad = +3$ lb/ft³ if $2.0 > d_{10} > 0.6$ mm (coarse sand size)
$\qquad = 0$ lb/ft³ if $0.6 > d_{10} > 0.2$ mm (medium sand size)
$\qquad = -5$ lb/ft³ if $0.2 > d_{10} > 0.06$ mm (fine sand size)
$\qquad = -10$ lb/ft³ if $0.06 > d_{10} > 0.02$ mm (coarse silt size)
$\quad \Delta\gamma_2$ = correction for gradation (size distribution),
$\qquad = +12$ lb/ft³ if CU = 5.0 (well graded)
$\qquad = 0$ lb/ft³ if CU = 2.0 (medium graded)
$\qquad = -8$ lb/ft³ if CU = 1.0 (poorly graded)
$\quad \Delta\gamma_3$ = correction for particle shape
$\qquad = +6$ lb/ft³ if $S = 0.7$ (high sphericity)
$\qquad = 0$ lb/ft³ if $S = 0.6$ (average sphericity)
$\qquad = -6$ lb/ft³ if $S = 0.4$ (low sphericity)

$$\gamma_{min} = 80 + \Delta\gamma_4 + \Delta\gamma_5 + \Delta\gamma_6 \qquad (4.4)$$

where γ_{min} = minimum dry unit weight, lb/ft³
$\quad \Delta\gamma_4$ = correction for particle size
$\qquad = +8$ lb/ft³ if $d_{10} > 2$ mm (gravel size)
$\qquad = +5$ lb/ft³ if $2.0 > d_{10} > 0.6$ (coarse sand size)
$\qquad = +1$ lb/ft³ if $0.6 > d_{10} > 0.2$ (medium sand size)
$\qquad = -5$ lb/ft³ if $0.2 > d_{10} > 0.06$ (fine sand size)
$\qquad = -10$ lb/ft³ if $0.06 > d_{10} > 0.02$ (coarse silt size)
$\quad \Delta\gamma_5$ = correction for gradation (size distribution)
$\qquad = +6$ lb/ft³ if CU > 5 (well graded)

$$= -2 \text{ lb/ft}^3 \text{ if } 5 > \text{CU} > 2 \text{ (medium graded)}$$
$$= -12 \text{ lb/ft}^3 \text{ if } 2 > \text{CU} > 1 \text{ (poorly graded)}$$
$$\Delta\gamma_6 = \text{correction for particle shape}$$
$$= +11 \text{ lb/ft}^3 \text{ if } S \simeq 0.7 \text{ (high sphericity)}$$
$$= +3 \text{ lb/ft}^3 \text{ if } S \simeq 0.6 \text{ (average sphericity)}$$
$$= -7 \text{ lb/ft}^3 \text{ if } S \simeq 0.4 \text{ (low sphericity)}$$

The above equations are for granular soils composed of the mineral quartz, which has a specific gravity of 2.65. For soils with other minerals present, the multiplication by a simple ratio of specific gravities will give a reasonably accurate estimate of γ_{max} or γ_{min}. Also, d_{10} = effective size of soil, CU = coefficient of uniformity, and S = sphericity (= 1 for a circle and 0 for a plate on end); see Appendix A.

As far as the ranking of soils with different relative densities goes, the following can be used:

For 0 percent $< D_R <$ 50 percent, the soil is considered *loose*.
For 50 percent $< D_R <$ 75 percent, the soil is considered *medium*.
For 75 percent $< D_R <$ 90 percent, the soil is considered *dense*.
For 90 percent $< D_R <$ 100 percent, the soil is considered *very dense*.

As this ranking indicates, the classification of the soils as to their relative densities and the amount of effort needed to achieve a given degree of compaction are not linear with respect to the values of D_R. For example. a dump truck dropping a load of sand onto the ground will produce a relative density of approximately 50 percent in that sand, which occurs without any compactive effort as such. Conversely, to increase D_R from 90 to 100 percent requires a tremendous amount of compactive effort. For these reasons other formulations of density have been suggested. Lacroux and Horn[3] approach the subject by defining two other relationships to augment the standard relative density equation. They are:

$$R_R = \frac{\gamma - \gamma_{min}}{\gamma_{max} - \gamma_{min}} \tag{4.5}$$

$$D_C = \frac{\gamma}{\gamma_{max}} \tag{4.6}$$

where R_R = density ratio
$\quad D_C$ = degree of compaction
$\gamma, \gamma_{min}, \gamma_{max}$ = dry unit weights as previously defined

Note that the degree of compaction D_C is numerically equal to the density ratio R_R divided by the relative density D_R. It is a very convenient term to use in the field control of projects involving soil compaction.

Whatever the method used for formulating the value of relative density, its significance in the behavior of engineering properties of granular soils is well established. Figure 4.1 shows how an increase in D_R from 22 to 83 percent increases both a soil's strength and its modulus of elasticity.[4] This can also be seen in Fig. 4.2, where Sowers and Sowers[5] present a number of relationships (ϕ, c, $\Delta H/H$) as functions of standard penetration resistance (N = blow count per foot of sampling spoon resis-

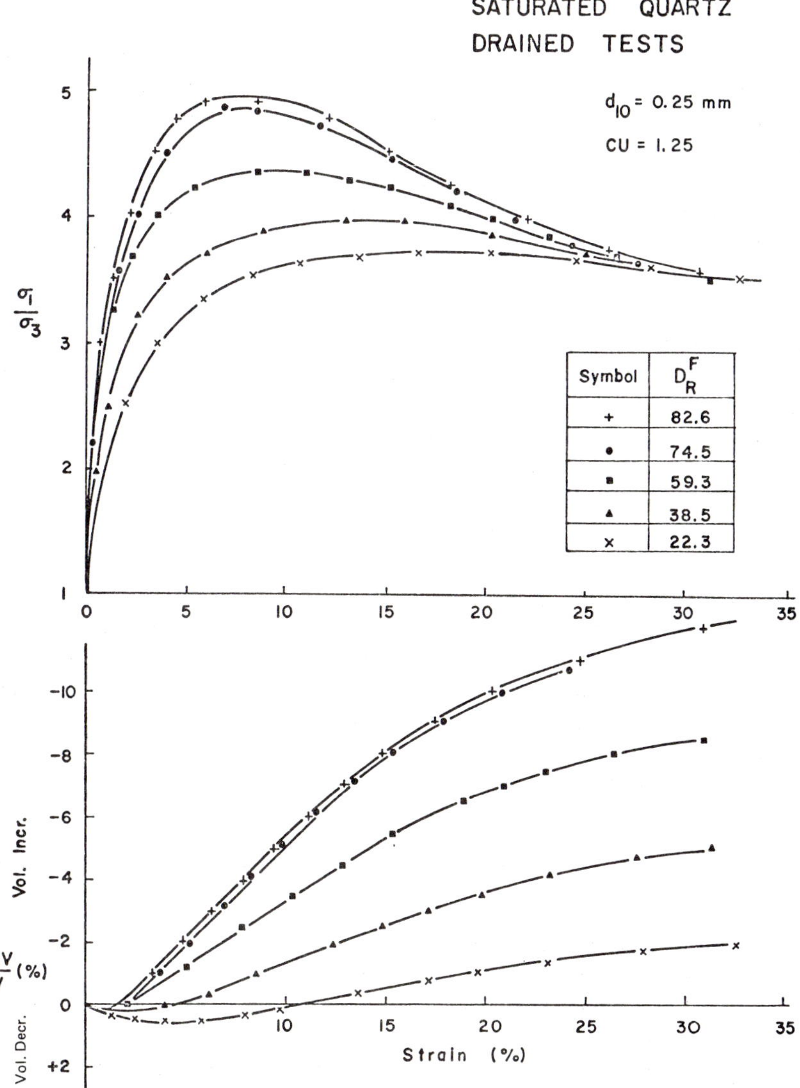

Symbol	D_R^F
+	82.6
●	74.5
■	59.3
▲	38.5
×	22.3

Figure 4.1 Typical drained triaxial test results on saturated quartz. *(After Koerner, Ref. 4.)*

tance). Also shown in Fig. 4.2 is the Gibbs and Holtz[6] relationship between relative density D_R and blow count N.

The rest of this chapter will describe the methods used in foundation engineering to increase the relative density of in situ granular soils. Each of the following four sections discusses the topic according to the method of compaction to be used (vibration or impact) and the location at which the compactive device is to be physically deployed (at ground surface or below ground surface).

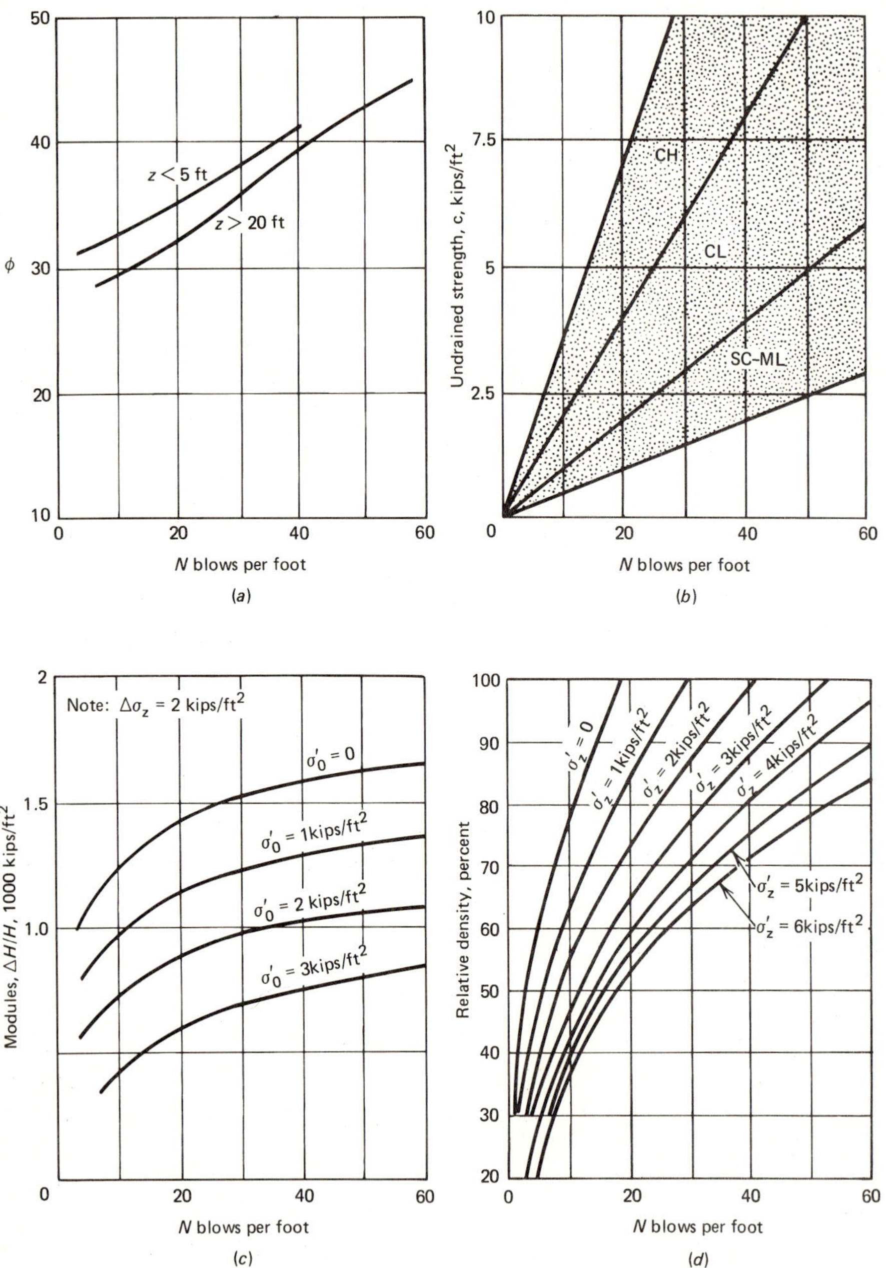

Figure 4.2 Relation of standard penetration resistance to soil properties for preliminary studies. *Note: All the relationships are approximations. (a) Cohesionless soil; (b) saturated clay, unified classification; (c) deformation modulus or one-dimensional compressibility; (d) relative density of dry sands. (After Sowers and Sowers, Ref. 5.)*

4.2 VIBRATION AT THE GROUND SURFACE

To properly address the topic of densification of granular soils by surface vibration, we must first note the number of construction devices available to choose from. They include the following, which are further elaborated upon in Table 4.1 (a nice historical account of their development is given by Foster[8]):

Smooth wheel rollers. Two types are available, one a three-wheeled roller (used primarily in the compaction of bituminuous pavements and base courses) ranging in weight up to 15 tons, the other a tandem roller ranging up to 20 tons in weight.

Rubber-tired rollers. These range in weight up to 200 tons and are either self-propelled or towed. Smaller models have about 10 tires on two axles with tire pressures up to 30 lb/in². Larger models have fewer, but considerably heavier, tires with tire pressures up to 150/in².

Vibratory rollers. Both smooth steel wheel rollers (up to 8 tons) and rubber-tired rollers (up to 30 tons) can be modified so they impart an impacting motion to the soil being compacted. Weights are placed eccentrically about a rotating shaft and timed so as to reasonably match the resonant frequency of the soil, between 15 and 30 Hz. These rollers are by far the predominant type used in the compaction of granular soils.

Sheepsfoot rollers. These steel wheel rollers, which have protruding feet completely around their surfaces, range up to 15 tons in weight, with foot lengths up to 9 in and foot areas up to 9 in². The feet, in piercing the soil, do the actual compaction, producing their maximum effect a few inches beneath the ground surface.

Grid rollers. These are intermediate between steel wheel and sheepsfoot rollers, with their rotating wheel made of a heavy wire mesh.

Tampers and rammers. These are small air- or gasoline-driven compaction devices used for densification of soils in confined areas.

Vibrating plate tampers. Usually manually pulled along the surface, these tampers can range up to 2 tons but are generally considerably smaller.

Numerous laboratory and field tests have been conducted on the behavior of granular soils when subjected to vibratory compaction, and the following items give some of the results of those efforts:

1. *Behavior related to granular soil type.* Obviously there is a tremendous variation even within the granular soil category. The ease of compaction seems to be related to the spread between limiting density states and the absolute value of the maximum density. Hilf[9] uses a compactibility function

$$F = \frac{e_{max} - e_{min}}{e_{min}} \tag{4.7}$$

   to evaluate the adaptability of granular soils to compaction. See Table 4.2.

2. *Behavior related to moisture content.* It is not nearly as critical to control moisture content in the compaction of granular soils as it is in the compaction of

Table 4.1 Compaction equipment and methods†

Equipment type	Applicability	Requirements for compaction of 95 to 100 percent standard Proctor maximum density					
		Compacted lift thickness, in	Passes or coverages	Dimensions and weight of equipment			Possible variations in equipment
				Soil type	Foot contact area, in^2	Foot contact pressures, lb/in^2	
Sheepsfoot rollers	For fine-grained soils or dirty, coarse-grained soils with more than 20 percent passing the No. 200 sieve. Not suitable for clean, coarse-grained soils. Particularly appropriate for compaction of impervious zone for earth dam or linings where bonding of lifts is important	6	4–6 passes for fine grained soil; 6–8 passes for coarse-grained soil	Fine grained soil $PI >$ 30‡ Fine-grained soil $PI <$ 30 Coarse-grained soil Efficient compaction of soils wet of optimum requires less contact pressures than the same soils at lower moisture contents.	5–12 7–14 10–14	250–500 200–400 150–250	For earth dam, highway, and airfield work, drum of 60-in diameter, loaded to 1.5 to 3 tons per lineal foot of drum, generally is utilized. For smaller projects 40-in diameter drum, loaded to 0.75 to 1.75 tons per lineal foot of drum, is used. Foot contact pressure should be regulated so as to avoid shearing the soil on the third or fourth pass.
Rubber tire rollers	For clean, coarse-grained soils with 4 to 8 percent passing the No. 200 sieve. For fine-grained soils or well-graded, dirty, coarse-grained soils with more than 8 percent passing the No. 200 sieve.	10 6–8	3–5 coverages 4–6 coverages	Tire inflation pressures of 60–80 lb/in^2 for clean, granular material or base course and subgrade compaction. Wheel load 18,000–25,000 lb. Tire inflation pressures in excess of 65 lb/in^2 for fine-grained soils of high plasticity. For uniform clean sands or silty fine sands, use large-size tires with pressures of 40–50 lb/in^2.			Wide variety of rubber-tire compaction equipment is available. For cohesive soils, light-wheel loads, such as provided by wobble-wheel equipment, may be substituted for heavy-wheel load if lift thickness is decreased. For cohesionless soils, large-size tires are desirable to avoid shear and rutting.

Smooth wheel rollers	Appropriate for subgrade or base course compaction of well-graded sand-gravel mixtures.	8–12	4 coverages	Tandem-type rollers for base course or subgrade compaction, 10–15 ton weight, 300–500 lb per lineal inch of width of rear roller.	3-wheel rollers obtainable in wide range of sizes. 2-wheel tandem rollers are available in the range of 1–20 ton weight. 3–axle tandem rollers are generally used in the range of 10–20 ton weight. Very heavy rollers are used for proof rolling of subgrade or base course.
	May be used for fine-grained soils other than in earth dams. Not suitable for clean, well-graded sands or silty uniform sands.	6–8	6 coverages	3-wheel roller for compaction of fine-grained soil; weights of 5–6 tons for materials of low plasticity, to 10 tons for materials of high plasticity.	
Vibrating baseplate compactors	For coarse-grained soils with less than about 12 percent passing No. 200 sieve. Best suited for materials with 4–8 percent passing No. 200 sieve, placed thoroughly wet.	8–10	3 coverages	Single pads or plates should weigh no less than 200 lb. May be used in tandem where working space is available. For clean, coarse-grained soil, vibration frequency should be no less than 1600 cycles/min.	Vibrating pads or plates are available, hand-propelled or self-propelled single or in gangs, with width of coverage of $1\frac{1}{2}$–15 ft. Various types of vibrating-drum equipment should be considered for compaction in large areas.
Crawler tractor	Best suited for coarse-grained soils with less than 4–8 percent passing No. 200 sieve, placed thoroughly wet.	10–12	3–4 coverages	No smaller than D8 tractor with blade, 34,500-lb weight, for high compaction.	Tractor weights up to 60,000 lb.
Power tamper or rammer	For difficult access, trench backfill. Suitable for all inorganic soils.	4–6 in for silt or clay, 6 in for coarse-grained soils.	2 coverages	30-lb minimum weight. Considerable range is tolerable, depending on materials and conditions.	Weights up to 250 lb, foot diameter 4–10 in.

†After NAVDOCKS, Ref. 7.

‡*PI* = plasticity index = liquid limit − plastic limit; see Appendix A.

Table 4.2 Compactibility F of cohesionless soils†

Classification	γ_{min}	γ_{max}	e_{min}	e_{max}	Maximum size	D_{10}‡	C_u‡	C_c‡	F§
SP-SM	90	108	0.54	0.84	#16	0.058	6.0	2.2	0.555
SM	75	97	0.83	1.36	$\frac{3}{4}$ in	0.0065	31	5.5	0.638
SP	92	112	0.48	0.80	#4	0.15	3.0	0.93	0.667
SP	93	113	0.46	0.77	$1\frac{1}{2}$ in	0.16	2.4	0.92	0.674
SP	95	116	0.43	0.74	#4	0.30	3.7	1.0	0.721
SP-SM	92	113	0.46	0.80	$\frac{3}{4}$ in	0.08	3.0	0.88	0.739
SP	85	107	0.54	0.94	#30	0.10	2.3	1.3	0.740
SP	97	118	0.40	0.70	$1\frac{1}{2}$ in	0.11	3.2	1.2	0.750
SP	99	120	0.38	0.67	$1\frac{1}{2}$ in	1.8	4.4	0.76	0.763
SM-ML	83	108	0.62	1.11	#4	0.012	8.3	1.5	0.790
SP-SM	79	103	0.60	1.08	#30	0.09	2.4	1.5	0.800
SP	103	124	0.33	0.60	$\frac{3}{8}$ in	0.17	5.0	0.75	0.818
SM	105	126	0.31	0.54	5 in	0.02	350	0.30	0.838
SP-SM	87	112	0.48	0.90	#4	0.08	3.0	1.3	0.875
SM	82	108	0.54	1.02	#16	0.023	6.5	1.4	0.889
SW-SM	95	119	0.39	0.74	3 in	0.05	10	1.4	0.897
SP	98	122	0.36	0.69	#4	0.37	5.1	1.2	0.917
SW-SM	98	125	0.34	0.71	3 in	0.07	6.8	1.0	1.088
SP-SM	97	124	0.33	0.70	$\frac{3}{4}$ in	0.10	5.0	1.4	1.121
SP-SM	84	115	0.44	0.97	$1\frac{1}{2}$ in	0.085	4.7	1.4	1.205
SP-SM	94	123	0.34	0.76	$1\frac{1}{2}$ in	0.12	4.4	1.3	1.235
SM	99	128	0.31	0.70	3 in	0.02	240	1.8	1.258
SP-SM	80	114	0.44	1.06	#16	0.07	3.7	1.6	1.409
SW-SM	80	116	0.42	1.07	$1\frac{1}{2}$ in	0.074	6.6	2.4	1.547
SM	83	120	0.38	0.99	#4	0.015	26	6.1	1.605
SM	102	134	0.23	0.62	$\frac{3}{4}$ in	0.01	120	1.9	1.695
GN-GM	113	127	0.31	0.47	3 in	0.14	86	1.2	0.517
GP-GM	112	129	0.32	0.52	3 in	0.03	200	0.50	0.625

GW-GM	116	133	0.26	0.44	5 in	0.17	171	2.2	0.692
GP-GM	110	128	0.30	0.51	3 in	0.11	191	15	0.700
GP-GM	117	133	0.24	0.41	5 in	0.125	160	4.0	0.708
GW-GP	111	130	0.27	0.49	3 in	0.20	105	7.5	0.815
GP	116	134	0.23	0.43	5 in	0.27	111	6.2	0.870
GW	119	139	0.24	0.45	3 in	0.51	45	2.2	0.875
GW	120	139	0.20	0.39	3 in	0.45	51	1.6	0.950
GW	119	139	0.21	0.41	3 in	0.18	94	1.1	0.952
GW	111	132	0.25	0.49	3 in	2.9	9.7	1.8	0.960
GP	115	136	0.22	0.44	5 in	0.38	29	0.61	1.000
GP	114	135	0.22	0.45	3 in	2.0	11	0.77	1.045
GW-GM	121	141	0.19	0.39	3 in	0.30	77	2.3	1.052
GM	122	141	0.17	0.36	$1\frac{1}{2}$ in	0.025	381	3.0	1.118
GW-GM	114	137	0.21	0.45	3 in	0.60	16	1.2	1.143
GW	112	138	0.20	0.48	3 in	2.0	12	1.3	1.400
GW	109	137	0.21	0.52	3 in	2.0	14	2.6	1.476
GP	114	140	0.18	0.45	3 in	1.7	10	0.76	1.500
GM	101	132	0.25	0.64	$1\frac{1}{2}$ in	0.03	260	12	1.560
GW-GM	111	139	0.19	0.49	3 in	1.8	13	2.3	1.578
GP	115	142	0.17	0.44	3 in	0.31	87	8.2	1.588
GW	123	146	0.13	0.34	3 in	0.21	124	1.1	1.615
GW-GM	110	139	0.19	0.50	5 in	0.42	43	2.1	1.631
GW-GM	115	142	0.17	0.45	3 in	0.15	133	1.1	1.647
GP-GM	112	140	0.18	0.48	3 in	0.42	26	4.2	1.667
GW-GM	112	140	0.18	0.48	5 in	0.25	56	1.0	1.667
GW-GM	114	142	0.16	0.45	3 in	1.2	15	1.7	1.812
GP	112	141	0.17	0.48	3 in	1.4	7.1	0.73	1.823
GW-GM	118	147	0.12	0.40	3 in	1.3	19	1.1	2.333

†After Hilf, Ref. 9.

‡D_{10} = effective size; $C_u = D_{60}/D_{10}$; $C_c = \dfrac{(D_{30})^2}{D_{10} \times D_{60}}$; see Appendix A.

§$F = (e_{max} - e_{min})/e_{min}$.

cohesive soils. Figure 4.3 shows the response curves typical of each soil group. It can be seen that the unit weight of granular soils decreases initially (surface tension forces are not easily overcome) until beyond 50-percent saturation, at which point the soils become relatively insensitive to water content variations and, in fact, return to their unit weight value when dry. A completely different phenomenon is seen with cohesive soils. Here massive particle reorientation occurs during the initial stage (up to the optimum water content), leading to a large density increase, followed by a density decrease as more water is added and the soil particles move further apart.

3. *Behavior related to lift thickness.* For average compaction equipment, the most efficient compaction occurs at a depth of about 2 ft, which is the greatest depth at which zero effective stress occurs during rebound of the soil. See Fig. 4.4*a* for results from D'Appolonia et al.[10] Thus 2 ft should be the absolute maximum lift thickness for granular soils being placed in a fill. More typically, lift thicknesses are 10 to 15 in.

4. *Behavior related to number of roller passes.* The same study by D'Appolonia et al. (Ref. 10) evaluated the effects of the number of roller passes on densification. Figure 4.4*b* shows the effects on dry unit weight for up to 45 passes. Both density and depth of maximim density increase, but the latter to only a minor degree.

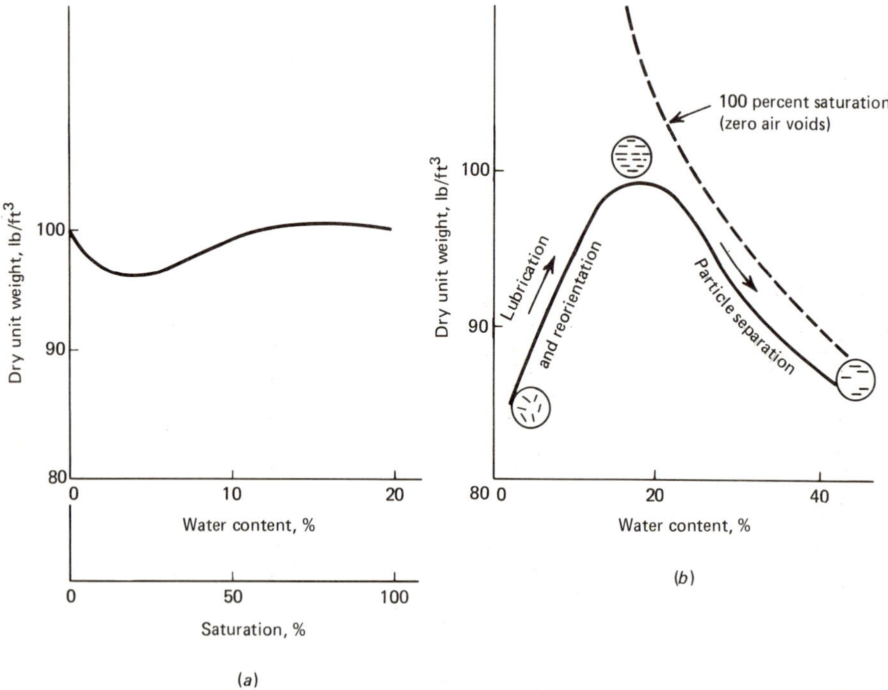

Figure 4.3 Comparison of water content versus dry unit weight responses for granular and cohesive soils; typical values are illustrated. (*a*) Compaction response for a well-graded quartz sand. (*b*) Compaction response for a medium-plasticity Kaolinite clay.

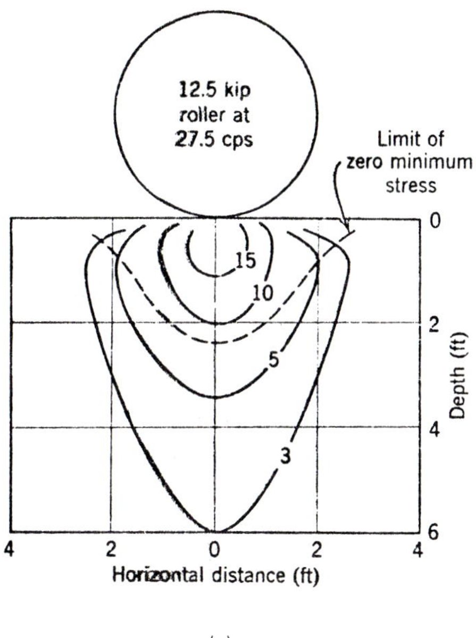

(a)

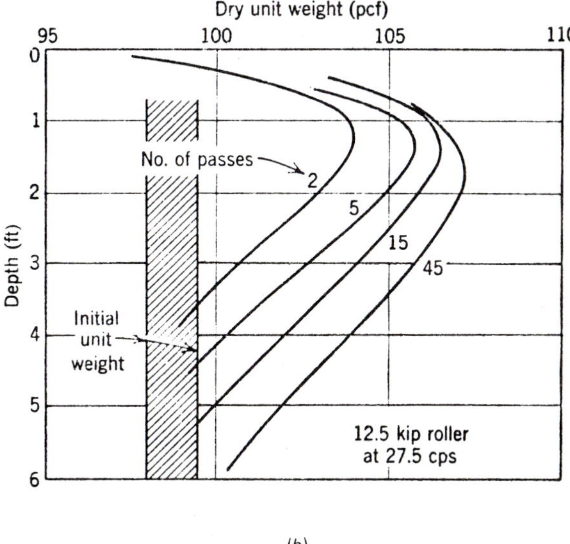

(b)

Figure 4.4 Field results from surface compaction tests. (*a*) Contours of maximum vertical dynamic stress beneath vibratory roller. (*b*) Densification by vibratory roller. *(After D'Appolonia et al., Ref. 10.)*

5. *Behavior related to heavy compactors.* Using a 50-kip vibratory roller (approximately four times the weight of the roller that produced the results shown in Fig. 4.4) which achieves a centrifugal force of 60 kips at 25 Hz, Moorehouse and Baker[11] conducted a study to evaluate the maximum depth of sand densification. As shown by the curves of Fig. 4.5a, significant relative density increases

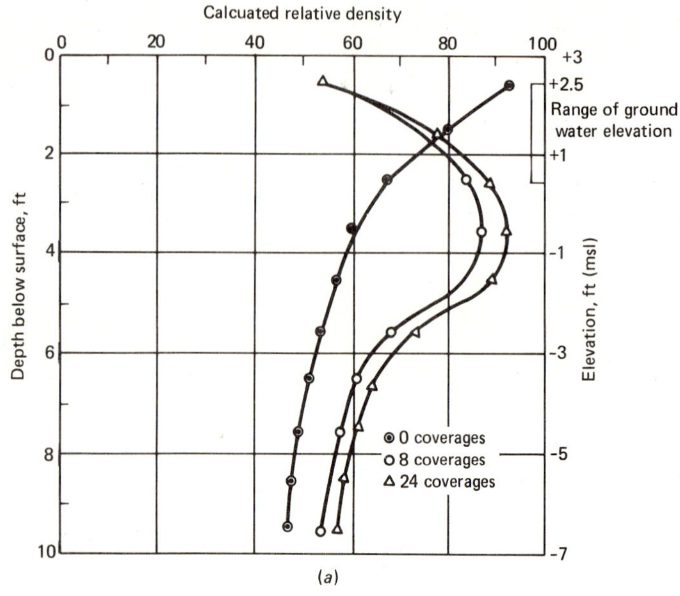

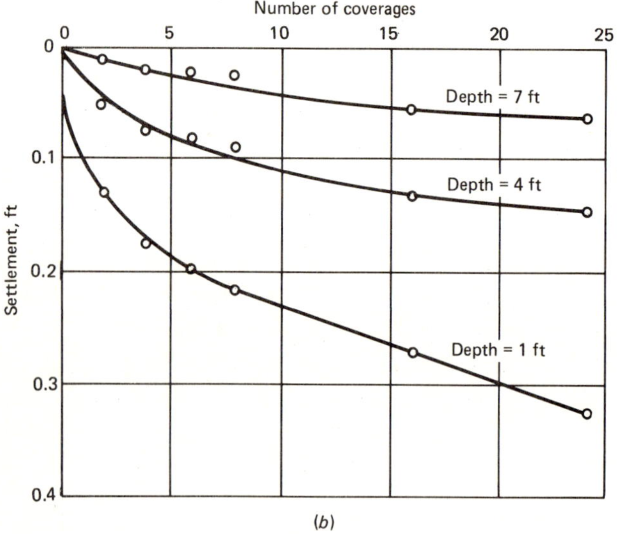

Figure 4.5 Field results from surface compaction tests. (*a*) Calculated relative density versus depth for 0, 8, and 24 coverages. (*b*) Settlement as a function of number of coverages. *(After Moorehouse and Baker, Ref. 11.)*

222

occurred at a depth of 4 ft with eight roller passes, and the influence was felt as deep as 7 ft. However, as seen in Fig. 4.5*b*, repeated roller passes beyond eight did little for increased densification at the greater depths. Thus, for all practical purposes, granular soils can be surface-compacted, using large vibratory rollers, to a depth of about 7 ft. Beyond that depth, other techniques are required.

4.3 IMPACT AT THE GROUND SURFACE (DEEP DYNAMIC COMPACTION)

The repeated dropping of a weight onto the ground's surface is one of the simplest and most basic methods of compacting loose soil. The technique was used long ago (see Schultze and Muhs[12]), and it has seen a rebirth recently with the use of heavier weights falling from higher heights. This rebirth can be traced to a series of patents granted in this country and others to L. Menard[13]: In the United States, it is patent number 3,898,844, dated Aug. 12, 1975.[14] The principal claims of the method as currently performed are the following:

1. Depth of compaction can reach 33 to 66 ft (10 to 20 m).
2. All soils (not just granular soils) can be compacted. This includes saturated fine-grained soils which liquefy under the imposed loads, thereby destroying whatever structure is present, and reorient at lower water contents and in a more stable manner.
3. The method produces equal settlements more quickly than do static (surcharge-type) loads.

The method, known as *deep dynamic compaction* or *deep dynamic consolidation,* uses a crane to lift a concrete or steel block, weighing up to 50 tons, up to 150 ft high, from which height it is allowed to fall freely back down to ground surface. Needless to say, it leaves its mark behind! The process is then repeated either at the same location or sequentially over other parts of the area to be stabilized. When the required number of repetitions is completed over the entire area, the compaction at depth is completed. The soils near the surface, however, are in a greatly disturbed condition. The upper 3 to 6 ft is then compacted in the same manner except that the weight used is somewhat lighter, has a greater area (and thus a lower stress), and is dropped over a smaller distance (which gives it a lower energy). Major variables in the process are:

Magnitude of the weight
Size (surface area) of the weight
Height of free fall of the weight
Number of drops per location
Distribution of drop locations over the site, i.e., continuously or only at specific spots
Nonhomogeneity of the soil
Strength and permeability anisotropy of the soil
Degree of saturation of the soil

While it would be ideal if a rigorous model of the process were available, taking into account all the variables mentioned, such is not the case. Only the rudimentary beginnings of an understanding of the process are available. These are that the weight, when it strikes the surface, produces P, S, and R waves. The P waves cause direct compression of the soil skeleton and (if the soil is at or near maximum saturation) induce pore water pressure. The S waves mobilize shear forces in the particles and also cause a densification. The R waves, and other surface waves, are potentially harmful in loosening surface soils and damaging adjacent structures. Thus it is seen that the P and S waves (the S waves travel at about half the speed of the P waves) do the work of compaction and produce a stress which has been estimated on the basis of elasticity theory. Scott and Pearce[15] have examined the problem in regard to both elastic and elasto-plastic models, and their results are shown in Fig. 4.6a and b. The rapidity of the entire process in achieving both maximum stress mobilization and mobilization of surface deflection can clearly be seen.

With regard to the depth of penetration of the compaction, only empirical information is available. The effective depth is related to the energy of the impact, and appears (see Fig. 4.7) to be within the following range[16]:

$$0.4 \sqrt{wh} < D < \sqrt{wh} \tag{4.8}$$

where D = effective depth, m
     w = weight being dropped, ton
     h = height of drop, m

Seeing that an analytic understanding of the situation is at a very formative stage, all deep dynamic compaction projects should have an initial pilot program and then be closely monitored during their construction. It is highly recommended that a small test section, at the site under consideration, be field-evaluated first. At this time instrumentation can also be evaluated and realistic goals set for the final densified soil mass. The actual work can be assessed during construction or afterward. A few case histories on different soil types will illustrate some of the assessment methods available:

Leonards et. al.[17] describe the deep dynamic compaction of a fine to medium sand where the cone penetration test was used to evaluate the soil's improved properties.

Charles et. al.[18] used deep dynamic compaction on a mine tailings site and assessed its performance with surface settlement monitoring and pore water pressure measurements.

Ramaswamy et. al.[19] compacted a saturated peaty clay and evaluated its performance by taking undisturbed samples before and after compaction and performing various tests on their physical, hydraulic, and mechanical properties.

Welsh[20] describes a number of case histories (including collapsing silts and sanitary landfills) where surface settlement monitoring, dynamic cone penetration, or standard penetration resistance methods were used for assessment purposes.

All of the above references cite evidence of significant improvement in soil properties after deep dynamic compaction. Indeed, it seems that this very basic and intu-

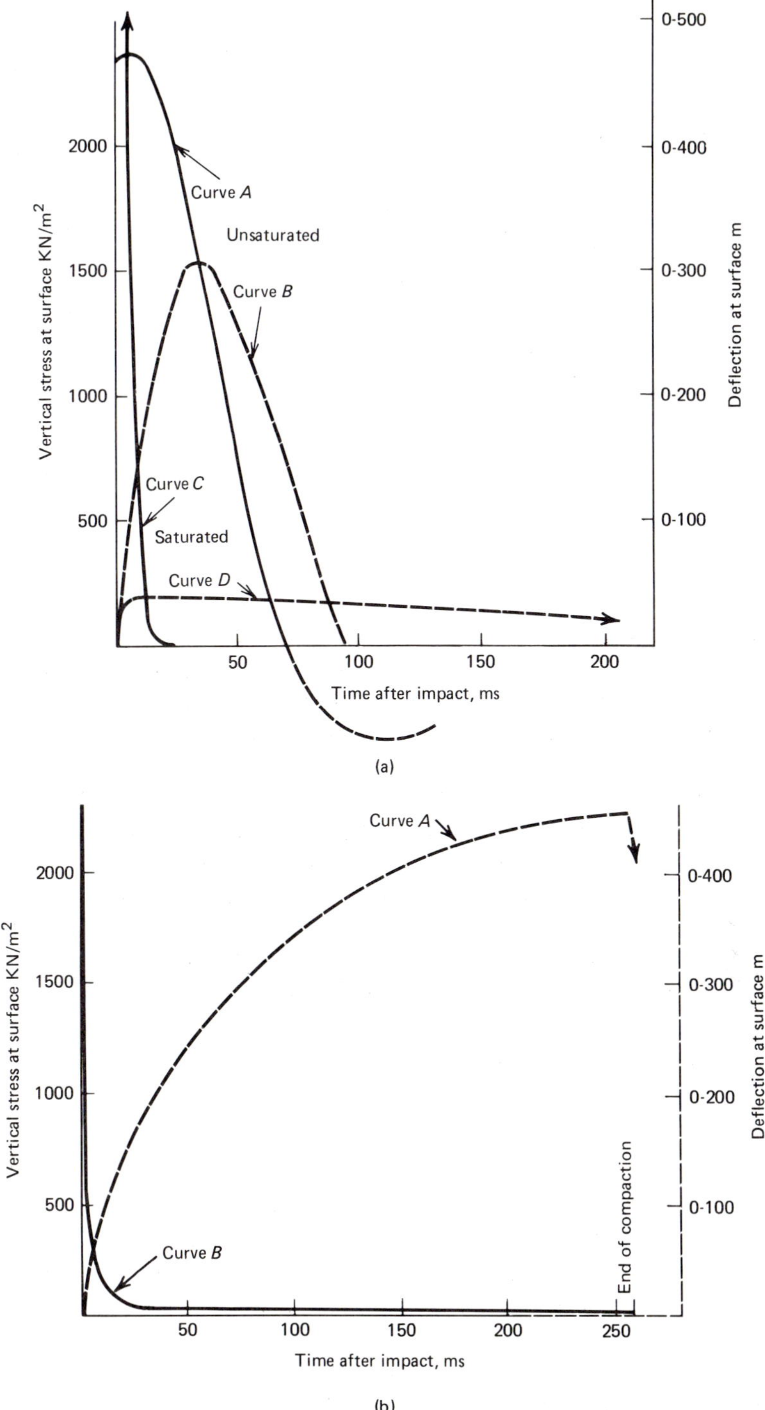

Figure 4.6 Stress and movement at the impact surface for elastic and elasto-plastic models. (*a*) Stress (*A* and *C*) and movement (*B* and *D*) at the impact surface; elastic model. (*b*) Stress (*B*) and movement (*A*) at the impact surface; one-dimensional elasto-plastic model. (*After Scott and Pearce, Ref. 14.*)

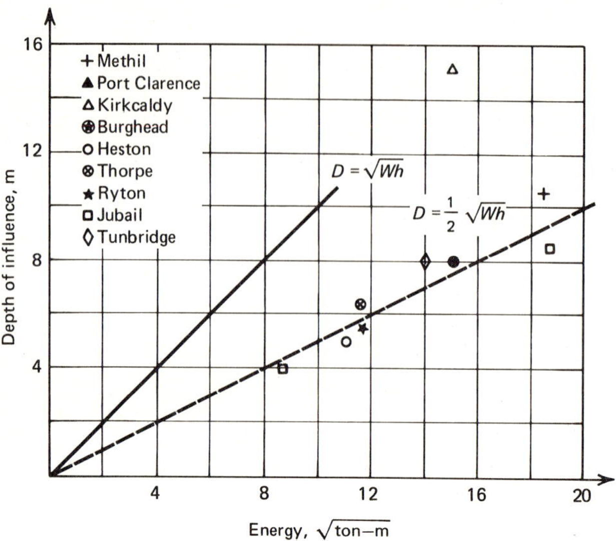

Figure 4.7 Comparison between depth of influence and delivered energy for deep dynamic compaction. *(After Dobson and Slocombe, Ref. 16.)*

itive method has been underutilized by the construction and geotechnical communities in the past. It certainly seems to be a cost-effective method for the improvement of a wide variety of soils in a wide variety of subsurface conditions.

4.4 VIBRATION AT DEPTH

One of the inherent problems associated with impacting surface compaction methods is ground vibration. (This comes about mainly from the R waves previously mentioned, which account for 67 percent of the total energy generated by the impacts.[21]) Usually measured on the basis of peak particle velocity at the point in question, the maximum level of vibration that is generally considered safe for structures is 2.0 in/s (see Bulletin 656 of the U.S. Bureau of Mines entitled "Blasting Vibrations and Their Effects on Structures," 1971). Shown in Fig. 4.8 is the approximate response of peak particle velocity to distance away from impact for deep dynamic compaction. In general, if a structure is located within 100 ft of the work site, an alternative method should be considered. Such alternatives are vibrocompaction and sand compaction piles, which we discuss in this section.

Vibrocompaction or *Vibroflotation*® is a depth technique used for the in situ densification of granular soils. As generally implemented, the procedure is to jet a probe, measuring about 18 in in diameter and 5 to 10 ft long, to the deepest portion of the ground to be densified. A built-in eccentric weight is then rotated around the central shaft of the probe. This compacts the soil radially. Granular soil backfill is added from the ground's surface to fill the anular space created by the compaction process. The probe is then slowly lifted toward the surface, compacting the soil strata as it

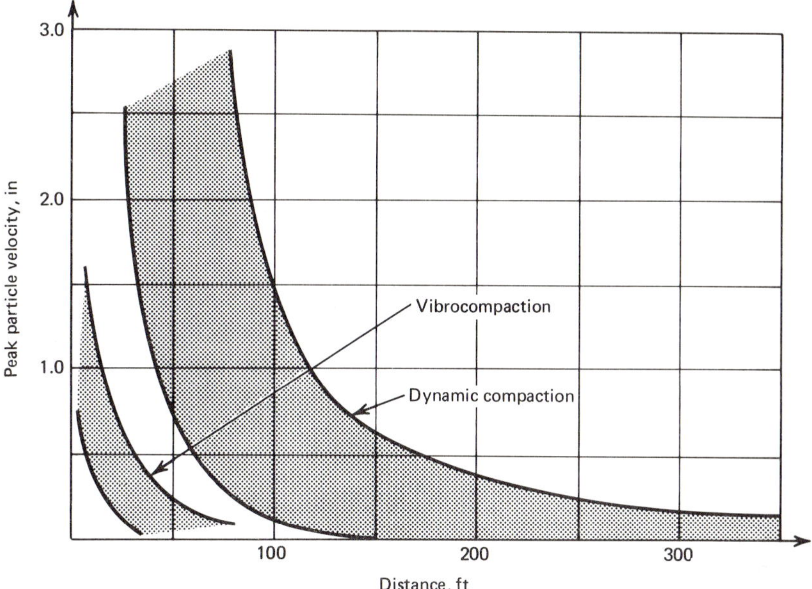

Figure 4.8 Comparison of general vibration levels induced by deep dynamic compaction and vibrocompaction. *(After Dobson and Slocombe, Ref. 16.)*

moves from the bottom upward. The cylinder of soil formed below the probe by its vibrating action is densified by periodically dropping the probe onto it as it proceeds out of the hole. The entire process is shown schematically in Fig. 4.9.

Vibrocompaction is controlled via the power consumption of the vibrator within the probe (reaching maximum current consumption means the soil is as dense as it can be under the particular conditions at a given site) and by the care that the operator gives to withdrawing from and backfilling the central portion of the hole. Withdrawal rates of 1 to 2 ft/min are customary. This last point is of concern since the density of the backfill zone must be assured. See Metzger and Koerner[22] for model test results which address this potential problem.

A significant project illustrating the use of vibrocompaction was the densification of approximately 13 m of ocean sand beneath 20 m of water during construction of the Oosterschelde Storm Surge Barrier in Holland.[23] A specially designed ocean-going barge, measuring $68 \times 33 \times 5.5$ m, was constructed to handle four separate probes which could be operated simultaneously. A typical response of the sand to densification is shown in Fig. 4.10, where both porosity decreases and cone penetration resistance increases can easily be seen.

The technique has been used for many projects, and the following items of interest have been discovered:

For underwater projects, the water jetted from the bottom of the probe during its placement causes soil liquefaction, which allows the probe to penetrate to its desired depth. The water is then redirected to the top of the probe where it aids

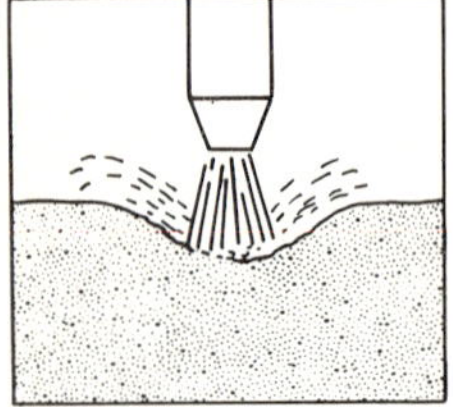

At start, lower jet is opened full.

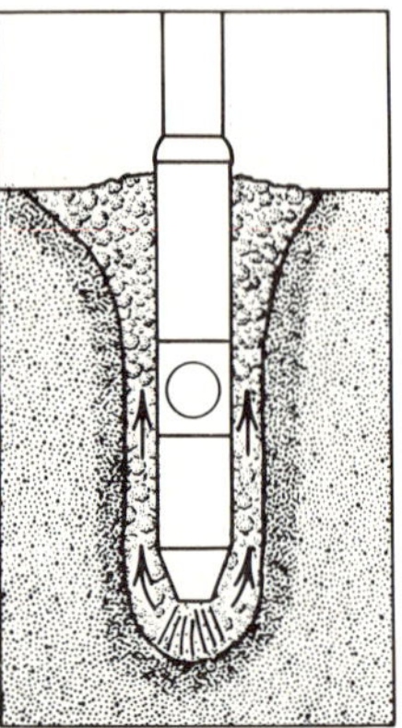

Water is introduced more rapidly than it can drain away. This creates a momentary "quick" condition ahead of the equipment which permits the vibrating machine to settle of its own weight to the desired depth.

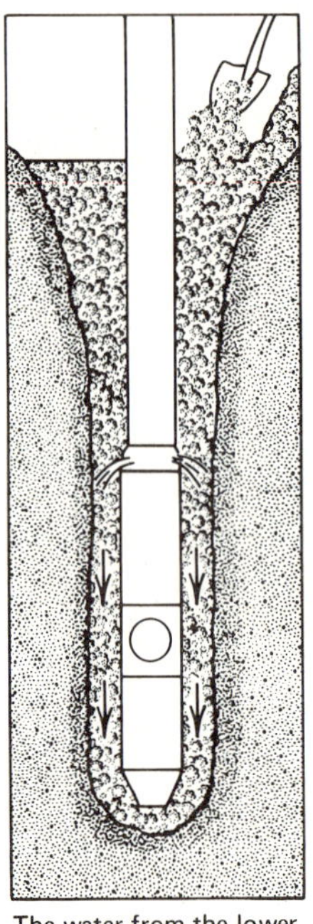

The water from the lower jet is transferred to the top jets, and the pressure and volume are reduced just enough to carry the sand to the bottom of the hole.

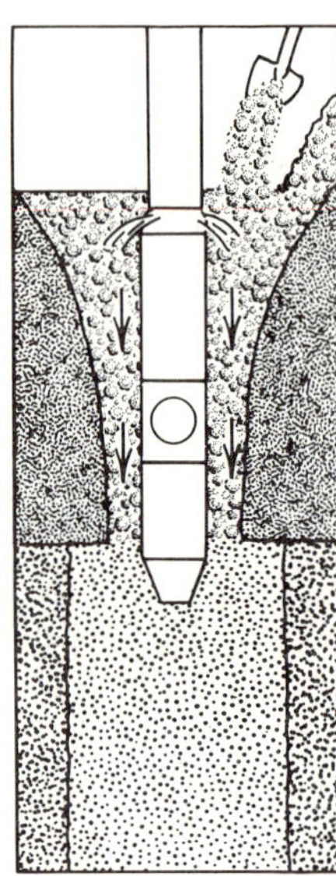

Actual compaction takes place during the intervals between the one-foot lifts which are made in returning the Vibroflot to the surface. The vibrator is first allowed to operate at the bottom of the crater until the desired density around the lower part of the machine is attained. By raising the vibrator step by step and simultaneously backfilling, the entire depth of soil is compacted.

Figure 4.9 Schematic diagram of vibrocompaction or vibroflotation densification of granular soils. *(Compliments of Vibroflotation Corporation, Pittsburg, PA.)*

in holding the sides of the hole open and in "puddling" the backfilling sand beneath the probe.

The radial vibratory motion has vertical components as well, which contribute to densification of the backfilling soil.

Ideal soils for the technique are cohesionless sands and gravels with a maximum of 10 percent of their particles passing the No. 200 sieve size (0.074 mm).

Maximum relative density attained is usually in the 80 to 90 percent range.

Maximum influence varies from about 7 ft in clean sands to 3 ft in sands containing more than 20 percent fines. Location of probe spacings should be adjusted accordingly.

The option of using the technique in fine-grained soils, where stone aggregate is used for backfill, is available. In this case, structural loads can be carried as with

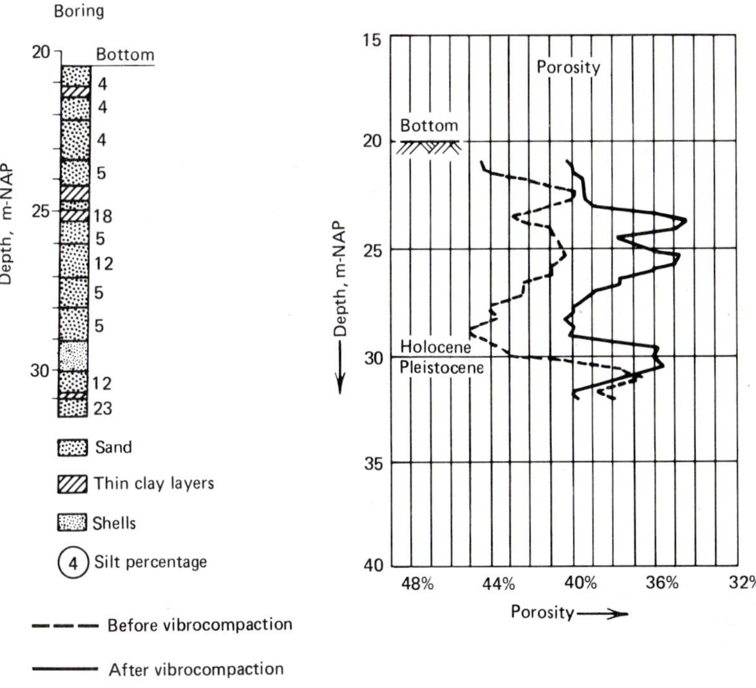

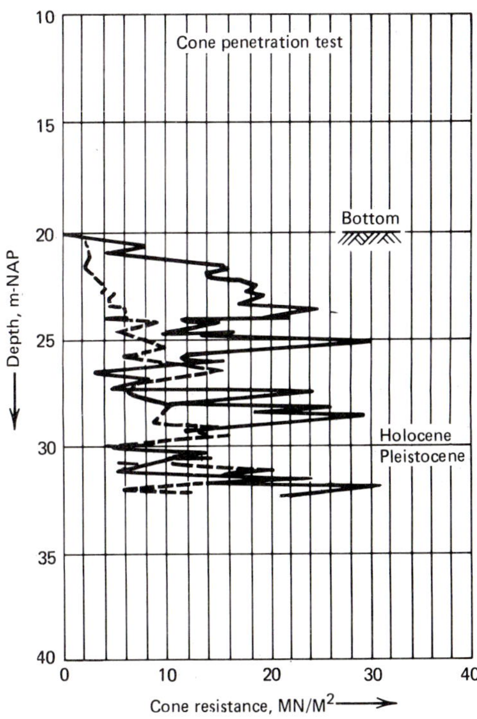

Figure 4.10 Porosity decrease and cone penetration resistance increases due to vibrocompaction of ocean sand. *(After Davis et al., Ref. 23.)*

standard deep foundation systems. The technique has also been used to stabilize slopes (see Mitchell and Katti for a complete review of the process[24]).

Densification results should be verified by some type of continuous monitoring, e.g., by cone penetration resistance or standard penetration resistance. Discrete monitoring has also been reported. The latter includes downhole nuclear methods and pressuremeter tests.

Locations for monitoring should be equidistant between vibrocompaction probe spots as well as directly at their centers.

Field inspection during construction is necessary. The inspection should record the ammeter movement during compaction, the type and amount of backfill sand used, and the surface movement that occurs.

The second technique to be discussed in this section is *sand compaction piles*. As the name implies, it consists of driving a hollow steel pipe with a hinged bottom plate down to the desired depth, filling it with sand, and withdrawing the pipe while air pressure is directed against the sand inside it. The hinged bottom plate opens during withdrawal and the sand backfills the void created earlier during the driving of the pipe. The in situ soil is densified while the pipe is being driven down, and the sand backfill prevents the soil surrounding the compaction pile from collapsing as the pipe is withdrawn. An interesting set of data, provided by Gupta[25] and shown in Fig. 4.11, illustrates the strong response of relative density to both the spacing and the size of the compaction piles.

An interesting variation of the above technique is the vibratory driving of a

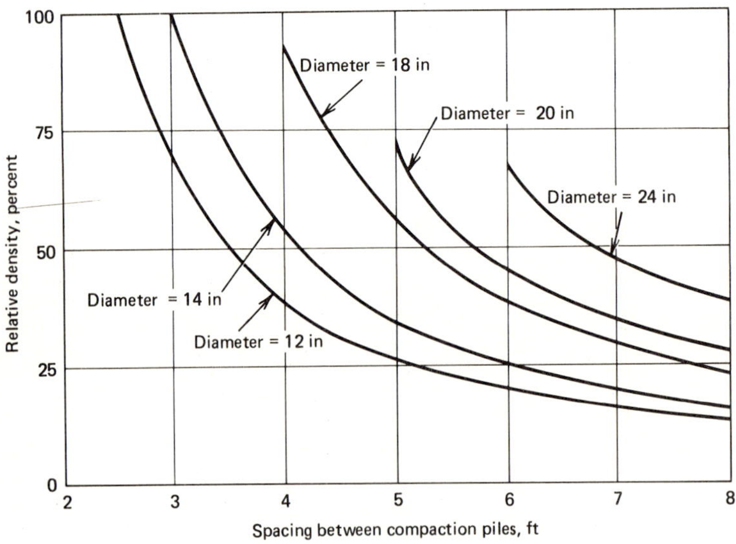

Figure 4.11 Effects of size and spacing of sand compaction piles on relative density of in situ soil. *(After Gupta, Ref. 25.)*

large-diameter hollow steel pipe with an open end. This method, called *Terra-Probe,*®
is described by Janes[26] and Anderson.[27] They use a 30-in-diameter pile vibrated in
"clean" sands or gravels. The maximum limits on the amount of fines that can be
present are 15 percent passing the No. 200 sieve (0.074 mm) and 3 percent passing
0.005 mm. These limits must be strictly adhered to, since the functioning of the
technique depends upon a lack of cohesion which allows the sand particles to freely
orient themselves both during penetration of the pipe and during its withdrawal. The
large diameter is also very important because no arching must occur within the pipe
during withdrawal inasmuch as it would cause a loosening condition.

A typical setup described used a Foster model 260 vibratory hammer, with a
30-in-diameter steel pipe 45 ft long attached to it. Probe spacings were quite close,
e.g., 3 to 4 ft between centers, which is not too disadvantageous because of the rapidity of the insertion-withdrawal cycle. One cycle of compaction took only 3 min to
perform.

As with vibrocompaction, field monitoring and job control can be carried out by
continuous or discrete testing before and after compaction, or by noting surface subsidence after densification. Since no additional backfill material is added by this particular technique, surface settlement is an excellent indicator of the amount of densification that has occurred.

4.5 IMPACT AT DEPTH

Blasting, through the use of buried, time-delayed explosive charges, has been used
to densify loose, granular soils. The sands and gravels must be essentially cohesionless with a maximum of 15 percent of their particles passing the No. 200 sieve size
and 3 percent passing 0.005-mm size. The moisture condition of the soil is also
important, for surface tension forces in the partially saturated state limit the effectiveness of the technique. Thus the soil, as well as being granular, must be dry or
saturated, which sometimes requires prewetting the site via construction of a dike
and reservoir system.

Using the technique requires careful planning and a relatively remote site. Theoretically, an individual charge densifies the surrounding adjacent soil and the soil
directly beneath the blast. It should not lift the soil situated above the blast, however,
since that upper soil should provide a surcharge load—or to put it another way, the
charge should not crater the soil.[28] Charge delays should be timed to explode from
the bottom of the layer being densified upward in a uniform manner. The uppermost
portion of the stratum is always loosened, but this can be surface-compacted by
vibratory rollers as previously described.

Information about the results of the technique is limited, but some data of a
quantitative nature are available. Hall[29] found that repeated blasts of small charges
are the most efficient way of achieving the desired results (see Fig. 4.12). Kummeneje and Eide[30] investigated the effects of successive blasts on settlement; their results
are shown in Fig. 4.13. In general, it is thought that the technique produces few
additional effects after three successive blasts in the same vicinity.

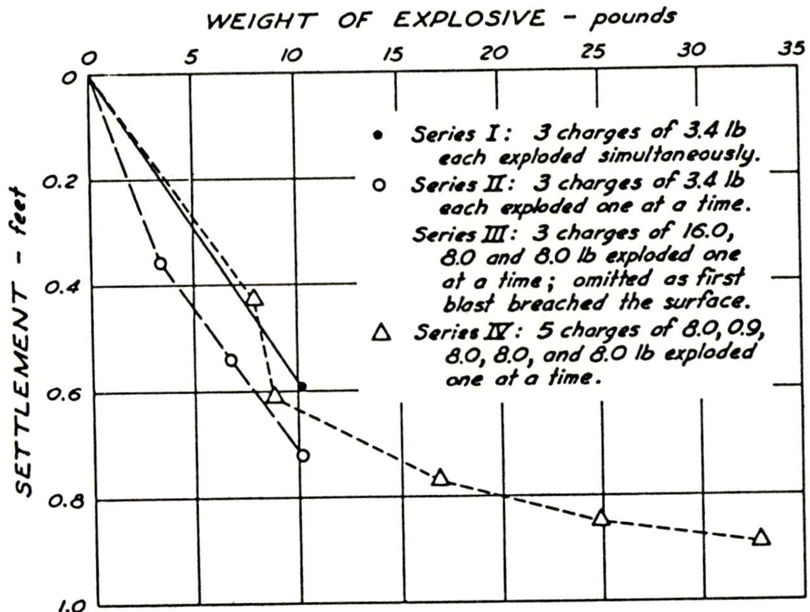

Figure 4.12 Results of field test using explosives as a soil compaction technique. *(After Hall, Ref. 29.)*

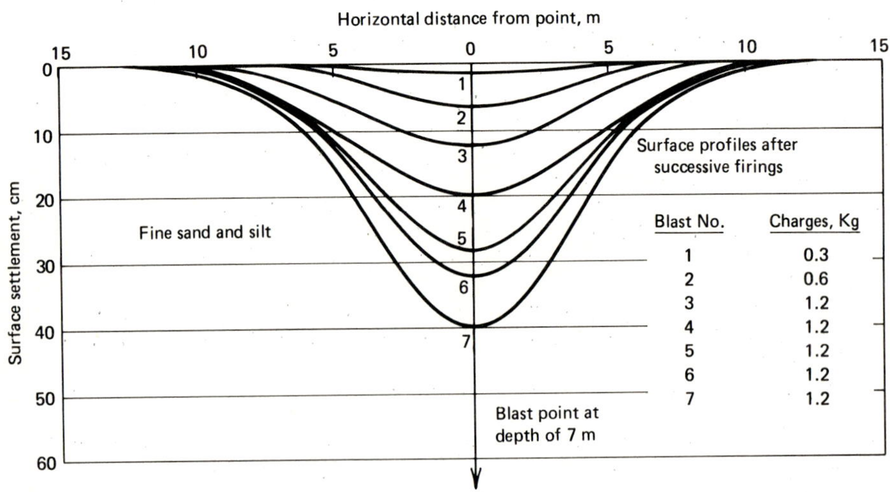

Figure 4.13 Results of field tests using explosives to assess the influence of successive firings on settlement. *(After Kummeneje and Eide, Ref. 30.)*

232

Lyman[31] has empirically related charge weight and sphere of influence as follows:

$$w = CR^3$$

where w = weight of explosive, lb

R = radius of sphere of influence, ft

C = constant (= 0.0025 for 60 percent dynamite)

The depth of the charge should be greater than R to avoid cratering. Charge sizes are from 0.5 to 8 lb (a typical explosive consists of 60 percent dynamite, 30 percent gelatine dynamite, and 10 percent ammonite), and three to five separate blasts are often used. A typical firing pattern is given by Mitchell[32] and is shown in Fig. 4.14.

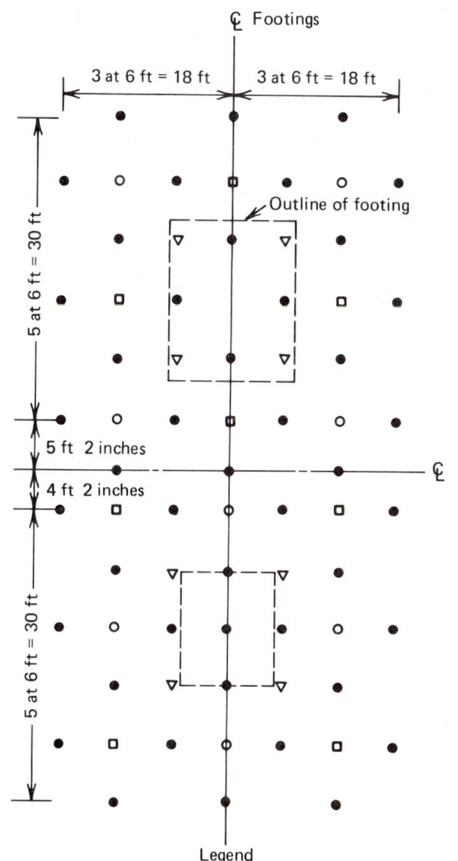

Figure 4.14 Typical firing plan for sand densification for spread footings. *(After Mitchell, Ref. 32.)*

In remote areas, where vibrations are of no concern, and under proper soil conditions, the technique can be most cost-effective. A specialty contractor with experience in the use of blasting to densify soils should perform the work. This technique, perhaps more than others described in this book, is not one for amateurs to engage in.

PROBLEMS

4.1. Prove that Eq. (4.1) is equal to Eq. (4.2), that is, prove

$$\frac{e_{max} - e}{e_{max} - e_{min}} = \frac{\gamma - \gamma_{min}}{\gamma_{max} - \gamma_{min}} \frac{\gamma_{max}}{\gamma}$$

4.2. Consult the ASTM book on standards, or other references on soils, and briefly describe:
    (*a*) The test for maximum density (from which γ_{max} and e_{min} are obtained)
    (*b*) the test for minimum density (from which γ_{min} and e_{max} are obtained)

4.3. Is it possible to have calculations showing that $D_R > 100$ percent? Or that $D_R < 0$ percent? If so, explain why.

4.4. The two soil properties most commonly determined in the field are its density and water content. Briefly describe the field tests for determining in situ density (sand cone, rubber-balloon, nuclear and drive cylinder methods) and water content (gravametric and nuclear methods). What is Soiltest's "speedy moisture tester?"

4.5. (*a*) Describe and illustrate how the sphericity of soil particles is measured and quantified. (Many geology texts have identification charts for sphericity.)
    (*b*) Define and illustrate the difference between a soil particle's sphericity and its angularity.

4.6. Determine γ_{max} and γ_{min} for each of the soil types given in the accompanying table using the empirical relationships of Eqs. (4.3) and (4.4).

Problem	Soil	D_{10} (mm)	C_u	S
(*a*)	Ottawa sand	0.42	2.0	0.70
(*b*)	Beach sand	0.21	8.0	0.45
(*c*)	River silt	0.035	20.	0.55
(*d*)	Bank gravel	2.1	6.5	0.64
(*e*)	Concrete sand	0.32	11.5	0.52

   Note: As earlier in the chapter, D_{10} = effective size; C_u = coefficient of uniformity ($= D_{60}/D_{10}$); S = sphericity ($= 1$ for a circle, $= 0$ for a plate on end).

4.7. (*a*) For the values of γ_{max} and γ_{min} obtained in Prob. 6, calculate the respective values of e_{min} and e_{max} based on a specific gravity of solids of 2.65.
    (*b*) Use the values found in part *a* to calculate a compactibility function F for each of the five soils [see Eq. (4.7)], and rank the soils accordingly.

4.8. The moisture sensitivity of granular soils, as far as dry unit weight is concerned, was shown to be relatively slight (recall Fig. 4.3*a*). This is not the case for their strength, where ϕ in the saturation range of 5 to 50 percent increases significantly over its value in dry and fully saturated granular soils. Explain the difference in this behavior.

4.9. Regarding the surface compaction of granular soils:

(*a*) Why is the use of vibratory rollers preferred?

(*b*) Why are sheepsfoot rollers essentially useless?

(*c*) Compare and contrast from a design point of view the differences between rollers used in tandem and individual rollers making separate passes.

(*d*) Answer part *c* from the contractor's point of view.

4.10. (*a*) Describe, on a phenomenological basis, how deep dynamic compaction densifies granular soils (use sketches in your discussion).

(*b*) How does water content influence this process? (Consider dry. partly saturated, and fully saturated conditions.)

(*c*) Menard's patent claims that the method can densify saturated, fine-grained soils, i.e., clays. How do you think it does this and what are your comments about the process?

4.11. In the discussion on deep dynamic compaction it was mentioned that the P and S waves do the actual work of compaction. But R waves (called surface or Raleigh-type waves) are also generated. Of what significance are these waves? (In your discussion describe their motion.)

4.12. (*a*) During vibrocompaction or Vibroflotation the main energy is transmitted radially from the vibrating device. How does the sand beneath the vibrator get compacted?

(*b*) What are the major variables involved in propagating this vibrational energy outward beyond the vibrator itself?

(*c*) Using Fig. P4.12, sketch a densification influence diagram for this response.

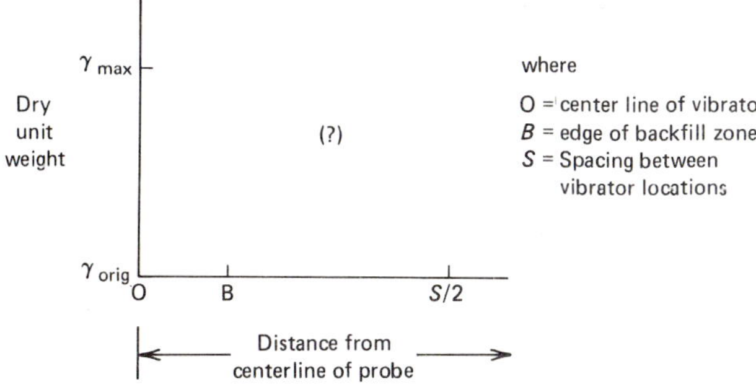

Figure P4.12

4.13. The vibrators used for vibrocompaction or Vibroflotation have also been used in cohesive soils to make a void which is then filled with 4- to 12-in-sized stone that is compacted by the vibrator as the probe is withdrawn and then used as a stone column for direct support of shallow foundations. Comment on this technique. How does it work mechanically? Give your assessment of it.

4.14. Evaluate the techniques of vibrocompaction (Vibroflotation), sand compaction piles, and Terra-Probe compaction according to the items listed in the accompanying table.

Item being evaluated or ranked	Vibrocompaction (Vibroflotation)®	Sand compaction	Terra-Probe®
Sophistication of equipment			
Direction of vibration			
Typical spacings of vibrator location			
Approximate cycle time			
Maximum relative density achievable			

4.15. Compare and contrast the use of explosives to densify granular soils with the methods discussed in Prob. 14.

4.16. (*a*) In discussing the use of explosives to densify granular soils below ground, it was mentioned that cratering is a potential problem. Explain what this phenomenon is and why it is undesirable.

(*b*) Cratering, however, can be a very positive and worthwhile goal. Explain the most obvious situation where this is so.

REFERENCES

1. Burmister, D. M., "The Importance and Practical Use of Relative Density in Soil Mechanics," *ASTM Proc.,* Vol. 48, 1948, pp. 1249–1268.

2. Koerner, R. M., "Limiting Density Behavior for Quartz Powders," *Power Technol.,* Vol. 3, 1969/70, pp. 208–212.

3. Lacroux, Y., and H. M. Horn, "Direct Determination and Indirect Evaluation of Relative Density and its Use on Earthwork Construction Projects," in *Proc. Conf. on Evaluation of Relative Density and its Role in Geotechnical Projects Involving Cohesionless Soils,* ASTM-STP 523, Los Angeles, CA, June 1972, pp. 251–280.

4. Koerner, R. M., "Effect of Particle Characteristics on Soil Strength," *J. Soil Mech. Found. Eng. Div., ASCE,* Vol. 96, No. SM4, July 1970, pp. 1221–1234.

5. Sowers, G. B., and G. F. Sowers, *Introductory Soil Mechanics and Foundations,* 3d ed., Macmillan, New York, 1970.

6. Gibbs, H. J., and W. G. Holtz, "Research on Determining the Density of Sands by Spoon Penetration Testing," in *Proc. Fourth Int. Conf. on Soil Mech. and Found. Eng.,* Vol. 1, London, 1957, pp. 35–39.

7. *Design Manual: Soil Mechanics, Foundations and Earth Structures,* NAVDOCKS DM-7, Dept. of the Navy, Bureau of Yards and Docks, Washington, DC, 1982.

8. Foster, C. R., "Field Problems: Compaction," Chap. 12 in *Foundation Engineering,* G. A. Leonards (ed.), McGraw-Hill, New York. 1962, pp. 1000–1024.

9. Hilf, J. W., "Compacted Fill," Chap. 7 in *Foundation Engineering Handbook,* H. F. Winterkorn and H. Y. Fang (eds.), Van Nostrand Reinhold, New York, 1975, pp. 244–311.

10. D'Appolonia, D. J., R. V. Whitman, and E. D'Appolonia, "Sand Compaction with Vibratory Rollers," in *Proc. ASCE Spec. Conf. on Placement and Improvement of Soil to Support Structures,* New York, 1968, pp. 125–136.

11. Moorehouse, D. C., and G. L. Baker, "Sand Densification by Heavy Vibratory Compaction," *J. Soil Mech. Found. Eng. Div., ASCE,* Vol. 95, No. SM4, July 1969, pp. 370–391.

12. Schultze, E., and H. Muhs, *Bodenuntersuchingen fur Ingenieur-Bauten,* Springer, Berlin, 1967.

13. Menard, L., and Y. Broise, "Theoretical and Practical Aspects of Dynamic Consolidation." *Geotechnique,* Inst. of Civil Engineers, London, Vol. 25, March 1975, pp. 3–17.

14. Method of Compacting Made-up Ground and Natural Soil of Mediocre Quality, U.S. Patent No. 3, 898, 844, August 12, 1975.

15. Scott, R. A., and R. W. Pearce, "Soil Compaction by Impact," in *Ground Treatment by Deep Compaction,* Inst. of Civil Engineers, London, 1976, pp. 19–44.

16. Dobson, T., and B. Slocombe, "Deep Densification of Granular Fills," in *Proc. Second Geotechnical Conf. on Design and Construction,* Las Vegas, NV, April 26–28, 1982, 21 pp.

17. Leonards, G. A., W. A. Cutter, and R. D. Holtz, "Dynamic Compaction of Granular Soils," *Geotech. Eng. Div., ASCE,* Vol. 106, No. GT1, June 1980, pp. 35–43.

18. Charles, J. A., D. Burford, and K. S. Watts, "Field Studies of the Effectiveness of the Ground Treatment Techniques—Dynamic Consolidation," in *Proc. Tenth Int. Conf. on Soil Mech. and Found. Eng..* Stockholm, June 1981, Vol. 3, pp. 617–622.

19. Ramaswamy, S. D., M. A. Aziz, R. V. Subrahamanyam, M. H. Abdul Khoder, and S. L. Lee, "Treatment of Peaty Clay by High Energy Impact," *J. Geotech. Eng. Div., ASCE,* Vol. 105, No. GT8, August 1979, pp. 957–967.

20. Welsh, J. P., "Dynamic Deep Compaction; a Ground Modification Technique," *Ground/Expo '82,* Houston, TX, March 22, 1982.

21. Woods, R. D.. "Screening of Surface Waves in Soils," in *ASCE Proc. Conf. on Placement and Improvement of Soil to Support Structures.* August 26–28, 1968, New York, pp. 407–436.

22. Metzger, G. V.. and R. M. Koerner, "Modeling of Soil Densification by Vibroflotation," *J. Geotech. Eng. Div., ASCE,* Tech. Note, Vol. 101, No. GT4, April 1975, pp. 417–421.

23. Davis, P., H. Nelissen, and A. Pladet, "Mytilus, a Soil Compaction Vessel," in Ref. 18, Vol. 3, pp. 641–644.

24. Mitchell, J. K., and R. K. Katti, "Soil Improvement—State-of-the-Art Report," in Ref. 18, Vol 4, pp. 567–575.

25. Gupta. S. N., "Discussion of 'Sand Densification by Piles and Vibroflotation,' by C. E. Basore and J. D. Boitano," *J. Soil. Mech. Found. Eng. Div.. ASCE,* Vol. 95, No. SM6, November 1969, pp. 224–226.

26. Janes, H. W., "Densification of Sand for Drydock by Terra-Probe," *J. Soil Mech. Found., Eng. Div., ASCE,* Vol. 99, No. SM6, June 1973, pp. 451–470.

27. Anderson, R. D., "New Method for Deep Sand Vibratory Compaction," *J. Constr. Div., ASCE,* Vol. 100, No. CO1, 1974, pp. 79–95.

28. Vesic, A. S., "Cratering by Explosives as an Earth Pressure Problem," in *Proc. Sixth Int. Conf. on Soil Mech, and Found. Eng.,* Montreal, 1965, vol 2, pp. 427–431.

29. Hall, C. E., "Compacting a Dam Foundation by Blasting," *J. Soil Mech. Found, Eng. Div., ASCE,* Vol. 88, No. SM3, June 1962, pp. 33–51.

30. Kummeneje, O., and O. Eide, "Investigations of Loose Sand Deposits by Blasting," in *Proc. Fifth Int. Conf. on Soil Mech. and Found. Eng.,* Vol. 1, Paris, 1961, pp. 491–496.

31. Lyman, A. K. B., "Compaction of Cohesionless Foundation Soils by Explosives," *Trans. ASCE,* Vol. 107, 1942, pp. 1330–1348.

32. Mitchell, J. K., "In-Plane Treatment of Foundation Soils," *J. Soil Mech. Found. Eng. Div., ASCE,* Vol. 96, No. SM1, January 1970, pp. 73–110.

FIVE

IN SITU DENSIFICATION METHODS IN COHESIVE SOILS

5.1 INTRODUCTION

Without question, the single most important feature of cohesive soils is their water content. (Compare this statement with the one that began Chap. 4.) As the water content of a cohesive soil increases, the following occur:

Its strength decreases.
Its compressibility increases.
Its creep deformation increases.
It becomes more difficult to work with in a general sense.
One begins to look for a different site to develop.

The behavior of these soils can be accurately visualized in terms of a graph of water content versus total volume, as shown in Fig. 5.1. Soil properties are most amenable to placement and densification when the soil is in its partially saturated state (recall Fig. 4.3*b*), but this is often found not to be the natural situation. The water content of many soils in their natural state lies above the saturation point, which is defined as the soil's *shrinkage limit*. And this region of saturation is so important that two other *consistency,* or *Atterberg, limits* are also related to it. These are the *plastic limit* and the *liquid limit*. For most engineering purposes the liquid limit, which is the upper value, separates any given soil in its plastic state from its viscous liquid state, the condition in which resistance to even the slightest load is absent. The plastic limit is intermediate between these two extremes. For typical values of these consistency limits see Table 5.1.

Table 5.1 Typical values of consistency limits for various cohesive soils†

Mineral	Exchangeable ion	Liquid limit, %	Plastic limit, %	Plasticity index, %	Shrinkage limit, %
Montmorillonite	Na	710	54	656	9.9
	K	660	98	562	9.3
	Ca	510	81	429	10.5
	Mg	410	60	350	14.7
	Fe	290	75	215	10.3
	Fe‡	140	73	67	
Illite	Na	120	53	67	15.4
	K	120	60	60	17.5
	Ca	100	45	55	16.8
	Mg	95	46	49	14.7
	Fe	110	49	61	15.3
	Fe‡	79	46	33	
Kaolinite	Na	53	32	21	26.8
	K	49	29	20	
	Ca	38	27	11	24.5
	Mg	54	31	23	28.7
	Fe	59	37	22	29.2
	Fe‡	56	35	21	
Attapulgite	H	270	150	120	7.6

†After Ref. 1.
‡After five cycles of wetting and drying.

Once the meaning of Fig. 5.1 is understood, it becomes obvious how cohesive soils must be dealt with in relation to foundation engineering: Their water content must be lowered as much as is feasible and practical. Every foundation technique described in this chapter has that concept as its goal. Which technique will be the most cost-effective way to solve a particular problem is a site-specific matter. But every method must be understood so that the appropriate one can be chosen for any given situation.

5.2 PRELOADING OR DEWATERING

As noted in Sec. 5.1, one must remove the water from a cohesive soil in order to make it more suitable for most foundation applications. One of the simplest, most straightforward, and most cost-effective methods to do this is by placing a surcharge fill on the ground surface and preloading before permanent construction begins. Usually a sand blanket or thick nonwoven geotextile is first placed on the ground, followed by successive layers of soil which constitute the preload. This increases the pore water pressure in the soil and, as the consolidation process occurs, an increase in the soil's effective stress takes place, accompanied by surface settlement. The process is completely described in all geotechnical engineering textbooks and is discussed

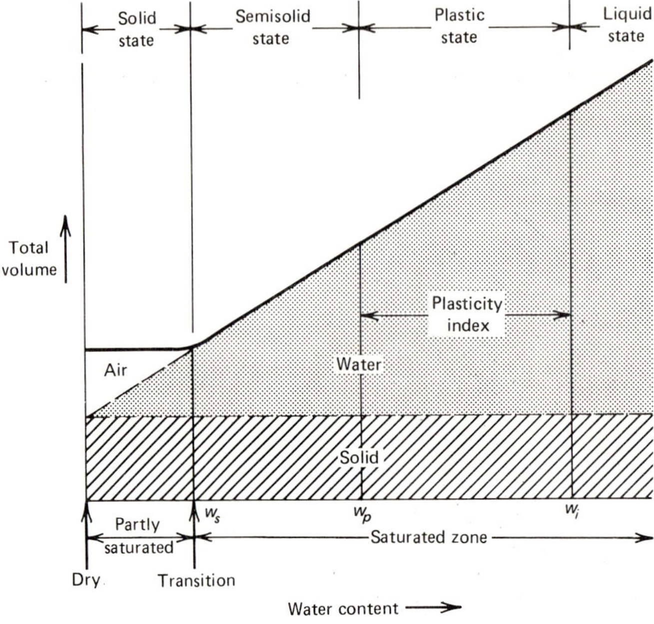

Figure 5.1 Volumetric behavior of fine-grained soils in relation to variations in water content. Plasticity index = $PI = w_l - w_p$. Flow index = FI = slope of liquid-limit curve (in percent). Toughness index = $TI = PI/FI$.

in Sec. 1.5.2 of this book. With adequate site development and planning, this is a most viable method of stabilizing fine-grained soils. It has two major drawbacks, however:

The time required for consolidation is usually long, particularly if the soil's permeability is low [k_v is proportional to c_v in Eq. (1.23)] or if the stratum is thick with only a single drainage.

Shear failures of soft soils are quite common, particularly if individual layers of preload are applied too quickly.

Increasing a soil's effective stress by dewatering the site can be brought about by another method—the use of *vacuum wellpoints*. See Fig. 5.2. This dewatering technique is usually carried out on granular soils (gravels, sands, and cohesionless silts) for the purpose of temporarily lowering the ground water table during a construction project. A typical construction site setup, along the lines described by Mooney,[2] is shown in Fig. 5.3. Both single-stage setups (able to lower the water to a maximum depth of about 18 ft) and multiple-stage setups (able theoretically to lower the water to any depth) are illustrated. Pumps are required to handle the discharge from the header connecting the individual wellpoints. The Dupuit equation is gen-

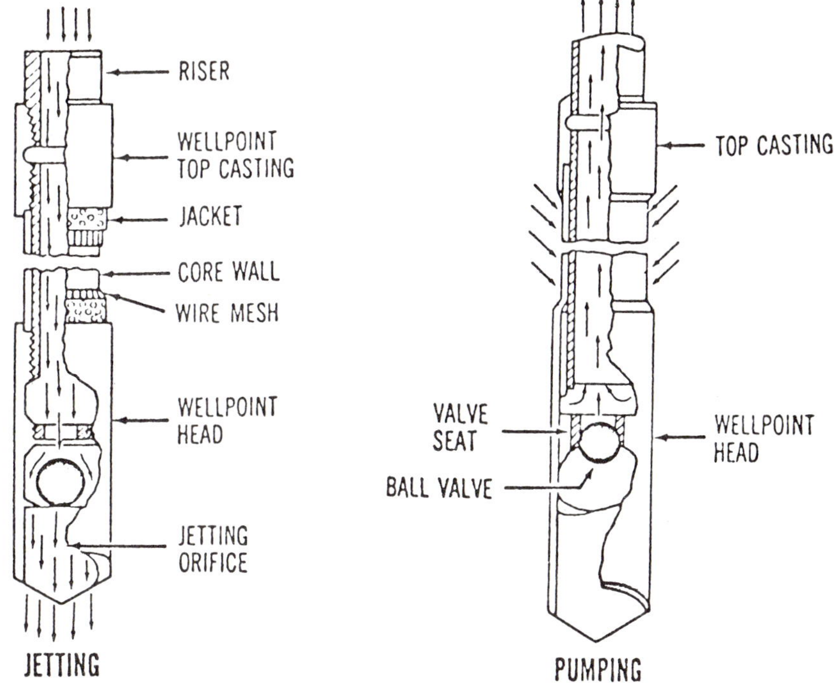

Figure 5.2 Details of typical wellpoints showing water flow during placement (by jetting) and during dewatering (by vacuum).

erally used to estimate the flow rate, and for a single wellpoint is as follows:

$$q = \frac{k(H^2 - h^2)}{\ln\,(R/r)} \tag{5.1}$$

where q = flow rate

k = coefficient of permeability

H = height of original water table above an impermeable stratum

h = height of the wellpoint above this stratum

R = radius of the area of influence (generally used as one-half the wellpoint spacing)

r = radius of the wellpoint

If the aquifer is confined between impervious strata under artesian pressure, the equation is modified to the following[3]:

$$q = \frac{2\pi\,kb(H - h)}{\ln\,(R/r)} \tag{5.2}$$

where b = thickness of the aquifer, and all other variables are as defined above.

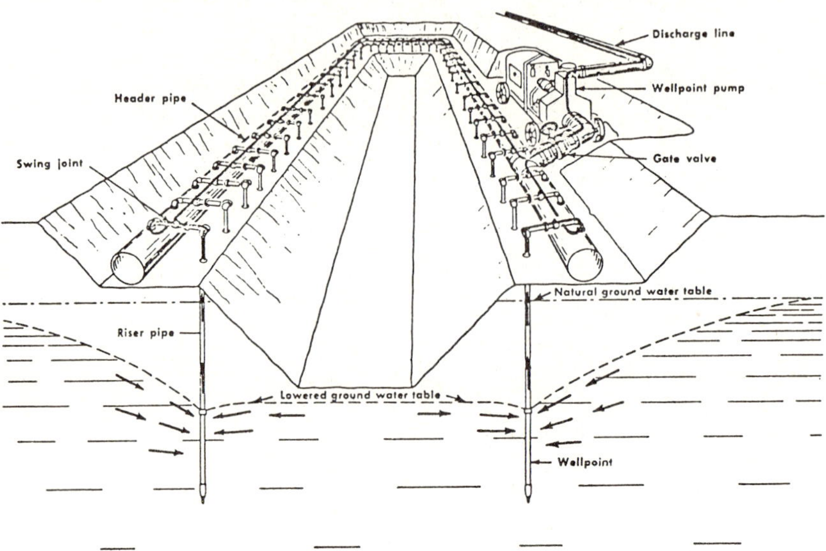

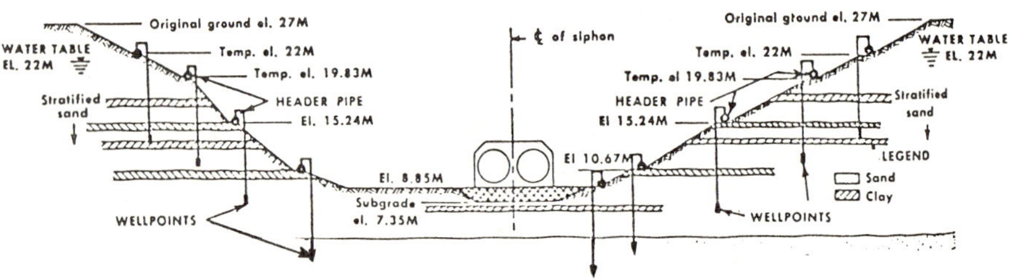

Figure 5.3 Typical construction site dewatering using vacuum wellpoints showing single- and multiple-level systems. *(After Mooney, Ref. 2.)*

Regarding wellpoint spacing in granular soils, Lancaster-Jones et al.[5] give a nomograph to use; see Fig. 5.4. Some further relevant comments about wellpoint dewatering in sands and gravels are given here:

The wellpoints are installed by jetting them to the desired depth.
Connections between the wellpoint riser and the header are valved and must be tight.
The suction pump (or pumps) is (are) usually placed below original ground level so as to reduce the head required for lifting the water.
Single-stage systems are most successful in areas where the water table must be lowered less than 15 ft.

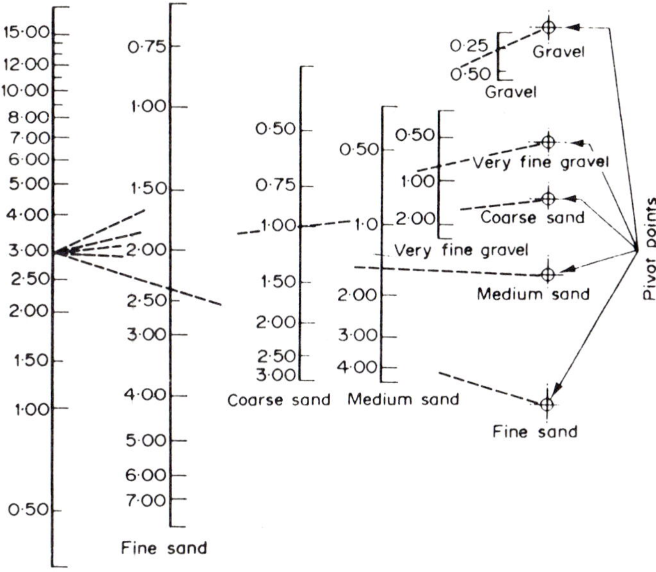

Figure 5.4 Nomograph for wellpoint spacings (in meters) for granular soils. *(After Lancaster-Jones, Ref. 3.)*

Clogged wellpoints can sometimes be freed by backflushing them with water.

Wellpoints that suck air can be recognized by the vibrations they induce on the lines they are connected to or by fluctuating pump gages; such wellpoints should be isolated from the rest of the system or suitably repaired.

Since wellpoints usually operate 24 h a day, the arrangement of the entire layout is critical insofar as labor costs are concerned.

Ejector pumps can be used to dewater at great depths, although flow rates are generally low and power costs are relatively high.

Submersible, or deep well, pumps can be placed into cased borings for dewatering by pushing the water up the casing; this no longer constitutes a vacuum system, but it does work in some situations.

There are no theoretical limitations on the depth at which a submersible pump can dewater.

Practical considerations in regard to the use of submersible pumps are their high cost of installation and heavy power consumption.

Example 5.1 What is the flow rate from a wellpoint located as shown in the illustration below if it is founded in a medium sand of $k = 0.012$-cm/s permeability? The wellpoint is 2.6 in (0.066 m) in diameter.

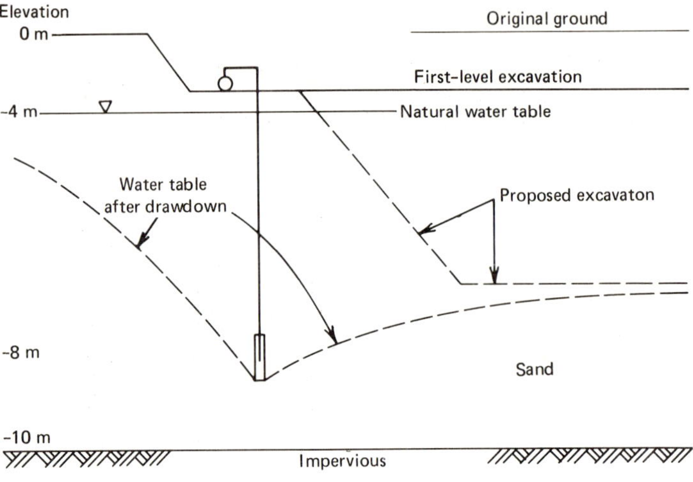

SOLUTION

$$q = \frac{k(H^2 - h^2)}{\ln(R/r)}$$

where R = 1.4 m (from Fig. 5.4)

$$q = \frac{(0.00012)(36 - 4)}{\ln(1.4/0.033)}(60)(60)$$

$$q = 3.7 \ \text{m}^3/\text{h}$$

So far this discussion about the use of wellpoints has focused on the dewatering of granular soils, i.e., sands and gravels, yet this chapter is about densification of cohesive soils. How can the wellpoint method be used for this latter purpose? The answer is that the wellpoints can be used to reduce the hydrostatic pressure of such soils, thereby increasing their effective pressure and consolidating them. A case history given by Partos and Koerner[4] illustrates this use of the technique. A soil profile for the area in which the research project was carried out is shown in Fig. 5.5, and it can be seen that the upper three strata were all compressible and saturated. Wellpoints were placed on various patterns in stratum 4 and vacuum pumping for 3 days produced the piezoemetric head elevations shown in Fig. 5.6. Settlements were immediately noted, but pumping was discontinued shortly thereafter due to the research nature of the project.

It is important to recognize that the use of wellpoint dewatering depresses the water table-piezometric levels for a considerable distance away from the work site. Figure 5.7 shows profiles of piezometric levels in two perpendicular directions at the research site just described. There it can be seen that the influence of the dewatering extended even beyond the range of the detection instruments. This type of depression results in an increase in the overlying unit weight of the soil (from buoyant to total unit weight), causing consolidation settlement, but also causing settlement beyond the desired limits of the construction site. To avoid this latter effect, one could con-

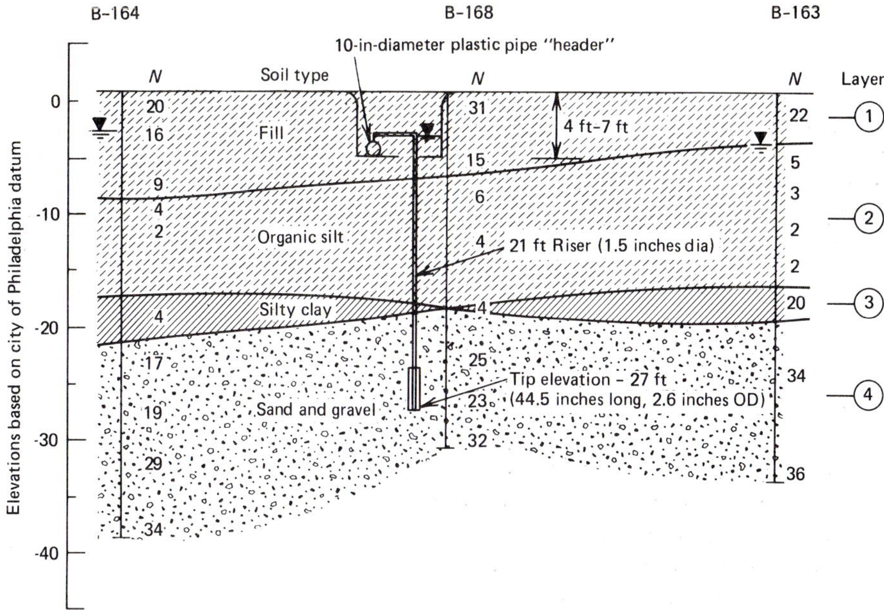

Figure 5.5 Soil profile showing various strata and typical wellpoint used in consolidating fine-grained soils. *(After Partos and Koerner, Ref. 4.)*

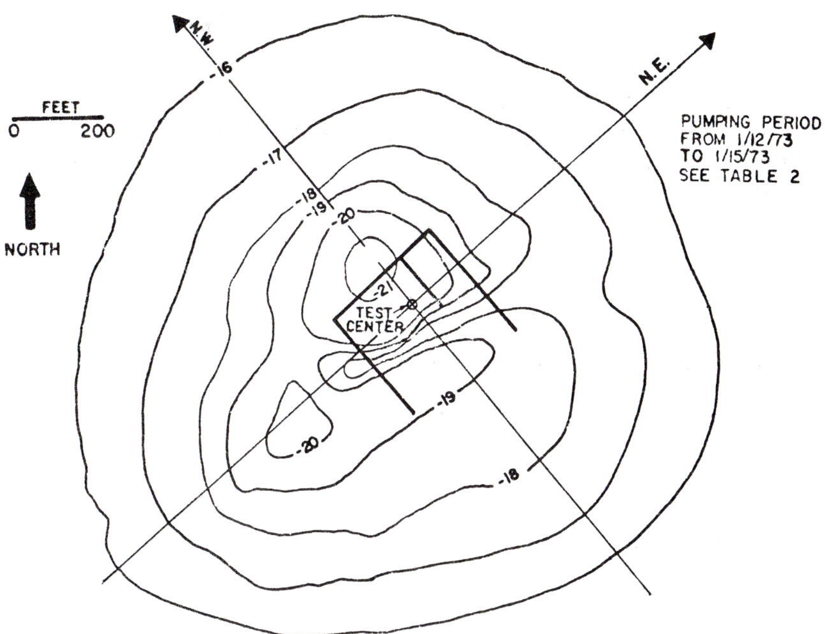

Figure 5.6 Contour plan of piezometric head elevations due to site dewatering in area indicated. *(After Partos and Koerner, Ref. 4.)*

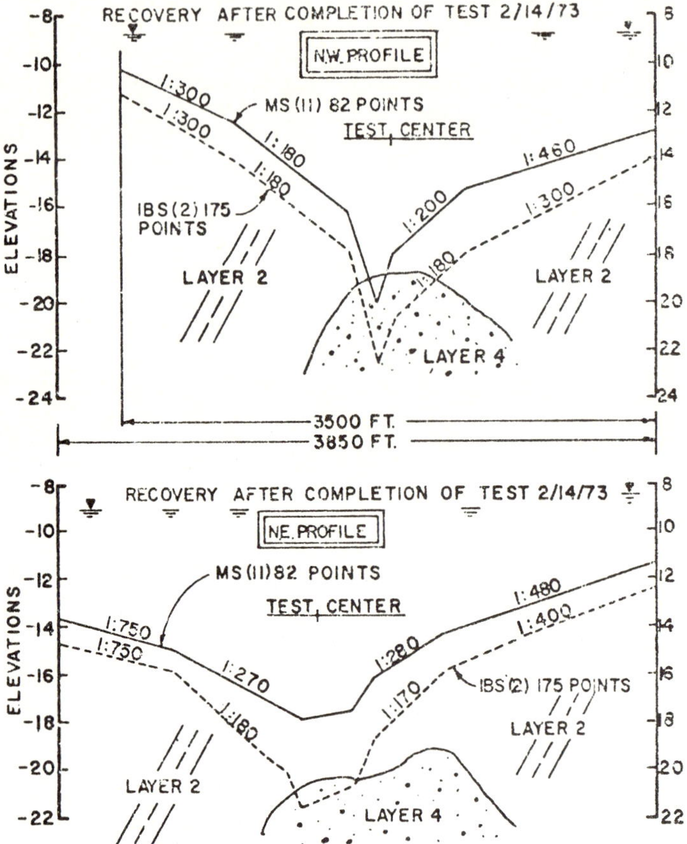

Figure 5.7 Profiles of piezometric head elevations for site shown in Figs. 5.5 and 5.6 showing extent of drawdown. *(After Partos and Koerner, Ref. 4.)*

struct a cutoff wall or could recharge the water table beyond the site limits, but both alternatives are expensive and far from foolproof. In general, the technique should not be used in urban areas where settlements could cause damage to adjacent structures.

5.3 RADIAL DRAINAGE VIA DRAIN WELLS

As noted in Sec. 5.2, the time it can take to consolidate relatively thick compressible strata can be long. Too long, in fact, for most situations where thicknesses exceed 20 ft and/or coefficient of consolidation values are less than approximately 0.001 cm^2/s. It can easily be seen by reviewing Eq. (1.23) that any technique which decreases the drainage path length and/or takes advantage of a higher coefficient of consolidation would be helpful in decreasing the time required for settlement to occur. If such a method were available, and used before permanent construction began, the site could be stabilized and postconstruction settlement limited to acceptable amounts. Such a technique does exist, and has been widely used over the past 40 years—the method of *radial drainage via vertical drain wells.* Prior to about 1975 it was generally implemented by installing vertical columns of free-draining sand in the soil to be consolidated, and then placing the permanent fill and/or a temporary preload on the ground surface (see Fig. 5.8). The drainage path length is now the distance between sand columns and quite often a coefficient of consolidation higher horizontally than vertically can be taken advantage of. Analogously to Eq. (1.23) for vertical flow, we now have one for radial flow:

$$t = \frac{T_h d_e^2}{c_h} \tag{5.3}$$

where t = time for consolidation

d_e = the center-to-center spacing between sand drains

T_h = the time factor for horizontal drainage

c_h = the horizontal coefficient of consolidation for horizontal drainage as mobilized by vertical compression

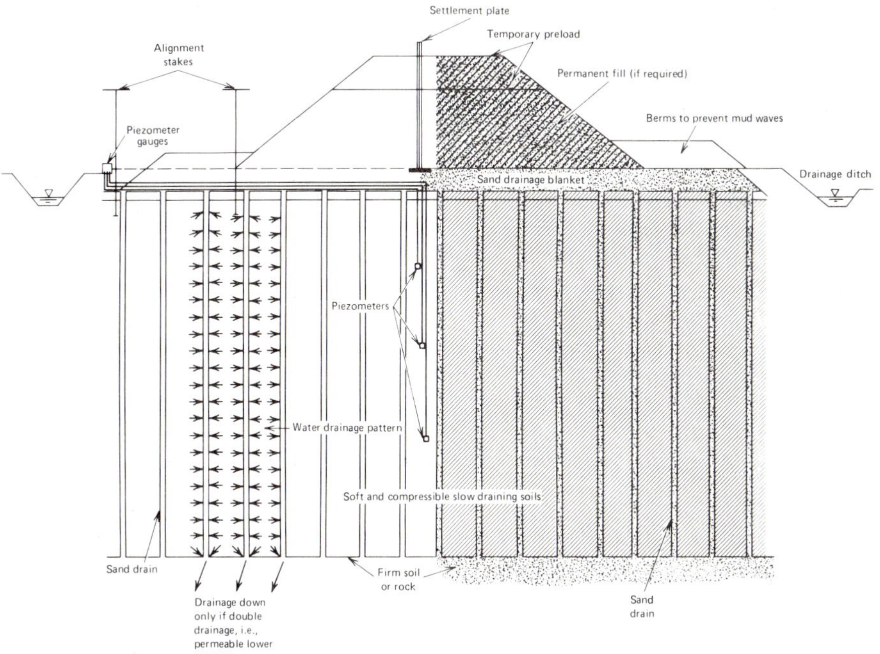

Figure 5.8 Typical layout of sand drain-preload installation showing drainage paths and recommended instrumentation.

Values of T_h have been solved for a wide variety of sand drain spacings d_e and sand drain diameters d_w, as shown in Fig. 5.9. c_h is sometimes considerably larger than c_v (particularly in sedimentary deposits of a continuous nature), but not always. Specific tests should be conducted to determine its value; see Johnson.[6] If no direct data are available, the value of c_v determined from standard consolidation tests should be the maximum value used. See Example 5.2, which shows how drastically soil con-

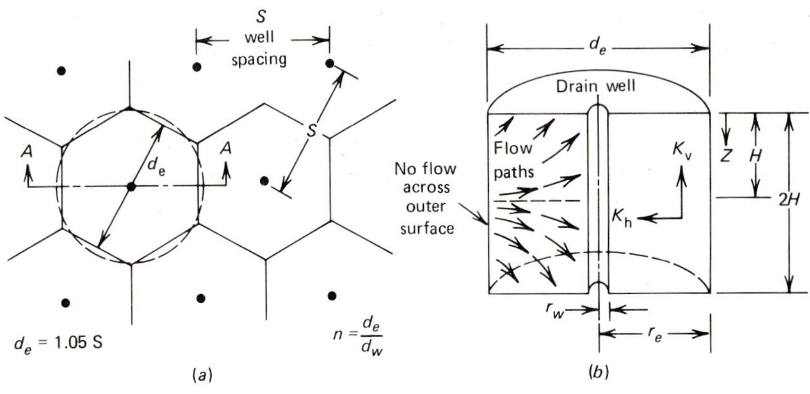

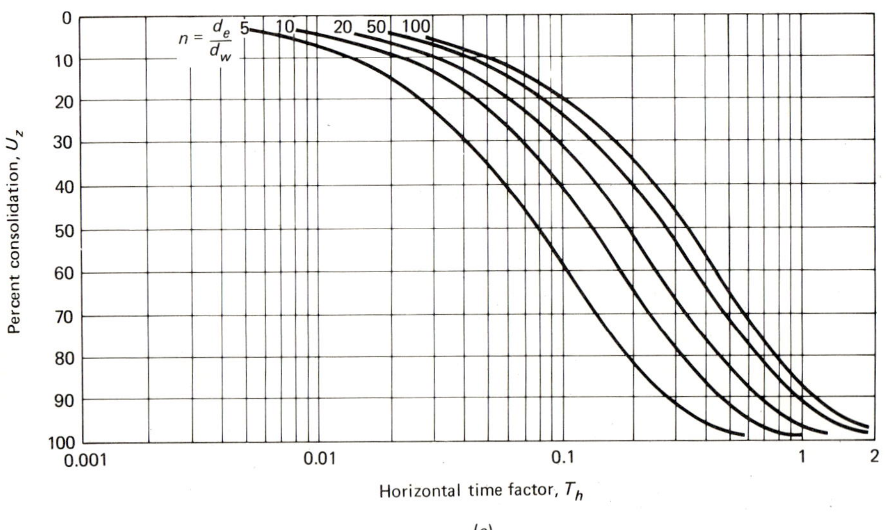

Figure 5.9 Theoretical results for radial consolidation to vertical drain wells. (*a*) Plan of drain well pattern. (*b*) Section *A-A*. (*c*) Values of T_h as a function of U_r for various ratios of sand drain spacing to sand drain size. (*After Barron, Ref. 5.*)

solidation time can be reduced when sand drains are installed. Obviously as the sand drains become larger in diameter and/or are placed closer to one another, the time needed for consolidation decreases.

Example 5.2: Sand drain design A site consists of 66 ft of organic clayey silt having high plasticity (OH soil) which is to be consolidated for eventual placement of large storage tanks. Soil consolidation tests indicate that $c_v = 0.005$ in^2/min and $c_h = 0.010$ in^2/min. Determine the time needed for 90 percent consolidation under the following conditions:

(*a*) Vertical drainage *without* sand drains, assuming an impermeable base layer

(*b*) Vertical drainage *without* sand drains, assuming a permeable base layer

(*c*) Radial drainage *with* sand drains based on various spacing patterns

SOLUTION For the cases where sand drains are not used, i.e., for vertical drainage, standard consolidation theory gives $T_v = 0.848$ for $U_z = 90$ percent. Thus

$$(a) \qquad t_{90} = \frac{T_v H^2}{c_v}$$

$$= \frac{0.848 \, (66 \times 12)^2}{0.005} \, \frac{1}{60(24)(365)}$$

$$= 203 \text{ years}$$

$$(b) \qquad t_{90} = \frac{T_v H^2}{c_r}$$

$$= \frac{0.848 \, (33 \times 12)^2}{0.005} \, \frac{1}{60(24)(365)}$$

$$= 51 \text{ years}$$

(*c*) For the case where sand drains *are* used, it is usual to assume a given diameter and then calculate, for a series of spacings, the consolidation times. For $d_w = 12$ in and $d_e = 25$ ft, $n = 25$. From Fig. 5.9, $T_h = 0.72$ and

$$t_{90} = \frac{T_h d_e^2}{c_h}$$

$$= \frac{(0.72)(25 \times 12)^2}{0.010} \, \frac{1}{60(24)(365)}$$

$$= 12.4 \text{ years}$$

For a range of spacings (and, for that matter, of diameters if they are desired), the accompanying table is provided.

d_w, ft	d_e, ft	n	T_h	r_e, ft	t_{90} Years	t_{90} Months	t_{90} Days
1	25	25	0.72	12.5	12.4	149	4460
1	15	15	0.66	7.5	4.0	48	1400
1	10	10	0.46	5.0	1.2	14	420
1	5	5	0.27	2.5	.18	2.2	66

This technique, properly called *vertical sand drains,* has been used in a wide variety of situations; see Table 5.2. A widely referenced study showing the beneficial effects of sand drains was undertaken in Sweden and reported on by Hansbo.[7] Figure 5.10*a* shows a plan view of the four test areas, which consisted of 10 to 14 m of very soft clay. Test area I was radially subdivided in three equal sections and 18-cm sand drains were installed at 0.9-, 1.5-, and 2.2-m spacings, respectively. The time-versus-settlement results from a 2.7-t/m^2 surcharge load are shown in Fig. 5.10*b*. It can easily be seen that for an equivalent settlement to occur, the closer spacings are significantly more beneficial. Results from test area II (sand drains at 1.5-m spacings

Table 5.2 Uses of vertical sand drains†

Field of application	Purpose Increase stability	Purpose Decrease postconstruction settlements
Highways		
Roadways	x	x
Structure approach fills	x	x
As alternative to structure	x	x
Airfields	x	x
Earth dams		
Foundations	x	x
Embankments	x	x
Warehouse floors	x	x
Buildings	x	x
Cellular cofferdams		
Stabilize fill-in cells	x	
Pile foundations		
Reduce "downdrag" or "negative" skin friction		x
Excavations		
Steepen allowable slopes	x	
Quay walls	x	x
Large-scale development of marginal areas	x	x

†After Johnson, Ref. 6.

under 2.7-t/m² surcharge), test area III (sand drains at 1.5-m spacings under 3.9-t/m² surcharge), and test area IV (no sand drains under 2.7-t/m² surcharge) are shown in Fig. 5.10c. There the sites with sand drains are seen to consolidate much more rapidly than the one without sand drains, and the amount of settlement for a given time after placement is seen to be greater for the more heavily surcharge-loaded site.

However, by no means all sand drain-surcharge installations have been successful. At one such site 18-in-diameter sand drains at 7.5-ft spacings were installed in a triangular pattern to consolidate 66 ft of organic, clayey, silty soil (see Koerner[8]). As surcharge fill was being placed, a shear failure in the foundation soil was mobilized. The failure plane eventually propagated up through the fill at one end, and well beyond the toe of the slope at the other end. Deformations were such that the sand drains were undoubtedly sheared off and the site had to be developed by alternative methods.

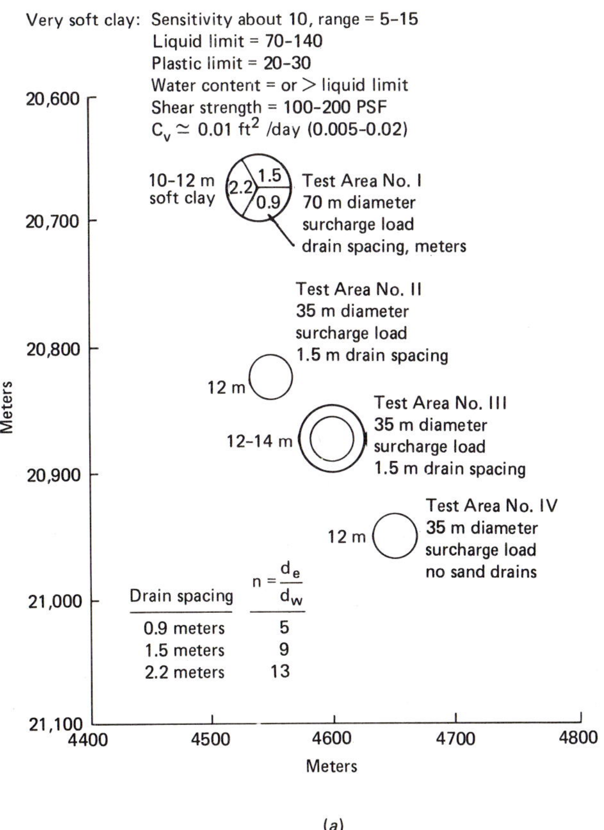

(a)

Figure 5.10 Swedish test site to evaluate effectiveness of vertical sand drains. (a) Plan of test areas at Ska-edeby. (b) Effect of sand drain spacing, test area I. (c) Effect of fill thickness in areas with and without sand drains, test areas II to IV. (*After Hansbo, Ref. 7.*)

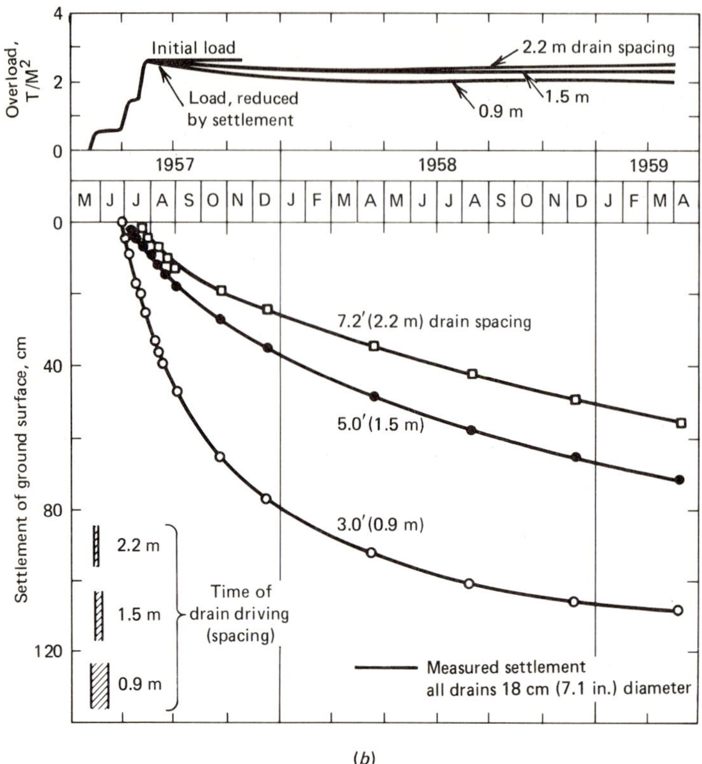

(b)

Figure 5.1 (*Continued*)

Some comments are in order about how sand drains are installed since their performance is influenced to some degree by the method of their installation. Three different techniques are common, although many others are available:

1. *Mandrel-driven.* In this method a closed-end hollow pipe (the mandrel) is driven to the bottom of the soil to be consolidated, it is filled with sand which is placed under approximately 100 lb/in² of air pressure, and then it is gradually lifted out of the ground. The closed end, being a hinged valve, opens, and the sand, under pressure, rushes out. The pipe continues to be lifted until it reaches the ground's surface. The entire process is then repeated for the next sand drain. The cycle takes 2 to 10 min, depending upon local conditions.

2. *Hollow stem-continuous flight auger.* Here a continuous flight auger of the same diameter as the intended sand drain is rotated into the soil to the specified depth. Since the auger is formed around a hollow pipe, sand, under pressure, can be introduced into it. As the auger is backrotated out of the hole, the sand is left behind and forms the sand drain. Cycle times for this installation method are 5 to 20 min.

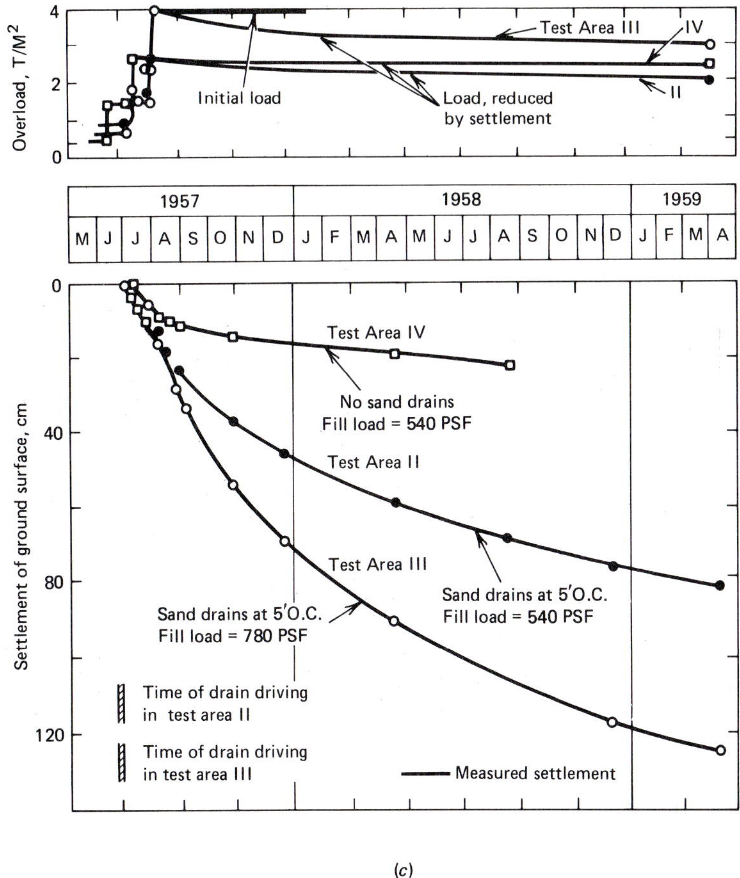

(c)

Figure 5.1 (*Continued*)

3. *Jetted hollow or closed-end pipes.* The driving into soil of either hollow or closed-end hinged pipes via water jetting is also used for sand drain installations. After pipe placement, the sand is introduced under pressure as the pipe is withdrawn. Cycle times are 5 to 20 min.

The problems of soil disturbance and smear around the periphery of the sand drain come up in regard to each of these construction methods. A list of those concerns is given in Table 5.3. Which method best avoids the problems is a subject still open for discussion—but the answer may already be merely academic, since a completely different style of drain using no sand at all, properly called a *drain wick,* has begun to dominate the market.

Drain wicks are usually fluted or nubbed plastic cores within geotextile sheaths that completely encircle those cores. They are usually 100 mm wide by 2 to 6 mm

Table 5.3 Disturbance effects from sand drain installations†

Disturbance effect	Remarks
Smear	Can be caused by any method of installation, even by a thin sharp knife drawn through soft clay
Soil displacement and remolding effects	Outward soil displacement caused by driven closed-end mandrels Inward soil displacement resulting from jetting methods or withdrawal of solid stem augers Either outward or inward soil displacement may result from hollow stem auger methods, depending on rotation and advance rates of auger
Grouting of thin sand layers	Caused by natural drilling mud formed by jetting methods
Thin film of mud on sides of drain	Caused by natural drilling mud formed by jetting methods
Contamination of sand backfill in drain	Possible in jetting methods if washing during jetting is insufficient; also possible when withdrawing driven mandrel if sand sticks in mandrel
Distortion of thin sand layers	Most likely with driven mandrel method, possible with solid or hollow stem augers; may be severe in varved or thinly bedded deposits

†After Johnson, Ref. 6.

thick, and as long as necessary; see Fig. 5.11a. Alternative styles resemble corrugated cardboard; see Fig. 5.11b. Constructionwise, a drain wick from a coil is inserted into a lance that is pushed or driven into the ground to the desired depth. Once inserted, the lance is removed, leaving the drain wick behind. The drain wick is then cut off with shears at ground level and the process is repeated at the next location. Cycle times are in the range of 1 to 5 min. The entire process resembles (and is somewhat analogous to) a large sewing machine, and is a vast improvement over the original installation methods reported by Kjellman[9] in 1948.

While the method of installing drain wicks appears to be established and some field data about their use are becoming available,[10] the design of drain wick projects is far from standardized. Most designs attempt to equate the diameter of the particular type of drain wick being considered to an equivalent sand drain diameter (see Example 5.2), and from that point on, at least, the procedure is standardized as illustrated in Example 5.2. To equate the two, one must multiply the percent of void space in the drain wick times its area to get an equivalent void circle, and then divide this number by the typical porosity of dense sand; see Example 5.3. The resulting value can probably be increased by 5 to 15 percent since there is no well resistance in drain wicks as there is in sand drains. That is, once the water enters the drain wick, its path is completely unhindered, which is not quite the case for sand drains where the path is obviously very tortuous.

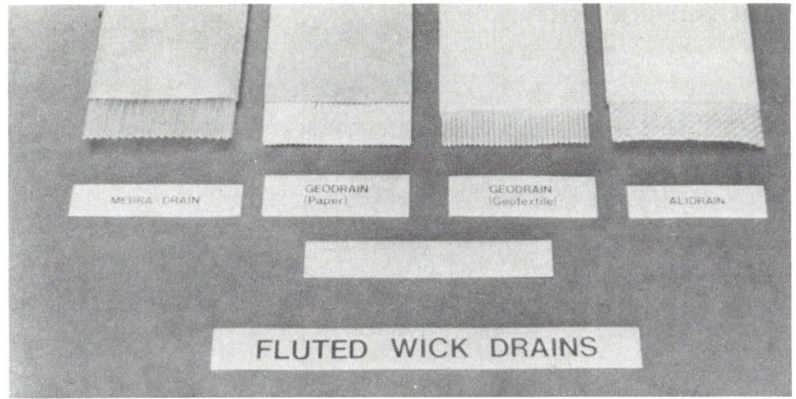

(a)

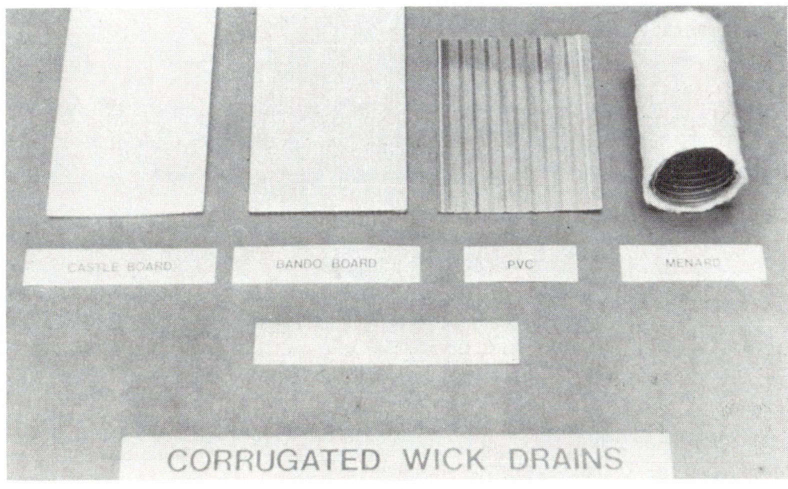

(b)

Figure 5.11 Photographs of drain wicks used to consolidate fine-grained soils. (*a*) Drain wicks constructed of plastic cores wrapped by geotextiles or paper sheaths. (*b*) Drain wicks where plastic is shaped like corrugated cardboard.

Example 5.3: Drain wick equivalency Given a drain wick with 85 percent open space and measuring 100 mm wide by 4.0 mm thick, what is the equivalent sand drain diameter using a porosity of 0.3?

SOLUTION Total void space in drain wick:

$$A = (100 \times 4.0)(0.85)$$

$$= 340 \text{ mm}^2$$

Equivalent diameter of a void circle:

$$d_v = \left[\frac{(340)(4)}{\pi} \right]^{1/2}$$

$$= 20.8 \text{ mm}$$

Equivalent diameter of a sand drain:

$$d_{sd} = \frac{20.8}{0.3}$$

$$= 69.3 \text{ mm}$$

Allowance for free drainage path (10 percent):

$$d_{allow} = 69.3 \times 1.10$$

$$= 76.2 \text{ mm}$$

$$= 3.0 \text{ in}$$

A more rigorous approach to drain wick design has been used in Europe and utilizes a combination of Kjellman's original work[9] and Barron's subsequent theory[5]. As developed by Hansbo,[11] the relevant design equations follow.

$$t = \frac{D^2}{8c_h} \left[\frac{\ln (D/d)}{1 - (d/D)^2} - \frac{3 - (d/D)^2}{4} \right] \ln \frac{1}{1 - U} \tag{5.4}$$

which can be simplified, since d/D is small, to

$$t = \frac{D^2}{8c_h} \left[\ln (D/d) - 0.75 \right] \ln \frac{1}{1 - U} \tag{5.5}$$

where t = consolidation time
$\quad c_h$ = coefficient of consolidation for horizontal flow
$\quad d$ = equivalent diameter of the wick drain ($\simeq$ circumference$/\pi$)
$\quad D$ = sphere of influence of the wick drain (for a triangular pattern use 1.05 times the spacing, for a square pattern use 1.13 times the spacing)
$\quad U$ = average degree of consolidation

5.4 ELECTRICAL METHODS

In this section we will treat two electrical methods used to densify in situ cohesive soils, each of which is related to the other. The first is *electro-osmosis,* used primarily for stabilization purposes; the second is an electrochemical method, used primarily for eliminating settlement problems. Both techniques require the soil to contain clay minerals, so that clays of either low or high plasticity and various clay mixtures are the target soil types. Obviously, the soil being densified is usually saturated, which is why densification is needed to begin with.

Electro-osmosis consists of placing in the soil to be stabilized a number of electrodes and then passing a direct current between them. The electric current induces a flow of water from the anode to the cathode. (This is because of the attraction of cations, and of the unbalanced, negatively charged, clay particles themselves, to the anode.) Simultaneously, water is expelled in the vicinity of the anode and it, with its free cations, migrates toward the cathode. The cathode is generally a perforated metal pipe which is used as a wellpoint for removing the water. The anode can be any type of metal rod. Typical electro-osmotic stabilization configurations are shown in Fig. 5.12. In general, both cathodes and anodes should be placed about 5 ft beneath the lowest elevation to be stabilized. Typical spacings of the cathodes (the wellpoints) are 20 to 30 ft apart, with the anodes being placed midway between them.

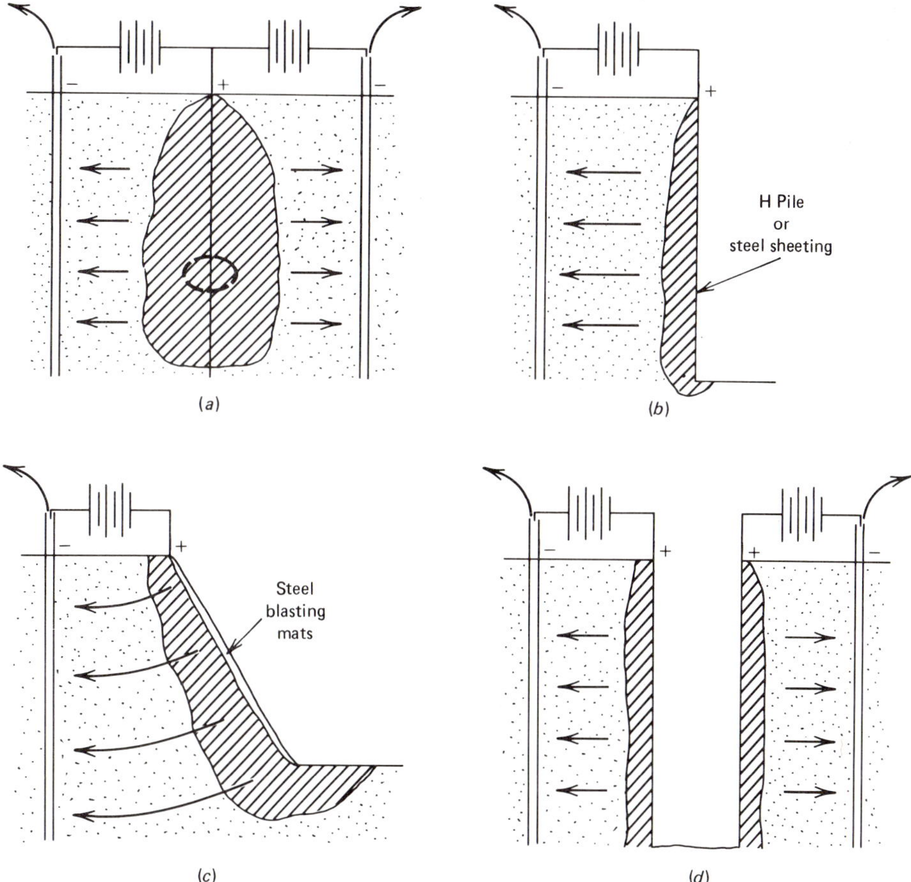

Figure 5.12 Typical stabilization configurations using electro-osmosis, where arrows refer to water flow and shaded areas to stabilized soil. (*a*) Tunnel-driving through soft soil. (*b*) Decreasing stress on temporary bracing. (*c*) Stabilizing soft slopes. (*d*) Sinking shafts in soft soils.

The flow rate to a cathode wellpoint can be estimated using a modification of Darcy's law (see Mitchell[12]) as follows:

$$q = k_e i_e A \tag{5.6}$$

where q = flow rate, m³/s
$\quad k_e$ = electro-osmotic coefficient of permeability based on voltage, 1×10^{-9} to 7×10^{-9} m/s per V/m
$\quad i_e$ = electrical potential gradient, V/m
$\quad A$ = cross-sectional area, m²

The flow rate can also be estimated using Eq. (5.7):

$$q = k_i I \tag{5.7}$$

where q = flow rate, m³/s
$\quad k_i$ = electro-osmotic water flow based on amperage, m³/s·A
$\quad I$ = current, A

In Eqs. (5.6) and (5.7), the values of k_e and k_i are related by the specific electrical conductivity σ as follows:

$$k_i = \frac{k_e}{\sigma} \tag{5.8}$$

where values of σ range from 0.02 S/m for low salt content soils to 0.30 S/m for high-salt-content soils.[12] Typical values of k_i are given in Table 5.4.[12]

The power consumption of an electro-osmotic stabilization project can be calculated from Eq. (5.9):

$$P = \frac{q \, \Delta V}{k_i} \times 10^{-3} \tag{5.9}$$

where P = power consumption, kWh
$\quad \Delta V$ = voltage drop, V

As interesting and intriguing as the electro-osmotic process is, it does have some serious drawbacks. Farmer[13] points out that with time its efficiency decelerates. That is, the very fact that the soil around the anode is stabilizing means it is drying out, and so the entire system becomes less efficient with time. It should also be noted that

Table 5.4 Typical values of electro-osmotic water transport coefficients (k_i)†

Soil type	Pore water salt concentration N	k_i, m³/(s·A)
Silty clay, kaolinite	10^{-3}	1×10^{-5}–5×10^{-7}
Silty clay, kaolinite	10^{-2}	5×10^{-8}–1×10^{-7}
Clay (illitic)	10^{-3}	3×10^{-8}–6×10^{-8}
Clay (illitic)	10^{-2}	2×10^{-8}–3×10^{-8}

†After Mitchell, Ref. 12.

the technique requires considerable power, with consumption varying from 1 to 10 kW/m^3 of stabilized soil.[13]

The second method to be described, *electrochemical* in nature, was apparently discovered by accident.[3] It was observed that during electro-osmotic stabilization, the constant flow of electricity began to erode the metal rods used as anodes, thereby requiring their rapid replacement. It was realized, however, that the metallic ions lost to the soil were partially cementing it, thus aiding in its stabilization and, even more important, reducing its compressibility. Currently, the electrochemical technique is used mainly to help prevent settlement problems.

The method consists of moving solutions through fine-grained soils via an electric field. (This is in sharp contrast to chemical grouting, the subject of Chap. 6, in which solutions are moved through soils by means of hydraulic pressures.) Since the electro-osmotic coefficient of permeability is relatively insensitive to soil type, a unit of electrical gradient is more effective in fine-grained soils than is a unit of hydraulic gradient.

The chemicals are introduced at the anode and are carried toward the cathode via the electric gradient. According to Mitchell[12] gradients of 50 to 100 V/m are necessary. Standard sodium silicates (and various combinations of retarders and accelerators) have been used. Also used are organic and inorganic compounds of aluminum or calcium, the most frequent being calcium chloride.[13]

The technique can also utilize direct electrolyte replacement to alter the characteristics of the clay being stabilized. For example, sodium and lithium montmorillonite clays have low shear strength and high compressibility. If an electrolyte solution containing calcium, magnesium, iron, or aluminum is electrochemically injected into them, the ionic replacement that results will produce a considerably more stable in situ structure.[3]

Both chemical additives and electrolyte replacement are effective in stabilizing in situ clay and reducing its compressibility. However, there are shortcomings. The technique requires a large amount of power, and a reduction of flow around the anode occurs as the additives begin to take effect. Thus the effectiveness of the method in stabilizing the clay diminishes as the process continues.

5.5 THERMAL METHODS

Of the two possible temperature extremes (hot or cold) that saturated, fine-grained soils may be subjected to, heat is only rarely used in construction activities. Certainly it is technically feasible to stabilize saturated clays by heat; for example:

Temperatures $\simeq$ 100°C cause drying and significant increases in the strength of clays, along with decreases in their compressibility

Temperatures $\simeq$ 500 °C cause permanent changes in the structure of clays resulting in decreased plasticity and moisture adsorption capacity

Temperatures $\simeq$ 1000°C cause fusion of the clay particles into a solid substance much like brick

A few case histories are available where heat has been used, e.g., those discussed by Beles[14] and Lancaster-Jones et al.[3] The latter cites the East European experience in using exhaust gases from burning fuel oil which are driven into boreholes 2 to 3 m apart. Both clays and loess have been stabilized in this manner. The economics of using heat, however, precludes its use in most construction projects, and only when a site is located near a large and inexpensive heat source should it be considered. On the other hand, *ground freezing* appears to be gaining in popularity during recent years. Conferences, numerous technical articles, and the appearance of several contractors who specialize in the technique are providing a good technical basis for its use.

Ground freezing is accomplished by bringing a refrigerant into the proximity of soil pore water that is stationary or moving at a rate less than 2 meters per day. The pore water around the refrigerant pipes begins to freeze and, with continued exposure, the ice layer expands until it comes into contact with the ice spreading out from adjacent refrigerant pipes. Eventually, a continuous wall of ice is formed. Stabilizing soil by this method provides a number of advantages:

High strength in the stabilized soil
Low permeability in the stabilized soil
A non-vibration-sensitive barrier to seepage flow or soil deformation
Its noiseless operation (except perhaps for the power plant needed to pump the refrigerant)
The wide range of soil types that it can stabilize

Numerous configurations for construction projects have been successfully designed; see Fig. 5.13.

In order to provide refrigerant to the soil, a number of schemes are possible; these are reviewed by Shuster[14] and shown in Fig. 5.14. Of those illustrated, two seem to be the most common. They are the use of expendable liquid refrigerants, like liquid nitrogen, liquid carbon dioxide, or liquid propane, and the in situ pumped-loop method via a secondary coolant.

The first of these is relatively simple. Freeze pipes are placed at approximately 1-m centers, adequately vented, and the liquid refrigerant is injected and allowed to boil (liquid nitrogen boils at atmospheric pressure). The ground freezes very rapidly, but the frozen zone is often very irregular and the method uses a great deal of energy. However, for a single project of short duration to solve a localized problem, the technique certainly has merit.

Much more popular is the technique that uses a primary plant and a pumped-loop, secondary circulating coolant. Originated by Poetsch in Germany nearly 100 years ago, the method is well suited to single installations of freeze pipe elements intended to provide maintenance freezing over a long period of time. The primary source of the refrigeration is a one- or two-stage ammonia or freon refrigeration plant. Heat transfer from soil to coolant occurs by convection, i.e., no phase change exists. The coolant distribution system consists of a closed-loop supply manifold connected to a number of parallel-connected freeze pipes placed in the ground and to a

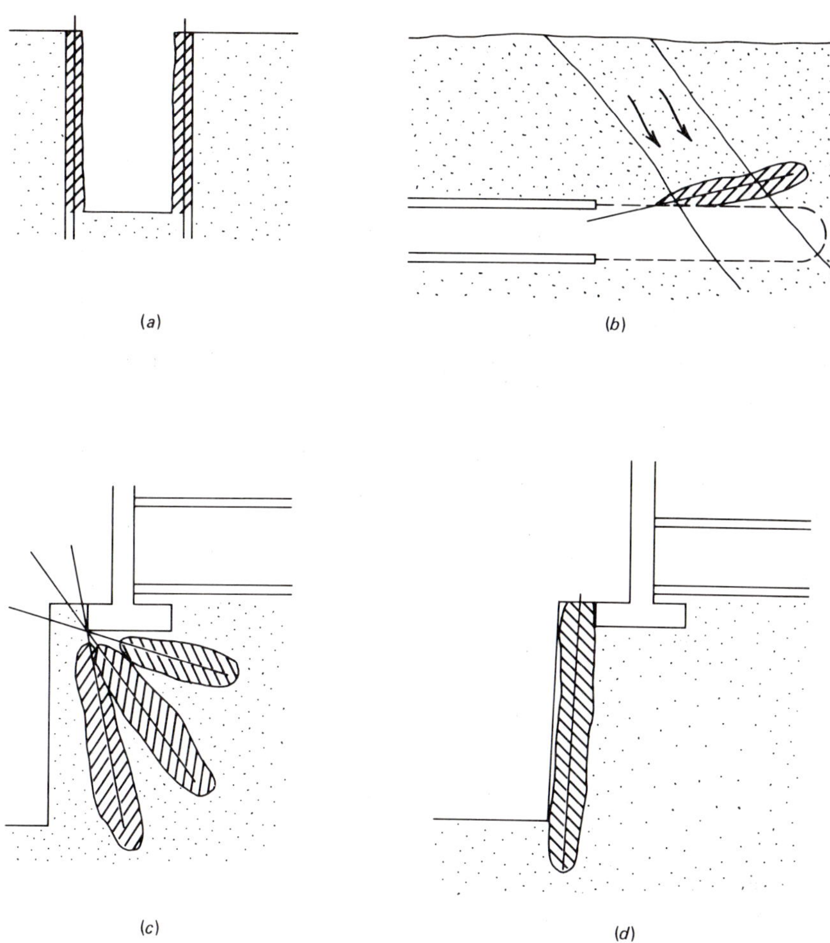

Figure 5.13 Typical construction configurations where ground freezing has been successfully used. (*a*) Sinking tunnel shafts. (*b*) Advancing tunnels in running ground. (*c*) Providing direct underpinning support. (*d*) Providing lateral restraint for excavations.

return manifold. The freeze pipes in the ground are typically 4 to 8 in in diameter with sealed lower ends. Smaller internal, or feed, pipes 1 to 3 in in diameter are inserted into the larger pipes, but not quite so far down so that they remain free of the bottom of the larger, capped pipe. Coolant is pumped into the inner feed pipes and returns through the annular space between the two. The proximity of the return coolant to the soil (they are separated only by the thickness of the larger pipe) is what produces the freezing. The usual coolant is brine, a mixture of sodium chloride and water with 10 to 23 percent NaCl, although diesel oil, propane, and glycol-water mixtures have also been used. A sketch of a typical freeze pipe and the method of its deployment for stabilizing soil in the construction of a 20-ft-diameter sewer in New York City is shown in Fig. 5.15.

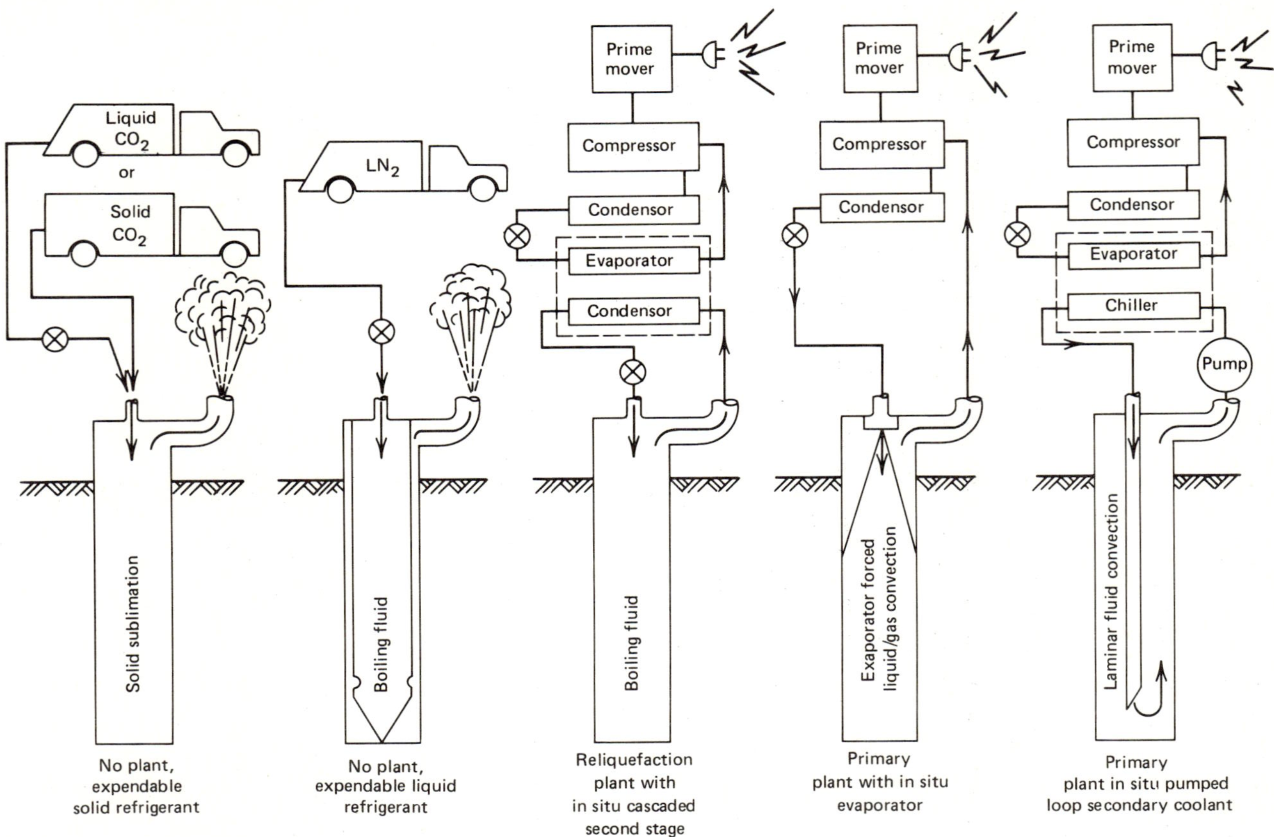

Figure 5.14 Alternative refrigeration methods for ground freezing. (*After Shuster, Ref. 14.*)

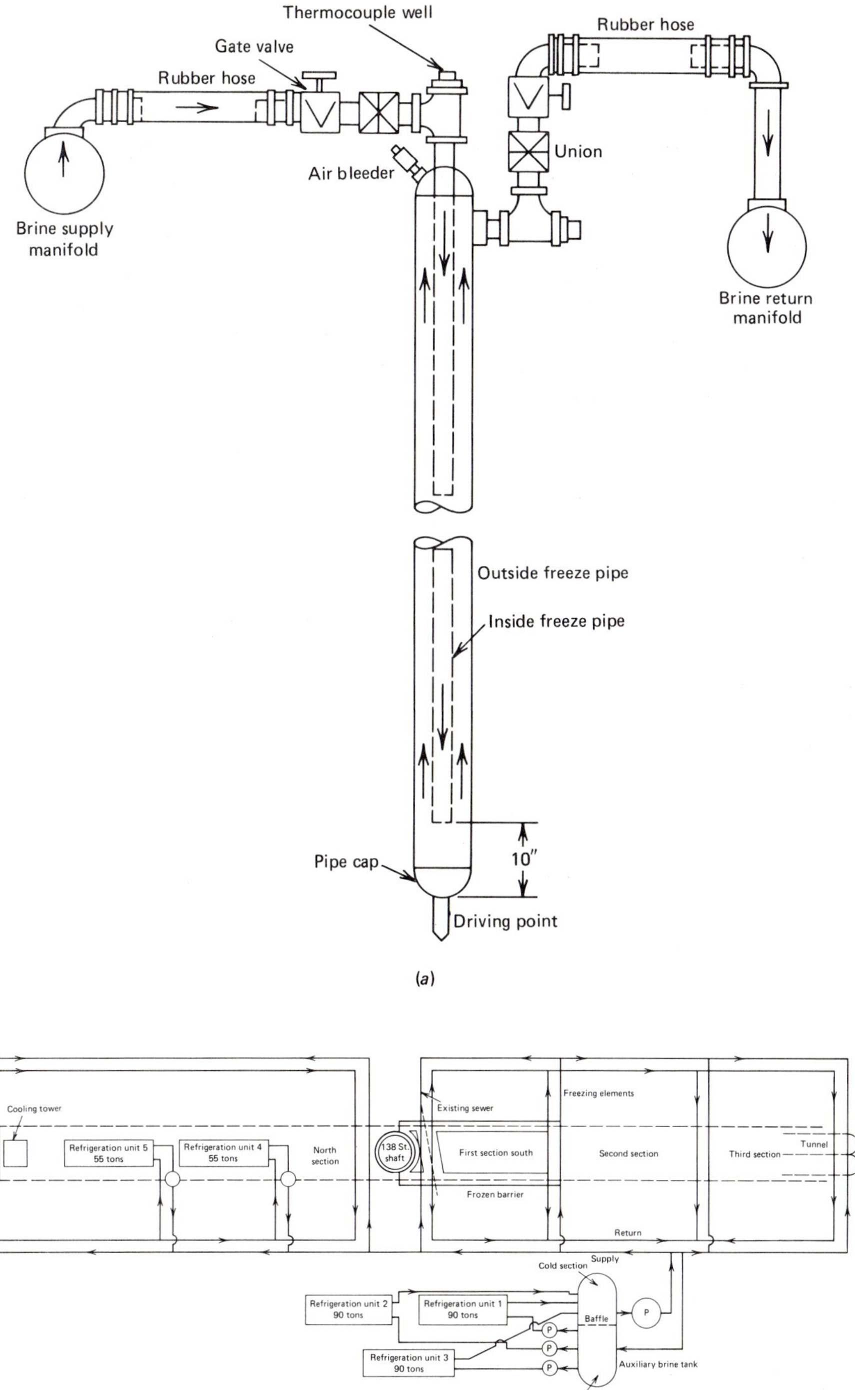

Figure 5.15 Typical freeze pipe and brine circuitry for stabilizing soil during construction of a large sewer tunnel. (*a*) General configuration of a typical freeze pipe. (*b*) Plan view of system layout. (*After Gail, Ref. 15.*)

Much of the design of ground freezing systems is done empirically, on the basis of past practice, but some analytic work is available. Mitchell[12] presents the following equations on heat content per unit volume of soil for unfrozen soil:

$$C_u = \gamma_d \left(C_m + \frac{C_w w}{100} \right) \qquad (5.10)$$

where C_u = volumetric heat for unfrozen soil
  γ_d = dry soil unit weight, kg/m^3
  w = water content, percent
  C_w = heat capacity of water [= 4186 J/(kg·K)]
  C_m = heat capacity of soil minerals [= 710 J/(kg·K)]
And for frozen soil:

$$C_f = \gamma_d \left(C_m + \frac{C_i w}{100} \right) \qquad (5.11)$$

where C_f = volumetric heat for frozen soil
  C_i = heat capacity of ice [= 2093 J/(kg·K)]
It should be recognized that the time required for a ground freezing system to take effect, i.e., for a continuous frozen wall to form between freeze pipes, is typically in the range of weeks or even months. However, this obviously depends on the type of coolant used, the temperature achieved, the freeze pipe size, and the spacing between pipes. The topic is discussed by Shuster,[14] and illustrated in Fig. 5.16.

The following miscellaneous comments about ground freezing are relevant to an understanding of where and when the technique should be considered for use:

The systems are conceptually relatively simple and straightforward, but are quite cumbersome to deploy in the field.

Once the frozen wall has formed, the energy required for its maintenance becomes constant and significantly lower than that required to produce it in the first place.

Sands and cohesionless silts (as well as clay soils) can be frozen as long as the groundwater in the vicinity is relatively stationary; see Fig. 5.17.

When the soil to be treated is partially saturated, or even dry, it can be prewetted before and wetted again during freezing.

Drilling for insertion of freeze pipes into the soil must be accurate because even a slight deviation will leave an unfrozen window of soil by the time the excavation bottoms out. Slope indicators have been used for monitoring purposes. See Chap. 9.

It is necessary to check the ground temperature before excavation by using thermocouples placed midway between the freeze pipes.

The frozen zone will generally be quite irregular in shape, reflecting the heterogeneous nature of the soil to begin with.

While the frozen wall of soil is indeed strong, it is creep-sensitive under sustained lateral loading; see Jessberger.[16]

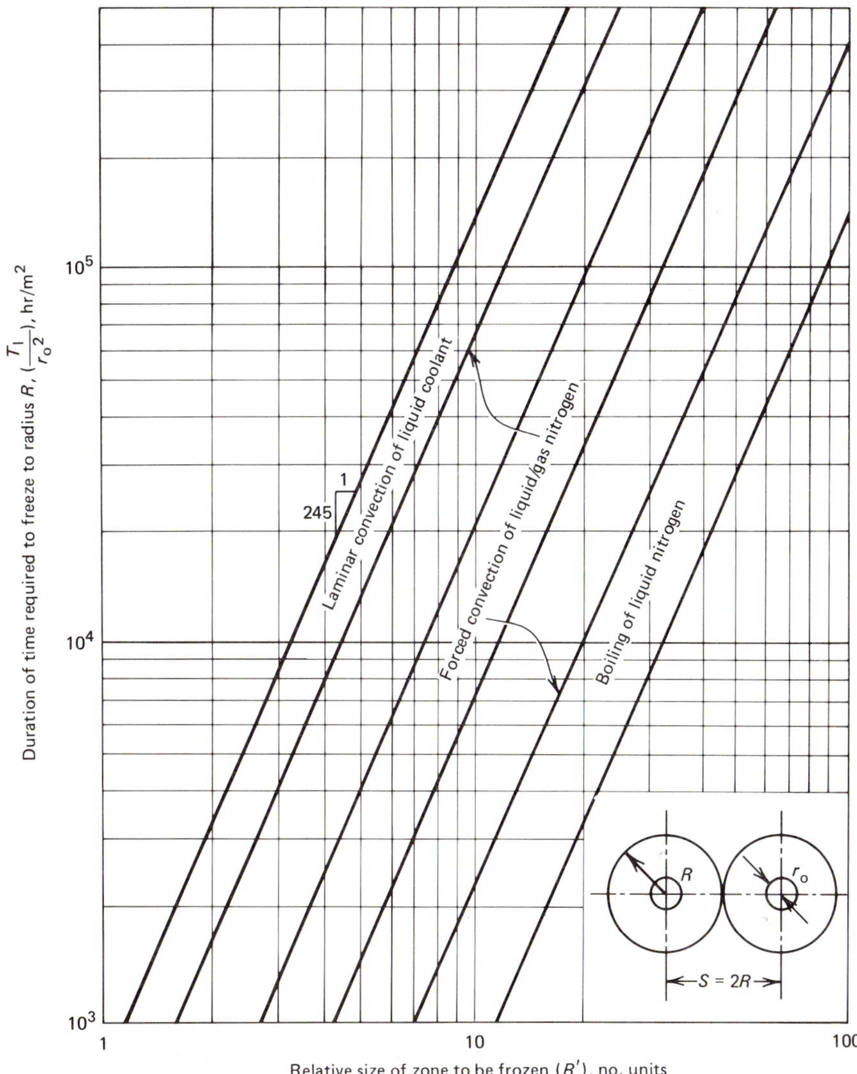

Figure 5.16 Required freezing time for a continuous wall to be formed between freeze pipes. Indicated bands represent normal range of observed field and laboratory results. However, results with forced convection of N_2 may vary more widely than is indicated owing to variables in control of the freezing process. $R' = R/r_0$.

Saturated soil expands during freezing and contracts during thawing. This can be important in underpinning applications.

Utility lines and/or shafts can be troublesome and must be taken into account before work proceeds.

A competent specialty contractor should perform the work. If the situation is critical, it is no place for amateurs!

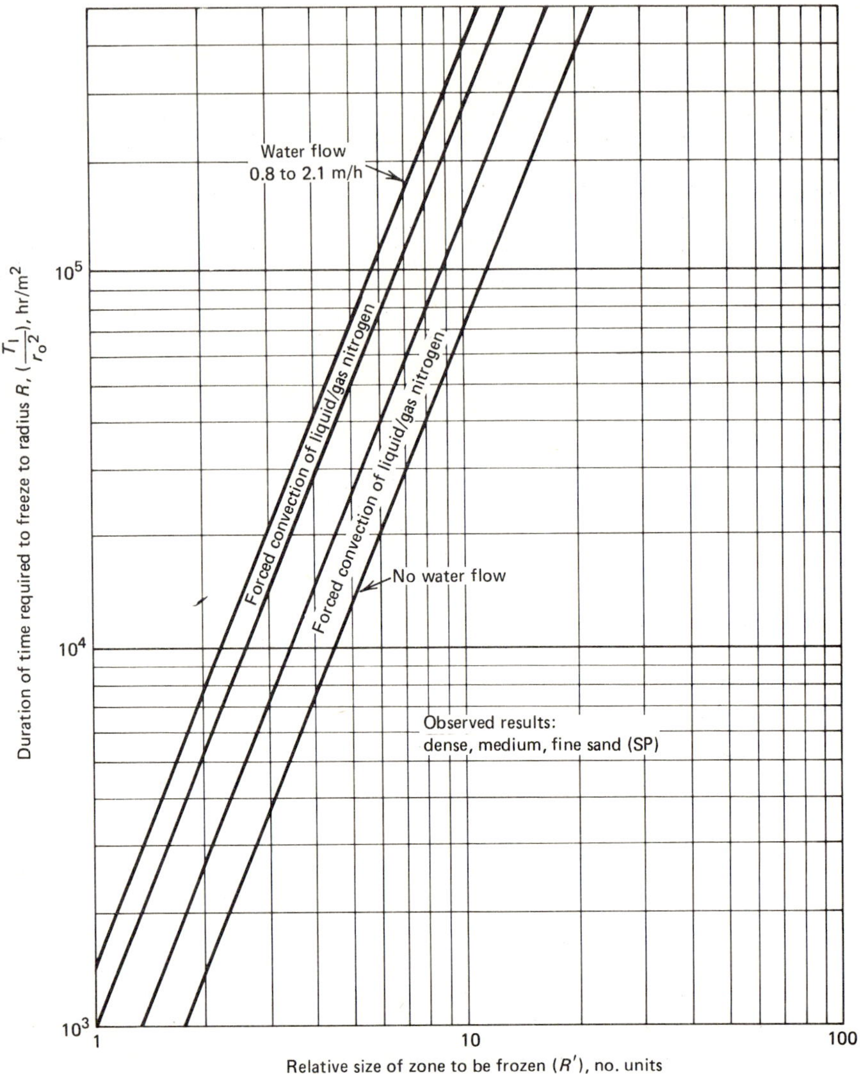

Figure 5.17 Influence of groundwater flow on required freezing time. *(After Shuster, Ref. 14.)*

PROBLEMS

1. (*a*) What are the ASTM definitions for shrinkage, plastic, and liquid limits of a soil?

   (*b*) Write a short description of the test used to evaluate each of these conditions.

   (*c*) When the tests are first described to a student (or when he or she first performs them), the general comments heard usually range from "unbelievable" to "nonsense." Can you think of a better way to evaluate the physical state of a fine-grained soil in terms of its variation in water content?

5.2. Table 5.1 lists four groups of cohesive soils, with their respective consistency limits. Note that these are all 100 percent clay soils.

(*a*) What influence would silt have if mixed in with the clay?

(*b*) What influence would sand or gravel have?

(*c*) What influence would organic material have?

(*d*) What influence would the addition of these factors have on the classification of fine-grained soils as shown on the plasticity chart in Appendix A?

5.3. (*a*) How is the preloading method expressed in equation form using the effective stress concept?

(*b*) Can water be used to preload soils? If so, how?

5.4 In evaluating soil consolidation times, the most critical soil variable is its coefficient of permeability.

(*a*) Considering that this is a low value for fine-grained soils, how is it usually evaluated?

(*b*) Under what geological conditions is the horizontal permeability of an in situ soil greater than its vertical permeability?

(*c*) How is the horizontal permeability sampled and/or measured?

(*d*) Is it reasonable to assume that using vertical permeability for design purposes is a conservative measure?

5.5. For a project involving the confined aquifer shown in Fig. P5.5, in which consolidation of the upper impervious stratum is the goal, what pumping rate can be anticipated? The wellpoints are 2.6 in in diameter at 5.0-ft spacings. The permeability of the sand is 0.22 cm/s.

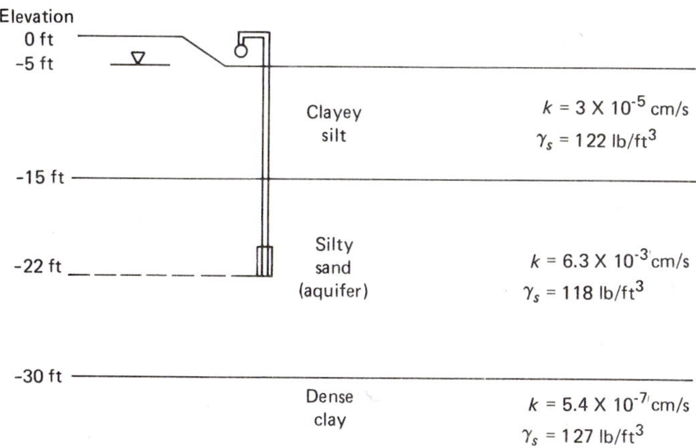

Figure P5.5

5.6 (*a*) What is the effective stress 10 ft below the surface in Prob. 5.5 before and after dewatering?

(*b*) What is the effective stress 20 ft below the surface before and after dewatering?

5.7. Using the data given in Fig. 5.7*a* for 82 wellpoints, and assuming that, for the situation described in Prob. 5.5, the water is drawn down to the bottom of the upper layer of clayey silt, how far beyond the wellpoints will the zone of influence be felt?

5.8. The usual radial drainage problem solutions (like that shown in Example 5.2) assume *only* radial drainage. However, pore water near the upper drainage blanket and near the bottom pervious layer (if one is present) surely must accept water in a vertical manner. What influence does this have on the following:

(*a*) The amount of consolidation settlement?

(*b*) The time rate of settlement?

5.9. Two problems connected with traditional sand drain design are the smearing of in situ soil that occurs during installation and the resistance the backfilled sand offers to the expulsion of pore water. Both decrease the effectiveness of the drain, although the effects of smear are considerably more important. How do you think these problems will be affected by the use of the newer style drain wicks?

5.10. (*a*) Consider a site where a thin sand layer covers a compressible silty clay, as shown in Fig. P5.10. What will be the time required to achieve 90 percent consolidation if the site is merely preloaded?

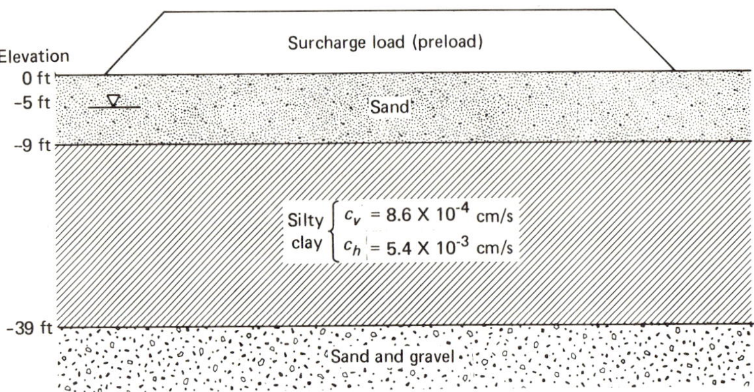

Figure P5.10

(*b*) If 18-in-diameter sand drains are placed at 10-ft spacings on a triangular grid pattern, what will be the time required to achieve 90 percent consolidation taking into account radial drainage alone?

(*c*) Develop a design graph showing time required to achieve 90 percent consolidation at this site in cases where sand drains of 6-, 12-, 18-, and 24-in diameters, respectively, are placed at spacings varying from 5- to 30-ft centers. Plot your results on log-log paper with time as the *x* axis and sand drain spacing as the *y* axis. (There should be four separate curves on the plot, one for each of the four sand drain diameters.)

5.11. In regard to the use of geotextile-encased plastic cores, the so-called "drain wicks," instead of sand drains, answer the following:

(*a*) Is there any difference in the shear strength of the foundation soil when drain wicks are used rather than sand drains?

(*b*) What influence would this difference have on the rate of surcharge fill placement?

(*c*) Discuss the possibility of the geotextiles which surround the plastic cores in the drain wicks shown in Fig. 5.11 becoming clogged.

5.12. Regarding electro-osmotic stabilization of fine-grained soils, answer the following questions:

(*a*) What different types of anodes can be used?

(*b*) What effect would the presence of salt water in the soil voids, as opposed to the presence of polluted groundwater, have on stabilization?

(*c*) During long-term stabilization does the electrical current have to be maintained 24 hours per day?

5.13. In areas where fine-grained soils have been polluted by hazardous materials such as PCBs or radioactive waste, it is conceivable that another chemical might be added which would neutralize, or stabilize, the hazardous material. Could electrochemical methods be used for this injection process? If so, discuss the technique you would use.

5.14. Regarding ground freezing:

(*a*) Why is there a limit to the maximum permissible groundwater flow rate?

(*b*) How does the rate of ground freezing vary with respect to the type of soil being stabilized, i.e., sand, silt or clay?

(*c*) How does the rate of ground freezing vary as the moisture content of the soil varies, say from 20 percent to fully saturated?

REFERENCES

1. *Final Report on Soil Stabilization Research,* Cornell University, Ithaca, New York, 1951.
2. Mooney, W. G., "Ground Water Control," *Civil Eng., ASCE,* March 1963, pp. 123–129.
3. Lancaster-Jones, P. F. F., E. McKeand, and F. G. Bell, "Ground Treatment," Chap. 14 in *Foundation Engineering in Difficult Ground,* F. G. Bell (ed.), Newnes-Butterworths, London and Boston, 1978, pp. 385–426.
4. Partos, A., and R. M. Koerner, "Monitoring Consolidation Settlements of Soft Soils Induced by Dewatering," in *Performance Monitoring for Geotechnical Construction,* ASTM-STP 584, 1975, pp. 111–126.
5. Barron, R. A., "Consolidation of Fine-Grained Soils by Drain Wells," *Trans. ASCE,* Vol. 113, 1948, pp. 718–754.
6. Johnson, S. J., "Foundation Precompression with Vertical Sand Drains," in *Proc. ASCE Spec. Conf. on Placement and Improvement of Soil to Support Structures,* 1968, pp. 9–39.
7. Hansbo, S., "Consolidation of Clay with Special Reference to Influence of Vertical Sand Drains," *Swedish Geotech. Inst.,* Proc. No. 18, 1960.
8. Koerner, R. M., "Consolidation of Compressible Soils by Sand Drain Installation and Surcharging," Master's thesis, Drexel University, Philadelphia, June 1963.
9. Kjellman, W., "Accelerating Consolidation of Fine Grained Soils by Means of Cardboard Wicks," in *Proc. Second Int. Conf. on Soil Mech. and Found Eng.,* Vol. 2, Rotterdam, 1948, pp. 302–305.
10. Seim, C., T. J. Walsh, and J. B. Hannon, "Wicks, Fabrics and Sawdust Overcome Thick Mud," *Civil Eng., ASCE,* July 1981, pp. 53–56.
11. Hansbo, S., "Consolidation of Clay by Band-Shaped Prefabricated Drains," *Ground Engineering,* July 1979, pp 16–25.
12. Mitchell, J. K., "Soil Improvement—State-of-the-Art Report," in *Proc. Tenth Int. Conf. on Soil Mech. and Found. Eng.* Vol. 4, pp. 509–565.
13. Farmer, I. W., "Electro-Osmosis and Electrochemical Stabilization," in *Methods of Treatment of Unstable Ground,* F. G. Bell (ed.), Newnes-Butterworths, London, 1975, pp. 26–36.
14. Shuster, J. A., "Controlled Freezing for Temporary Ground Support," in *Proc. First N. Amer. Rapid Exch. and Tunneling Conf.,* Chicago, June 1972, pp. 863–894.
15. Gail, C. P., "Tunnel Driving Using Subsurface Freezing," *Civil Eng., ASCE,* May 1972, pp. 37–40.
16. Jessberger, H. L., "A State-of-the-Art Report: Ground Freezing: Mechanical Properties, Processes and Design," *Eng. Geol.* (Amsterdam), Vol. 18, 1981, pp. 5–30.

SIX

GROUTING

6.1 OVERVIEW

Grouting is a process whereby fluidlike materials, either in suspension or solution form, are injected into subsurface soil or rock for one, or more, of the following purposes:

To decrease permeability
To increase shear strength
To decrease compressibility

The suspension-type grouts include soil, cement, lime, asphalt emulsion, etc., while the solution-type grouts include a wide variety of chemicals. Each type will be described in detail later in the chapter.

The basic function the grouting is intended to serve should always be kept in mind when considering different aspects of the subject. Among those basic functions, shown in Fig. 6.1, are the following:

Permeation or penetration. Here the grout flows into the soil voids or rock seams with only minimal effect on the original structural arrangement.
Compaction or controlled displacement. In this situation the grout remains more or less intact as a mass and exerts pressure on the soil (usually) or rock (in some cases), to compact and thereby densify the weaker zones of the nearby areas.
Hydrofracturing or uncontrolled displacement. When the grouting pressures are greater than the tensile strength of the soil or rock being grouted, the latter material fails, and the grout rapidly penetrates into the fractured zone. Thus lenses of nearly solid grout are projected into densified regions of the in situ soil or displaced rock masses.

270

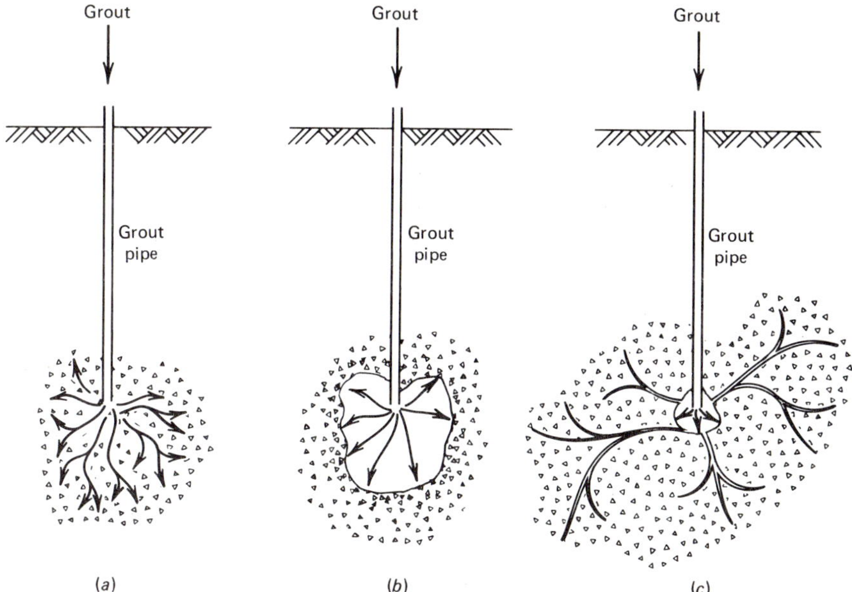

Figure 6.1 Various functions involved in soil and rock grouting. (*a*) Permeation grouting (penetration). (*b*) Compaction grouting (controlled displacement). (*c*) Hydrofracturing (uncontrolled displacement).

The technology of grouting, as with many other foundation methods, is far from new, yet it is in a constant state of flux, with new materials and construction techniques being developed constantly. Modern grouting began in the mining industry, with its concern over seepage and strength control in mine tunnels and shafts, then was taken up by civil engineering, where it was used in the construction and maintenance of subways, deep building foundations, and large dams. It is now in use in all the fields of foundation engineering, the most recent being to provide for the cutoff of seepage from, and the containment of, hazardous waste landfills (see Chap. 8). It has long been tagged "an art, not a science," but this somewhat negative image is changing as recent research and development work broadens our knowledge of the area. Key references are books by Cambefort,[1] Bowen,[2] Tallard and Caron,[3] Karol,[4] various conference proceedings,[5-6] and an extensive bibliography that has been compiled on the subject.[7] As much detail as possible from these will be included in this discussion, as well as from other relevant references in the literature of foundation engineering.

6.2 SUSPENSION GROUTS

Suspension grouts are made up of solid particles (soil, cement, lime, asphalt emulsion, etc.) carried in water. When these suspensions are injected into soil for the purpose of permeating the voids of that formation, the relationship between grout particle size and soil void size must be considered. In regard to this, the following relationship has been proposed as a rough guide[8]:

$$GR = \frac{d_{15}\ (\text{formation})}{d_{85}\ (\text{grout})} > 20 \tag{6.1}$$

Where GR = groutability ratio (seen in Fig. 6.2)

d_{15} = particle size at which 15 percent of the soil is finer (of the formation being grouted)

d_{85} = particle size at which 85 percent of the soil is finer (of the grout being injected)

This criterion basically limits the use of suspension grouting to permeation of sands and gravels. When this is the case, it appears as though the travel distance from the grout pipe varies as one-third the power of pumping pressure, pumping rate, soil permeability, and soil porosity. This has been quantified in the following manner[8]:

$$r = 0.62 \left(\frac{Rgt}{n} \right)^{1/3} \tag{6.2}$$

where r = grout penetration radius, ft

t = grout gel time, min

n = porosity of soil formation

g = acceleration due to gravity, ft/s^2

R = ratio of viscosity of water to viscosity of grout

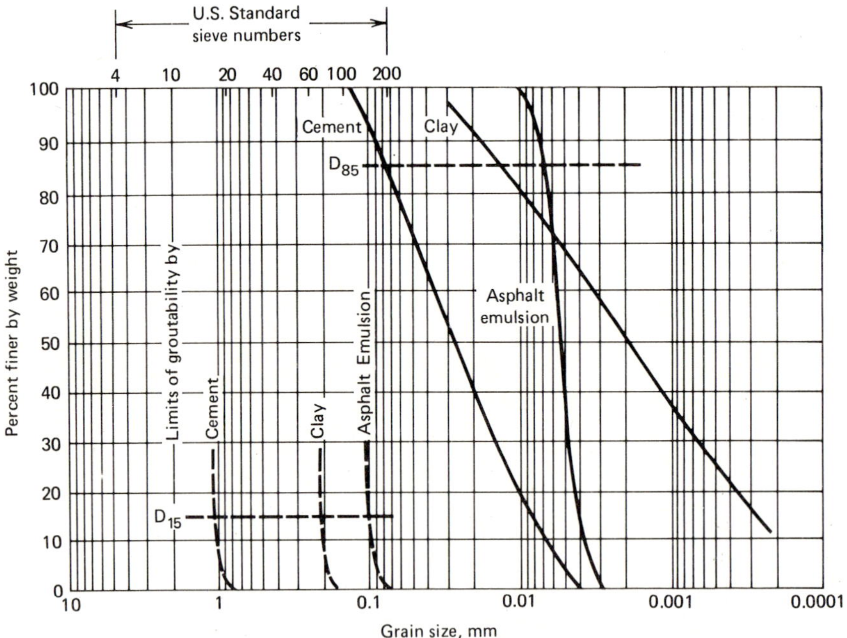

Figure 6.2 Groutability of soils by various suspension grouts. *(After Kravetz, Ref. 8.)*

Other considerations that must be taken into account in grouting design are the grout's setting time (if it sets too quickly it may harden in the grouting equipment, and if too slowly it can get washed away by moving groundwater), and its stability (the suspended particles should not settle out before the grout is in its final position).

6.2.1 Grouting with Soil Itself

In dealing with very coarse, in situ soil formations, soil itself can be used to fill up (permeate) some of the void volume. However, even fine sands and silts used for this purpose settle out quite quickly after being injected. Therefore, the grout is usually composed of a very-fine-grained soil—bentonite clay being a commonly used material. The structure of bentonite clay is such that water is readily absorbed into its surface, and its viscosity, strength, and flow properties can be controlled reasonably well. Jefferies[9] has evaluated some of these properties and has shown how sensitive they are to various mixing procedures; see Fig. 6.3. (More detail regarding clay soils and their permeability characteristics will be given in Chap. 8 in the discussion of landfill liners and covers.)

When the water-to-soil ratio is kept very low, e.g., is kept equivalent to a zero slump concrete, almost no permeation of the grout into the soil voids occurs. Pressure is then exerted by the grout against the soil mass from the outlet of the grout pipe, which causes densification and the movement of adjacent areas (so long as they are not constrained from moving). This latter application was originally known as "mud-jacking" and was used to raise pavement slabs, to underpin shallow building foundations which had settled, etc. Currently, it is also used to densify in situ soils and, in a liberal sense, to form soil compaction piles. Obviously, the use of a more stabilized material would do better in this regard. Such materials are bentonite-cement mixes and/or cement itself.

6.2.2 Grouting with Bentonite-Cement Mixes

The addition of cement to bentonite results in a suspension which has interesting synergistic properties and has been widely used as a permeation grout. The area of its greatest use has probably been to grout beneath dams—an application we will treat more fully in Sec. 6.2.3 on cement grouting.

A carefully monitored clay-cement grout project reported by Terzaghi and Lacroix[10] used a number of separate injections to create an impervious cutoff beneath an existing dam. Permeabilities were measured before and after each injection and a ratio k_{before}/k_{after} was calculated and plotted as seen in Fig. 6.4. The clay-cement grout gradually decreased the permeability but after the sixth application a more fluid clay soil grout was introduced which was very effective. The entire process reduced the permeability approximately 500 times.

Grouts will have different properties depending on the amounts of soil, cement, and water they contain, just as they will differ according to the type of soil and type of cement. Laboratory testing using sequential variation of the parameters involved is necessary to determine those properties accurately. Figure 6.5 shows the type of

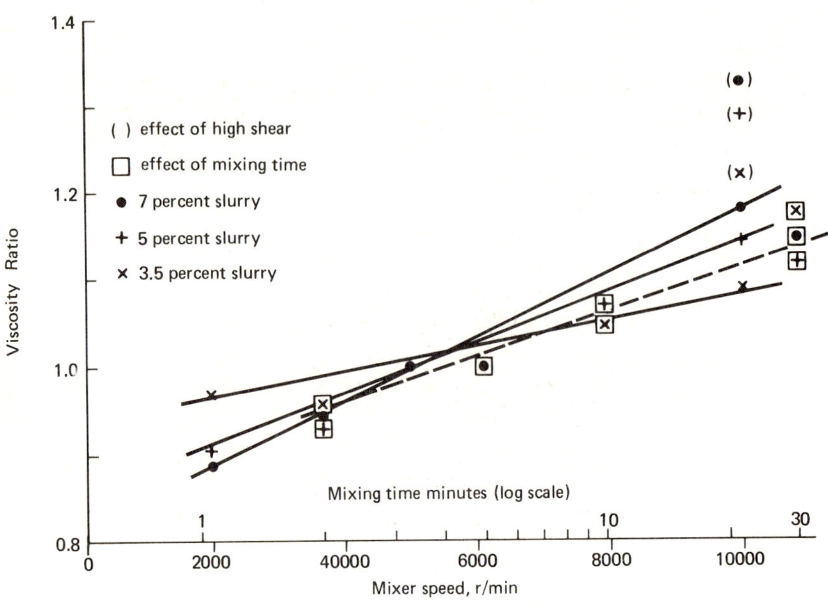

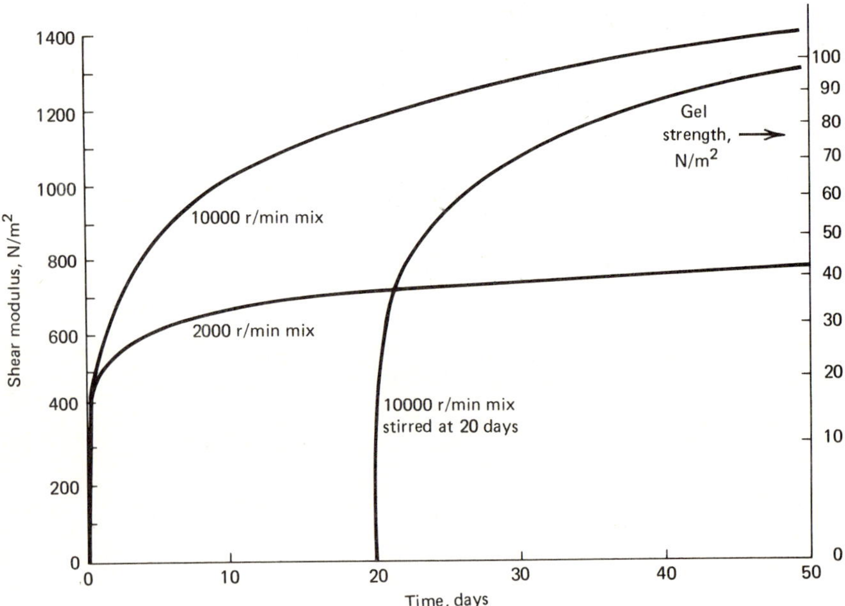

Figure 6.3 Effect of mixing time and procedure on properties of bentonite clay grout. *(After Jefferies, Ref. 9.)*

274

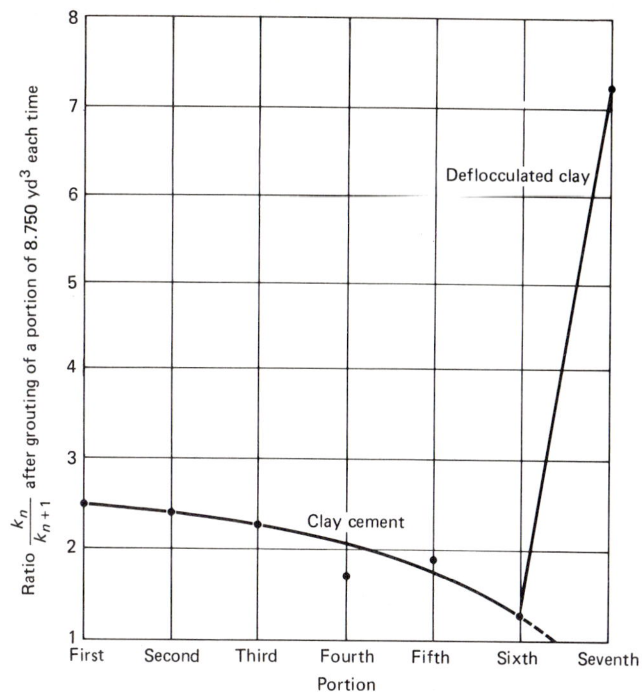

Figure 6.4 Grout curtain permeability before and after a number of separate clay-cement grouting injections. *(After Terzaghi and Lacroix, Ref. 10.)*

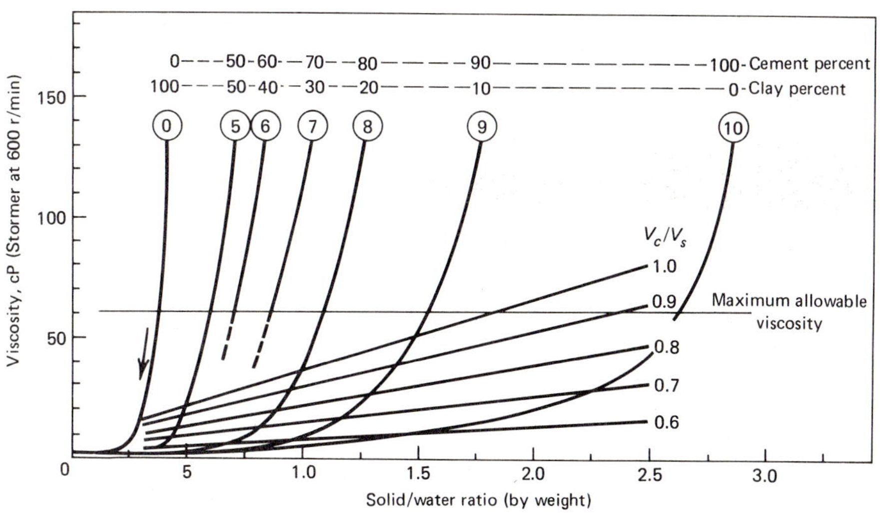

Figure 6.5 Curves required to describe accurately the basic properties of clay-cement-water grouts; shown here are curves for viscosity. *(After Kravetz, Ref. 8.)*

information that is required for one such property: viscosity. Changes in viscosity are shown for various solid-to-water ratios and for different cement-to-clay ratios. It can easily be seen that the viscosity varies widely as the mix changes, and similar variations also occur in the permeability and strength of these materials.

Low-water-content soil-cement mixtures, like soil alone, can serve as displacement grouting. The advantage of these mixtures over soil is permanence of the grout; their disadvantage is increased cost. The advantages and disadvantages of other aspects such as equipment, pressures they can exert, and pumping rates, however, are roughly equivalent.

6.2.3 Grouting with Cement Mixes

The literature on cement grouting is simply enormous. More has been written on its use in seepage cutoff beneath dams than on any other subject in grouting. Cement grouts are usually formed from Type I (standard) Portland cement and water. Figure 6.6 shows the effects of differing water/cement ratios on compressive strength, shear (flow) strength, and bleeding. *Bleeding* is where the cement particles come out of suspension before complete curing of the grout has occurred, and is obviously more pronounced the greater the water content and fluidity of the grout. Other ingredients sometimes used in cement mixes are fine sand, clay (as discussed in Sec. 6.2.2), fly

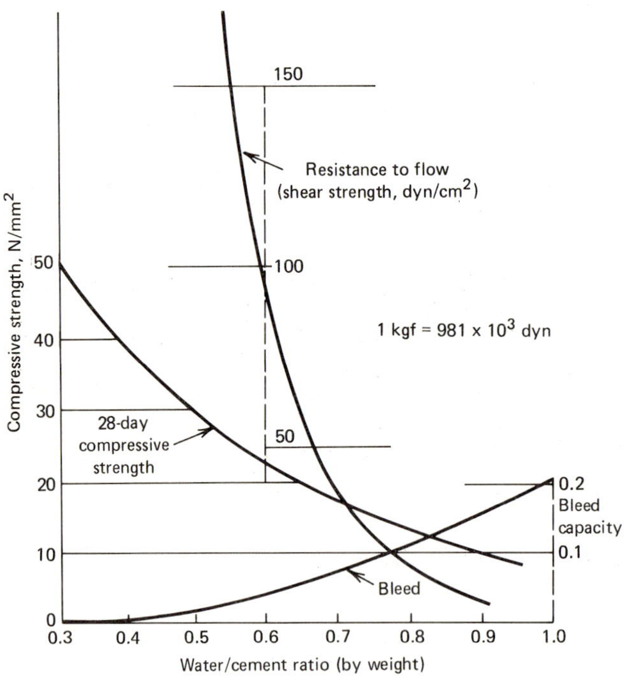

Figure 6.6 Effects of water content on cement grout properties. *(After Littlejohn, Ref. 11.)*

ash, fluidizers, accelerators, and retarder or expansion additives; Littlejohn[11] discusses the effects of each.

Regarding the individual constituents in cement grouts:

Water should be of drinking quality, although seawater has been used successfully where there is no contact between the grout and steel reinforcing.

Cement is usually Type I Portland, in which approximately 99 percent of the particles pass 44 to 100 μm. This particle size is adequate for soils that allow a grout flow of 0.05 cm/s or for rock that contains openings about 160 μm in size. Finer soils or openings must use Type III Portland cement, which is adequate to penetrate soils that allow a flow of about 0.001 cm/s and rock that contains openings 20 μm in size. These are the practical limits of cement grouting.

Clay is considered a filler in cement grouts (i.e., it is present in amounts $\leq$ 3 percent by weight) and can be used to reduce bleeding by holding the cement in suspension while it is setting.

Sand is a true filler in cement grout and is used to reduce the overall cost of the material.

Pozzolans, such as fly ash or ground slag, are sometimes used as fillers but their properties and their reactions to the grout being used should be determined through laboratory testing before actual field work commences.

Additives or *admixtures* are meant to impart controlled features to the final composition of the grout. Table 6.1 lists a number of them.

When selecting a particular cement grout for use, one would obviously like to know its final strength, flow rate, set time, shrinkage, permeability, and durability. Some information on these properties can be found in textbooks on concrete and cement grouts, e.g., in books by Popovics,[12,13] but laboratory testing of the proposed material is always recommended.

6.2.4 Grouting with Asphalt Emulsions

The oil refining industry can process its residues in an emulsifying plant in such a way as to produce nearly circular droplets of asphalt suspended in a water matrix (a small amount of emulsifying agent is added, as are, sometimes, other manufacturing controls). By choosing the proper emulsifying agent, one can obtain negatively-charged anionic asphalt globules (for maximum flow and penetration) or positively charged cationic asphalt globules (to attach to negatively charged clay soils). Furthermore, manufacturing techniques can allow for rapid setting (RS), medium setting (MS), or slow setting (SS) varieties in both anionic and cationic types. Thus the final products result in globular spheres 1 to 2 μm in diameter,[14] which can be injected into the soil voids or rock cracks and can "break down" at controlled rates.

Soils into the medium silt range have been grouted using emulsified asphalt, and so have rock masses with cracks as fine as 10 μm is size. The type of emulsion setting needed, RS, MS, or SS, depends on the flow of groundwater in the voids. The use of

Table 6.1 Common additives to cement grout used to impart specific properties to the final product†

Admixture	Chemical	Optimum dosage, % cement wt	Remarks
Accelerator	Calcium chloride	1–2	Accelerates set and hardening
	Sodium silicate	0.5–3	Accelerates set
	Sodium aluminate	0.5-3	Accelerates set
Retarder	Calcium lignosulphonate	0.2–0.5	Also increases fluidity
	Tartaric acid	0.1–0.5	
	Sugar	0.1–0.5	
Fluidiser	Calcium lignosulphonate	0.2–0.3	
	Detergent	0.05	Entrains air
Air entrainer	Vinsol resin	0.1–0.2	Up to 10% of air entrained
Expander	Aluminum powder	0.005–0.02	Up to 15% preset expansion
	Saturated brine	30–60	Up to 1% postset expansion
Antibleed	Cellulose ether	0.2–0.3 (for $w <$ 0.7)	Equivalent to 0.5% of mixing water
	Aluminum sulphate	Up to 20% (for $w < 5$)	Entrains air

†After Littlejohn, Ref. 11.

slow-setting emulsions are generally recommended since they can travel the greatest distance into the material being grouted.

Although asphalt emulsions can permeate soils down into the silt range, medium to fine sands and silts are usually grouted with solution grouting, the topic of Sec. 6.3.

6.3 SOLUTION GROUTS

The solution grouts are indeed numerous. Tallard and Caron[3] use the following classification (which is a bit too detailed for our purposes here, so that we will discuss only the more commonly used of them):

A. Aqueous solutions
   1. Silicate derivatives
   2. Other mineral gels
   3. Lignosulfite derivatives
   4. Other plant derivatives
   5. Polyacrylamides
   6. Phenoplasts
   7. Aminoplasts
   8. Combinations of the above
B. Collodal solutions
   1. Organic solutions
   2. Mineral solutions
C. Nonaqueous solutions
   1. Synthetic resins
   2. Vulcanizable oils
   3. Bitumen and other heated materials
   4. Solvent systems
D. Emulsions
   1. Bituminous
   2. Others
E. Products reacting with the ground
   1. Reaction with ground or groundwater salts
   2. Reaction with groundwater
F. Combined systems

Karol and Welsh[15] and Karol[16] discuss items A.1, A.3, A.5, A.6, A.7, and C.1, and since they are the most commonly used solution grouts, we will comment on them briefly also. Before doing so, however, it should first be noted that solution grouts can generally permeate finer soils than can suspension grouts. See Fig. 6.7 and compare the responses shown there with the curves given in Fig. 6.2 for suspension grouts.

6.3.1 Silicate Derivatives

Sodium silicate (Na_2SiO_4, also called "water glass") is commercially available as a relatively inexpensive (compared with other chemicals, that is) aqueous solution. Insofar as grouting goes the basic chemical reaction with which we are concerned is between it and carbonic acid (which is itself produced by water and carbon dioxide):

$$H_2O + CO_2 \rightleftharpoons H_2CO_3 \text{ (carbonic acid)}$$

$$Na_2SiO_4 + H_2CO_3 \rightarrow Na_2CO_3 + H_2SiO_4$$

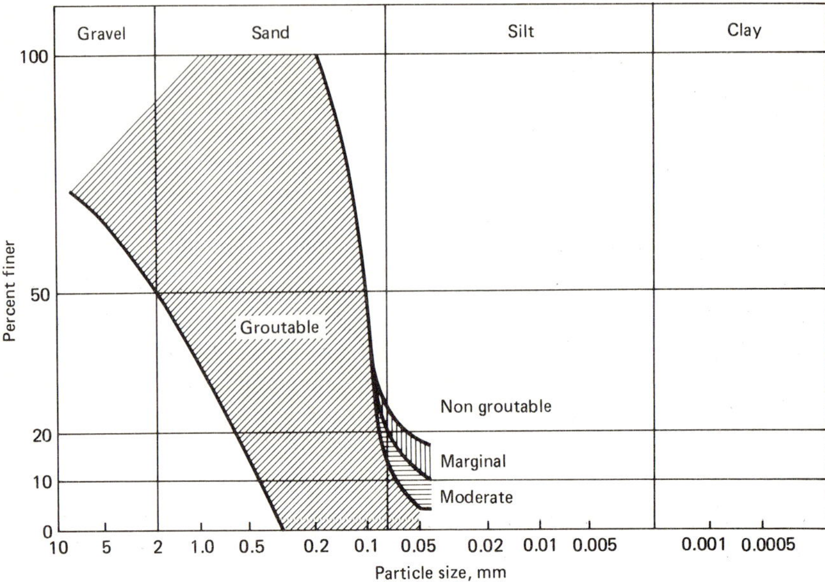

Figure 6.7 Groutability of soils by various solution grouts. *(After Baker, Ref. 17.)*

In the above reaction, salt$_1$ (the sodium silicate) and acid$_1$ (carbonic acid) react to form salt$_2$ (sodium carbonate, a solid) and acid$_2$ (silicic acid, a liquid). Silicic acid further breaks down to

$$H_2SiO_3 \rightarrow H_2O + SiO_2$$

where the silicon dioxide is a solid precipitate. Thus the two original liquids react to form two solids which remain in the soil voids or rock cracks as the grout, as well as a small amount of residual water which causes some shrinkage (called *syneresis*) that is of concern.

The above reaction, however, is quite slow, and so a salt, like calcium chloride, can be added, which greatly increases the speed of reaction. Then the primary solid precipate becomes calcium silicate ($CaSiO_3$), and reaction times are so fast that two separated injection pipes are required so that the mixing is done in situ as close as possible to the area targeted for stabilization.

The method just described, known as the *two-shot Joosten process,* was developed in the 1930s but is rarely used today. Instead, a delayed and controlled reaction has also been developed using an organic reagent, usually ethyl acetate or formamide, which slowly breaks down the acid and produces a collodal silica gel. The gellation period can be varied from several minutes to a few hours, which allows for greater penetration of the area to be stabilized by the grout, more homogeneous conditions in the grout, and greater overall economy of operation. The common silicate-based grout used today contains sodium silicate, a reactant (formamide in the United States), an accelerator (one of several possible salts), and a water carrier in proportions roughly of 40, 10, 10 and 40 percent respectively.

Work by Warner[18] on a number of patented grout types and by Clough et al.[19] on variations of the above formula has led to a reasonably good understanding of grout properties. For example, the work by Clough et al. used proportions consisting of 30 to 70 percent silicate, 6 to 12 percent formamide, 0 to 9.4 g/l of calcium chloride, and 18 to 64 percent water. This range of variables resulted in gel times from 1 to 12 h, triaxial stress-strain responses as shown in Fig. 6.8a, secant modulus

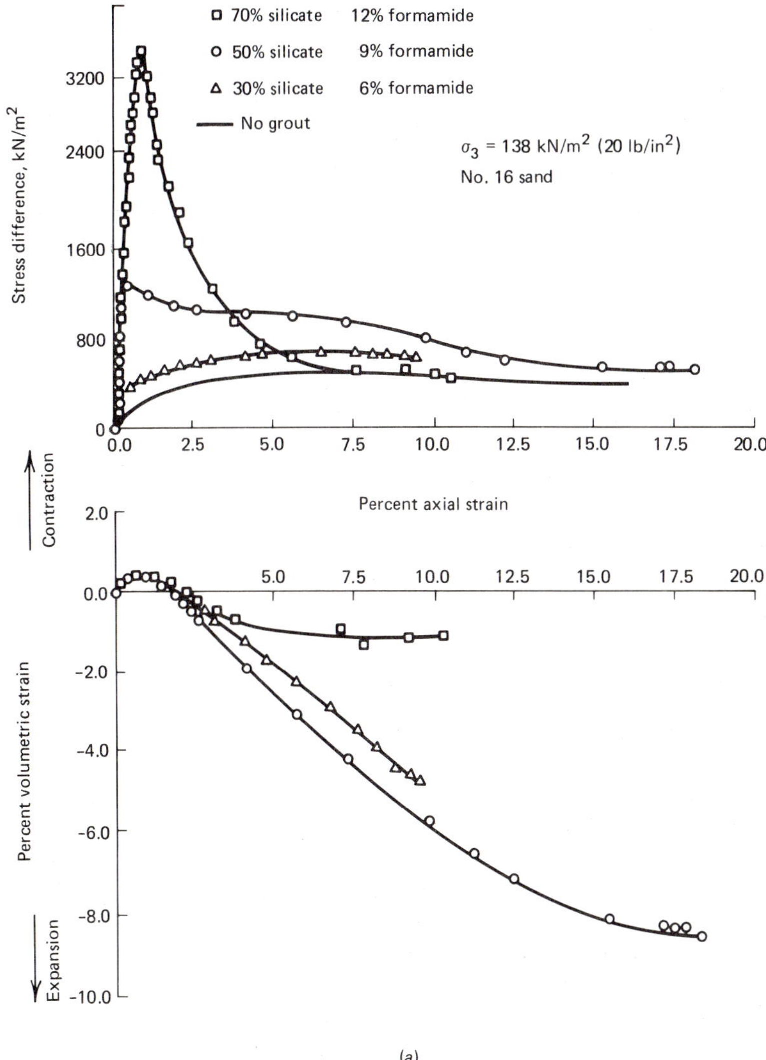

Figure 6.8 Results of silicate-stabilized sands in triaxial shear and creep modes. (a) General triaxial shear response. (b) Modulus response to confining pressure. (c) Accelerated creep behavior. (d) Long-term creep behavior. *(After Clough et al., Ref. 19.)*

behavior as shown in Fig. 6.8*b*, and accelerated and long-term creep behavior as shown in Fig. 6.8*c* and *d*, respectively. These response curves show that silicate-grouted soils can indeed be quite strong and of high modulus, but are somewhat sensitive to long-term sustained loads, i.e., they can undergo creep.

Recently, formamide has been labeled by the Environmental Protection Agency (EPA) as a possible carcinogenic agent and a replacement is being actively sought after by many organizations. The general direction of research seems to be into the use of organic reactants without accelerators. Reference 6 contains several papers on such materials.

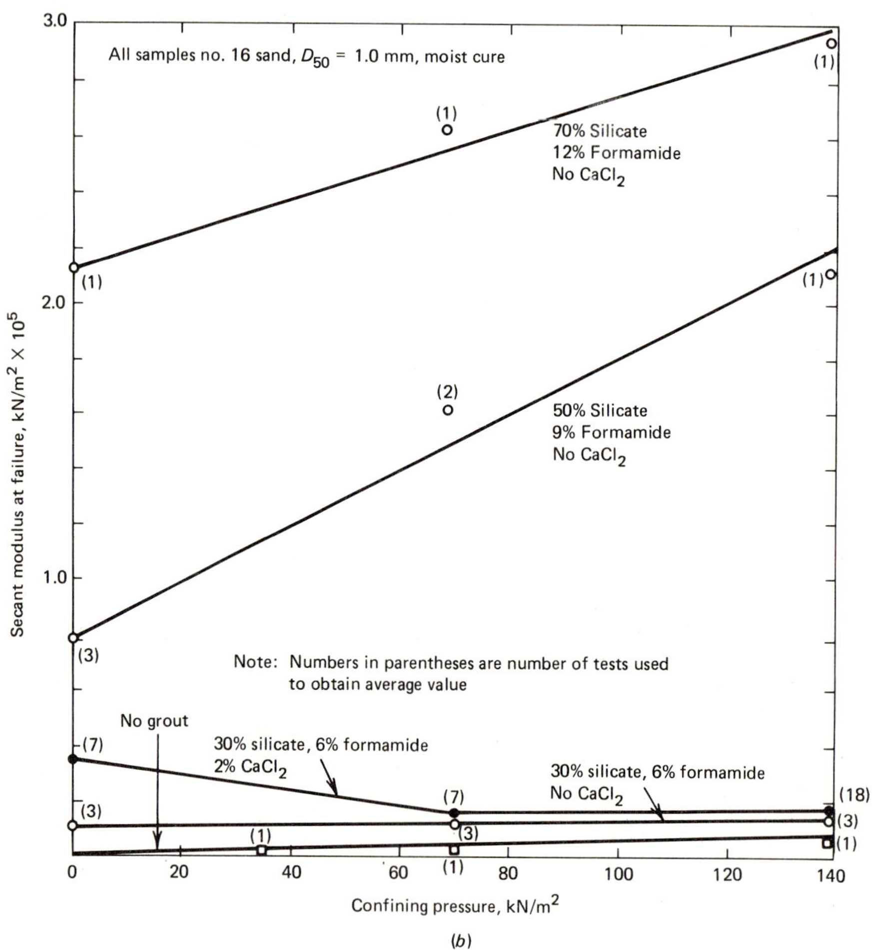

Figure 6.8 (*Continued*)

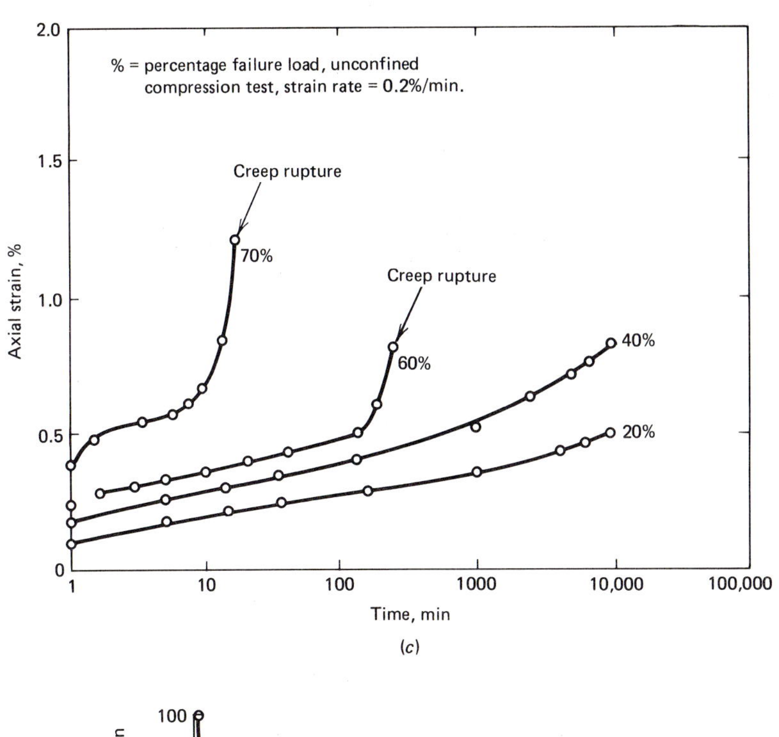

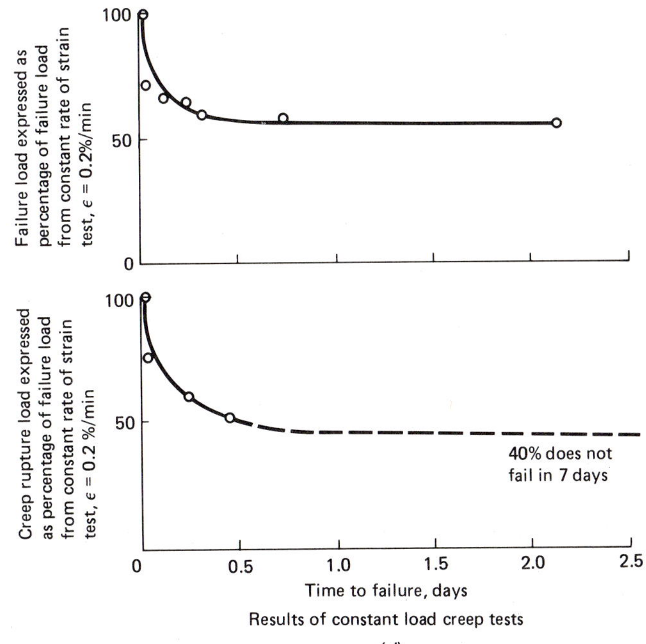

Figure 6.8 (*Continued*)

6.3.2 Lignosulfate Derivatives (Chromelignins)

The wood processing industry (e.g., paper mills) produces a waste liquor, known as *lignin,* which, if dumped into creeks, would be a major source of water pollution. Lignin, however, is useful in grouting, and the Clean Streams Act of 1972, which restricted such dumping, was a major stimulant to developing it as a grout material. The raw lignin can be used directly, or as a dried, catalyst-mixed powder to which other additives are sometimes introduced before injection. This introduction of additives is important for control of gel time and uniformity in performance of the grout because the lignin itself is an extremely variable material.

The grouts are usually made up of lignosulfates and a hexavalent chromium compound, e.g., calcium lignosulfate and sodium dichromate. Combined with an acid, the chromium ion changes valence from $+6$ to $+3$, thereby oxidizing the lignosulfate into a gel. Acids or acid salts are sometimes added to control the grout's pH.

Properties, as described by Karol,[4] are the following:

Composition: 75 percent lignosulfate, 20 percent sugar, 5 percent ash
Viscosity: 3 to 8 cP
Set time: variable
Strength: 50 to 200 lb/in^2 (unconfined compression)
Creep limit: 25 to 50 percent of the above strength

While the use of a waste byproduct is indeed noble, it must be noted that hexavalent chromium (the catalyst present in all lignosulfate grouts) is highly toxic. Its toxicity is lessened by its reaction with the lignosulfate, but the reaction is not usually complete and so a residual amount of toxicity always remains.

6.3.3 Polyacrylamides

There are a number of acrylamide grouts, most of which are characterized by constant viscosity, good penetrability, good control of gel time, and adequate strength. In general, Karol[16] describes them as two organic monomers: acrylamide and a cross-linking agent, such as methylene-bisacrylamide, in a ratio of approximately 20:1. They are catalysed with a two-component redox system. One part, the catalyst or initiator, is a peroxide or persalt; the other part, the accelerator or activator, is usually triethanolamine (TEA). There are many variations on the above composition available (or formerly available) on the commercial market, some of which are listed here:

AM-9 (American Cyanamid Co. product withdrawn in 1978)
Nitto SS (Japanese product banned in Japan in 1974)
Rocagil BT (French product of Rhone Progil Co.)
Rocagil 1295 (French product of Rhone Progil Co.)
Sumisoil (Japanese product of Sumatomo Chem. Co. Ltd., withdrawn)

Q-Seal (distributor's trade name for AM-9)
PWG (distributor's trade name for AM-9)
Terragel (American product withdrawn in 1979)

The reason for withdrawing, and in one case banning, several products is that acrylamide in solution or in powder form is neurotoxic, although the gel is nontoxic. By following the manufacturers' recommendations, however, the health hazard can be eliminated. Acrylamide grouts should not be used near potable water sources.

Properties of acrylamide grouts, as described by Karol,[16] are as follows:

Gel time: 5 s to 12 h (controlled by temperature, catalyst, activator, and inhibiter)
Strength: 50 to 200 lb/in^2 (unconfined compression)
Creep limit: 25 to 50 percent of the above strength
Permanence: excellent

6.3.4 Phenoplasts

Phenoplast grouts are formed from the reaction of phenol to an aldehyde. At ambient temperatures, they require an acid to react. The phenoplast commonly used for grouting is resorcinal in reaction with formaldehyde, where sodium hydroxide is used as a catalyst. Its primary advantage is a very low viscosity (1.5 to 3 cP), which means it can penetrate fine soils through grout pipes spaced relatively far apart.

As described by Karol,[16] there are three commercially available materials:

Rocagil (Rhone Progil Co.)
Geoseal (Borden Chem. Co.)
Terranier (ITT Rayonier Co.)

It should be noted that all of them contain phenol, a formaldehyde, and an alkaline base, and thus are health hazards (i.e., are toxic) and potentially environmental pollutants.

6.3.5 Aminoplasts

The major ingredients in aminoplast grouts are urea and formaldehyde. They require an acid environment to complete their reaction. The grout has low viscosity (10 to 20 cP), produces a stabilized soil strength of about 200 lb/in^2,[15] and exhibits creep limits higher than the other solution grouts so far discussed.

The following materials are available commercially:

Rocagil (Rhone Progil Co.)
Cyanaloc (American Cyanamid Co.)
Herculox (Halliburton Co.)
Diarock (Diamond Shamrock Co.)

Although toxicity and environmental pollution are not so problematic with aminoplasts as with other solution grouts, they give off ammonia in reaction, which precludes their use in confined or poorly ventilated areas.

6.3.6 Water-Reactive Grouts

Materials which gel or polymerize with water have been evaluated as possible grouts. Two products are available, both polyurethanes[17]:

TACSS (Japanese product)
CR 250 (used primarily as a sewer sealant)

They have low viscosity, but exhibit the toxicity and environmental problems associated with the other grouts just reviewed.

6.3.7 Water-Reactive Foam Grouts

A series of grouts which foam in reaction with water have been evaluated by Vinson and Mitchell.[20] They investigated many such grouts and found that polyurethane holds the most promise. Catalysts are used to control the reaction rate and the nature of the foaming process. Surface-active agents also control the foam's bubble size. In general, they found that polyurethane-foamed plastics for grouting have the following properties:

Viscosity about 13 cP
Reaction times > 1 min
Strength 1400 to 6000 lb/in^2 (unconfined compression)
Permeability $\simeq 10^{-6}$ cm/s

This approach to soil grouting seems quite intriguing although it has not yet seen a great deal of use.

6.3.8 Summary

Karol[21] has presented an interesting comparison between the relative degrees of toxicity, viscosity, and strength of the various solution grouts described in this section, which we summarize in Table 6.2. When that information is coupled with the graphic data shown in Fig. 6.7, one has a reasonable basis on which to select the best solution grout for a given soil.

We are now also in a position to compare solution and suspension grouts in regard to how the grain sizes of formation soils limit their use. Little[22] gives one such comparison and in Fig. 6.9, based on that reference, it can be seen that the order of grout penetrability, with grouting materials ranked according to their ability to pen-

Table 6.2 Relative ranking of solution grouts as to their toxicity, viscosity, and strength†

Grouts	Corrosivity or toxicity	Viscosity	Strength
Silicates			
Joosten process	Low	High	High
Siroc	Medium	Medium	Medium high
Silicate-bicarbonate	Low	Medium	Low
Lignosulfates			
Terra Firma	High	Medium	Low
Blox-all	High	Medium	Low
Phenoplasts			
Terranier	Medium	Medium	Low
Geoseal	Medium	Medium	Low
Aminoplasts			
Herculox	Medium	Medium	High
Cyanaloc	Medium	Medium	High
Acrylamides			
AV-100	High	Low	Low
Rocagel BT	High	Low	Low
Nitto-SS	High	Low	Low
Polyacrylamide			
Injectite 80	Low	High	Low
Acrylate			
AC-400	Low	Low	Low
Polyurethane			
CR-250	High	High	High
CR-260			

†After Karol, Ref. 21.

etrate soils of decreasing particle size, is as follows:

Cement
Clay-cement
Clay
Silicates (and other chemicals)
Asphalt emulsions

6.4 GROUTING EQUIPMENT AND METHODS

Because of the different functions of grouting, the differences between grouts themselves, and the differences between materials to be grouted, generalizations about grouting equipment and methods are difficult to make. Nevertheless, an overview is important.

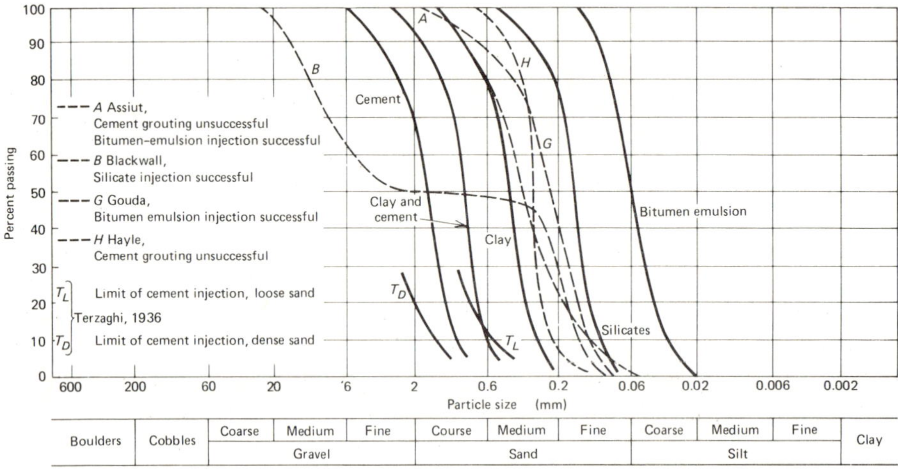

Figure 6.9 Limit of penetrability of different grouts (both suspension and solution). *(After Little, Ref. 22.)*

6.4.1 Drilling of Grout Holes

Most grout holes are drilled with an EX bit, which results in diameters of approximately $1\frac{1}{2}$ in. The holes are usually drilled with rotary drilling rigs, and diamond bits are used for rock. Percussion drilling equipment is sometimes used in soil and casing is always needed. Since samples of the rock or soil cannot be taken using these methods, one can specify that a small percentage of the hole be cored or cased so that samples can be retrieved.

The holes are usually drilled vertically unless a specific objective requiring another orientation is involved. Two of the common objectives for which other orientations are needed are angling beneath building foundations and intersecting fracture planes.

In rock, the grout holes are always washed out and pressure-tested. The volume of water inflow is measured and plotted against a log of the hole. A grouting plan is then made after inspecting the adjacent holes also. The "Lugeon unit" is sometimes used, 1 Lugeon unit being equal to a water take of 1 liter per meter of hole per minute at 10 bars of pressure. Since this is about 150 lb/in² of pressure and normal grouting pressures are much lower, a refined value is used as follows[23]:

$$\text{Lugeon value} = \text{water take } [l/(m \cdot min)] \frac{10 \text{ (bars)}}{\text{actual pressure (bars)}}$$

$$= 0.0107 \text{ water take } [ft^3/(ft \cdot min)] \frac{150 \text{ (lb/in}^2)}{\text{actual pressure (lb/in}^2)}$$

("Take" is actually the water outflow from the grout hole and is a commonly used grouting term.) For a sense of proportion 1 Lugeon $\simeq 1.3 \times 10^{-5}$ cm/s ($\simeq 10$ ft/h). Materials allowing only 1 Lugeon of water take are solid enough as foundations that grouting is not necessary, materials allowing 10 Lugeons generally need grouting, and materials allowing 100 Lugeons have open joints and large voids that must be grouted.

6.4.2 Injection Methods

Caron et al.[24] describe several types of grout injection methods:

1. Bottom-up grouting, shown in Fig. 6.10*a*, drives the casing to full depth, withdraws it to a specified elevation, and then injects the grout. The process continues upward to the surface of the material being grouted.
2. Sealed-in sleeve pipe injection (with tube à manchettes), shown in Fig. 6.10*b* and in more detail in Fig. 6.11, allows for several injections at the same elevation. Injection is done by a double packer which is inserted in the sleeve pipe to the desired location. Although the method is more complicated than bottom-up grouting and the initial cost during drilling is greater, it allows for independent drilling and grouting operations and the selection of different grout viscosities as the grout take varies at different locations.
3. Simultaneous drilling and grouting, shown in Fig. 6.10*c,* allows for injection during rotary drilling. At a predetermined distance, the drill rod is withdrawn and grout is injected into the soil or material through a separate drill rod. The process continues from the top downward. It is only suitable for pervious, granular soils.

6.4.3 Grout Plant Equipment

The mixing plants and delivery systems for suspension and solution grouts differ mainly in their storage and mixing configurations. A typical cement (suspension) grouting plant is shown in Fig. 6.12. In essence it consists of a mixer, an agitator, a pump, and piping connected to the grout holes. Two systems are shown for the piping. One is a circulating type where unused grout is returned to the agitator, the other a single-line type where the grout not pumped into the hole is wasted.

For solution grouts, separate ingredients are stored in stationary tanks or tank trucks and metered out (on a flow volume basis), mixed at junction points, and brought to the intended grout pipe. The grout pipes are connected in a manifold system, but each is separately valved so that complete control is obtained over their flow.

6.4.4 Grout Injection Measurements

The basic information to be measured during the grouting process is weight for suspension grouts or volume for solution grouts, and grout flow rate and pressure for both.

Accurate measurements of weight or volume are required to confirm mix proportions. In solution grouting, Baker[17] recommends positive displacement meters for this purpose.

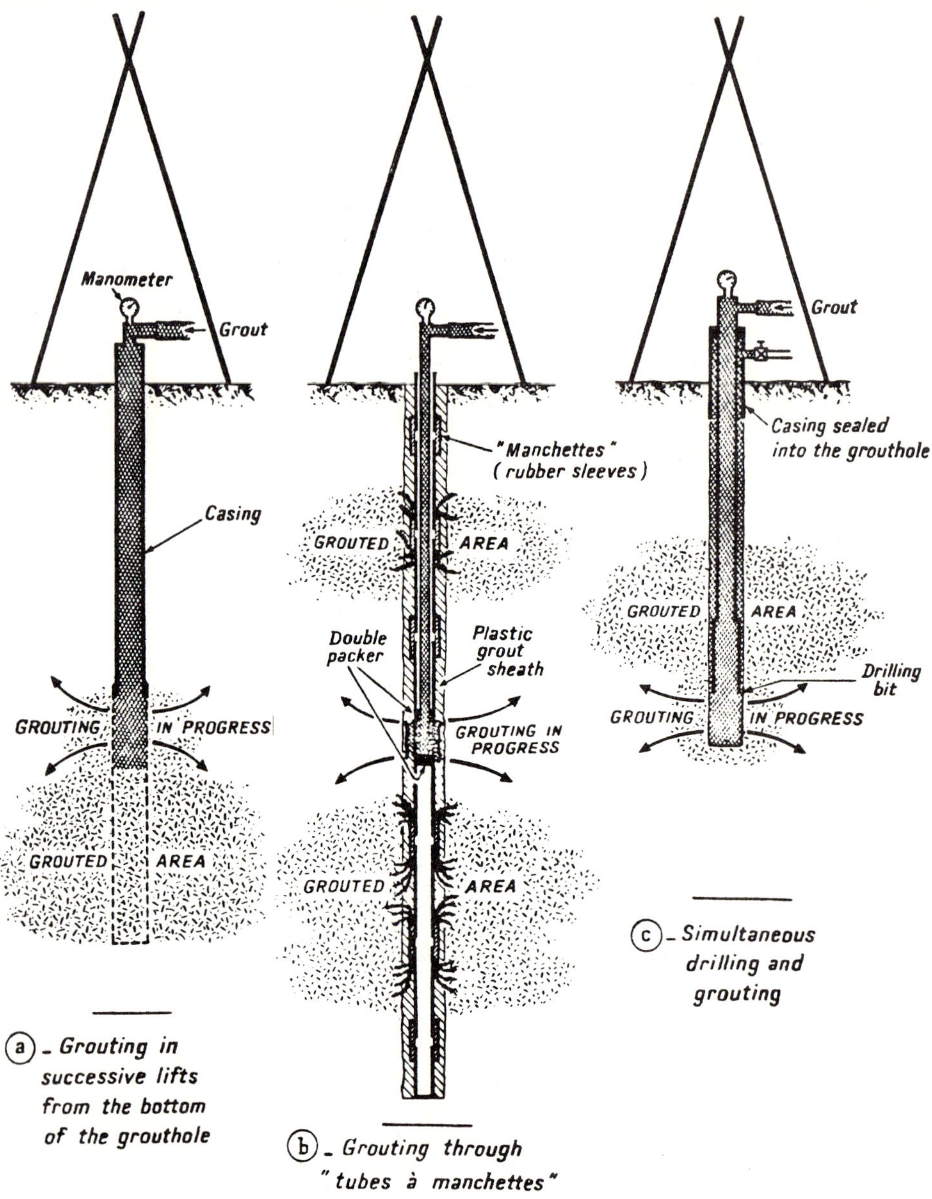

Figure 6.10 Various types of grout injection methods. *(After Caron et al., Ref. 24.)*

During injection, the flow rate of the properly mixed grout should be constantly monitored and plotted against the grout pressure to understand what is happening below ground. A number of different types of flow meters are available.[17] If the equipment is stationary, the readings should be recorded continuously; see Jefferies et al.[26]

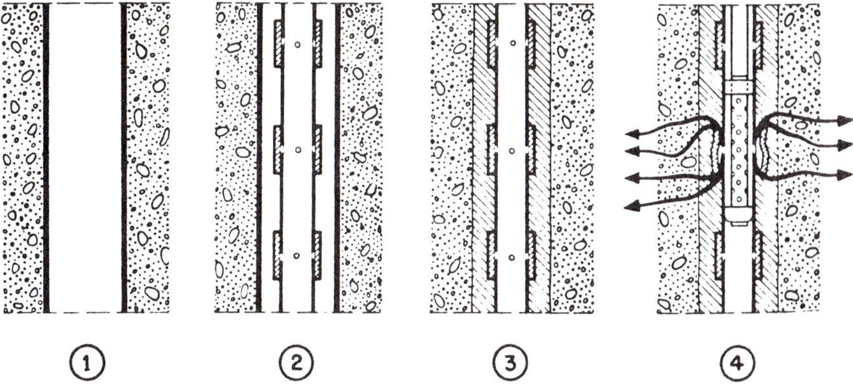

Figure 6.11 Details of installation of sleeve pipe injection method (tube à manchettes). (1) Boring and casing; (2) inserting the tube à manchettes; (3) sealing in tube à manchettes and withdrawing casing; (4) injection by means of a double packer. *(After Caron et al., Ref. 24.)*

Pressures should also be recorded continuously,[26] but are often monitored periodically on a Bourdon tube type of gauge too. These gauges usually take a dreadful battering on most grouting jobs, and yet they provide such valuable information that they are worth taking care of. Most have rated accuracies of 1 percent of their full face value, which for a 100-lb/in^2 gauge means an accuracy within 1lb/in^2. However, in lower-pressure ranges, these gauges often function poorly. It should be noted that pressure is monitored at the grout station. The pressure at the grout pipe is often much less because of friction losses that occur in long lengths of hose leading to the pipe. Additionally, there is a large pressure drop at all fittings and manifolds.

Typical response curves of flow rate versus pressure and flow rate versus time are shown in Fig. 6.13. By means of such curves, one can follow on a particular job how the soil or rock is responding to the grout being injected. Each basic function of grouting listed at the beginning of this chapter, and illustrated in Fig. 6.1, is shown by these curves:

a. For permeation, or penetration, grouting, flow rate-versus-pressure and flow rate-versus-time curves are parallel to each other (resulting in a nearly linear pressure-versus-time response) at flow rates which depend upon the porosity of the soil or the sizes of the crack openings in the rock to be grouted. The flow rate, of course, always decreases with time since the voids are being gradually filled with the grout. The weight or volume of grout injected depends on the structure of the material being grouted.

b. For compaction, or controlled displacement, grouting, the response is very similar except that generally lower flow rates are to be expected. As with permeation grouting, those rates decrease with time as the in situ formation being grouted becomes densified within the limits of the equipment and grout being used.

c. For hydrofracturing, or uncontrolled displacement grouting, the responses are markedly different than for the other two functions. Here a pressure is always

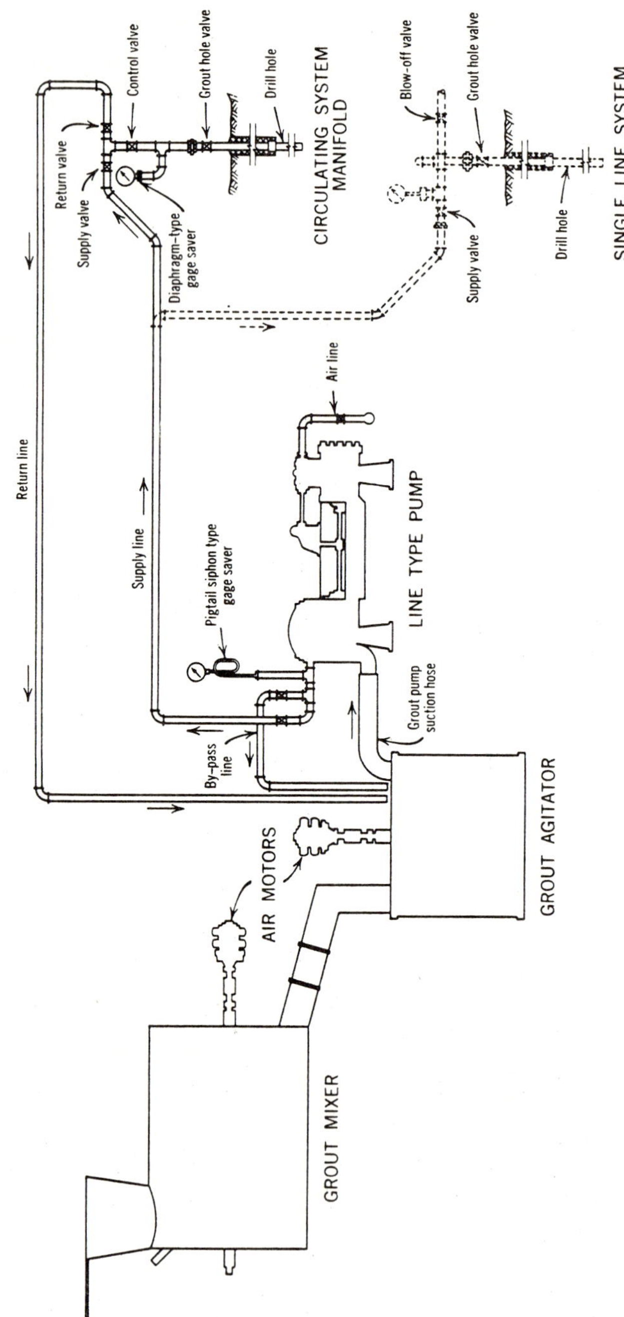

Figure 6.12 Typical cement (suspension) grouting plant. (*After Ref. 25.*)

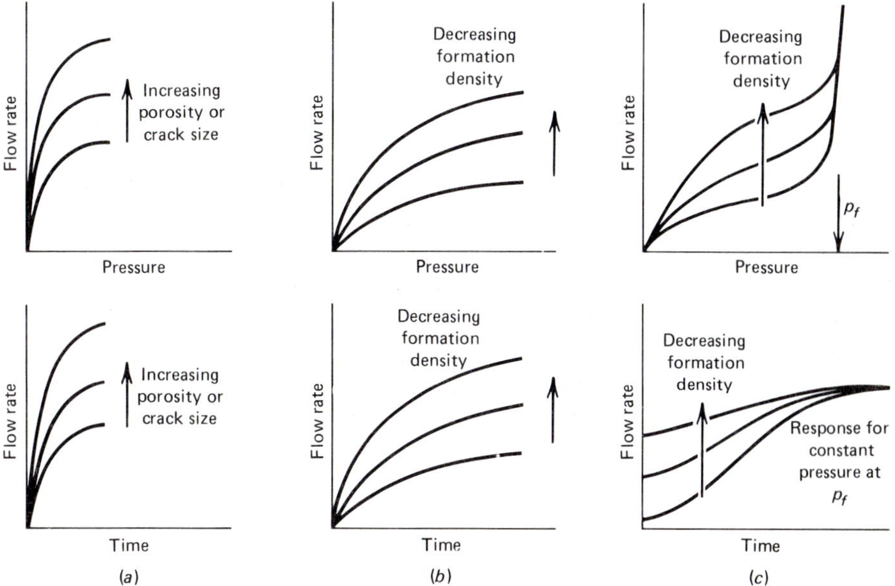

Figure 6.13 Grout project control curves showing how soil or rock is responding to injection. (*a*) Permeation (penetration) grouting; (*b*) compaction (controlled displacement) grouting; (*c*) hydrofracturing (uncontrolled displacement) grouting.

arrived at where the tensile strength of the soil or rock fails and a lens of grout shoots into the vacated space; recall Fig. 6.1*c*. The values of hydrofracture pressures in terms of their magnitudes and orientations are being actively investigated. Theoretically, they can be determined by means of the following equation:

$$p_f = \bar{\sigma}_v \nu(1 + \sin \phi) \tag{6.3}$$

where p_f = hydrofracture pressure
σ_v = effective vertical stress
ν = Poisson's ratio
ϕ = angle of shearing resistance
Practically, Cambefort[27] has found the following relationship:

$$p_f = \frac{q_f}{2\pi k} \frac{\gamma_g V_g}{H V_w} \ln \frac{r}{r'} \tag{6.4}$$

where p_f = hydrofracture pressure, lb/ft^2
q_f = rate of grout flow, ft^2/s
γ_g = unit weight of grout, lb/ft^3
k = permeability, ft/s
V_g = viscosity of grout, cP
V_w = viscosity of water, cP
H = thickness of zone being grouted, ft

r = radius of zone being grouted, ft

r' = radius of grout hole, ft

Certainly an understanding of hydrofracture pressure (and how to determine it accurately) would go a long way toward clearing up the ongoing arguments about what pressures to use in various situations that require grouting. Current practice in the United States (particularly in relation to cement grouting for seepage control beneath dams) is to use 1 lb/in^2 of pressure per foot of depth below ground surface. This is generally thought to be suitable for permeation grouting involving nominal amounts of controlled displacement. European practice is to use much higher grout pressures. Those higher pressures may, or may not, cause hydrofracturing, but such an occurrence, it is felt, is not particularly undesirable since it must have taken place along a zone of weakness which by means of it has now been stabilized or been made less permeable. Zones adjacent to the hydrofracture lenses are also permeated and/or displaced in the process. The proper method to use, however, has still not been agreed on.

6.5 GROUTING DESIGNS AND LAYOUTS

Depending upon the material to be grouted, (i.e., whether it is rock or soil), its quality (i.e., in terms of fracturing or density), and the purpose for grouting (i.e., for seepage control or strengthening), a number of grouting designs and layouts are possible. Some of the more common situations and their grouting needs will be reviewed here.

6.5.1 Seepage Control in Rock under Dams

This area, the "granddaddy of grouting," has had considerable exposure. Depending upon the degree of fracturing in the rock near its surface, a shallow blanket grout is placed, through which a deep curtain grout is injected; see Fig. 6.14a. Note that this cutoff is an extension of the clay corewall of the dam and the interface between the two is critical so that a potential seepage path will not be formed. Details on this topic can be found in Refs. 25, 28, and 29. It is significant to note here that the failure of the Teton Dam on June 5, 1976, was attributed to this aspect of its construction. Quoting from the report of the independent panel that reviewed the causes of the dam's failure[30]:

> The records show that great effort was devoted to constructing a grout curtain of high quality, and the Panel considers that the resulting curtain was not inferior to many that have been considered acceptable on other projects. Nevertheless, the Panel's on-site tests and other field investigations showed that the rock immediately under the grout cap, at least in the vicinity of Stas. 13 + 00 to 15 + 00, was not adequately sealed, and that additional unsealed openings may have existed at depth in the same locality. The leakage beneath the grout cap was capable of initiating piping in the key trench fill, leading to the formation of an erosion tunnel across the base of the fill. The Panel considers that too much was expected of the grout curtain, and that the design should have provided measures to render the inevitable leakage harmless.

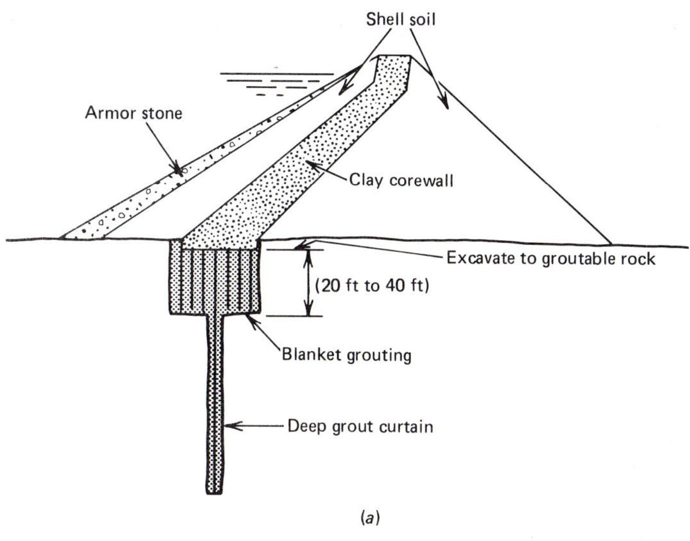

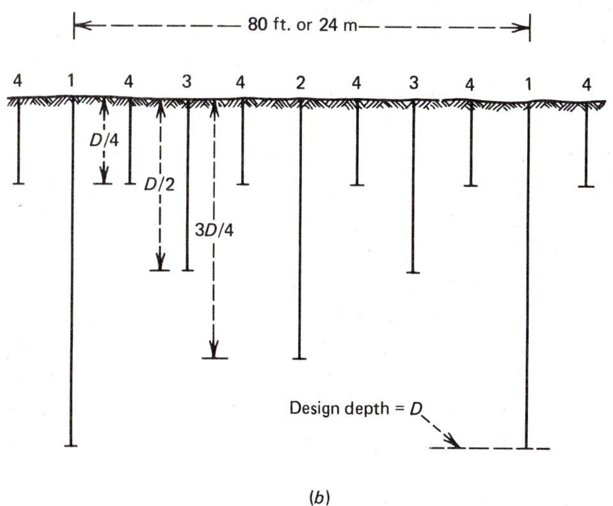

Figure 6.14 Blanket and curtain grouting for seepage control beneath dams on rock foundations. (*a*) Grout pattern typically used in fractured rock where upper surface is highly fractured. (*b*) Conventional closure pattern for curtain drilling and grouting. Numbers are sequence of drilling and grouting. *(After Wahlstrom, Ref. 29.)* (*c*) Schematic locations of pattern blanket and curtain grout holes in bedrock of an earth dam of moderate size. *(After Wahlstrom, Ref. 29.)*

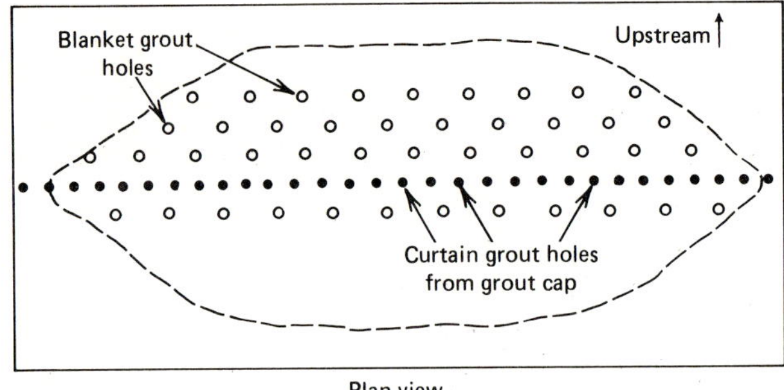

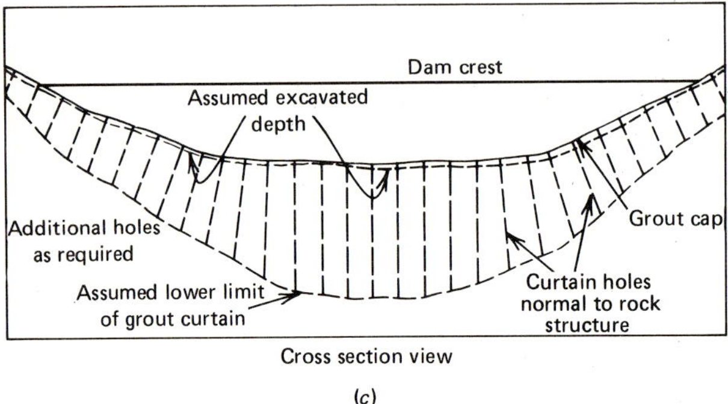

Figure 6.14 (*Continued*)

The Teton Dam, like almost all dams constructed for similar purposes on comparable sites, used cement suspension grouting.

Wahlstrom[29] provides the information about the spacing and depth of the curtain grouting shown in Fig. 6.14*b*. This pattern varies with location along the axis of the dam, as shown in Fig. 6.14*c*.

Each situation is, of course, unique and many unusual patterns can be designed. Wahlstrom[29] covers a wide range of dam types and geologic conditions, including the following:

Rock-fill dam with impermeable concrete face
Zoned earth and rock-fill dam, type 1
Zoned earth and rock-fill dam, type 2
Earth dam
Concrete gravity dam with "C" holes intersecting a fault zone
Concrete gravity dam with double-grout curtain and "C" holes intersecting a permeable fault zone
Concrete gravity dam with special-purpose "C" holes
Concrete gravity dam with a porous filter to collect seepage water
Hollow concrete gravity dam with possible flow of seepage water past grout curtain

Concrete buttress dam with possible seepage past grout curtain
Concrete arch dam, type 1
Concrete arch dam, type 2

Several of the above configurations are shown in Fig. 6.15.

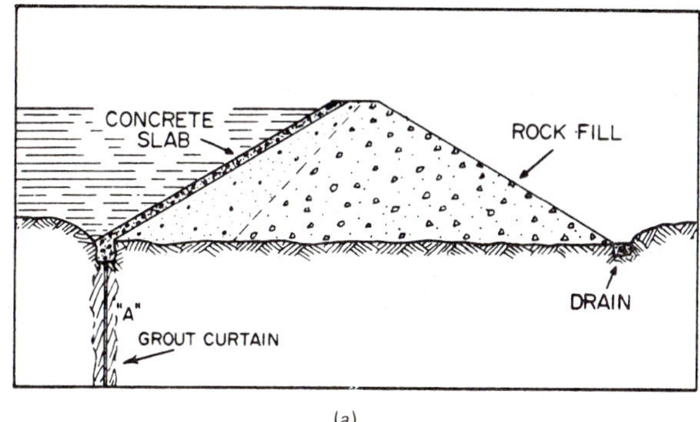

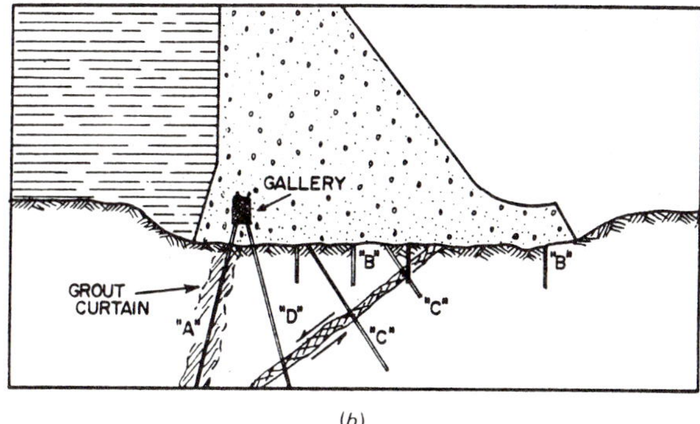

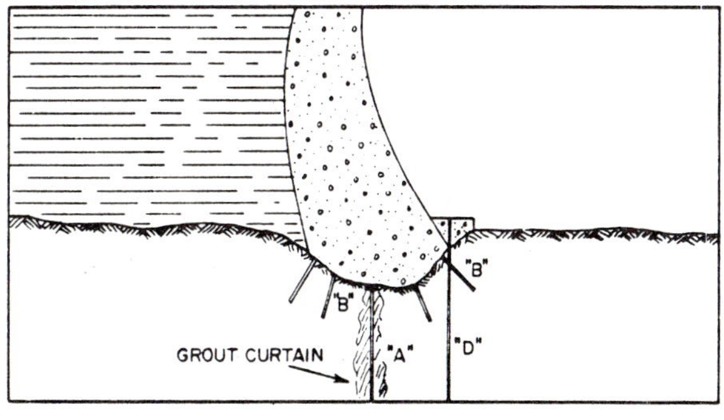

Figure 6.15 Grout curtain configurations for seepage control of rock foundations. (*a*) Rock fill dam with concrete-lined face. (*b*) Concrete gravity dam with intersecting fault zone. (*c*) Concrete arch dam. (*After Wahlstrom, Ref. 29.*)

6.5.2 Seepage Control in Soil under Dams

This particular topic is an extension of what was just discussed in Sec. 6.5.1, with the following differences: Parallel rows of curtain grout lines are customary, a wider variety of grouts have been used for seepage control in soil than in rock, and the grouts are usually placed much more deeply. The largest grout curtain ever built lies beneath the Aswan High Dam, Egypt, built in 1967, and seals 1,650,000 m³ of alternating strata of sands, gravels, silts, and sandstones. The grout curtain used was made up of sodium silicate and sodium aluminate, and it was placed to a depth of 255 m![31]

As another example of a deep grout curtain used to prevent (although the word "limit" is probably more accurate) seepage in soil, see Fig. 6.16. In that situation, flexibility of the grout curtain was desirable, and hence silicate solution grouting or

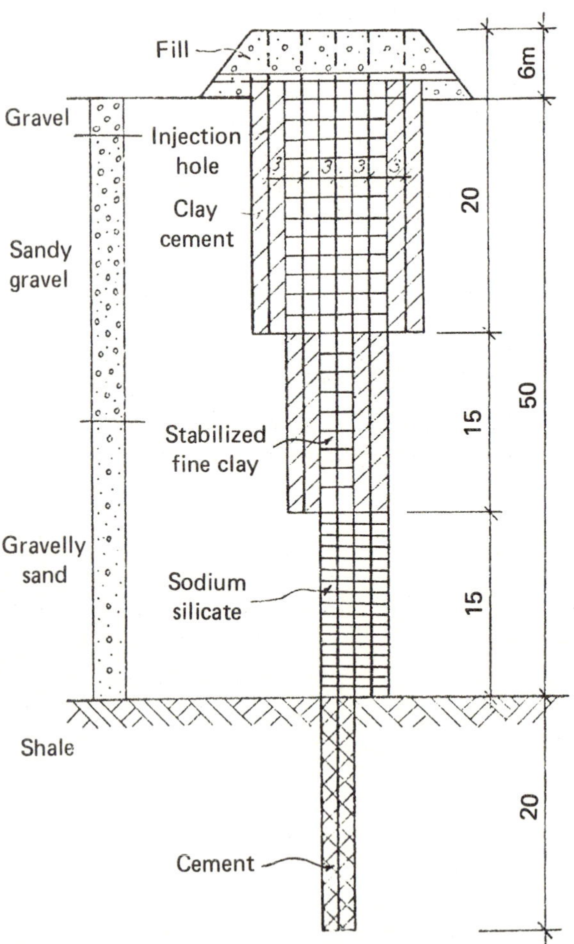

Figure 6.16 Deep grout curtain (of various types of grout) beneath earth dam, France (all values in meters). (*After Bonazzi, Ref. 32.*)

clay suspensions were used in some places, while in other areas watertightness was the main objective and clay-cement suspension grouts were used. The five rows of holes shown were drilled from a 6-m-high working platform which made it possible to grout right up to the original ground surface.

6.5.3 Seepage Control in Soil for Advancing Tunnels

When transit or utility tunnels are to be placed beneath the water table and the soils encountered have permeabilities greater than approximately 0.001 cm/s, water inflow problems can be expected. And along with this water inflow, soil can be eroded into the tunnel face (called "running ground" by tunnelers), resulting in piping collapses and adverse surface settlements. The long-existing solution to this problem has been to drive tunnels in such problematic soil under compressed air, but that is a slow, costly, personally dangerous, and overall objectionable practice in today's construction world.

An effective alternative is to grout the soil in advance of tunneling. A typical pattern of grout holes is shown in Fig. 6.17a. The grouting operation is carried out ahead of the tunnel's progress up to the limit of the water table. The sleeve pipe method, which uses double packers (as shown in Figs. 6.10b and 6.11), is the recommended technique. If the water inflow into the area to be tunneled is not sufficiently decreased, the grout holes can be reentered so that additional grout can be injected. Chemical solution grouts, with relatively quick setting times, are recommended.

6.5.4 Seepage Control in Soil for Cutoff Walls (Jet Grouting)

The containment of hazardous materials and toxic waste landfill leachates is a problem of enormous proportions. (Chapter 8 will concentrate on this particular topic.) And with over 30,000 abandoned toxic waste sites in the United States, it cannot be avoided. To isolate an existing pit from its surrounding environment, a row of closely spaced grout holes is drilled and continuous grouting is performed from their bottoms upward; recall Fig. 6.10a. Chemical grouts with low viscosities are usually used so that the grout curtain is tight and so that drilling costs are minimized.

However, a new technique, called *jet grouting,* has now been introduced which is well suited to address the problem of toxic waste containment. It seems to have been developed in Japan, but its current use in England and France, as well as in the United States, makes its origin somewhat unclear. Whatever its history, though, its deployment is similar from one country to another.

A high-speed jet of water released from a 1- to 2-mm nozzle at the end of the grout pipe, at pressures of 200 to 5000 kg/cm^2, cuts through the soil (or soft rock). When a slot has been formed, a chemical solution grout is substituted for the water, and that stabilizes in place as a solid mass. The grout pipe is lifted upward from the bottom of the hole as the process continues. When the grout is level with the ground's surface, the "wall panel" is complete and work is started on an adjacent area; see

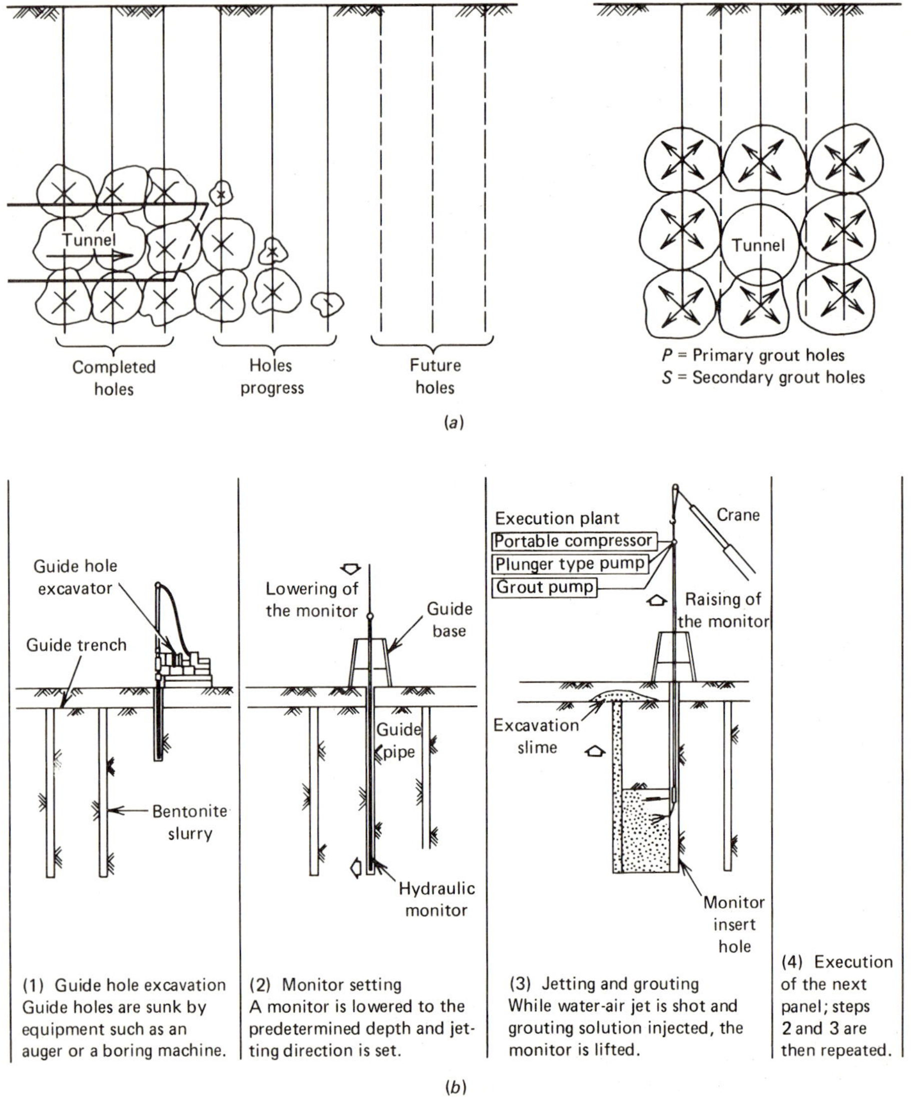

Figure 6.17 Seepage containment designs using chemical solution grouts. (*a*) Cross section and front view of grouting in advance of tunneling to prevent seepage. (*b*) Jet grout formation of seepage cutoff wall. (*After H. Baker Co.*)

Fig. 6.17*b*. The technique forms relatively thin walls, but it seems able to assure continuity between injection borings.

A slight variation on the jet grouting method is to rotate the jet 360°, forming a disk at the desired elevation. By forming a series of these disks at the same elevation, a horizontal layer of solid grout can be formed. This technique has been used to demonstrate in situ placement of a grouted bottom below an existing landfill; see Brunsing and Grube.[33]

6.5.5 Stabilization Grouting in Soil in Advance of Tunnels

The construction of transit and utility tunnels can cause settlement and stability problems for adjacent buildings and structures located above them. An interesting study by Verfel[34] shows how grouting was used for the Prague subway in order to eliminate such problems; see Fig. 6.18. Different grouts were used in different soil strata. The pressure of the grouting raised adjacent buildings between 12 and 17 mm. Grout take was about 35 percent of the grouted volume of the original soil.

6.5.6 Stabilization Grouting for Underpinning

Rather than stabilize *around* whatever is being constructed, it might prove more efficient to stabilize *the structure itself.* Obviously, this approach would be useful when the structure is settling for reasons not easily amenable to the usual grouting

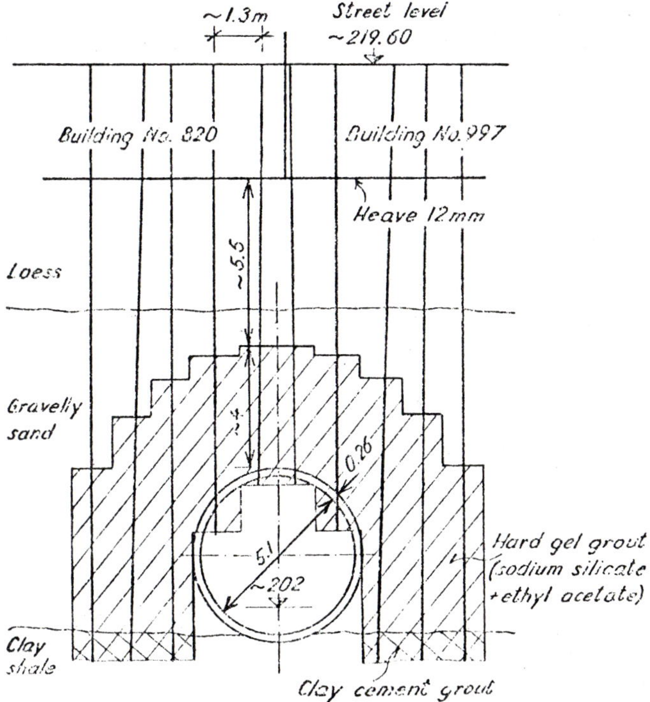

Figure 6.18 Different grout types and how they were used to stabilize buildings located adjacent to and above site of construction of Prague subway (values in meters). *(After Verfel, Ref. 34.)*

methods.[35] Grout pipes are often angled (see Fig. 6.19*a*), however, and at sharp angles create problems for drilling, and are often placed in very irregular patterns (see Fig. 6.19*b*). All kinds of grout, including stiff cement grouts where lifting is desirable, can be used to stabilize structures. Compaction grouting of this type is difficult and must be closely monitored, but has been successfully carried out (see Stilley[36]).

6.6 GROUT MONITORING SCHEMES

Since most grouting occurs at subsurface elevations where its behavior can usually only be postulated (i.e., often only guessed at!), the normal way of monitoring a grouting program has been to measure the quantities of materials injected (actually this is done for payment purposes since most grout projects are bid for on the basis of quantities to be injected), flow rates, and grout pressures[26]. The use of devices to measure these processes was described in Sec. 6.4.4. By "monitoring" we mean making a positive assessment of the results of the injected grout. This can be accomplished in traditional ways by taking soil borings or rock cores[37] and then testing them for strength, permeability, compressibility, etc., using standard laboratory methods. The problem here is that one must be selective about the locations at which to take the test borings, as well as the depths from which they are taken. Obviously there will be limitations imposed both by time and money.

A better approach is to use a geophysical or nondestructive testing (NDT) method which will provide a continuous trace of conditions before and after grouting either along the ground's surface or within adjacent grout pipes. Huck and Waller[38] have assessed the situation on a unified basis; see Table 6.3.

In general, the following *surface-deployed methods* have been used to assess the completeness and adequacy of grouting:

Standard penetration testing
Cone penetration testing
Seismic refraction
Electrical resistivity
Ground-probing radar
Inductance surveys

In addition, the following *borehole-deployed methods* have also been used:

Piezometers
Pressuremeter testing
Cross-hole seismic testing
Cross-hole radar

These methods will be discussed in Chap. 9, "Foundation Control Techniques." At the very least, one must be cognizant of the adjacent ground and structures founded

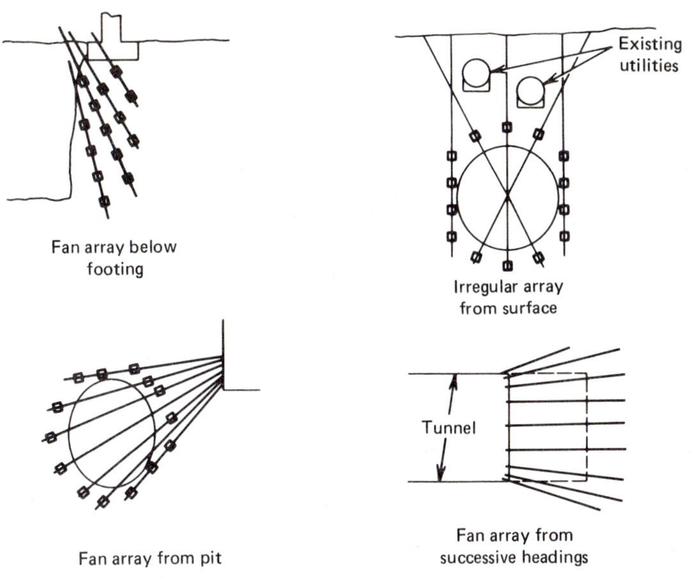

Figure 6.19 Different grouting patterns for use of stabilization grouting in soil. (*a*) Details of angled drilling. (*After Baker, Ref. 17.*) (*b*) Complicated drilling pattern in industrial plant. (*After Hinkle and Yen, Ref. 35.*)

Table 6.3 Grout monitoring procedures at various times during the grouting process†

Grouting activity	Minor monitoring effort only	Intense monitoring effort
Prior to grouting	Inspect equipment Set elevation survey points Establish monitoring plan and procedures	Inspect equipment Set elevation survey points Establish monitoring plan and procedures
During drilling		Conduct pregrout radar and cross-hole acoustic surveys
Grout materials	Certificates of compliance Trial grout mixes	Certificates of compliance Trial grout mixes Independent laboratory tests
During grout	Monitor injection pressure and flow rate Grout samples for gel time and storage Plot grout-take log Heave measurements on survey points	Monitor and record injection pressure and flow rate (strip chart) Grout samples for gel time and storage Plot grout-take log Heave measurements on survey points In situ deformation measurements as appropriate (inclinometer and/or in situ strain) Pore-pressure data In situ resistivity Acoustic emission monitoring for hydrofracturing
After grouting	Final heave survey Final review and signoff	Final heave survey Postgrout radar and acoustic surveys Final review and signoff

†After Huck and Waller, Ref. 38.

thereon. Benchmarks and position surveys should always be made, *before grouting begins* and during the process, to determine if vertical or horizontal movement has occurred.

In closing this section on monitoring, it should be noted that grouting is a relatively violent phenomenon accompanied by noise emissions. These noises, called *acoustic emissions,* can be picked up with the appropriate equipment, and when sufficient pickup points are involved can be used (by means of source location techniques similar to those employed in seismology to locate earthquake epicenters) to pinpoint the subsurface events giving rise to them. By displaying the progression of such events, it should be possible to develop a subsurface "picture" of what is happening. Work in this area is in progress; see Koerner et al.[39]

PROBLEMS

6.1. The usual contractual agreement when working with a grouting subcontractor is to pay for the work on a unit price basis for material used, e.g., dollars per gallon injected, etc. Why is this the case when alternative methods like paying a lump sum for a specified job are possible?

6.2. Although some large contractors will do their own grouting, mostly it is done by specialty grouting subcontractors. Why is this the case?

6.3. One of the major problems with suspension grouts is that the particles being carried in suspension often settle out before they reach their intended position in the soil. Rank the following suspension grouts according to their "set times" from fastest to slowest: sand, silt, clay, cement, fly ash, asphalt emulsion, and a bentonite-cement mix.

6.4. For almost all types of grouts, a critical property is their viscosity. Define the term and the units of measure connected with it, and describe a viscosity test.

6.5. Considering that water is attracted to bentonite clay by as much as a 20:1 ratio, and water forms a chemical reaction with cement, hydrating it at a ratio of about 0.7:1, how does a bentonite-cement grout mixture behave in the presence of water?

6.6. In working with cement grouts, when would a Type III Portland cement be preferable to the Type I usually used?

6.7. What are the advantages of using fly ash or ground slag in cement grouting rather than naturally occurring soils?

6.8. At what stage in the process of refining of crude oil can an emulsifying plant be used to produce an emulsified asphalt?

6.9. Solution grouts do indeed produce stabilized soils, as the curves shown in Fig. 6.8a indicate. In certain circumstances, however, the problem of creep has arisen. What mechanism allows creep to occur since the soils being grouted are usually sands or coarse silts which are generally not considered to be creep-sensitive? (Explain on the basis of effective stresses.)

6.10. Considering the groutability of various types of soils (see Figs. 6.2, 6.7, and 6.9), what method, if any, could be used to grout fine-grained soils such as silty clays and clays?

6.11. All of the soil types recommended for grouting, i.e., those shown in Figs. 6.2, 6.7, and 6.9, have uniformly graded particle sizes. What happens to groutability characteristics when the in situ soil formation exhibits a gap-graded particle size distribution curve?

6.12. Evaluate the characteristics listed in the following table for the various types of grout injection methods illustrated in Fig 6.10.

Characteristic being evaluated	Bottom-up grouting (Fig. 6.10a)	Sleeve pipe injection (Fig. 6.10b)	Top-down grouting (Fig. 6.10c)
(a) Initial drilling cost			
(b) Cost per unit of injected grout			
(c) Suitability for suspension grouting in rock			
(d) Suitability for solution grouting in soil			
(e) Suitability for resumption of grouting at a later time without redrilling			

6.13. (a) For suspension grouting, as shown in Fig. 6.12, list the advantages and disadvantages of a circulating manifold over a single-line manifold system.

  (b) Are these same two alternatives available for solution grouting?

6.14. In American grouting procedures, hydrofracturing of the in situ material to be grouted is not thought to be a suitable practice.

  (*a*) Why is this so for designers?

  (*b*) Why is this so for contractors?

  (*c*) Why do the European designers and contractors seem to prefer high-pressure grouting and the hydrofracturing that usually accompanies it?

6.15. To assess in the design stage of constructing a dam the amount of underlying rock to be grouted and the degree to which it might fracture, the older method of percent core recovery gives less information than the newer concept of rock quality designation (RQD). Describe these two methods and state why RQD is the preferred way of carrying out the assessment.

6.16. When considering how to design an earth or earth-rock dam to be built on a soil foundation having permeable strata beneath it, how far upstream and downstream should the soil profile be investigated? Why is it necessary to do so?

6.17. When evaluating the in situ hydraulic characteristics of a rock foundation, what is the advantage of using a Lugeon measurement rather than a standard falling or constant head permeability test? (Note that these tests are generally performed by means of a double-packer sleeve pipe arrangement inserted to a specific elevation in the borehole, as shown in Figs. 6.10*b* and 6.11).

6.18. What are some of the considerations that go into determining the "design depth" of a deep grout curtain as shown in Fig. 6.14?

6.19. For deep grout curtains in soil beneath earth or earth-rock dams, as shown in Fig. 6.16, vertical alignment during drilling is critical. Why is this the case and how can the verticalness of the alignment be checked during the drilling operation?

6.20. When driving a tunnel in saturated sand, water and soil inflow problems are invariably encountered. There are three major ways of handling them that are usually considered:

Use of a closed shield (or a slurry shield) at the face of the tunnel
Driving the tunnel under compressed air, which balances the external hydrostatic and soil pressures
Grouting in advance of the tunnel, as shown in Fig. 6.17*a*

Evaluate these three methods in terms of the characteristics listed in the accompanying table:

Characteristic being evaluated	Closed shield	Compressed air	Advance grouting
(*a*) Initial cost			
(*b*) Anticipated progress of tunnel advancement			
(*c*) Risk to personnel			
(*d*) Likelihood of damage to adjacent properties			
(*e*) Quality of finished job			

6.21. The use of jet grouting to form vertical seepage cutoff walls as illustrated in Fig. 6.17*b*, is in direct competition with the use of the slurry construction method to form cutoff walls, as described in Sec. 3.4.3. Evaluate these techniques in terms of the characteristics listed in the following table:

Characteristic being evaluated	Jet grouting	Slurry construction
(*a*) Cost of installation		
(*b*) Type of material used		
(*c*) Width of trench		
(*d*) Maximum depth that can be reached		
(*e*) Terrain limitations		
(*f*) Reliability of seepage "tightness" of final barrier		

6.22. The use of stabilization grouting to support structures in advance of tunneling wo k, as shown in Figs. 6.18 and 6.19, is in obvious competition with the use of direct contact underpinning methods (see Figs. 3.31 and 3.32) and lateral support underpinning methods (see Fig. 3.15). Evaluate these techniques in terms of the characteristics listed in the following table:

Characteristic being evaluated	Grouting	Direct concrete underpinning	Lateral support underpinning
(*a*) Probable initial cost			
(*b*) Likelihood of damage to structure during underpinning construction			
(*c*) Likelihood of damage to structure after underpinning construction			
(*d*) Likelihood of damage to structure after all construction ceases			

REFERENCES

1. Cambefort, H., *Injection des Sols,* Vol. 1: *Principles and Methods*; Vol. 2: *Applications*, Eyrolles, Paris, 1964.
2. Bowen, R., *Grouting in Engineering Practice,* 2d. ed., Applied Science and John Wiley and Sons, New York, 1981.
3. Tallard, G. R., and C. Caron *Chemical Grouts for Soils: Available Materials,* Vol. 1, and *Engineering Evaluation of Available Materials,* Vol. 2, Federal Highway Administration, U.S. Dept. of Transportation, Washington, DC, June, 1977.
4. Karol, R.H. *Chemical Grouting*, Marcel Dekker, New York, 1982.
5. *Grouts and Drilling Muds in Engineering Practice,* Institute of Civil Engineers, Butterworth, London, 1963.
6. W. H. Baker (ed.), *Proc. Conf. on Grouting in Geotech. Eng.,* New Orleans, February 1982.
7. *Bibliography on Grouting,* Miscellaneous paper No. C-78-8, U.S. Army Corps of Eng. Waterways Experiment Station, Vicksburg, MS., 1978.
8. Kravetz, G. A., "Cement and Clay Grouting of Foundations—The Use of Clay in Pressure Grouting," *J. Soil Mech, Found. Eng. Div.,* ASCE, Vol. 84, No. SM1, February 1958, 1546-1 to pp. 1546-30.
9. Jefferies, S. A., "Effects of Mixing on Bentonite Slurries and Grouts," in Ref. 6, pp. 62–76.
10. Terzaghi, K., and Y. Lacroix, "The Mission Dam," *Geotechnique,* Vol. 14, March 1964, pp. 13-50.
11. Littlejohn, G. S., "Design of Cement Based Grouts," in Ref. 6, pp. 35–48.
12. Popovics, S., *Concrete Making Materials,* McGraw-Hill, New York, 1979.
13. *Fundamentals of Portland Cement Concrete: A Quantitative Approach,* Vol. 1, *Fresh Concrete,* John Wiley and Sons, New York 1983.
14. Nicholls, R., *Composite Construction Materials Handbook,* Prentice-Hall, Englewood Cliffs, N.J. 1976.
15. Karol, R. H., and J. P. Welsh, "Chemical Grouts," in *Encyclopedia of Chemical Technology,* Vol. 5, 3d ed., Kirk-Othmer (ed.) John Wiley and Sons, New York, 1979, pp. 368–374.
16. Karol, R. H., "Chemical Grouts and Their Properties," in Ref. 6, pp. 359–377.
17. Baker, W. H., "Planning and Performing Structural Chemical Grouting," in Ref. 6, pp. 515–539.
18. Warner, J., "Strength Properties of Chemically Solidified Soils," *Soil Mech. Found. Eng. Div.,* ASCE, Vol. 98, No. SM11, November 1972, pp. 1163–1185.
19. Clough, G. W., W. M. Kuck, and G. Kasali, "Silicate-Stabilized Sands," *J. Geotech. Eng. Div.,* ASCE, Vol. 105, No. GT1, January 1979, pp. 65–82.
20. Vinson, T. S., and J. K. Mitchell, "Polyurethane Foamed Plastics in Soil Grouting," *J. Soil Mech. Found. Eng. Div.,* ASCE, Vol. 98, No. SM6, June 1972, pp. 579–602.

21. Karol, R. H., "Seepage Control with Chemical Grout," in Ref. 6, pp. 564–575.

22. Little, A. L., "Groundwater Control by Exclusion," in *Methods of Treatment of Unstable Ground,* F. G. Bell (ed.), Newnes-Butterworths, London, 1962, pp. 37–68.

23. Houlsby, A. C., "Cement Grouting for Dams," in Ref. 6, pp. 1–34.

24. Caron, C., T. F. Herbst, and P. Cattin, "Injections," Chap. 9 in *Foundation Engineering Handbook,* H. F. Winterkorn and H.-Y. Fang (eds.), Van Nostrand Reinhold, New York, 1975, pp. 337–353.

25. *Pressure Grouting,* U.S. Bureau of Reclamation Tech. Memo. No. 646, Denver, June 1957.

26. Jefferies, M. G., B. T. Rogers, and D. W. Reades, "Electronic Monitoring of Grouting," in Ref. 6, pp. 769–781.

27. Cambefort, H., "The Principles and Applications of Grouting," *Quart. J. Eng. Geol.,* Vol. 10, 1977, pp. 44–63.

28. Sherard, J. L., R. J. Woodward, S. F. Gizienski, and W. A. Clevenger, *Earth and Earth-Rock Dams,* John Wiley and Sons, New York, 1963.

29. Wahlstrom, E. E., *Dams, Dam Foundations and Reservoir Sites,* Elsevier, New York, 1974.

30. *Failure of Teton Dam,* summary and conclusion report to U.S. Dept. of the Interior and the State of Idaho, W. L. Chadwick (Chair of report committee), Idaho Falls, December 1976.

31. Bowman, W. G., "Record Grout Curtain Seals Nile's Leaky Bed," *Eng. News Rec.,* Vol. 180, February 28, 1968, pp. 22–24.

32. Bonazzi, D., "Alluvium Grouting Proved Effective on Alpine Dam," *J. Soil Mech. Found. Eng. Div., ASCE,* Vol. 91, No. SM6, 1965, pp. 81–93.

33. Brunsing, T. P., and W. E. Grube, Jr., " A Block Displacement Technique to Isolate Uncontrolled Hazardous Waste Sites," in *Proc. Nat. Conf. on Control of Hazardous Waste Sites,* Hazardous Materials Control Research Institute, Washington, D.C. 1982, pp. 249–253.

34. Verfel, J., "Soil Strengthening by Grouting for the Metro of Prague" (in Czech), in Z. Bazant, *Methods of Foundation Engineering,* Elsevier, Amsterdam, 1979.

35. Hinkle, R. D., Jr., and B. C. Yen, "Grouting Halts Chemically Induced Settlement," in Ref. 6, pp. 974–986.

36. Stilley, A. N., "Compaction Grouting for Foundation Stabilization," in Ref. 6, pp. 923–937.

37. Ballwy, G., and P. Niemants, "New Laboratory Techniques to Test Grouted Fractured Rock," in Ref. 6, pp. 809–821.

38. Huck, P. J., and M. J. Waller, "Quality Control for Grouting," in Ref. 6, pp. 781–792.

39. Koerner, R. M., W. M. McCabe, P. J. Huck, and J. P. Welsh, "Hydrofracture Monitoring Using the Acoustic Emission Method," in *Proc. Conf. on Geotech. and Environ. Aspects of Geopressure Energy,* S. K. Saxena (ed.), Engineering Foundation, New York, 1980, pp. 153–164.

GEOTEXTILES (FILTER FABRICS)

7.1 OVERVIEW

The impact of *geotextiles* (also called *construction fabrics, synthetic fabrics, filter fabrics, road rugs,* or simply *fabrics*) on geotechnical and construction engineering within the past 5 to 10 years is nothing short of awesome. Sales have moved from a few thousand square yards per year in the 1960s and early 1970s, to about 6,000,000 yd^2 in 1976, to 15,000,000 yd^2 in 1977, then to over 100,000,000 yd^2 in 1981. These figures are for North America only; European use is comparable, and a reasonable estimate of current worldwide use is approximately 250,000,000 yd^2 per year. At an average installed price of somewhat over \$1 per square yard, the industry currently does over \$250,000,000 worth of business per year.

Geotextiles are porous fabrics manufactured from synthetic materials such as polypropylene, polyester, polyethylene, nylon, polyvinyl chloride, glass, and various mixtures of these (see Fig. 7.1). They come in thicknesses ranging from 10 to 300 mils, in widths up to 30 ft, in roll lengths up to 2000 ft, and with permeabilities comparable in range from coarse gravel to fine sand. They can be constructed in a variety of ways, the most common methods being:

Woven, made from continuous monofilament fibers
Woven, made from continuous slit-film fibers
Nonwoven, made from continuous or staple fibers laid down in a random pattern and then mechanically entangled into a relatively thick, feltlike fabric by means of punching with barbed needles
Nonwoven, made from continuous fibers laid down in a random pattern and then joined at the fiber crossover points by thermal or chemical bonding and pressed through rollers (calendered) into a relatively thin fabric

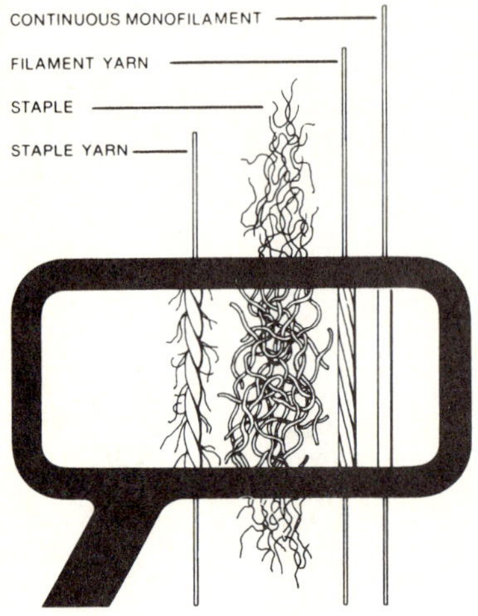

Figure 7.1 Various fibers and fiber yarns used to make geotextiles. *(After Ref. 1.)*

Grid, made from a sheet of polymer, usually polyethylene, that is punched with holes (closely spaced, varying in diameter from 0.5 to 2.0 in) in a uniform pattern and then elongated in at least one direction

Hybrid combinations of any of the above materials and techniques, usually manufactured for specific applications

Each of the above forms is shown in Fig. 7.2.

With the variety of materials and processes available to manufacture such fabrics, it should come as no surprise that a wide selection of geotextiles is available. On the one hand, this is advantageous in that the market is very active and strongly competitive, offering a wide choice of geotextiles for a given job; on the other, the situation is confusing, mainly because of the range of physical, mechanical, hydraulic, and durability properties that are offered. For example, there are the following:

1. Physical properties:
   a. Mass: 3 to 30 oz/yd^2
   b. Thickness: 10 to 300 mil
   c. Specific gravity: 0.9 to 1.4
   d. Percent open area (POA): nil to 36 percent
   e. Equivalent opening size (EOS): 30 to 300 U.S. Sieve No.
2. Mechanical properties:
   a. Grab strength: 50 to 5000 lb/in
   b. Grab elongation: 20 to 200 percent
   c. Modulus (10 percent secant): 100 to 10,000 lb/in
   d. Mullen burst: 50 to 2000 lb/in^2
   e. Trapezoidal tear: 20 to 300 lb

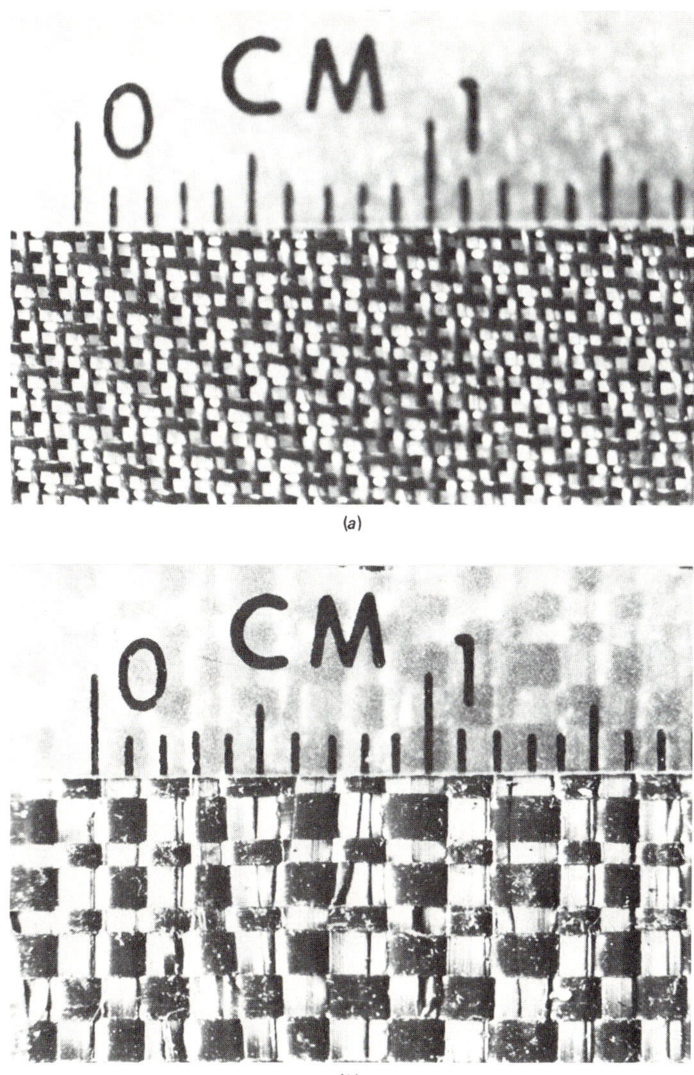

Figure 7.2 Photographs of various styles of geotextiles and geogrids. (*a*) Woven: continuous monofilament fibers; (*b*) woven: slit film fibers; (*c*) nonwoven, continuous fibers: needle-punched; (*d*) nonwoven, continuous fibers: spun-laid, heat-bonded; (*e*) geogrid: holes punched in a polymer sheet and then elongated for strengthening; (*f*) composite geotextiles.

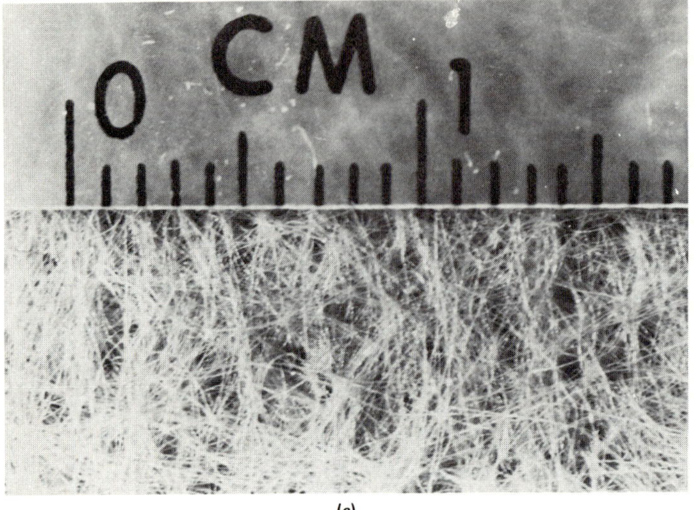

(c)

(d)

Figure 7.2 *(Continued.)*

3. Hydraulic properties:
   a. Cross-plane permeability: 0.01 to 5 cm/s
   b. In-plane permeability: nil to 2.0 cm/s
4. Durability properties:
   a. pH resistance: 3 to 11 (excellent)
   b. Biological resistance: generally excellent
   c. UV stability: poor to reasonable
   d. Moisture absorbency: nil to 3 percent

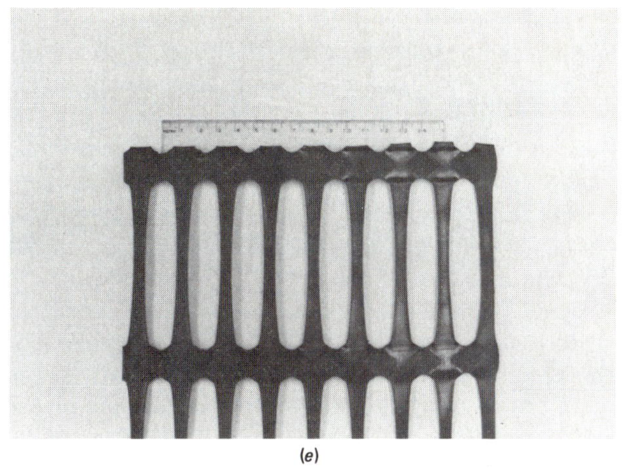

(e)

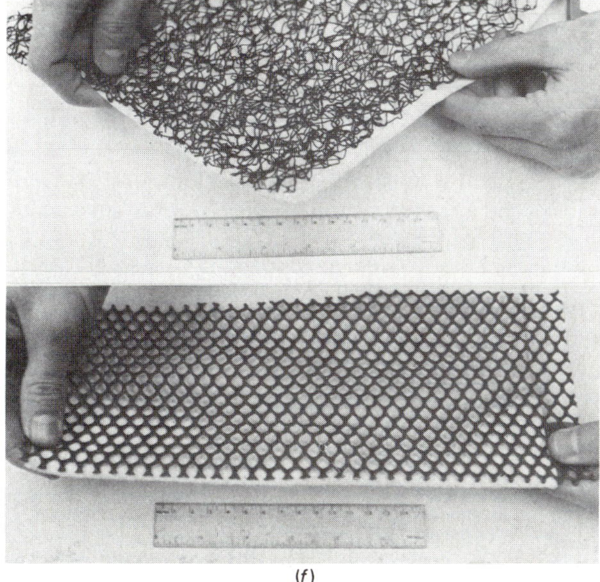

(f)

Figure 7.2 *(Continued.)*

The accuracy of the values claimed for these properties in relation to a given geotextile is validated by laboratory testing carried out according to standard procedures which have been, or are in the process of being, developed by many standards organizations. In the United States, the American Society for Testing and Materials (ASTM) has a joint Textile and Geotechnical Engineering Committee (D13/D18) that investigates the reliability of the various tests. Brief descriptions of such tests can be found in two texts on the subject by Koerner and Welsh[2] and Rankilor.[3] More complete test descriptions and actual values of the various properties associated with specific geotextiles can be found in the literature most manufacturers of geotextiles put out on their products; see Appendix E for names and addresses of the manufacturers, as well as a list of many geotextiles with some of their properties.

In selecting the proper geotextile for a specific project, i.e., in the design process, one must first determine the major function (or sometimes functions) that the geotextile is intended to perform. Those functions usually have to do with soil separation, reinforcement, filtration and drainage, control of erosion, and the use of geotextiles as flexible formwork for the placement of cement and grout. These topics will form the focal points of Secs. 7.2 through 7.6. As much information on design as possible will be incorporated into the discussions, but it should be realized that the geotextiles used in most projects are selected on the basis of past experience. These experiences are often summarized for future use by way of guidelines or specifications giving minimum performance values for a particular function (a typical guideline will be given in Sec. 7.7), but the selection is never cut and dried.

7.2 GEOTEXTILES AS SEPARATORS

A basic tenet of geotechnical and construction engineering is that soils with vastly different particle sizes and particle size distributions cannot be placed together. This is shown graphically in Fig. 7.3a, where it is apparent that the smaller-sized soil will eventually migrate into the voids of the larger-sized soil. Similarly, if the strength of the finer soil is low in relation to the stresses the larger-sized material can exert against it, the larger-sized soil will penetrate into the finer material. Taken collectively, the "intrusions" of the two soils into one another can spell disaster for the soil system where it occurs. Composite strength is reduced, drainage can be completely blocked off, compressibility usually is increased, the lifetime of the system is shortened, and the likelihood of failure for any engineering project connected with it is greatly increased.

In such a case, one solution is to use a material having particle size characteristics intermediate between the two soils. Figure 7.3b shows a "filter" soil useful for this purpose, but in reality, of course, two or even three intermediate soils may be necessary to separate the original soils successfully. Thus an incremented series of filter soils may be required for a given project, and in each case careful control of the filter soil is required.

The obvious alternative to the use of such a filter soil, or soils, is the use of a geotextile. The advantages of using geotextiles as separators over using single- or multiple-soil filters are the following:

Simplicity in construction
Less excavation required in many cases
Less weight placed on underlying soils
Less time required for construction

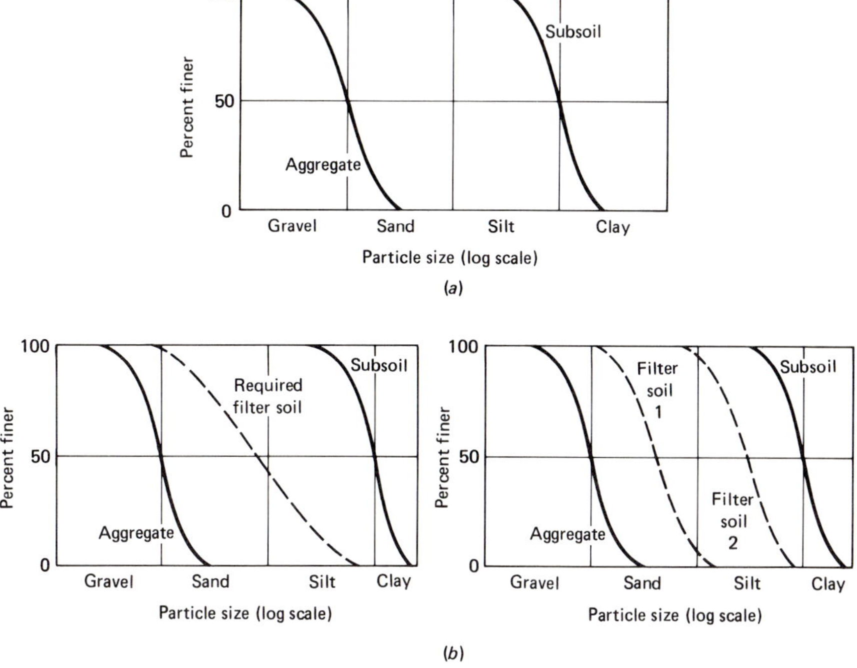

Figure 7.3 Problems caused by soils whose particle sizes are extremely dissimilar being placed adjacent to one another, and standard use of soil filters to solve them. (*a*) Typical particle size distributions. (*b*) Single-soil filter (left) and multiple-soil filter (right).

Less chance for problems and/or errors to arise
Lower project cost

Applications where geotextiles have been used in place of filter soils are the following:

1. Between natural foundation soils and stone aggregate for paved and unpaved roads, railroads, parking lots, staging areas, sidewalks, etc.; see Fig. 7.4.
2. In earth dams to separate rip-rap from shell soil, shell soil from core wall soil, core wall soil from drainage soils; see Fig. 7.5.
3. As a waterproofing agent to prevent cracks in existing asphalt pavements from propagating through the asphalt overlays. Geotextiles have found significant use in this area of preventing crack reflection. The general procedure is as follows:
   a. Clean the existing road surface.
   b. Fill existing small cracks (<0.25 in) with crack filler, and the larger cracks (>0.25 in) with asphalt, hot mix, or cold patch.
   c. Spray an asphalt sealant uniformly over the existing pavement, in amounts

Figure 7.4 Geotextiles used as separators between in situ soil and stone aggregate. *(Photos compliments of Crown Zellerbach Corp., Washougal, WA 98671.)*

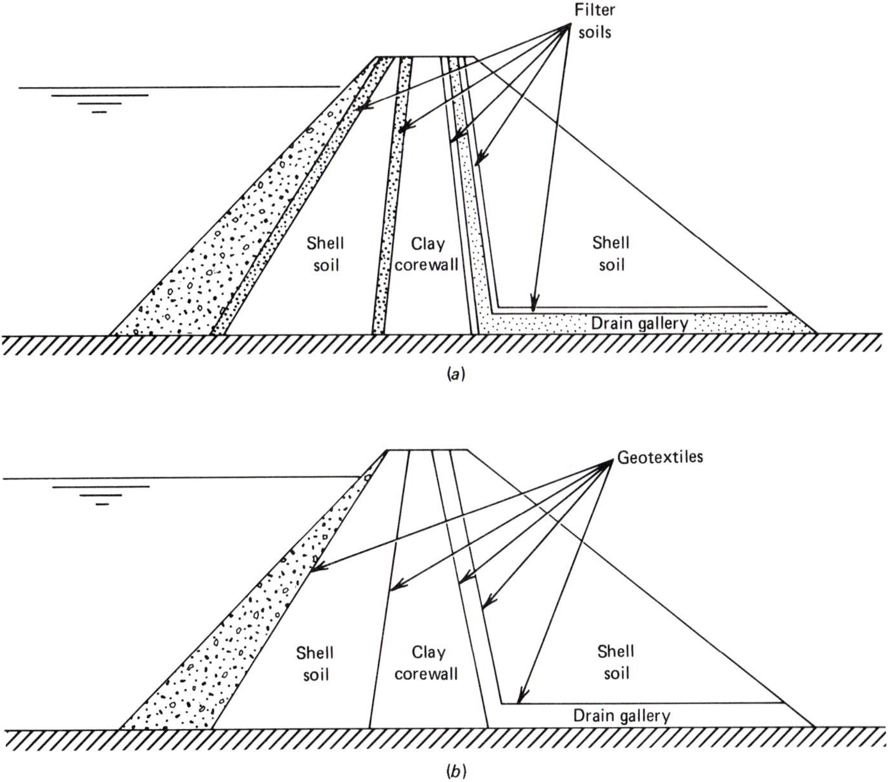

Figure 7.5 Use of geotextiles in zoned earth dams to replace filter soils. (*a*) Typical section of zoned earth dam with required filter soils. (*b*) Similar zoned earth dam using geotextiles to eliminate filter soils.

ranging from 0.15 to 0.60 gal/yd^2 depending on the porosity of the pavement and the absorbency of the fabric.

d. Place the fabric on the prepared surface (free of wrinkles or folds and with joint overlap of 1 to 3 in).

e. Place a hot mix overlay on the fabric as soon as possible. (A design procedure for determining the required overlay thickness, without and with fabric, is available; see Majidzadeh et al.[4])

f. Roll and compact the overlay.

In general, the technique has proved to be valuable and cost-effective, but there have been problems with it in areas of severe freeze-thaw cycles and in local zones where the existing pavement has structural failures; see Koerner and Welsh[2] for further details.

4. In many situations a separator is needed to break the continuity of an existing, and usually detrimental, process. Examples are frost heave zones[5] and salt migration.[6] Both phenomena occur within the soil's capillary zone where water from

below either freezes, causing the ground surface to heave, or brings salts to the surface, killing vegetation. Geotextiles can be used to encapsulate a sand layer, or a geotextile-fluted, plastic insert–geotextile composite can be used as a capillary break. The technique appears to be very effective in breaking the continuity of water flow, thereby eliminating the feed water causing the heave or the upward movement of the water bearing the salts. The method will be discussed further in Sec. 7.4.

7.3 GEOTEXTILES AS REINFORCEMENT

The composite interaction between steel and Portland cement in reinforced concrete is a well-known situation where each constituent material contributes its strength to the whole, thus minimizing the other's weakness. In the same manner, a geotextile with good tensile strength can contribute to the load-carrying ability of soil which is poor in tension but good in compression. This fortunate situation has provided the major impetus for geotextile use to date. The most important such use has been the construction of unpaved roads over soft, subgrade soils, where the geotextile is laid down over the soil and stone or rock aggregate is placed directly on it. But geotextiles are also used to reinforce walls and embankments. Both use categories will be described below.

Geotextiles placed between a natural soil subgrade below and a stone or rock aggregate surfacing above in unpaved roads, parking lots, staging areas, etc., certainly function as separators. As the topic of separation was discussed in Sec. 7.2, however, we need not consider it here. But when geotextiles are significantly deformed, as is the case when they are placed over soft subsoils, their tensile stress is substantially mobilized, and thus they serve as reinforcers also. In fact, the more geotextiles are deformed, as happens when they are placed over extremely soft subsoils, the greater the load they can carry and therefore the less stone is required as a support above, and so the reason for their usefulness is readily apparent. In order for geotextiles to function really significantly as reinforcers, the natural soil subgrade should have a California Bearing Ratio (CBR) of 2.5 or less. This is equivalent to an unconfined compressive strength of about 10 lb/in² (69 kPa) or a cone penetration value of 150. And soils with CBR values of 0.5 and less have also been made serviceable using geotextiles in this manner.

The proper placement procedure for fabrics being used to reinforce road subgrades obviously depends on many factors, which make each job somewhat unique. Nevertheless, some generalized comments, applicable to most situations, can be made:

The site should be cleared of debris, sharp objects, rocks, trees, tree stumps (which should at least be cut off at ground level), and other material that could puncture the fabric when fill is placed upon it.
If the site is such that construction machinery can be moved onto it, it should be at least rough-graded, with holes being filled in and high spots leveled.
After the site is cleared and leveled, the fabric should then be rolled out over it, by

means of construction equipment if feasible or by hand if the job is small or equipment cannot maneuver. A front-end loader can be easily fitted with a sling to support a pipe run through the core of the fabric roll, if machinery can be used. Caution must be exercised, however, so that the wheels or tracks of the equipment do not tear the fabric. Thus it is preferable to use the soil subgrade as the working platform, not already laid fabric.

The end of each fabric roll and the beginning of the next must be overlapped, as well as the sides of adjacent fabric sections. The amount of overlap needed depends on the compressibility of the soil involved, the nature of the traffic that will be supported, the thickness of the aggregate that will be placed on the fabric, and the deformation characteristics of the fabric itself. Thus a firm rule about overlapping is difficult to state. Typical overlap widths are 12 to 48 in, depending on the status of the conditions just mentioned. Some manufacturers state their recommended overlap dimensions for different ground conditions as a percentage of roll width.

An alternative to overlapping which should be considered in regard to softer soils, where both economics and large deformations are factors, is sewing the fabric ends together. Both air-operated and electric sewing machines for geotextiles are available. Seam strengths should be capable of carrying at least 80 percent of the strength of the geotextile itself. Adhesive strips have also been used to attach the ends.

When the aggregate is placed on the fabric, caution should again be exercised. The stone or rock should not be dumped directly onto the geotextile, nor should the dump trucks run over it. Instead the aggregate should be dumped on previously placed aggregate and spread by a bulldozer or front-end loader. And the blade or bucket of the spreader should be kept sufficiently high so that the aggregate is not pulled over the fabric but is dropped onto it from a minimum height above. Naturally, the quality of the aggregate will dictate how much care is required in this part of the construction sequence.

After the aggregate has been placed, a road grader should rough-grade the site, then a loaded truck should rut it. The ruts should then be filled with additional aggregate and the entire surface compacted, followed by the final grading.

A wearing surface, if required, can then be placed, following standard procedures.

See Fig. 7.6, which illustrates the above procedure by means of schematic sketches. This particular geotextile application has been very successful in part because of the many design methods available that allow one to calculate the aggregate thickness (i.e., the amount of aggregate) which will be needed both with and without the use of fabric, so that comparative costs and savings can be measured. It is generally cheaper to use a geotextile than not to use one. This is particularly the case when the job site is more than 10 mi from the quarry supplying the stone aggregate. Beyond this distance, the cost of transporting the stone goes up significantly, whereas the cost of a geotextile is relatively insensitive to transportation distances.

Most geotextile manufacturers have a design procedure for calculating required aggregate thickness in terms of subgrade soil strength alone, without the use of a geotextile, as well as the required thickness when any of their particular styles of

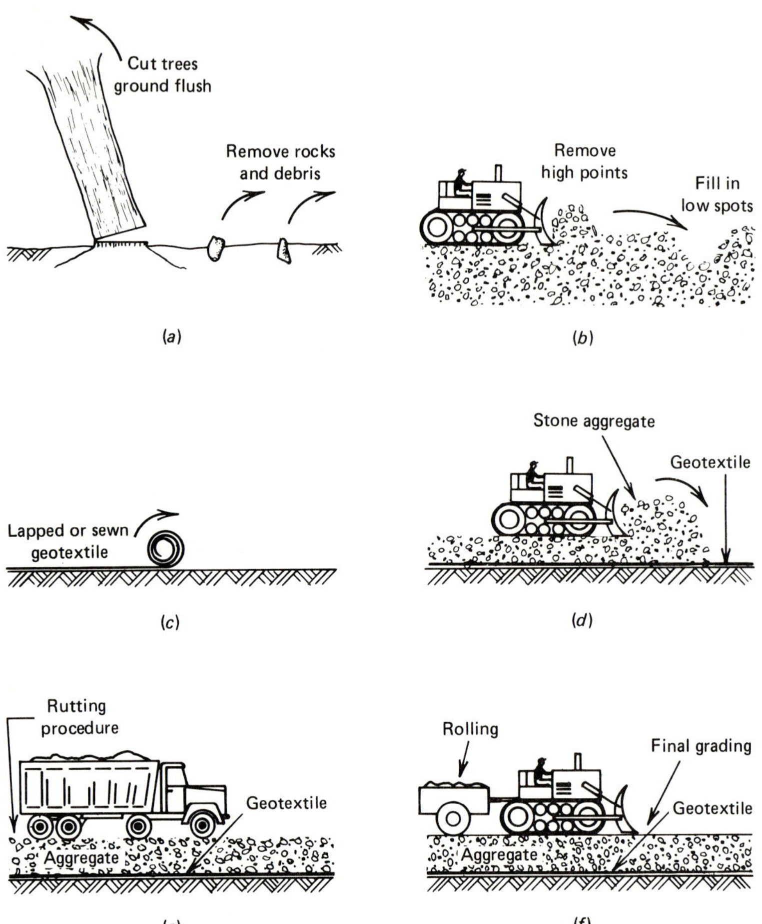

Figure 7.6 General procedures for using geotextiles in the construction of an unpaved road. (*a*) Clear trees, stumps, and large brush. (*b*) Level ground surface as much as possible. (*c*) Roll out geotextile by hand or machine. (*d*) Spread aggregate, working from previously placed aggregate. (*e*) Rut aggregate and geotextile system with truck. (*f*) Fill in ruts, grade aggregate, and roll finished surface.

geotextiles is used. The majority incorporate into this procedure the largest type of vehicle, maximum magnitude of axle load, and greatest tire pressure that the final road will be able to accommodate. Most such procedures are only good for any particular manufacturer's geotextiles and cross referencing or generalizing to those produced by others is basically impossible. However, *one* procedure is available that allows for a comparison between different geotextiles, given the same design conditions for each fabric being considered—a procedure based on the initial stiffness, i.e., the modulus, of the geotextiles.

Giroud and Noiray[7] begin that procedure using geometrical relationships to describe identical situations without and with the use of geotextile reinforcement. Then, introducing the hypothesis that soil strength increases from π to $\pi + 2$ times the undrained shear strength when geotextiles are used, they arrive at the following equations. For the case where a geotextile is not used, the result for the unknown aggregate thickness h_0 for on-highway trucks is

$$c_u = \frac{P}{2\pi(\sqrt{P/p_c} + 2h_0 \tan \alpha_0)(\sqrt{P/2p_c} + 2h_0 \tan \alpha_0)} \tag{7.1}$$

Similarly, for off-highway trucks,

$$c_u = \frac{P}{2\pi(\sqrt{P\sqrt{2/p_c}} + 2h_0 \tan \alpha_0)(\sqrt{P/2p_c\sqrt{2}} + 2h_0 \tan \alpha_0)} \tag{7.2}$$

where c_u = cohesion of subgrade soil, Pa
    P = axle load, N
    p_c = tire inflation pressure, Pa
    h_0 = thickness of aggregate layer, m
    α_0 = angle of load distribution pyramid, deg

For the case where a geotextile *is* used, the contrasting equation for the unknown aggregate thickness h is

$$(\pi + 2)c_u = \frac{P}{2(B + 2h \tan \alpha)(L + 2h \tan \alpha)} + \frac{1}{a\sqrt{1 + (a/2s)^2}} \tag{7.3}$$

where the various constants are defined by Giroud and Noiray.[7] These equations have been plotted by Giroud and Noiray, resulting in a family of curves showing the aggregate thickness required when no geotextile is used and the savings in aggregate thickness when various geotextiles *are* used; see Fig. 7.7 for an example of such curves. Values needed for the analysis are:

Soil strength expressed as CBR or unconfined compression strength.
Geotextile modulus. This should equal the secant modulus at 10 percent strain, which is conservative considering the other values that might be used, e.g., initial or offset tangent moduli.
Characteristics of vehicles that will use the finished road, such as their weight, tire inflation pressure, and the number of anticipated vehicle passes.
An estimate of anticipated rut depth, for it is this value which mobilizes the geotextile's initial strength. These ruts exist only immediately after construction, for they are filled in with stone after the first few truck passes.
An example of this procedure follows.

Example 7.1: Unpaved road design Using the Giroud and Noiray[7] design procedure, determine the stone aggregate necessary for an unpaved road that must be able to carry vehicles having an 80-kN axle load, a tire inflation pressure of 480 kPa, and a rut depth of 0.3 m, if the soil CBR = 1 and the number N of vehicle passes = 340. Solve the problem for a series of moduli and plot the results showing the required aggregate thicknesses versus geotextile stiffness.

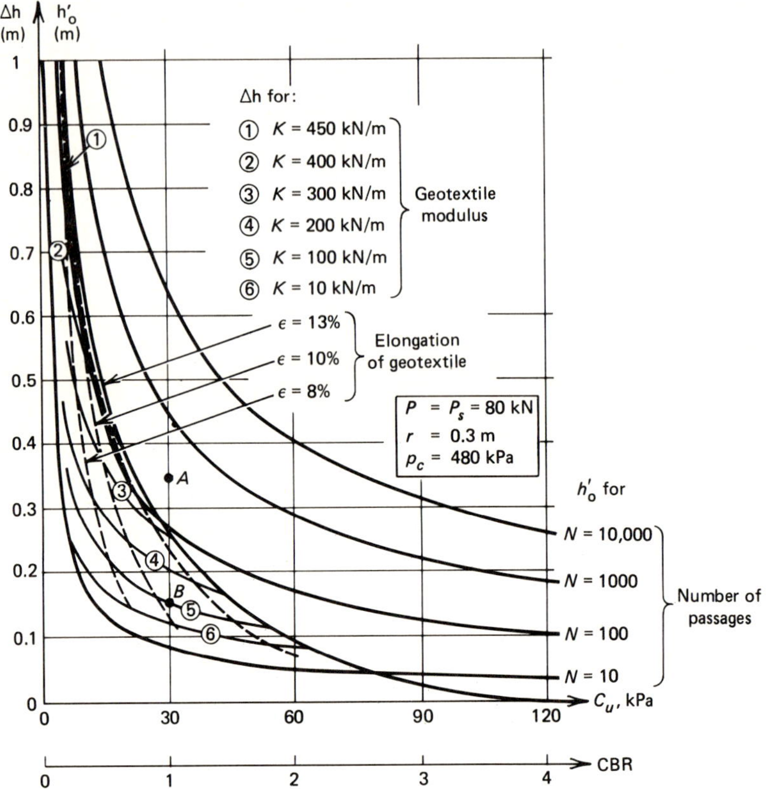

Figure 7.7 Plots of required aggregate thickness h_0' for a road not reinforced by a geotextile and possible reduction in required aggregate thickness Δh resulting from the use of a geotextile versus subgrade soil cohesion. Chart assumes traffic over road of an on-highway truck having a standard axle load. *(After Giroud and Noiray, Ref. 7.)*

SOLUTION Point A on Fig. 7.7 directly shows the required aggregate thickness when no geotextile is used—0.35 m. Now, selecting geotextile moduli of 450, 400, 300, 200, 100, and 10 kN/m (corresponding to curves 1 through 6, respectively, in Fig. 7.7), the data given in the accompanying table and chart result:

Geotextile modulus, kN/m	Aggregate savings, m	Aggregate thickness with geotextile, m
450	0.27	0.08
400	0.26	0.09
300	0.25	0.10
200	0.20	0.15
100†	0.15†	0.20
10	0.12	0.23

†Point B on Fig. 7.7.

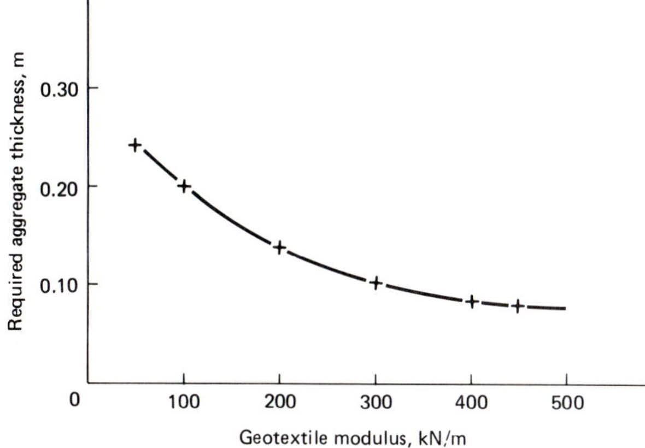

Another major way in which geotextiles can be used as reinforcement is in the construction of fabric-reinforced walls and embankments. This technology is borrowed from the standard technology for reinforced earth walls, insofar as both its methodology[8] and its design[9] go. The concept behind reinforced earth walls is to embed thin metal strips, or mesh, attached to wall facing elements, in soil. The elements form the face of the wall and the friction and adhesive interaction between soil and strip or mesh provides its stability. These types of walls have enjoyed widespread popularity over the past 10 to 20 years; recall Sec. 3.5.4. Recently, however, geotextiles have also been used to form such walls, and are able to provide both facing and stability simultaneously. The construction procedure is illustrated in Fig. 7.8, and is described here:

1. Start with an adequate working surface and staging area.
2. Lay a geotextile sheet of proper width on the ground's surface with 4 to 7 ft at the wall face draped over a temporary wooden form.
3. Backfill over this sheet with soil. Granular soils or soils containing a maximum of 30 percent silt and/or 5 percent clay are customary. Fatty clays of low permeability should not be used.
4. Construction equipment must work from the soil backfill and be kept off the unprotected geotextile. The spreading equipment should be a wide-tracked bulldozer that exerts little pressure against the ground on which it rests. Rolling equipment likewise should be relatively light-weight.
5. When the first layer has been folded over, the process should be repeated for the second layer, with the temporary facing form being extended from the original ground surface (in which case it must be braced from below) or the wall being stepped back about 6 in so that the form can be supported from the first layer. In the latter case, the support stakes must penetrate the fabric.
6. This process is continued until the wall reaches its intended height.
7. A surfacing, e.g., pavement, can be placed on top of the upper layer of the wall. Obviously, if this is done, the surfacing and its loading must be considered as a surcharge load in the wall's design.

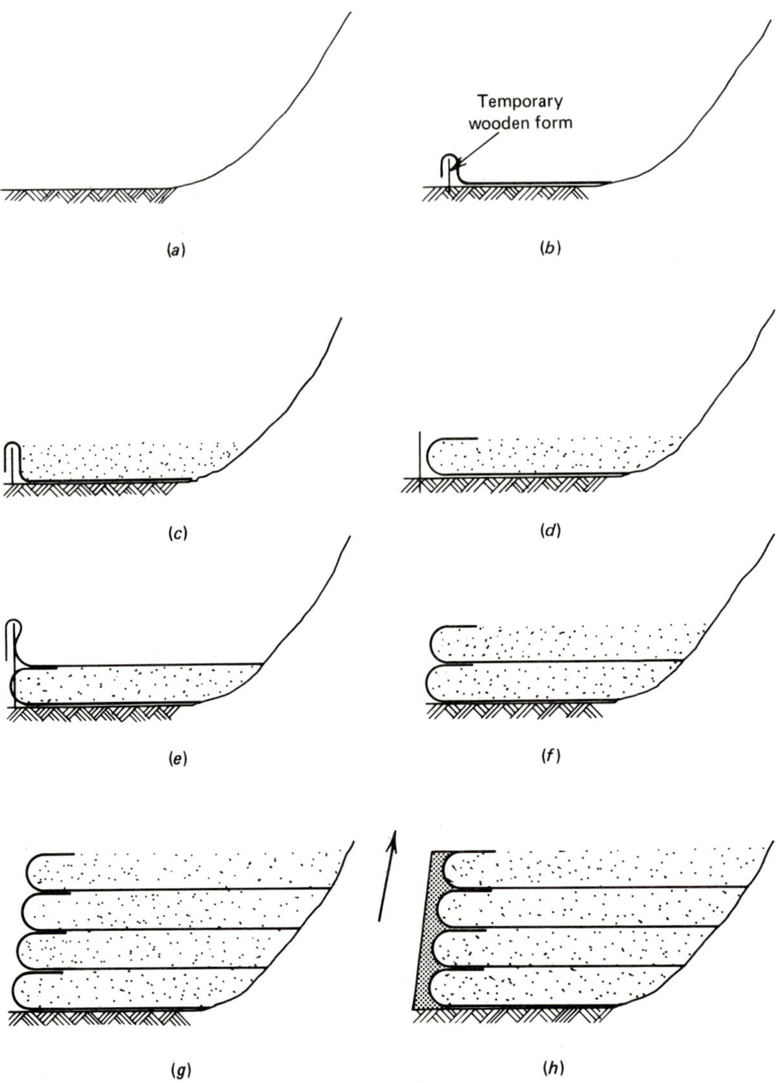

Figure 7.8 General construction procedures for using geotextiles in fabric wall construction. (*a*) Prepare working surface and staging area. (*b*) Place initial geotextile on ground surface. (*c*) Backfill over geotextile, grade, and compact. (*d*) Place loose geotextile end over backfill. (*e*) Place second geotextile sheet over backfill, as with step *b*. (*f*) Backfill, grade, compact, and lay loose end over backfill. (*g*) Continue to top of wall. (*h*) Place protection on wall face.

8. For protection against ultraviolet light and safety against vandalism, the faces of such walls must be protected. Both shotcrete and gunite have been used for this purpose. See Fig. 7.9 for typical completed structures.

The design of geotextile-reinforced walls is similar in principle to that of reinforced earth walls. The major design steps are as follows[10-12]:

Figure 7.9 Photographs of site preparation and actual construction of geotextile-reinforced walls. *(Photos compliments of Crown Zellerbach Corp., Washougal, WA 98671.)*

Assessment of the earth pressure that will be exerted against the wall (active Rankine conditions are assumed) versus the tensile resistance of the fabric. This analysis results in the vertical spacing needed between layers.

Determination of the maximum pressure that will be exerted at the face of the wall so that the designer can provide against any bursting of the fabric.

Assessment of the adhesive resistance of the individual fabric elements in relation to the backfill soil. This analysis results in determination of the fabric length.

Calculation of long-term creep deformation of the fabric-soil system to assess long-term horizontal deformations.

Calculation of overall stability of the wall and its foundation to avoid a rotational failure.

Calculation of foundation deformation to determine estimates of vertical settlement.

7.4 GEOTEXTILES IN FILTRATION AND DRAINAGE

Geotextiles can be used in filtration and drainage systems in much the same way as soil filters and drains; their advantages over soils, however, are numerous. Depending upon the particular situation, those advantages can include:

Less excavation required
Less soil to dispose of
Faster installation
Greater system stability
Tensile strength of the geotextile added to the system
Lighter load placed on subsoils
Less technical detail in planning and construction
Generally lower costs

Geotextiles have been used successfully in a wide variety of situations. A number of those are shown schematically in Fig. 7.10. In each, the geotextile is being used as a filter (for cross plane flow), a drain (for in plane flow), or both. The distinction between the two functions is as follows:

Geotextile filtration occurs in fabrics where water flow brings some of the finer particles of the soil being protected into the structure of the geotextile itself. This soil modification above the fabric and void modification within the fabric eventually reach stability, or equilibrium, and from then on only clear water passes through. For relatively thin geotextiles (woven and heat-set), most of the filtering takes place within the soil upstream from the fabric.

Geotextile drainage occurs either cross plane, when the above described processes have reached stability, or in plane, when water is transmitted within the geotextile structure itself. In this latter case, a bulky geotextile or a composite system is needed; see Figs. 7.2(c) and 7.2 (f).

How effective these functions are in a given system can be checked by means of long-term soil-geotextile flow tests. The general response to these tests by a wide range of soil types and geotextiles, as reported by Koerner and Ko,[13] is shown in Fig.

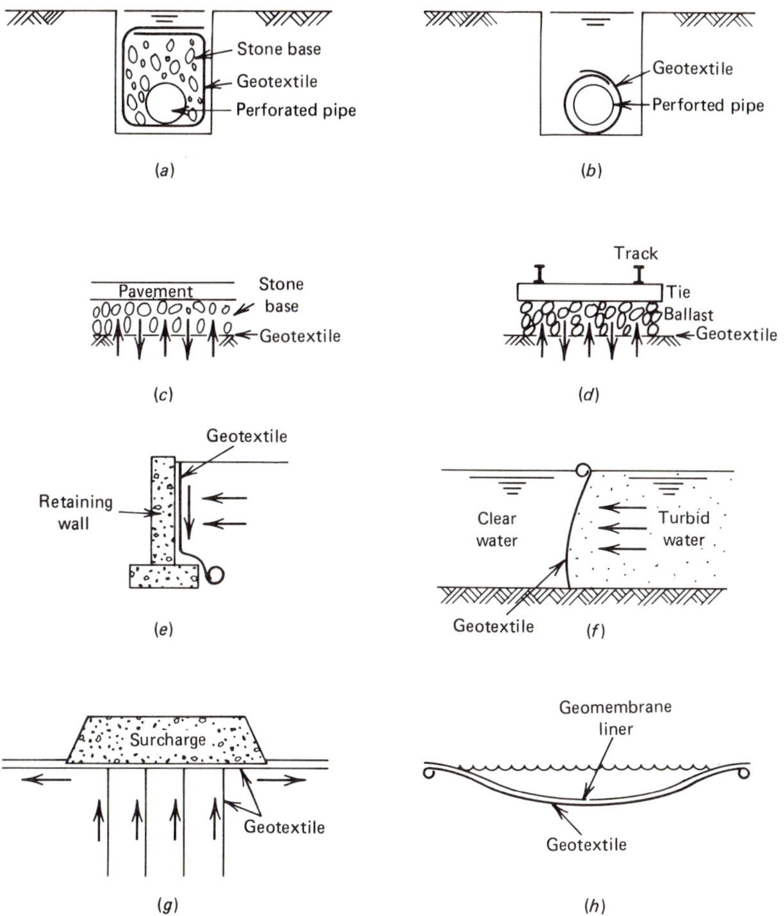

Figure 7-10 Situations where geotextiles have been used for filtration and drainage. (*a*) Highway and railroad underdrains. (*b*) Field and agriculture wrapped drain pipes. (*c*) Dissipating pore water pressure and allowing for drainage. (*d*) Eliminating pumping in railroads. (*e*) Providing drainage behind walls. (*f*) Silt curtain filter. (*g*) Vertical and horizontal drainage of soft subsoils. (*h*) Geomembrane underliner to vent gases and remove water.

7.11. Critical in this regard is the slope of the final portion of the curve. Once stability has been established within the soil and the geotextile, slope m_f should be equal to zero. If it continues to decrease, the geotextile or upstream soil is probably clogging and the problem in the long term might result in a completely cut-off flow of water. But as the amount of time required to reach stability indicates, it could take weeks of testing to determine just what the actual situation is.

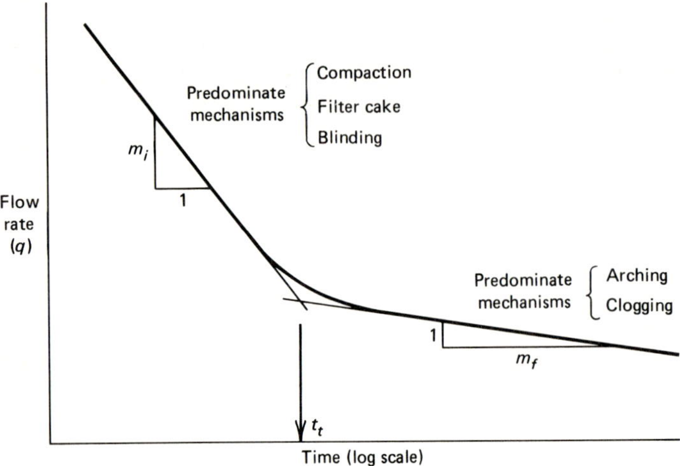

Figure 7-11 Generalized long-term flow response of soil-geotextile systems. t_t = transition time (minutes for sands, hours for silts, days for clays); m_i = initial slope = function of soil compaction, filter cake buildup, and blinding over fabric voids; m_f = final slope = function of soil arching over fabric voids and clogging within fabric. *(After Koerner and Ko, Ref. 13.)*

An alternative to the testing procedure noted above is to empirically relate the geotextile's void size to the particle characteristics of the soil being protected. Giroud[14] has offered a set of criteria relevant to the prevention of piping (soil passing through the geotextile), which we give in Table 7.1. However, it should be noted that it is difficult to obtain O_{95}, the apparent opening size of the geotextile, a value not generally given by the fabric manufacturers and not easy to determine. It is often assumed to be the equivalent opening size (EOS) of the fabric; see Refs. 2 or 3.

Table 7.1 Criteria for preventing soil piping in one-way flow situations†

Relative density	$1 < \text{CU} < 3$	$\text{CU} > 3$
Loose ($D_R < 35\%$)‡	$O_{95} < (\text{CU})(d_{50})$§	$O_{95} < \dfrac{9}{\text{CU}} d_{50}$
Intermediate ($35\% < D_R < 65\%$)	$O_{95} < 1.5(\text{CU})(d_{50})$	$O_{95} < \dfrac{13.5}{\text{CU}} d_{50}$
Dense ($D_R > 65\%$)	$O_{95} < 2(\text{CU})(d_{50})$	$O_{95} < \dfrac{18}{\text{CU}} d_{50}$

†After Giroud, Ref. 14.

‡ $D_R = \dfrac{\gamma - \gamma_{\min}}{\gamma_{\max} - \gamma_{\min}} \left(\dfrac{\gamma_{\max}}{\gamma} \times 100 \right)$

§ O_{95} = apparent opening size of geotextile; d_{50} = soil particle size corresponding to 50 percent finer than maximum size; $\text{CU} = \dfrac{d_{60}}{d_{10}}$ = soil coefficient of uniformity

In regard to drainage, the *cross-plane permeabilities* of all the geotextiles currently in use are sufficient for most situations. Permeabilities range from 0.01 to 5 cm/s, which is usually sufficient to drain soils. A possible exception is when geotextiles are used beneath erosion control structures (rip-rap or concrete blocks). Here the overlying material can block a sizable portion of the geotextile, and when rapid dissipation of increased pore water pressure is necessary, failures (i.e., blowouts of the surfacing) can result. It should be noted that the values given for the permeabilities of geotextiles are not particularly meaningful since geotextiles vary in thickness under load. It is better is to use the value of k_n/h where k_n is the permeability normal to the plane of the fabric and h is the fabric's thickness. This term is called *permittivity*.

Example 7.2 illustrates the use of a geotextile behind a small gabion wall (gabions are wire baskets filled with stone), and shows that for common applications most geotextiles have sufficient permittivity to pass water.

Example 7.2: Design for a gabion wall that uses a geotextile Select a geotextile to be placed behind a 7-ft-high gabion wall whose permittivity will allow it to accommodate full head loss. The backfill soil is a silty sand with $k = 0.01$ cm/s $(=3.3 \times 10^{-4}$ ft/s$)$.

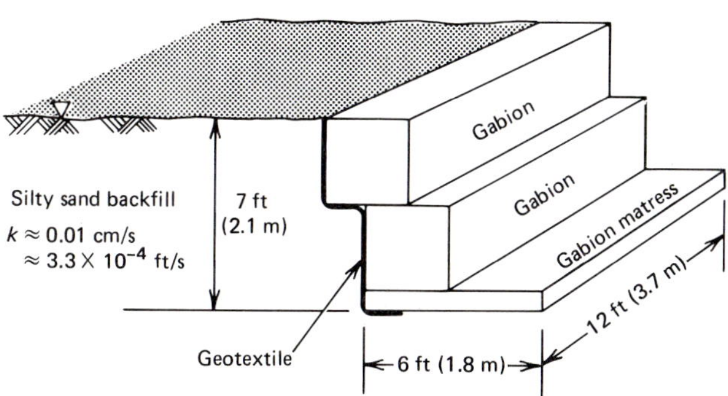

SOLUTION Maximum flow to geotextile:

$$Q = kiA$$

$$= k\frac{\Delta h}{\Delta s}A$$

$$= (3.3 \times 10^{-4})(\tfrac{7}{7})(1 \times 7)$$

$$= 23.1 \times 10^{-4} \text{ ft}^2/\text{s/ft wall}$$

Calculate required permittivity k_n/h:

$$Q = kiA$$

$$= k_n \frac{\Delta h}{h} A$$

$$\frac{k_n}{h} = \frac{Q}{\Delta h A}$$

$$= \frac{23.1 \times 10^{-4}}{7(7 \times 1)}$$

$$\left.\frac{k_n}{h}\right|_{\text{req'd}} = 4.6 \times 10^{-4}\,\text{s}^{-1}$$

Check candidate geotextiles. Try Bidim C-22:

$$h = 60 \text{ mil}$$

$$= 0.0015 \text{ m}$$

$$k_n = 0.003 \text{ m/s}$$

Therefore
$$\frac{k_n}{h} = \frac{0.003}{0.0015}$$

$$= 2.0 \gg 4.6 \times 10^{-4} \qquad \text{OK}$$

Try Typar 3401:

$$h = 15 \text{ mil}$$

$$= 3.8 \times 10^{-4} \text{ m}$$

$$k_n = 0.02 \text{ cm/s}$$

$$= 2 \times 10^{-4} \text{ m/s}$$

Therefore
$$\frac{k_n}{h} = \frac{2 \times 10^{-4}}{3.8 \times 10^{-4}}$$

$$= 0.53 \gg 4.6 \times 10^{-4} \qquad \text{OK}$$

Check the selected geotextiles in terms of their susceptibility to soil piping or clogging following criteria listed in Table 7.1.

The other aspect of flow through geotextiles that should be discussed is *in-plane drainage,* which occurs mainly within bulky, needle-punched fabrics. These fabrics can be as thick as 300 mil ($=0.3$ in $= 0.76$ cm) and contain anywhere from 60 to 95 percent void space, depending on the normal stresses imposed upon them.[15] The term which best describes this characteristic is *transmissivity* (symbolized by θ), which is equal to the in-plane permeability k_p times the geotextile thickness h. As with permittivity, transmissivity is the recommended value to work with because it includes the fabric thickness, a dimension strongly affected by the normal stresses placed on the fabric.

The equations for these two terms (permittivity and transmissivity) which describe a geotextile's ability to conduct water are as follows:

$$\psi = \frac{k_n}{h} \tag{7.4}$$

where ψ = permittivity
 k_n = coefficient of permeability "normal" to the fabric
 h = thickness of fabric under load

and
$$\theta = k_p h \tag{7.5}$$

where θ = transmissivity
 k_p = coefficient of permeability "in the plane" of the fabric
 h = thickness of fabric under load

Figure 7.12 illustrates results from planar transmissivity studies carried out by Koerner and Sankey[16] for a number of bulky, needle-punched geotextiles.[17] Here it can be seen that the residual transmissivity of these fabrics is from 0.004 to 0.010 ft^3/min per foot of fabric.

The in-plane flow of water in a geotextile has a number of possible applications (some were illustrated in Fig. 7.10). Nowhere, however, is the usefulness of this property so evident as in the application of geotextiles to chimney drains and drainage galleries within zoned earth dams.[18] Those drainage systems are intended to carry whatever water gets through the clay core wall to beyond the downstream toe of the slope, without causing a buildup of excessive pore water pressure within the dam itself. The functioning of these systems is critically important to the stability of the dam. Example 7.3, adapted from material presented by Giroud,[19] shows how geotextiles can be used in place of conventional granular soil chimney drains. Note that conventional granular soil chimney drains need to be protected from clogging by graded filter soils, a feature which is eliminated by the use of a proper geotextile. Geotextiles are also potentially more cost-effective and easier to deal with during construction than the conventional drains.

The application of geotextiles to such drainage systems is but one example of instances where the property of in-plane flow of water could be useful. In regard to gravity drainage, others are:

Retaining walls
Fin drains (flow interceptors)
Insulation between pavement above and a geomembrane below

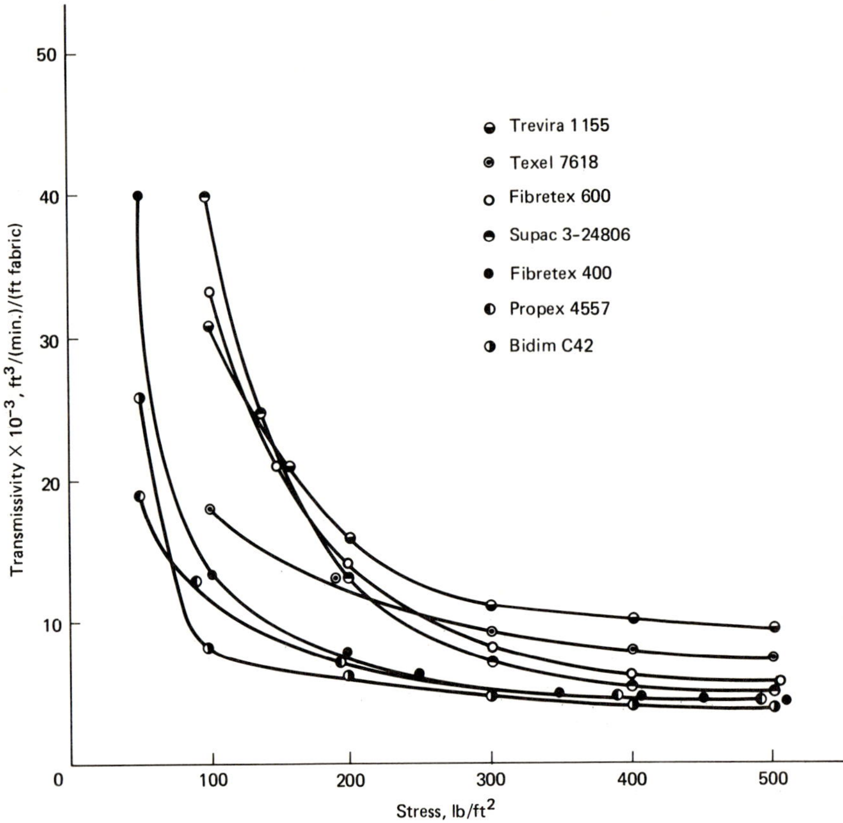

Figure 7.12 Transmissivity response to applied normal stress for various nonwoven, needle-punched geotextiles. *(After Koerner and Bove, Ref. 17.)*

In regard to pressure drainage:

Vertical drain wells
Reinforced earth walls
Geotextile-encapsulated soil
Horizontal drains in a dam
Horizontal drain beneath an embankment

Example 7.3: Design for geotextile chimney drain Design a geotextile chimney drain for the zoned earth dam shown in the accompanying illustration.

SOLUTION Calculate the flow rate q of the water coming to the geotextile drain (calculations are always based on a unit width of dam, and in this case we will use a per-meter-of-width basis):

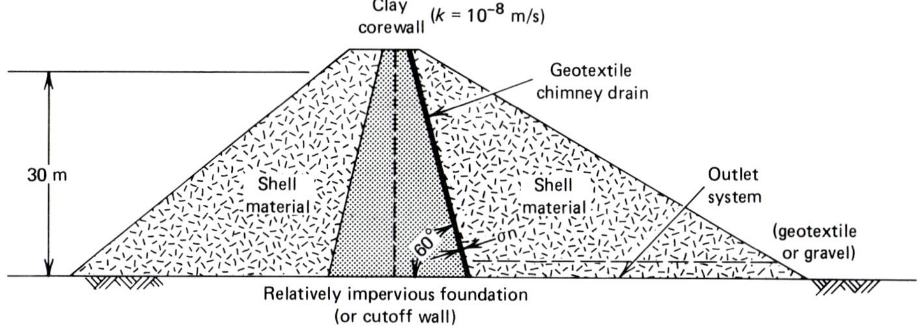

$$q = kh\,\frac{F}{N}$$ (standard flow net analysis which results in 7 flow lines per two equipotential head drops)

$$= (10^{-8})(30)(\tfrac{7}{2})$$

$$= 1.1 \times 10^{-6}\ \mathrm{m^3/s}$$

Determine the flow gradient i in the geotextile. Since the geotextile exerts no pressure of its own (i.e., we are dealing only with gravity flow), the gradient is merely the slope of the drain:

$$i = \sin 60°$$

$$= 0.87$$

Now calculate the required geotextile transmissivity $\theta_{req'd}$:

$$\frac{q}{L} = \theta_{req'd}i$$

$$\theta_{req'd} = \frac{q}{Li}$$

$$= \frac{1.1 \times 10^{-6}}{(1)(.87)}$$

$$= 1.3 \times 10^{-6}\ \mathrm{m^2/s}$$

This value of $\theta_{req'd}$ is based on FS = 1.0.

If $FS_{req'd} = 5$, then $\theta_{req'd} = 6.5 \times 10^{-6}\ \mathrm{m^2/s}$. Now select a geotextile that exhibits the above transmissivity at the pressures which will be exerted against it. (*Note:* This analysis is based on the maximum pressure; higher up in the drain where the pressures are lower, the required transmissivity will be lower.)

Note that the minimum residual value of transmissivity shown in Fig. 7.12 for the nonwoven, needle-punched geotextiles that were tested is 0.004 ft³/min per foot of fabric. This converts to $6.2 \times 10^{-6}\ \mathrm{m^3/s}$ per meter of fabric and is very close to the required transmissivity. Thus any of the geotextiles listed in Fig. 7.12 would be adequate from a permeability point of view.

As in Example 7.2, the candidate geotextile(s) must now be checked in terms of susceptibility to soil piping or clogging according to the criteria given in Table 7.1.

7.5 GEOTEXTILES IN EROSION CONTROL

The prevention of the erosion of land surfaces (see Fig. 7.13) represents a formidable challenge to many areas of geotechnical and construction engineering. And in the construction of structures used to control erosion, geotextiles can be used to great advantage. In our discussion of erosion control, it will be obvious that geotextiles serve a number of different functions, e.g., separation, reinforcement, filtering, drainage, and combinations of these. The topic is covered in a section all its own, however, since it is itself such a specific problem in need of a solution.

7.5.1 Use of Geotextiles Alone

The traditional way to protect slopes from sheet or gully erosion is by planting them with a quick-growing vegetative cover, e.g., crown vetch. However, for the 2 to 6 weeks required by the plant seeds for germination (a time frame also influenced by the amount of rainfall that occurs), the seed itself must be protected from washing away. Hay, with an asphalt emulsion tack coat or a PVA spray,[20] has often been used to prevent this from occurring, but geotextiles present an interesting alternative protection. Geotextiles laid over, and pinned or stapled to, the ground not only can protect the germinating seed, but can be made from biodegradable paper of varying

Figure 7.13 Typical erosion pattern on exposed soil surface.

lifetimes. At least two types of geotextiles, both made out of a loose-knit fabric with biodegradable paper strips that last from 2 months to 1 year woven throughout, are used for this purpose.

For long-term protection against erosion caused by rainfall runoff, there is an interesting system involving a stiff nylon mesh having a very open structure protected on the bottom by a geotextile; see Fig. 7.14. This type of system has been used successfully to line diversion ditches for canals and highway slopes. Its cost-effectiveness over concrete-lined trenches and other systems is quite impressive; see Hoffman and Adamsky.[21]

Figure 7.14 Geotextiles used as ground cover to prevent soil erosion; upper figure as constructed, lower figure after 1-year's growth. *(Photos compliments of American Enka Co., Enka, NC 28728.)*

7.5.2 Geotextiles beneath Rock Rip-Rap

The need for a filter between the natural soil being protected against erosion and the rock rip-rap surface doing the protecting is easily justified on the basis of the many failures sustained by protected slopes, jetties, groins, and other waterfront structures that did not have filters.[22] Barrett,[23] back in the mid-1960s, was the first to use geotextiles instead of graded soil filters for this application; see Fig. 7.15 for typical configurations.[24] (As an historical note, this was the first type of application for which synthetic geotextile fabrics were used and marks the beginning of the geotextile industry as we know it today. The fabrics originally used were almost exclusively woven monofilament polypropylene with a mass of about 8 oz/yd^2 and about 20 percent open area. Many are still serving as intended without exhibiting any degradation, excessive clogging, or other kind of deterioration after periods of over 15 years.)

The general procedures for placing a geotextile underliner beneath rock rip-rap are as follows:

First, if it is at all possible to do so, remove any water from the immediate vicinity of the site where the fabric is to be placed. This can be done by means of a small cofferdam, by diverting the water, by dewatering the area, or by working at low flow or low tide. The result will be a more productive effort and a better job when completed.

Prepare the slope over which the fabric is to be laid as much as possible. At a minimum, remove debris, rocks, stumps, and so on, and fill in large voids or slope irregularities.

Compact the slope, if possible. Even hand-rolling is helpful in minimizing future differential movement of the completed erosion control structure.

If working in the dry, excavate the toe trench at the intended lower elevation of the fabric. Then spread the fabric out on the slope and place and anchor it into the toe trench. Anchorage can be achieved by means of pins or stakes, or by placing soil or rock on top of the fabric in a sausagelike configuration. Proceed up the slope from the toe trench in this manner.

If working in the wet, the toe trench must be excavated and simultaneously the fabric must be placed in it and anchored. Thus you must take into account two factors at once, the presence of the water and having to proceed up the slope. It is important that the fabric be anchored to the slope along its entire underwater area, since water below the fabric can lift it out of place and wave action can tear it or rip it out of the toe trench. In cold climates, ice formation on the fabric can be very troublesome.

The amount of overlap needed between fabric ends and sections depends on the nature of the in situ slope soil (mainly its compressibility), on the type of erosion control structure to be placed above the fabric (mainly its weight), and on the deformation characteristics of the fabric itself (mainly its modulus and creep behavior). Lap distances of 2 to 5 ft are usual. If the fabric ends are to be joined

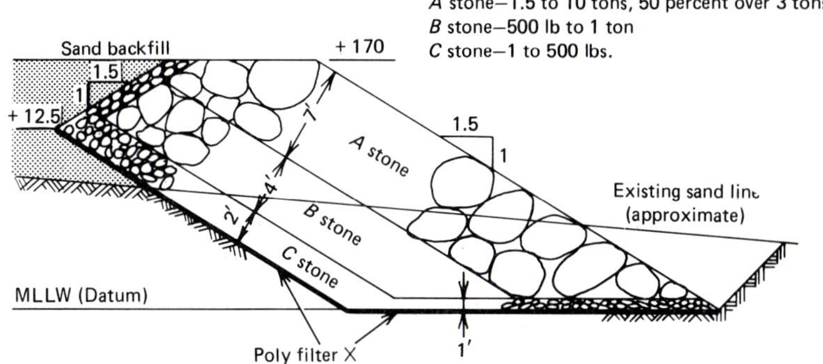

Note: Extend filter cloth at least 1 ft past edges and lap 2 ft at seams.
Lay loosely, do not stretch.

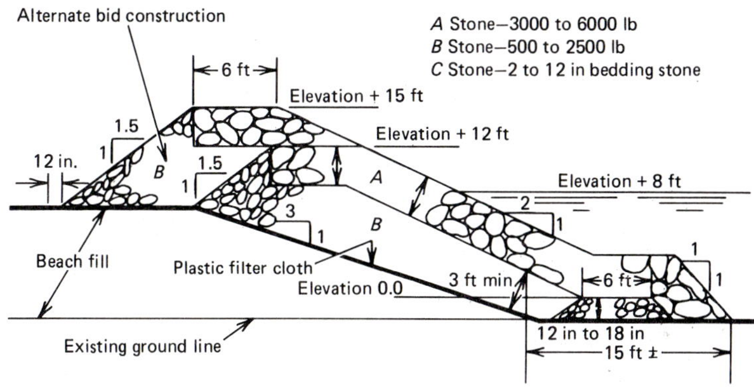

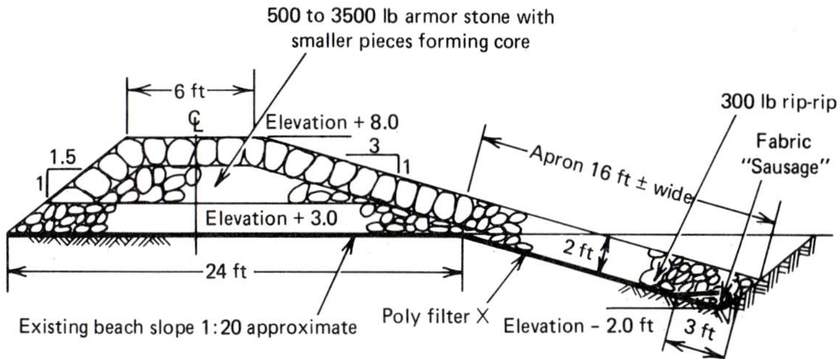

Figure 7.15 Typical cross sections of waterfront erosion control systems. *(After Dunham and Barrett, Ref. 24.)*

by bonding or sewing, the joint strength must be at least 80% of the strength of
the fabric itself.

Secure the upper end of the fabric into a trench or other suitable site for anchorage.
Surface water should not be able to get beneath it and it should be protected
against vandalism and prolonged exposure to the elements. No more than a minimum amount of the fabric should ever be left exposed, particularly in warm,
sunny climates since the fabrics are usually made from ultraviolet-sensitive
material.

If the rip-rap that will rest on the fabric is heavy, protect the material from puncture
by placing a 3- to 5-in layer of sand or run-of-bank gravel between them. Start
at the toe and work up the slope laying this protective layer, then follow the same
procedure for constructing the final erosion control structure.

Any geotextile used for this purpose must be resistant to soil migrating through
it (i.e., to soil piping), as that would result in an instability of the rip-rap. The appropriate geotextile should be selected according to the criteria listed in Table 7.1. The
fabric must also be able to pass very large quantities of water in many situations, so
its permeability must be adequate. And for such critical applications, the percent
open area of the fabric should be quite large, e.g., approximately 6 percent, or larger
depending upon the specific conditions.

Depending on the characteristics of the rip-rap and the control exercised during
its placement by the contractor, other fabric properties can be significant—for example, tensile strength, creep resistance, impact resistance, puncture resistance, and
tear resistance. Unfortunately there is a lack of field and laboratory data, and hence
of a precise design methodology at this time.

7.5.3 Geotextiles beneath Precast-Articulated Concrete Blocks

Aside from manually constructed geotextile–rock rip-rap systems, a number of fac-
tory-constructed erosion control systems are available. These systems (many of
which are patented) usually have a high-strength geotextile that serves as a sub-
strate, onto which are bonded small precast concrete blocks that are hinged together.
Sometimes the concrete blocks are cabled together ahead of time and placed as a
unit on the geotextile. The hinging provides for flexibility (articulation) between the
blocks, and separates them from one another to allow the passage of water from
above and below. Since the concrete blocks cover a high percentage of the geotextile's
surface, however, one should be careful about using such systems in situations where
rapid dissipation of pore water pressure is necessary, e.g., in areas with large tidal
zones or where heavy ship traffic produces large waves.

7.5.4 Erosion Control Mattresses

Two layers of fabric separated by a drop stitch or sewn together only at intervals of
about 12 in have been used as a flexible form into which cement grout is pumped, to

form an erosion control mattress. Figure 7.16 shows typical cross sections of such mattresses, which can be used to build a very aesthetic looking channel or bank lining system. As with the erosion control systems discussed in Sec. 7.5.3, the rapid dissipation of excess pore water pressure in the soil beneath a mattress can be troublesome. In those cases, the use of a filter point system is necessary, with the filter points being closely spaced and having a high permeability.

A typical construction sequence for installing mattresses along the banks of a channel or stream (many of the systems are patented) is as follows:

Remove stumps, boulders, and brush from the site. Grade sufficiently to provide a slope that is stable in the absence of erosive forces. In general, an average slope steeper than 1:1.5 is not recommended. Cut an upper toe trench to prevent undercutting of the completed structure in the event of heavy water runoff.

Place the shop-assembled fabric panels (usually from 2000 to 3000 ft^2 per section) over the embankment, with their seams straight and preferably angled perpendicular to the shoreline. The fabric must be flexible and light-weight, and usually must be placed by hand, beginning upstream. Guide ropes to the opposite shore or small boats may also be used to assist in fabric placement.

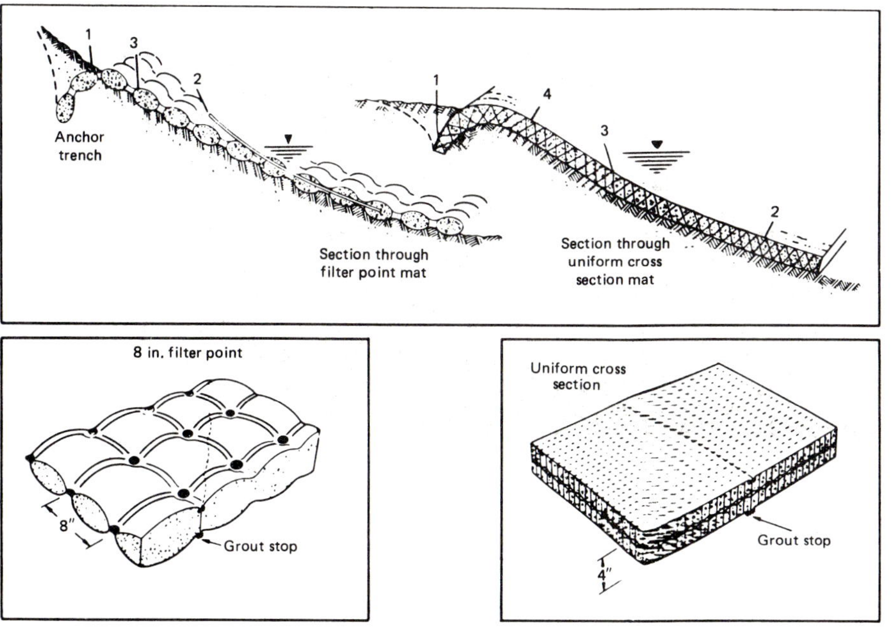

Figure 7.16 Typical cross sections and schematics of erosion control mattresses. Numbers in upper sketches refer to sequence of cement grout injections. *(After Construction Techniques, Inc., Cleveland, OH 44120.)*

Once placed, sew or otherwise join the fabric panels together to create a monolithic structure of the required length and width.

Inject ready-mix mortar into the fabric envelope with a mortar pump; a pump with a capacity of from 10 to 20 yd^3/h is usually sufficient. The fabric in the anchor trench is filled first so that it can serve to position the rest of the envelope. Fill the underwater portion of the mat next. Then fill any remaining sections. The lateral spread of the fluid concrete can be controlled by shop-installed strips, which serve as grout stops. Production rates as high as 5000 ft^2 of mat filled per 8-h shift are regularly achieved on large projects.

Backfill the toe trench at the top of the embankment.

7.5.5 Geotextiles as Silt Fences

The passage by Congress in 1972 of the Clean Streams Act had a significant impact on heavy construction practices. Previously the topsoil was completely stripped at the beginning of a project, stored long-term in a stockpile, and then used for the final grading of the site when the project was completed; since 1972, however, an "erosion control and sedimentation" plan is required for all projects involving streams. This means the topsoil must be stripped sequentially, stored in nonactive areas within each stripped area, and replaced section by section as soon as possible. Thus the impact of soil erosion on downstream waterways is greatly decreased. Generally each individual parcel being worked on has a small retaining structure built around it which stops the flow of turbid water, allows the sediments to settle out, and permits clear water to pass on. Those retaining structures can take the form of small earth dams, bales of hay or brush tied together, or geotextile silt fence.

Two types of geotextile silt fences are available. In one, the geotextile is strung between vertical posts, while in the other, mesh wire is attached to vertical posts to which the geotextile is then fastened. In both cases the posts must be adequately embedded in the ground, since they act like cantilever beams, and the geotextile must be anchored into a ground trench to prevent the runoff from flowing beneath it; see Fig. 7.17.

These types of geotextile silt fences meet most governmental regulations, are certainly cost-effective, and have sufficient ability to withstand ultraviolet exposure for the lifetime of the project. Moreover, complete design methodology has been developed for their use; see Example 7.4, based on work done by Bell and Hicks (Ref. 25).

Example 7.4: Silt fence design The estimated runoff Q_0 from a construction site is 30 cm^3/s, and it contains particles of sediment with diameters D as small as 0.005 cm. The ground slope α in the vicinity of the fence is 15°. The fabric being considered for use has thickness t of 25 mil (0.064 cm), a coefficient of

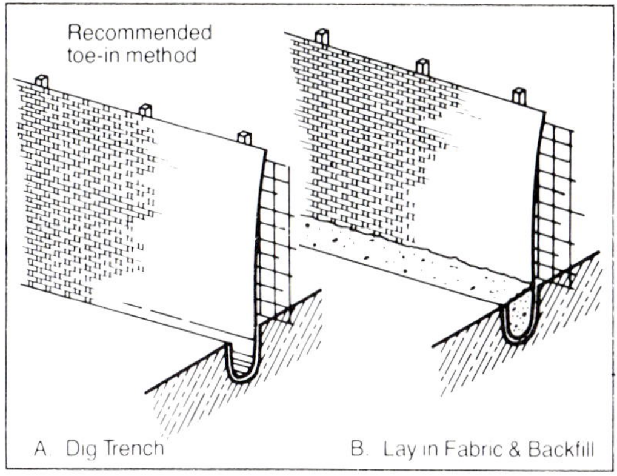

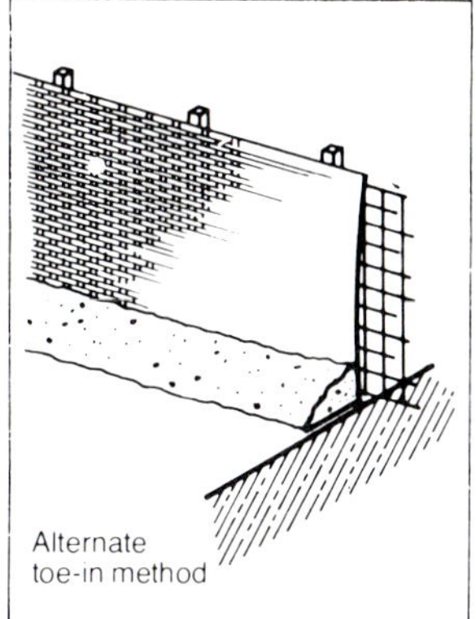

(a)

(b)

Figure 7.17 Geotextile silt fence construction using wire mesh attached to fence posts. (*a*) Schematic diagrams of recommended techniques. (*b*) Photos of completed geotextile silt fences. (*After Mirafi, Inc., Charlotte, NC 28224.*)

permeability k_f of 0.05 cm/s, a threshold pressure head h_2 of 3 in (7.6 cm), a fabric mobilization head h_1 of 9 in (estimated) [22.9 cm (estimated)], a tested wide-width tensile strength T_{test} of 160 lb/in, and a modulus E of 1200 lb/in.

Determine (*a*) the required height of the fence, (*b*) the factor of safety (FS) for the geotextile used without mesh support (i.e., when it is supported only by posts placed 4 ft apart center to center), and (*c*) the factor of safety for the geotextile when it is supported by wire mesh 12 in wide by 6 in high attached to the posts.

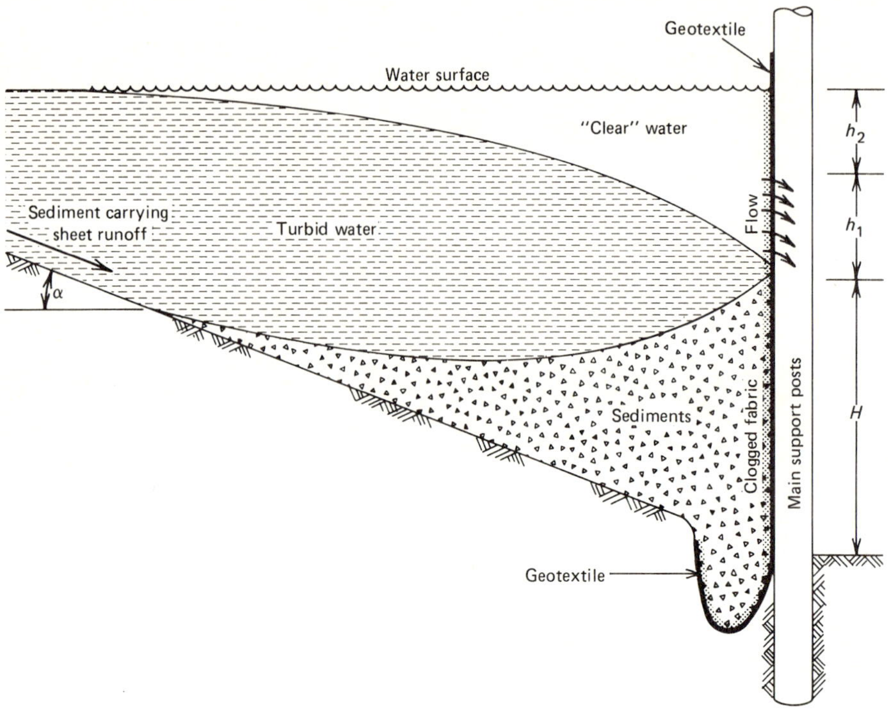

SOLUTION (a) The required height X is found as follows:

$$X = \frac{Q_0 \tan \alpha}{8915 D^2} \left[1 + \frac{1}{-1 + \sqrt{1 + 2tQ_0/k_f h_2^2}} \right]$$

$$= \frac{(30)(0.268)}{(8915)(0.005)^2} \left[1 + \frac{1}{-1 + \sqrt{1 + (2)(0.064)(30)/(0.05)(7.6)(7.6)}} \right]$$

$$= 36.07(1 + 1.9)$$

$$= 104.6 \text{ cm}$$

$$= 3.43 \text{ ft}$$

Allowing for freeboard, use a silt fence 4.0 ft high.

(b) The factor of safety FS for a geotextile used without mesh support, when the posts are placed 4 ft apart center to center, is equal to its tested wide-width tensile strength T_{test} divided by the maximum tensile strength T_{max} it will be called upon to exhibit in the given actual situation. Therefore, we find the FS is as follows:

$$T_{max} = \left[\frac{(K_A \gamma_s' + \gamma_w)(H)}{r} \right] \left(\frac{S^2}{8} + r^2 \right)$$

where K_A = (coefficient of active earth pressure) = 0.40
γ_s' = (effective unit weight) = $100 - 62.4 = 37.6$ lb/ft^3 = 0.0218 lb/in^3
γ_w = (unit weight of water) = 62.4 lb/ft^3 = 0.0361 lb/in^3
H = (sediment height) = $X - h_1 - h_2 = 104.6 - 22.9 - 7.6 = 74.1$ cm (29.2 in)
S = (post spacing) = 4 ft = 48 in
E = 1200 lb/in

$$r = \frac{1}{2.75} \left[\frac{(K_A\gamma_s' + \gamma_w)HS^4}{E} \right]^{1/3}$$

$$= \frac{1}{2.75} \left[\frac{[(0.40)(0.0218) + (0.0361)](29.2)(48)^4}{1200} \right]^{1/3}$$

$$= \frac{1}{2.75}(17.95) = 6.53$$

Therefore
$$T_{max} = \left\{ \frac{[(0.40)(0.0218) + (0.0361)](29.2)}{6.53} \right\} \left(\frac{48^2}{8} + 6.53^2 \right)$$

$$= (0.2004)(330.6)$$

$$= 66.3 \text{ lb/in}$$

and
$$\text{FS} = \frac{T_{test}}{T_{max}}$$

$$= \frac{160}{66.3}$$

$$= 2.4 \qquad \text{OK}$$

(*c*) The factor of safety for a geotextile used with mesh support measuring 12 in wide by 6 in high behind it can be found as follows:

$$T_{max} = \left[\frac{(K_A\gamma_s' + \gamma_w)(H - b/2)}{r} \right] \left(\frac{a^2}{8} + r^2 \right)$$

where $K_A, \gamma_s', \gamma_w, H$, and E are as given in part *b*, and
a (width of mesh) = 12 in
b (height of mesh) = 6 in

$$r = \frac{1}{2.75} \left[\frac{(K_A\gamma_s' + \gamma_w)(H - b/2)a^4}{E} \right]^{1/3}$$

$$= \frac{1}{2.75} \left\{ \frac{[(0.40)(0.0218) + 0.0361](29.2 - 3)(12)^4}{1200} \right\}^{1/3}$$

$$= \left(\frac{1}{2.75} \right)(2.72)$$

$$= 0.989$$

$$\text{Therefore} \qquad T_{max} = \left[\frac{[(0.40)(0.0218) + (0.0361)](29.2 - 3)}{0.989} \right]$$

$$\left[\frac{(12)^2}{8} + (0.989)^2 \right]$$

$$= (1.187)(18.98)$$

$$= 22.5 \text{ lb/in}$$

$$\text{and} \qquad FS = \frac{T_{test}}{T_{max}}$$

$$= \frac{160}{22.5}$$

$$= 7.1 \qquad OK$$

7.6 GEOTEXTILES AS FLEXIBLE FORMS

The use of wooden or metal forms to support poured concrete or cement grout until it has cured sufficiently for a permanent set is, of course, a standard procedure in construction work. Yet in many cases the use of a geotextile as a "flexible" form into which the concrete or grout is poured can be quite advantageous. Indeed, the method of using two geotextiles separated by a drop stitch or joined at discrete points as erosion control mattresses falls into this category (recall Fig. 7.16) and could have been discussed here as well as in Sec. 7.5.4. The primary functions of the geotextile when used as a flexible form are containment, and filtering, and these aspects will be illustrated in the following descriptions of specific applications.

7.6.1 Providing for Uniform Bearing

When precast footings, piers, or caissons, are to be placed on an uneven surface, as exposed rock often is, the bearing pressure becomes very high at the few points of contact. This situation can be avoided by attaching a geotextile to the bottom of the footing before placement, then after placement filling the space between the footing and the fabric with a high-strength concrete or cement grout. The result is an extension of the original footing whose bottom is the exact inverse replica of the uneven rock surface. Koerner and Welsh[2] describe this technique, as well as a related one where a geotextile bag is used as a sheetpile cutoff within an earth dam.

7.6.2 Remedying Scour beneath Existing River Piers

Somewhat similarly to the circumstances just described, the bearing capacities of bridge piers in rivers can be severely threatened by river scour. When this occurs, geotextile tubes can be used to form a perimeter ring around the pier as shown in

Fig. 7.18. The tubes are filled with concrete or cement grout, forming a barrier to the remaining voids beneath the pier. After curing, pipes connecting to the interior space beneath the pier are used for injection and venting purposes in the placement of high-strength concrete. This provides for the renewed bearing capacity of the pier. Of necessity the work must be done by divers, with all machinery for the concrete and grout delivery systems being operated from the bridge deck or a river barge.

7.6.3 Providing Support and Stability for Underwater Pipelines

Underwater pipelines, used, for example, to connect offshore oil drilling towers to delivery stations or to land, suffer from underwater scour in very unpredictable ways. Quite often divers find long lengths of pipe, originally placed on the ocean bottom, floating unsupported in the water, and such sections are obviously prone to high bending stresses and to breaking. To provide support for them, a number of systems have been devised. In general, they all use geotextiles as flexible forms that are filled with concrete or cement grout. Various configurations have been developed[27,28]:

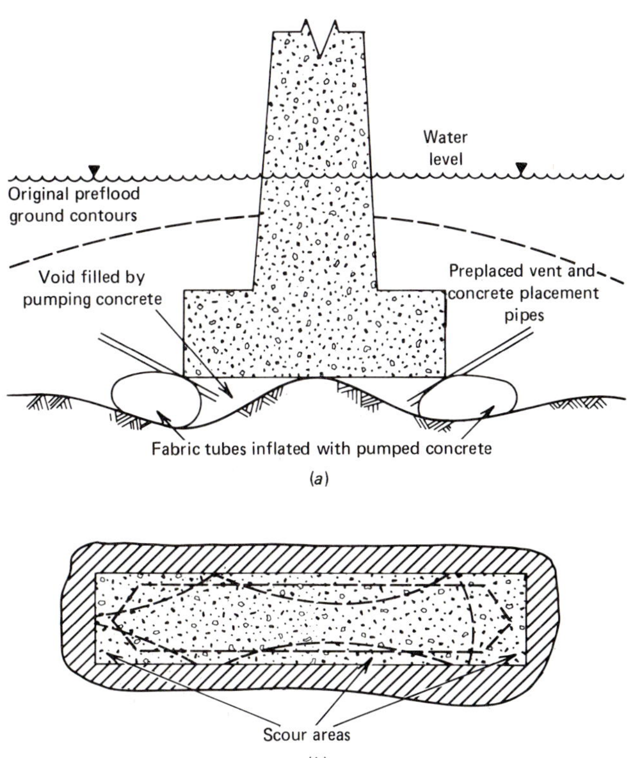

Figure 7.18 Repairing bridge piers after river scour using geotextile tubes filled with grout then grouting the interior. (*a*) Elevation view; (*b*) plan view. (*After Welsh, Ref. 26.*)

A single geotextile bag filled with a continuous concrete or grout mass and placed beneath the section of pipe when scour depth is 3 ft or less

A number of bags filled with grout and forming a pyramid type of support for scour depths greater than 3 ft; see Fig. 7.19*a*

Systems for repairing damaged underwater pipelines; see Fig. 7.19*b*

A complete pipe saddle providing support, resistance to buoyancy, and protection against boat anchor dragging; see Fig. 7.19*c*

A protective dome within which intricate underwater repairs or connections can be made; see Fig. 7.19*d*

7.6.4 Pile Jacketing

All piles in marine environments suffer deterioration at varying rates. The deterioration is caused by normal marine exposures, wet-dry cycles, freeze-thaw cycles and chemical, industrial, and sanitary wastes. Wood, concrete, and steel piles are all affected, and the location of the deterioration is always the same—in the tidal and splash zones. Customarily this is a distance of 5 to 15 ft along the pile.

A number of pile restoration techniques are available, but the use of geotextiles in a flexible form system is extremely cost-effective. The procedure is shown graphically in Fig. 7.20 and is described here:

Clean loose material from the pile by means of a wire brush, air jet, or water blasting.

Place wire mesh reinforcing around the pile, with double layers where it is most severely deteriorated (in some cases reinforcing might not be necessary).

Place a prefitted fabric jacket (usually made of woven monofilament nylon) around the pile, close the jacket's industrial zipper, and clamp its lower part around the pile. Note that this clamp may have to be placed slightly below the mudline, which will require some underwater digging.

Adapt the top of the jacket into a pipe for accepting grout from the top of the structure the pile supports or from a floating barge.

Grout the jacketed pile. A typical mix uses 8 to 11 bags of type I or II Portland cement, 2200 to 2500 lb of concrete sand, 50 to 60 gal of water, and a pumping aid designed to impart 4 to 6 percent air entrainment and thus eliminate bleeding of the concrete.

After curing, the clamps can be removed and the jacket stripped off, although it cannot be reused. See Koerner and Welsh[2] for additional details.

Figure 7.19 Various applications of geotextiles used as flexible forms filled with grout to support underwater pipelines. (*a*) Grout inflation sequence to support undermined pipeline. (*b*) Method of repairing damaged underwater pipeline. (*c*) Saddlebag weight coating with support beneath pipeline also. (*d*) Protective dome formed in situ using flexible framework. (*After Littlejohn et al., Refs. 27 and 28.*)

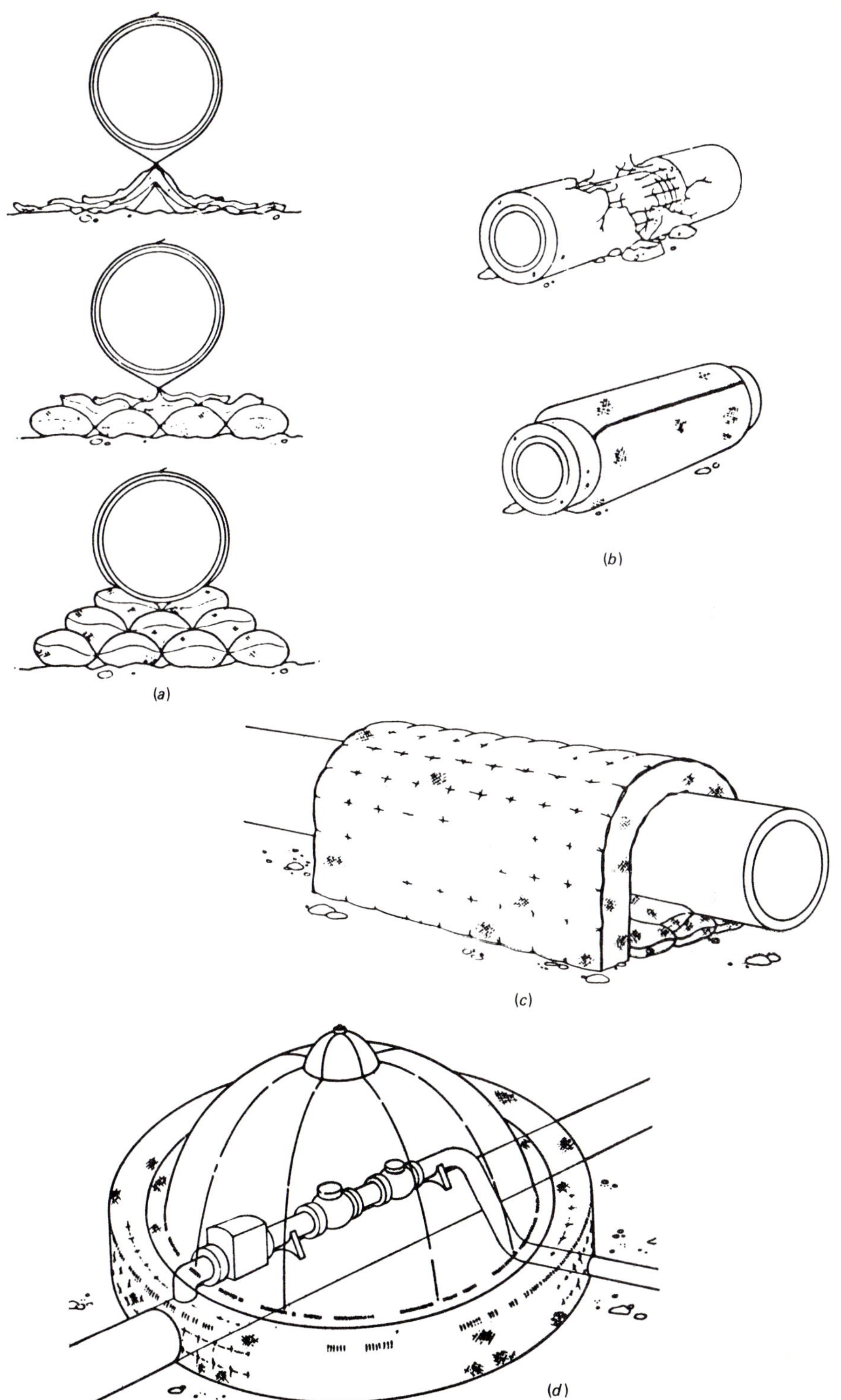

(a)
(b)
(c)
(d)

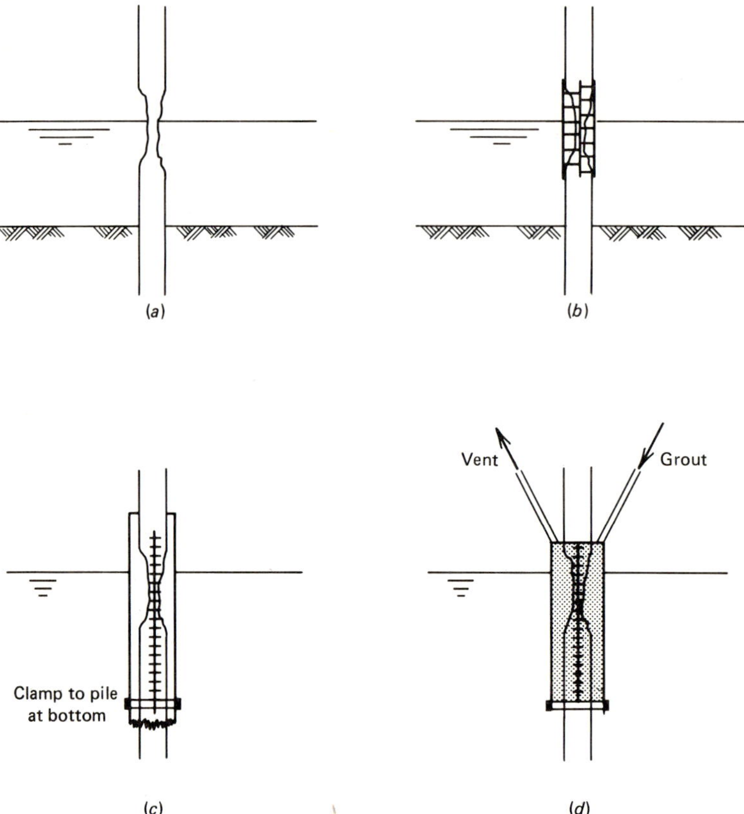

Figure 7.20 Construction sequence for restoring deteriorated piles in marine environments using geotextiles as flexible forms. (*a*) Clean loose material from deteriorated pile. (*b*) Place wire mesh reinforcement around pile. (*c*) Jacket pile with fabric. (*d*) Inject grout into fabric jacket. (*e*) Jacketed pile ready for grouting. (*f*) Finished piles restored to original strength.

7.6.5 In Situ Columns for Providing Mine and Cavern Stability

Construction on sites located above abandoned mines and known or suspected caverns or systems of caves can be very problematic. What is needed is an accurate assessment of the susceptibility of the cap rock above those large voids to subsidence or total collapse. However, this area of subsidence prediction is far from fully developed in a quantitative manner; recall Fig. 1.13 and the text of Sec. 1.2.7.

To provide a measure of insurance in such instances, core borings approximately 5 in in diameter can be made over the entire site, into which geotextile stockings (usually made of woven monofilament nylon) 12 to 24 in in diameter are inserted to the full depth of the boring. A grout pipe is then inserted into the stocking and grout is injected under pressure from the bottom upward. In regions where the rock is sound, the inflation of the stocking is resisted by the surrounding rock. Where a void

(e)

(f)

Figure 7.20 (*Continued*)

exists, however, the stocking is inflated to its full diameter by the grout, forming an enlarged section which extends along the length of the void. At the upper end of the void, it "shoulders" directly beneath the cap rock, thus providing an in situ support column. One system of such columns is shown schematically in Fig. 7.21.

As with many other ways of using geotextiles in flexible form systems, methods of creating in situ support columns are patented, and a patent search should be conducted before work commences; see Koerner and Welsh.[2]

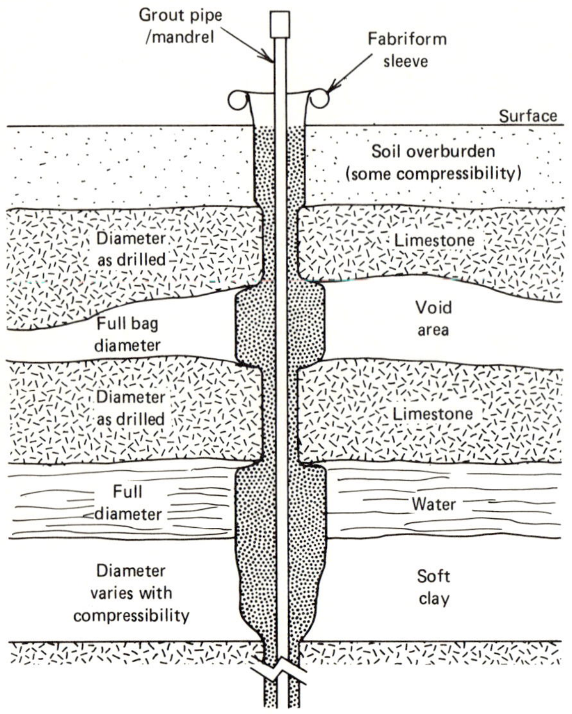

Figure 7.21 Idealized sketch of cross section of completed fabriform mortar piles for aiding in mine and cavern stability. (*After B. A. Lamberton, Intrusion Prepakt, Cleveland, OH 44114.*)

7.7 SUMMARY

As noted in the Overview of this chapter, the geotextile area is experiencing rapid growth and is in a state of rapid transition. This is true in regard to types of geotextiles, applications, testing, and design. The two books on the subject listed as Refs. 2 and 3, the proceedings of the two conferences listed as Refs. 4 and 5, several reports to the Federal Highway Administration (by Bell[25] and Haliburton et al.[29]) and brochures from geotextile manufacturers form the bulk of the available literature which should be consulted when beginning work in this area.

We have discussed in some detail in the examples the design of projects that can use geotextiles, but in only a few instances have those design methods been field-tested. Certainly such testing is needed to simplify the selection of what geotextile(s) to use in a given situation. But for the present what are available are numerous guidelines that match geotextile properties to particular applications. Those guidelines, made by federal, state, and local agencies in the United States, agencies of some foreign governments, manufacturers, and individuals, are often in the form of specifications. The data given in Table 7.2, for example, were obtained from such a specifications guideline issued by the Canadian Ministry of Transportation and Communications.[30] It should be emphasized, however, that such guidelines should be used only for the selection of geotextiles to be used in ordinary circumstances and for the applications stipulated. Whenever any unusual circumstances are anticipated in

Table 7.2 Guidelines used by Canadian Ministry of Transportation and Communications for geotextile selection†

Strength properties	Test method	Applications 3, 4, 5, 7, 9‡	Applications 1, 2, 5a, 6, 8‡
Tensile strength of unaged fabric in any principal direction, lb	CGSB Standard 4-GP-2 9.2	150	75
Tensile strength of fabric in any principal direction after testing in weatherometer, lb	ASTM Standard G23	110	55
Bursting strength of unaged fabric, lb/in^2	CGSB Standard 4-GP-2 11.2	250	125
Puncture strength of unaged fabric, lb/in^2	MTC method	60	30
Abrasion resistance in any principal direction, lb	ASTM Standard D1175	55	30

†After Barsvary and Korgemagi, Ref. 30.
‡The applications are defined as follows:

1. A filter cover around perforated drainage pipes
2. Lining around the perimeter of french drains
3. Between the subgrade and an open graded layer of the pavement structure to prevent infiltration by fines
4. Along the interface between subsoil and embankment to eliminate contamination and to assist in obtaining a more even load distribution on soft subsoil
5. Erosion control under rip-rap, behind gabions, and along highway slopes
5a. Erosion control under rip-rap, behind gabions, and along highway slopes where drop height for placing material on the fabric is zero or where sand cushion protection is provided
6. Silt fences and beneath sediment basins
7. A preventive measure against piping in excavations
8. Vertical and horizontal drains to increase rate of soil consolidation
9. A reinforcing membrane in reinforced earth embankments

regard to a particular project, its design in terms of geotextile selection must be based on a prior assessment of the critical function, or functions, the geotextile will be called on to fulfill followed by a careful selection of the fabric to be used based on that analysis. This way of designing from "first principles" to required geotextile properties has been the guiding philosophy behind the design methodology presented in this chapter, and it is the direction the author hopes this exciting field will follow in the future.

PROBLEMS

7.1 Describe in words and illustrate by sketches the differences between:
    (*a*) A geotextile
    (*b*) A geogrid
    (*c*) A geomembrane (see Sec. 8.5.3)

7.2 When water is moving through a soil-geotextile system, both drainage and filtration usually occur. Explain the differences between these two primary functions of geotextiles (use sketches in your answer).

7.3 (*a*) The ability of a geotextile to drain water is often expressed by percent open area (POA) and/or equivalent opening size (EOS). Explain each of these concepts. (The latter is a Corps of Engineers term; both, however, have been adopted by ASTM.)

   (*b*) Based on the descriptions of POA and EOS you just gave, which of them best characterizes the drainage capability of the following types of geotextile (see Fig. 7.2):

1. Woven, monofilament fiber geotextile
2. Woven, slit film fiber geotextile
3. Nonwoven, needle-punched geotextile
4. Nonwoven, heat-bonded geotextile
5. Geogrid
6. Composite geotextile

7.4 In evaluating the drainage capability of geotextiles, the two terms "permittivity" (ψ) and "transmissivity" (θ) were introduced. Why is this necessary for geotextiles, when in soils engineering we use k_n and k_p for the cross-plane and in-plane permeabilities, respectively?

7.5 (*a*) The stability of geotextiles upon their exposure to ultraviolet light (UV) was mentioned as being "poor to reasonable." Why are polypropylene and polyester (the most commonly used polymers) so poor in this regard? (It will be necessary to investigate the chemical structure of these materials to answer this question.)

   (*b*) The addition of carbon black to the polymer feedstock during fiber preparation greatly retards its susceptibility to degradation by ultraviolet light. Why? (Explain in terms of chemical structure.)

   (*c*) In view of the answers to parts *a* and *b*, what philosophy of construction should be adopted when using geotextiles?

7.6 Geotextile manufacturers list their respective fabrics by width and roll length. Why is this information necessary for both designers and contractors?

7.7 When the edge of one geotextile section meets the edge of another, a seam is required. There are three ways of handling this situation: overlap, sewing, and adhesive strip. List the advantages and disadvantages of each of these approaches.

7.8 Does the amount of geotextile overlap from one section to the next depend on the relative stability of the foundation soils involved? If so, how? What are typical overlap values?

7.9 (*a*) When coarse aggregate is placed on top of in situ soil, two phenomena are likely to occur: "soil pumping" and "aggregate intrusion." Show by sketches what these situations look like.

   (*b*) Soil pumping into aggregate negatively influences both the permeability of the aggregate and its strength. How?

   (*c*) Aggregate intrusion into the subsoil improves the subsoil's strength. How does this occur?

   (*d*) What is wrong with using aggregate to improve in situ soil strength?

7.10 (*a*) If the overall cost of using a geotextile for a given project is about $1.00 per square yard what thickness of stone aggregate is it necessary to use to offset that cost if the stone weighs 140 lb/ft^3 and costs $5.00 per ton.

   (*b*) What thickness is necessary as the cost of the stone increases in increments of $5.00 per ton up to $70.00 per ton (its price in Alaska)? Plot those incremental costs against the amount of aggregate needed to offset the cost of using a geotextile and comment on the results.

7.11 (*a*) What is frost heave (recall Secs. 1.2.1 and 1.2.2) and how do geotextiles, or better, composite geotextile systems, work to eliminate the problem?

   (*b*) In arid regions, what causes upward salt migration and how does the use of geotextiles and geotextile systems help eliminate it?

7.12 In using geotextiles to stabilize unpaved roads over soft soils, it is sometimes recommended that two-thirds of the aggregate thickness be placed, then a loaded truck be driven over the area, the area be regraded, and finally the remaining one-third of the aggregate thickness be placed. What is the reasoning behind this procedure?

7.13 In constructing an unpaved haul road to serve a construction site, determine the thickness of aggregate saved, if the road must last for 10,000 vehicle passages, by using a geotextile. The soil subgrade has an average CBR of 0.5 and there are a variety of geotextiles available. Solve the problem for geotextile moduli of 450, 400, 300, 200, 100, and 10 kN/m and plot the results of moduli (x axis) versus thickness of aggregate saved (y axis). Use Fig. 7.7 and follow the procedure given in Example 7.1.

7.14 Geotextiles can be tested for a number of moduli of elasticity, where such tests always involve a strip of fabric being clamped and stretched between two jaws of a testing machine. The resulting response is usually presented as a graph of strain (x axis) plotted against load per unit width (y axis), which exhibits the general elastic-plastic behavior of the fabric. With this type of response in mind:

(*a*) Should the tangent modulus of a fabric or its secant modulus be used for design purposes?

(*b*) If the secant modulus is used, at what percent strain should it be calculated? (*Note:* Geotextiles fail at 40 to 100 percent strain.)

(*c*) Would the width of the actual test specimen matter, even though the results are standardized to a unit width calculation? (*Note:* ASTM has a strip test using a width of 1 in, a grab test using a width of about 2 in, and a wide-width test using a width of 8 in.)

7.15 The generalized long-term flow response of water passing through a soil-geotextile system was shown in Fig. 7.11. The two slopes there were labeled as being controlled by compaction, filter cake, blinding, arching, and clogging. Explain those terms in words and sketches.

7.16 If some soil particles remain within the structure of a nonwoven, needle-punched geotextile, does this mean that it cannot drain water? If not, what is the explanation for this situation?

7.17 Table 7.1 is intended for use in selecting the proper geotextile to prevent piping in a given soil. It gives the apparent opening size of the geotextile (O_{95}) needed in terms of the average particle size of the soil (d_{50}) and the coefficient of uniformity of the soil (CU).

(*a*) What is meant by "piping"?

(*b*) Explain how to evaluate O_{95}.

(*c*) To what physical or hydraulic properties of the geotextile is O_{95} related?

7.18 Using a geotextile as an underliner for an erosion control system requires a knowledge of the friction that can be mobilized between the soil and the geotextile. If that friction is insufficient, the structure will become unstable and possibly will fail.

(*a*) What kind of laboratory test would you suggest to measure this property?

(*b*) If that friction were compared with the soil's frictional strength, an "efficiency" would result that could have a maximum value of 1. Taking a silt of low plasticity (ML) as the in situ soil, rank the following geotextiles with regard to efficiency:

Woven, monofilament fibers
Woven, slit film fibers
Nonwoven, needle-punched
Nonwoven, heat-set
Geogrid

7.19 When using stone rip-rap placed on top of a geotextile in erosion control work, why is it usually recommended to place a 3- to 5-in layer of sand between the fabric and the rip-rap?

7.20 In using geotextiles as underliners in erosion control structures (see Fig. 7.15), what is the danger in covering most of the surface of the geotextile with facing material?

7.21 (*a*) In using geotextiles as flexible forms for pile restoration work (see Fig. 7.20), why are woven monofilament nylon fabrics generally used?

(*b*) Does the final shape of the restored section always remain circular? If not, why?

(*c*) What effect does the water which is within and surrounds the jacket have on the freshly placed concrete?

(*d*) What effect does the water have on the curing of the concrete?

7.22 In the use of flexible forms to create columns for supporting sites over abandoned mines and possible large caverns, as shown in Fig. 7.21, what geotextile properties are critical, and why?

REFERENCES

1. Information supplied by Man Made Fibers Producers Assoc., Inc. Education Dept., 1150 Seventeenth St., NW, Washington, DC, 20036.
2. Koerner, R. M., and J. P. Welsh, *Construction and Geotechnical Engineering Using Synthetic Fabrics,* John Wiley and Sons, New York, 1980.
3. Rankilor, P. R., *Membranes in Ground Engineering,* John Wiley and Sons, New York, 1981.
4. Majidzadeh, R., M. S. Luther, and H. Skylut, "A Mechanistic Design Procedure for Fabric-Reinforced Pavement Systems," in *Proc. Second Int. Conf. on Geotextiles,* Vol. 2, Las Vegas, August 1–6, 1982, pp. 529–534.
5. Roth, H., "Filter Fabric for Improving Frost Susceptible Soils," in *Proc. Int. Conf. on the Use of Fabrics in Geotechnics,* Vol. 1, Paris, April 20–22, 1977, pp. 23–28.
6. Clough, I. R., and W. J. French, "Laboratory and Field Work Relating to the Use of Geotextiles in Arid Regions," in Ref. 4, Vol. 2, pp. 447–452.
7. Giroud, J. P., and L. Noiray, "Geotextiles—Reinforced Unpaved Road Design," *Geotech. Eng. Div., ASCE,* Vol. 107, No. GT9, September 1981, pp. 1233–1254.
8. Bell, J. R., A. N. Stilley, and B. Vandors, "Fabric Retained Walls," in *Proc. Thirteenth Geotech. and Soil Eng. Symp.,* Moscow, ID, April 1975, pp. 271–287.
9. Lee, K. L., B. D. Adams, and J. J. Vagneron, "Reinforced Earth Retaining Walls," *J. Soil Mech. Found. Eng. Div., ASCE,* Vol. 99, NO. SM10, October 1973, pp. 745–764.
10. Holtz, R. D., and B. B. Broms, "Walls Reinforced by Fabrics—Results of Model Tests," in Ref. 5, Vol. 1, pp. 113–118.
11. Nichols, R. L., "Comparison of Fabric Sheet-Reinforced Earth Design Methods," *Can. Geotech. J.,* Vol. 18, 1981, pp. 585–592.
12. Murray, R. T., "Fabric Reinforced Earth Walls: Development of Design Equations," *Ground Eng.,* Vol. 13, No. 7, October 1980, pp. 1–22.
13. Koerner, R. M., and F. K. Ko, "Laboratory Studies of Long Term Drainage Capabilities of Geotextiles," in Ref. 4, Vol. 1, pp. 91–95.
14. Giroud, J. P., "Filter Criteria for Geotextiles," in Ref. 4, Vol. 1, pp. 103–108.
15. Rollin, A. L., R. Denis, L. Estaque, and J. Masounave, "Hydraulic Behavior of Synthetic Non-Woven Filter Fabrics," *Can. J. Chem. Eng.,* Vol. 60, April 1982, pp. 226–234.
16. Koerner, R. M., and J. E. Sankey, "Transmissivity of Geotextiles and Geotextile/Soil Systems," in Ref. 4, Vol. 1, pp. 173–176.
17. Koerner, R. M., and J. A. Bove, "In-Plane Hydraulic Properties of Geotextiles," accepted for publication ASTM *J. Geotech. Test.,* May 1983.
18. List, F., "Use of Geotextiles in Dam Construction," in Ref. 4, Vol. 1, pp. 189–192.
19. Giroud, J. P., "Designing with Geotextiles," *RILEM Mater. Struct. J.* Vol. 14, No. 82, July/August 1981, pp. 257–272.
20. Koerner, R. M., and T. A. Okrasinski, "Erosion Control of Granular Soils Using PVA," *J. Constr. Div., ASCE,* Vol. 104, No. CO3, September 1978, pp. 279–294.
21. Hoffman, G. L., and R. Adamsky, "Nylon Erosion Control Mat," in *Proc. Sixty-First Ann. Mtg. of the TRB,* Washington, DC, January 1982; available from Pennsylvania Dept. of Transportation, Materials and Testing Div., Harrisburg, PA 17120.
22. Miller, S. P., *Bank Distress of Low Water Wiers in Big Creek, LA,* U.S. Army Corps of Engineers, Waterways Experiment Station, Vicksburg, MS, 1978.
23. Barrett, R. J., "Use of Plastic Filters in Coastal Structures," in *Proc. Tenth Int. Conf. on Coastal Eng.,* Chap. 62, Tokyo, September 1966, pp. 1048–1067.
24. Dunham, J. W., and R. J. Barrett, "Woven Plastic Cloth Filters for Stone Seawalls," *J. Waterways, Harbors, Coastal Eng. Div., ASCE,* Vol. 100, No. WW1, February 1974, pp. 13–22.
25. Bell, J. R., and R. G. Hicks, *Evaluation of Test Methods and Use Criteria for Geotechnical Fabrics in Highway Applications,* Final report to the Federal Highway Administration, Contract No. DOT-FH-119353, Oregon State University, Corvallis, OR, February 1982.
26. Welsh, J. P., "Penn DOT Utilizes a New Method for Solving Scour Problems Beneath Bridge Structures," *Highway Focus,* Vol. 9., No. 1, May 1977, pp. 72–81.

27. Littlejohn, G. S., "Grouting of Platforms and Pipelines Offshore," in *Proc. Symp. on Geotech. Aspects of Offshore and Nearshore Structures,* Asian Inst. of Tech., Bangkok, December 14–18, 1981.
28. Littlejohn, G. S., R. M. Connor, and H. Leuenberger, "Support and Protection of Pipelines Offshore," in *Proc. Conf. on Offshore Insp., Repair and Maintenance,* Edinburgh, 1981, pp. 1–7.
29. Haiiburton, T. A., J. D. Lawmaster, and V. C. McGuffey, *Use of Engineering Fabrics in Transportation-Related Applications,* Interim Report to the Federal Highway Administration, Contract No. DTFH 61-80-C-00094, October 1981.
30. Barsvary, A. K., and P. Korgemagi, "The Geotechnical Use of Filter Fabrics in Highway Construction," Report No. EM-13, 4th ed., Engineering Materials Office, Ministry of Transportation and Communications, Ontario, Canada, September 1979.

EIGHT

LANDFILLING, LINERS, AND COVERS FOR WASTE DISPOSAL

8.1 OVERVIEW

The obvious question before beginning this chapter is, Why discuss waste disposal at all in a book on geotechnical and construction methods of foundation engineering? The answer, quite simply, is that this area is interdisciplinary by its very nature and can be handled by geotechnical engineers as well as by engineers of any other discipline. However, there is a danger that in approaching the topic oversimplification of many complex phenomena will be made. Often, a group effort is necessary, or at least consultation with other engineers and scientists having expertise in areas of related interest. The areas which come to mind are the following:

Construction	
Geotechnical	
Chemistry—organic, inorganic, and physical	
Sanitary and environmental engineering	Technical
Clay mineralogy	
Polymer and liner subdisciplines	
Public health	
Economic planning and policy	
Land use planning	Social/economic
Attorneys and other legal professionals	
Public relations	

356

By virtue of its historical development, the field of landfilling can be subdivided into the categories of industrial, municipal, and hazardous wastes. However, this seems somewhat artificial since the philosophy for all three is in essence identical—to take the waste material, place it in a suitably lined area, backfill around it as it is being placed, and eventually cover it over in an environmentally safe and secure manner. Obviously in dealing with a highly toxic material, more care and greater precautions must be exercised than in handling something like building demolition waste. But the particular procedures required to handle different types of wastes will be noted in the course of the discussion in this chapter.

The magnitude of the problem of waste is simply enormous. No matter what alternative strategies for waste treatment there are at present or might be developed in the future (e.g., incineration, composting, recycling, etc.), the need for safe and secure landfills will always be present. In regard to the nature and magnitude of the problem, Table 8.1 shows that 148 million tons of municipal solid waste were generated in the United States in 1977 alone. Table 8.1 projects 175 million tons of such waste being produced by 1985, while some predictions have gone as high as 201 million tons (4.67 lb per person per day) by then and 225 million tons (5.00 lb per person per day) by 1990.

Hazardous waste, the disposal of which is very difficult, has been defined as follows (Ref. 2):

A "hazardous waste" is any waste or combination of wastes of a solid, liquid, contained gaseous, or semisolid form which because of its quantity, concentration, or physical, chemical or infectious characteristics, may (1) cause or significantly contribute to an increase in mortality or an increase in serious irreversible or incapacitating reversible illness; or (2) pose a substantial present or potential hazard to human health or the environment when improperly treated, stored, transported or disposed of, or otherwise managed.

Fitting into this general category, albeit at varying levels of toxicity, were nearly 40 million metric tons of materials produced in the United States in 1980; see Table 8.2.[3] In this regard, the actual decision as to whether a waste is to be classified as hazardous or not is far from trivial. Figure 8.1 presents an attempt to arrive at such a decision.[4]

As indicated by the title of this chapter, we will be discussing disposal of waste (industrial, municipal, hazardous, or other types) by means of landfilling procedures only; furthermore, we will discuss the design and construction of landfills only in terms of the most recent technological developments in the field.

8.2 PROPER LANDFILLING PROCEDURES

After selecting the proper site for a landfill (a subject in itself of great complexity; see Refs. 5 and 6), a proper design must be drawn up for disposing of the waste. (*Note:* The area chosen must be above the maximum level of the local water table. If it is not, the situation must be corrected by backfilling or creating a very shallow landfill. This is obviously a tremendous drawback which in all but a few cases will

Table 8.1 Generation and recovery of municipal solid waste in the United States from 1969 to 1985†
(In thousands of tons, as-generated wet weight)

Material	1960	1970	1977	1985
Ferrous metals				
Source separation	. . .	. . .	35	50
Magnetic separation‡	50	150	200	200
Mixed-waste processing	. . .	. . .	50	400
Total	50	150	285	650
Aluminum				
Source separation	. . .	10	140	225
Mixed-waste processing	. . .	. . .	. . .	5
Total	. . .	10	140	230
Paper				
Source separation	5,575	7,115	10,180	12,150
Glass				
Source separation	100	160	500	865
Mixed-waste processing	. . .	. . .	. . .	5
Total	100	160	500	870
Rubber				
Source separation	330	255	160	170
Total materials recovery				
Source separation	6,005	7,540	11,015	13,460
Magnetic separation‡	50	150	200	200
Mixed-waste processing	. . .	. . .	50	410
Total	6,055	7,690	11,265	14,070
Energy recovery from combustibles	. . .	. . .	750	9,400
Total recovery	6,055	7,690	12,015	23,470
Total gross discards	87,000	131,000	148,000	175,000
Percent recovered	7	6	8	13
Percent source separated	7	6	7	8

†After Ref. 1.

‡Includes systems magnetically separating ferrous scrap, but doing no other resource recovery.

result in the selection of another site.) There are three methods of disposal available (see Fig. 8.2):

Trench methods are used for disposing of relatively small waste volumes when adequate backfill material is present and the site has a relatively high water table. Large scrapers or bulldozers excavate a starter trench equal to a few machine widths (approximately 20 ft), 100 to 500 ft in length, and 3 to 5 ft in depth. The in situ soil must have sufficient cohesion to ensure that side slopes will not collapse [e.g., an unconfined compression strength of 15 lb/in^2 (103 kPa) or greater]. Wastes are dumped from the sides of the trench into the hole or from within the hole itself, spread into layers (whose thicknesses depend on the type

**Table 8.2 Quantities of potentially hazardous wastes
in the United States, 1974†**

(In millions of metric tons)

Industry	Dry basis	Wet basis
1. Batteries	0.005	0.010
2. Inorganic chemicals	2.000	3.400
3. Organic chemicals, pesticides, explosives	2.150	6.860
4. Electroplating	0.909	5.276
5. Paints	0.075	0.096
6. Petroleum refining	0.625	1.757
7. Pharmaceuticals	0.062	0.065
8. Primary metals	4.454	8.335
9. Leather tanning and finishing	0.045	0.146
10. Textile dyeing and finishing	0.048	1.770
11. Rubber and plastics	0.205	0.785
12. Special machinery	0.102	0.162
13. Electronic components	0.026	0.036
14. Waste oil re-refining	0.057	0.057
Totals	10.763	28.755

†After Ref. 3.

of material), and simultaneously compacted by the spreading bulldozer. Final height of the fill should be reached at the end of each day, for the full trench width, athough the length of trench utilized can vary. Backfill is obtained by extending the length of the trench or by starting a new, and parallel, trench. For this method, as for the other two also, each day's deposit of waste must be completely covered before work ends for the day.

The *area method* is used where large volumes of waste are to be landfilled, in soil which is not stable enough to maintain well-defined slopes. The wastes are dumped on the ground, then spread by bulldozer and compacted into layers of 18 to 30 in each. When the height of each layer reaches 5 to 10 ft, cover material is brought in using bulldozer-loaded or self-loading scrapers. The area covered per day depends on the size of the operation. Each day's covered lift is called a *cell*. As shown in Fig. 8.2b, cells can be stacked upon one another. Adequate compaction is needed in these situations. When filling a natural ground depression, ravine, or canyon, this method of successive cells is usually used.[8] Excellent compaction is needed when the number of cells stacked vertically is more than three.

The *ramp method* of landfilling is an adaptation of the area method and is used where a very low water table exists and large volumes of wastes are anticipated. Usually a single cell is constructed to final height but the cover is taken from directly in front of the working face. Cells are therefore thicker and shorter than in the area method.

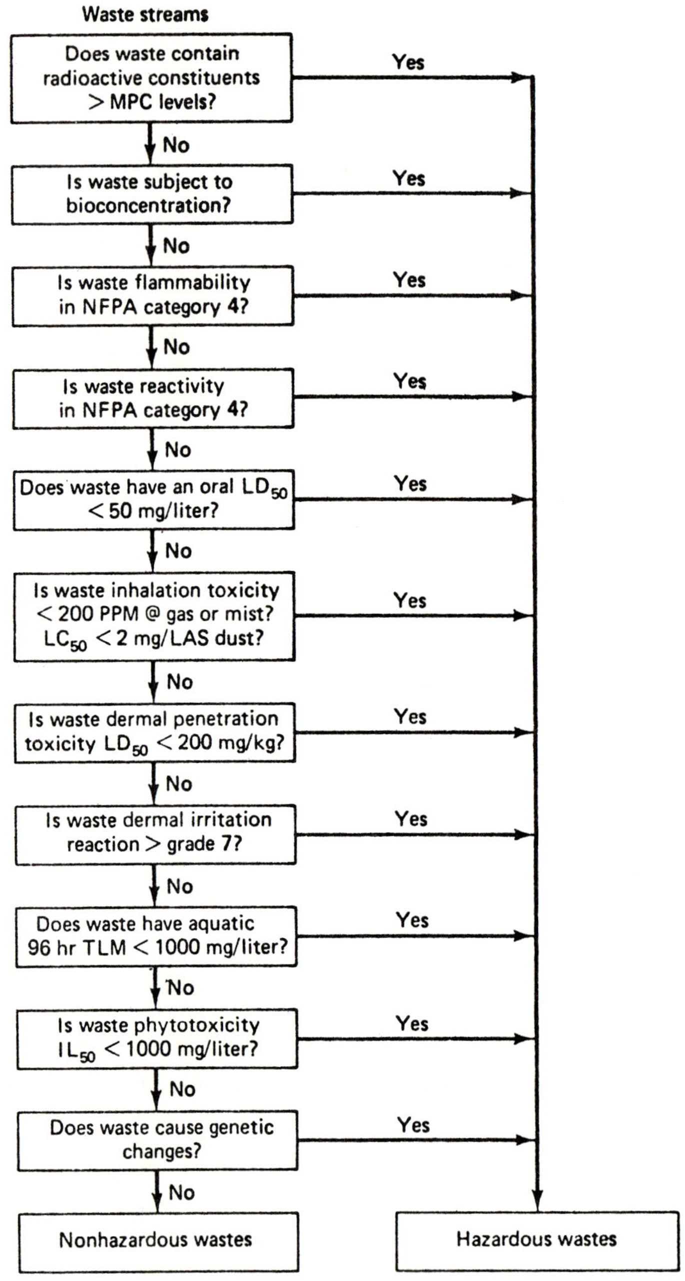

Figure 8.1 Model for deciding whether a material is hazardous. *(After Ref. 4.)*

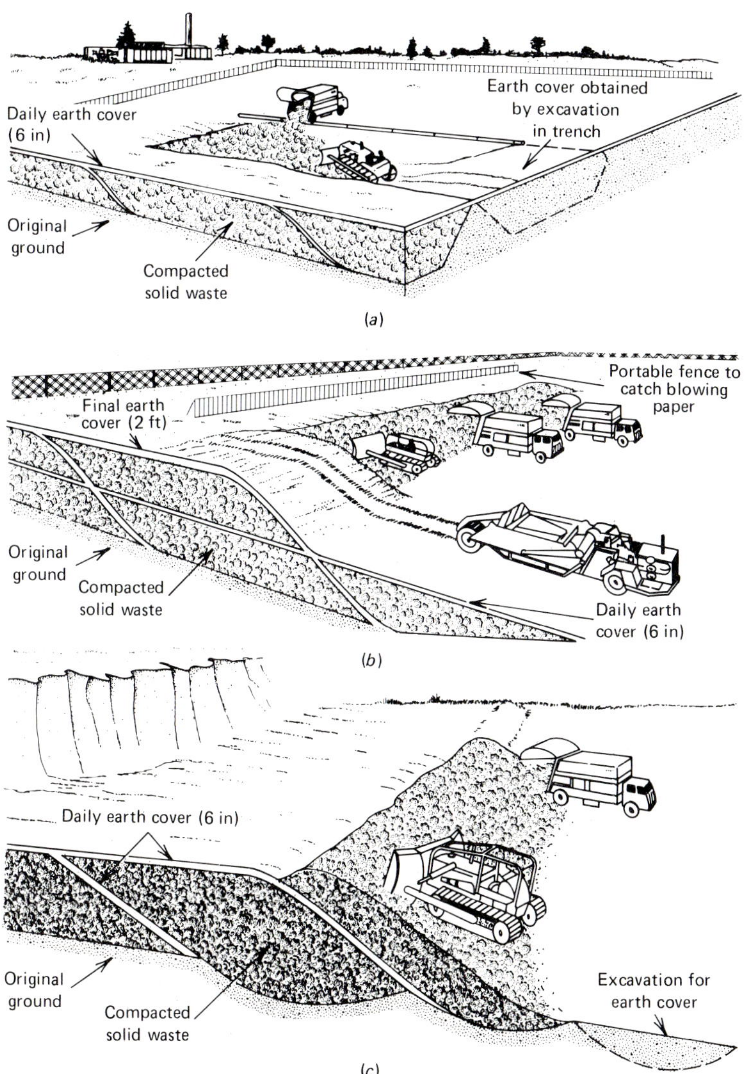

Figure 8.2 Typical sanitary landfill, showing three different methods of operation: (*a*) trench method, where trucks deposit the loads into a trench and a bulldozer spreads and compacts, and covers them over at the end of the day; (*b*) area method, where a bulldozer spreads out the refuse and a scraper hauls the cover material; and (*c*) ramp method, where the refuse is compacted as before but the cover is excavated from directly in front of the working face. (*After Brunner and Keller, Ref. 7.*)

The nature of the soil used for covering the waste materials is extremely important. Regarding the suitability of different soils see Table 8.3.[7] Obviously at many sites the available borrow is often distinctly less than the optimum material one could wish to use. In such cases, the daily earth cover over each cell can be obtained from on site excavations, while the final cover over the uppermost cell can be an imported borrow or a blend of on site and imported borrow that improves the local soil's characteristics. Keeping down costs, of course, is of major concern in regard to what soil or blend of soils will be used.

Waste materials disposed of by the above three methods will have dry densities varying from 10 to 40 lb/ft^3, moisture contents of 10 to 35 percent, and quite low bearing capacities, e.g., 400 to 700 lb/ft^2. Baling or shredding the waste materials before deposition improves their in situ properties, but at greatly increased costs.

Self-compaction or sometimes compaction induced with the aid of a surcharge fill, can lead to the better utilization of a landfilled site. And because of the large portions of most wastes that are combustible, settlements of up to 40 percent of the original thickness, occurring within 1 to 4 years after disposal, are not uncommon. However, during this settling process two phenomena—sometimes having disastrous consequences—occur. These are the generation of gases within the degrading landfill, and the creation of a polluted liquid, called *leachate,* which can flow gravitationally down into the soil and subsurface water table. These topics are discussed in Secs. 8.3 and 8.4, respectively.

8.3 LANDFILL GAS GENERATION AND MOVEMENT

The gases produced at most landfill sites are methane and carbon dioxide. Methane tries to rise, and so if an impermeable cover is placed over the entire landfill, it will eventually move laterally beyond the landfill boundaries to a permeable soil where it can escape to the atmosphere. The distances involved can be thousands of feet. In one situation in southeastern Pennsylvania, where the upper ground surface was frozen, the methane could only escape into the unfinished basements of private residences. Several cases of brain damage and other health disorders were reported. Some of the basements were in homes located up to 4000 ft from the edge of the landfill.

To avoid such problems, the methane gas must be properly vented. Each cell cover should be shaped so that a vent is located at the uppermost slope of the bottom of the cover. The vent itself either wastes the gas into the atmosphere or collects it for energy utilization; see Fig. 8.3 for a typical vent at a sanitary landfill.[5]

The carbon dioxide produced, being 1.5 times denser than air and 2.7 times denser than methane, separates and moves to the bottom of the landfill. Since it is readily soluble in water, it will enter the leachate that has formed there. If the landfill liner is inadequate, the gas will escape (generally along with the leachate, although it can escape independently because of its higher diffusion characteristics) and enter into the groundwater. Once dissolved, it lowers the water's pH, causing an increase in hardness and mineral content. (*Note:* Carbon dioxide and water form carbonic

Table 8.3 Suitability of general soil types for use as landfill cover material†

Function	Clean gravel	Clayey-silty gravel	Clean sand	Clayey-silty sand	Silt	Clay
Workability, placement of cover	Good	Excellent	Excellent	Good	Fair	Poor
Prevent rodents from burrowing or tunneling	Good	Fair to good	Good	Poor	Poor	Poor
Keep flies from emerging	Poor	Fair	Poor	Good	Good	Excellent‡
Minimize moisture entering fill	Poor	Fair to good	Poor	Good to excellent	Good to excellent	Excellent‡
Minimize landfill gas venting through cover	Poor	Fair to good	Poor	Good to excellent	Good to excellent	Excellent‡
Provide pleasing appearance and control blowing paper	Excellent	Excellent	Excellent	Excellent	Excellent	Excellent
Grow vegetation	Poor	Good	Poor to fair	Excellent	Good to excellent	Fair to good
Be permeable for venting decomposition gas§	Excellent	Poor	Good	Poor	Poor	Poor

†After Brunner and Keller, Ref. 7.
‡Except when cracks extend through the entire cover.
§Only if well drained.

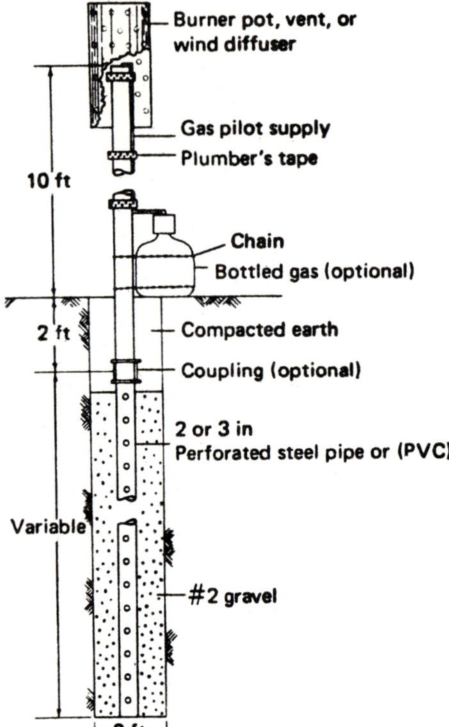

Figure 8.3 Typical well-type waste gas burner used at sanitary landfills. *(After Tchobanoglous et al., Ref. 5.)*

acid, which reacts with limestone to form soluble calcium carbonate. This is the exact process which occurs in the formation of sinkholes in limestone areas as described in Sec. 1.2.9.) Similar reactions occur with magnesium carbonates.[5] To collect the carbon dioxide gas, a sand-gravel drainage layer above the bottom of the landfill liner is needed, or a bulky geotextile of adequate transmissivity can be used, along with a perforated-pipe collection system for gathering the gas (and leachate) for proper disposal at the ground's surface.

The use of vertical cutoff walls to contain the laterally moving gases should not be considered as a primary containment strategy. Rather their use should be considered only if problems occur in regard to nonengineered landfills, for example, or if cracks or leaks develop in the liner of a lined landfill. These walls will be considered separately in Sec. 8.7.

8.4 LANDFILL LEACHATE GENERATION AND MOVEMENT

The generally high moisture content of a landfill to begin with (10 to 35 percent values are not uncommon), any rainfall or snow melt which gets through the landfill cover, and lateral moisture movement from beyond the landfill into it will all react with the waste materials themselves to form a polluted liquid called *leachate,* which often has dreadful characteristics. Table 8.4 shows typical compositions for leachates from a number of different landfills,[9] although variations in composition can be even greater than indicated there.

Leachate should be properly collected above the primary liner at the bottom of the landfill, adequately treated, and disposed of in accordance with accepted environmental engineering principles. (Landfill liners will be discussed in more detail in Sec. 8.5.) If, however, no liner is present (as is the case for many existing landfills) or if the liner leaks, the leachate will escape from the cell into the surrounding soil. Here any of a number of situations can occur:

If the surrounding soil is nearly dry and the leachate volume is small, it will be held within the soil voids for an indefinite period of time, i.e., vadose zone containment.

At higher natural water contents for the soil and larger leachate volumes, the soil voids reach their field capacity, at which point the liquid phase becomes continuous. The somewhat diluted leachate now moves gravitationally downward with some amount of lateral spread occurring also.

Still further along, the soil voids become saturated with the diluted leachate, and hydrodynamic principles can be used for predicting its movement. In this regard a number of sophisticated finite-element models have been used (see Hwang and Koerner,[10] for example).

As time continues, the leachate contained within the moving water migrates in an ever enlarging plume of polluted water. Depending on downstream conditions, the results can be academic or disastrous. It is here that hazardous landfill leachates are of great concern, particularly if downstream water users exist (i.e., users whose wells or water intakes are affected by the leachate pollution).

The basic strategy, however, is to keep the leachate from escaping the landfill site to begin with. Both primary and secondary liners can be used to achieve this end. If these methods fail, the use of a containment wall should be considered. These topics will be discussed in Secs. 8.5 and 8.7.

8.5 LANDFILL LINERS INCLUDING GEOMEMBRANES

Landfill liners for both bottom and side use are constructed from a wide range of materials. (See Appendix F for a list of manufacturers, suppliers, and contractors dealing with various types of liners.) The classification we will follow here is given below (Ref. 11 covers the subject in greater detail, but from a somewhat different perspective).

1. Rigid
   a. Shotcrete or gunite
   b. Concrete

Table 8.4 Composition of different leachates from various sanitary landfills†

All figures in milligrams per liter except specific conductance, which is in microsiemens per centimeter, and pH, in pH units

Source parameter (1)	SHWRL	Drexel	Georgia Tech	Georgia Tech recirculate	Olin Ava. landfill	Dupage LW5B	Dupage MM63	Dupage MM61	Dupage LW6B
Time Span, in years	1	2	3	3	3.5	5.5	8.5	14.5	16.0
COD	16,000–22,000	1,000–51,000	4,320–12,000	4,280–9,288	2,700–10,650	8,000	2,940	360	40
BOD_3	7,500–10,000		2,500–11,000	2,750–6,900	1,550–8,450	14,000‡	4,560‡	125‡	225‡
TOC			1,230–5,000	256–2,798					
pH	5.2–6.4	3.7–8.5	5.2–5.6	4.8–5.4	5.9–8.1	6.3			7.0
TS	10,000–14,000	0–42,000	2,442–12,500	1,627–6,918	4,028–7,790				
TDS	10,000–14,000					6,794	5,910	1,104	1,198
TSS	100–700	10–26,500	34–610	12–385					
Specific conductance	6,000–9,000								
Alkalinity ($CaCO_3$)	800–4,000	0–9,700	558–2,280	302–1,370	1,800	5,810	4,720	1,630	2,250
Hardness ($CaCO_3$)	3,500–5,000	0–5,500	450–1,940	370–1,040	1,400	2,200	2,250	690	540
Total P	25–35	0–130	2.8–26	0.65–22	12	1.2§	0.17§	0.5§	8.9§
Ortho-P	23–33								
NH_4-N		0–482	56–187	68–114					
$NO_3 + NO_2 - N$	0.2–0.8					0.7¶	0.5¶	0.14¶	1.6¶
Ca	900–1,700		125–750	60–435	80	308	44	156	102
Cl	600–800	4.7–2,340	98–385	91–248	400	1,330	946	205	135
Na	450–500	0–7,700	64–143	62–109		810	615	63	74
K	295–310					610	220	85	100
Sulfate	400–650	25–450	81–156	12–138		2	1	1	2
Mn	75–125		3–10	4–65	20	0.06	0.09	0.24	0.06
Mg	160–250		26–75	17–63		450	725	110	90
Fe	210–325	0–1,716	9–95	4–110	330	6.3	12	106	0.6
Zn	10–30	0–167			5.5	0.4	0.05	0.10	4.5
Cu		0.9–9				<0.5	<0.3	<0.5	<0.5
Cd						<0.05	<0.05	<0.05	<0.05
Pb						0.5	1.0	1.0	1.0

Source parameter (1)	Win-netka LW17	Win-netka LW58	Win-netka LW13	Riverside bin no. 1	Mission Canyon landfill, March 18 1969	March 24 1971	Sonoma, CA, Control	Recirculate	West Virginia University cylinder C	Range of all values
Time Span, in years	2.1	12.5	16.1	1.5	>1	>6	2	2	0.33	
COD	4,260	415	153		76,800	3,042	3,260–22,500	22,700–89,320		40–89,520
BOD$_3$	6,400‡	250‡	105‡	81–33,100	10,900	908	2,250–19,300	15,500–33,600	24,200–33,360	81–33,360
TOC										256–28,000
pH				5.6–7.6	5.75	7.4	4.7–5.4	4.6–6.5	5.25–5.88	3.7–8.5
TS									36,250–59,200	0–59,200
TDS	2,306	994	584		44,900	13,409	724–14,080	14,600–21,010		584–44,900
TSS					172	220	16–200	22–600		10–700
Specific conductance										2,810–16,800
Alkalinity (CaCO$_3$)	3,370	3,080	1,450	730–9,500	9,860	3,677	240–3,920	1,680–7,900	16,200–20,850	0–20,850
Hardness (CaCO$_3$)	1,180§	970§	710§	650–8,120	22,800	8,930			8,350–10,950	0–22,800
Total P	2.6	2.75	1.3	0.2–29	0.24	0.65	0–2.8	1.4–79.2	87–128	0–130
Ortho-P										6.5–85
NH$_4$ − N				0.2–890	0	270	0–81	880–194	828–1,106	0–1,106
NO$_3$ + NO$_2$ − N	0.5¶	0.43¶	0.2¶				0–0.90	0–4.7¶		0.2–10.29
Ca	100	109	109	115–2,570	7,200	216	20–1,082	561–1,800	2,405–2,790	60–7,200
Cl	429	701	70	96–2,350	660	2,355	56–181	920–1,565	1,465–2,310	4.7–2,467
Na	298	348	34	85–1,805	767	1,160	80–338	408–1,010	871–1,439	0–7,700
K	168	220	39	29–1,860	68	440	48–148	260–910	2,970–3,770	28–3,770
Sulfate	13.6	1	5		1,190	19	20–250	257–1,040	320–768	1–1,558
Mn	1.14	0.1	0.2							0.09–125
Mg	198	200	75	64–410	15,600	8,714	2.6–608	316–672	326–420	17–15,600

367

Table 8.4 Composition of different leachates from various sanitary landfills† (*Continued*)

All figures in milligrams per liter except specific conductance, which is in microsiemens per centimeter, and pH, in pH units

Source parameter (1)	Winnetka LW17	Winnetka LW58	Winnetka LW13	Riverside bin no. 1	Mission Canyon landfill March 18 1969	Mission Canyon landfill March 24 1971	Sonoma, CA Control	Sonoma, CA Recirculate	West Virginia University cylinder C	Range of all values
Fe	96	13.6	11	6.5–305	2,820	4.75	22–1.050	165–300	546–860	0–2,820
Zn	1 .5	0.6	0.1				0.23–9.0	85–95		0–370
Cu	<0.5	<0.5	·<0.5				0.2–0.44	<0.2–0.4		0–9.9
Cd	<0.05	<0.05	<0.05				0	<0.05–0.16		0.03–17
										<
Pb	1.0	0.5	1.0				<0.5–181	<0.0–2.0		<0.10–2.0

After Chain and DeWalle, Ref. 9.

‡Based on 20-day BOD.

§Blackwell value calculated from Mg and Ca concentration.

¶Using only NO_3 analysis. NO_2 not determined.

   c. Soil cement
   d. Bituminous concrete (asphalt)
   e. Bituminous panels
2. Flexible (natural materials)
   a. Compacted soils
   b. Chemically treated soils
   c. Bentonite clays
3. Flexible (synthetic materials)
   a. Geomembranes (plastics and elastomers)

Before beginning with specific details of each of these lining systems, however, the importance of soil subgrade preparation must be commented upon. No liner, no matter how rigid or highly reinforced it is, can withstand large differential settlement without eventually leaking or possibly failing completely. Adequate site preparation and compaction is simply mandatory for a lining system to perform adequately. Each situation is somewhat unique in regard to this stipulation. And certainly geotechnical engineering methods of stabilizing soils, such as those discussed in Chaps. 4 and 5, are at the heart of adequate site preparation. Specific failure mechanisms can come about from the following:

Insufficient soil compaction
Nonuniform soil compaction
Improper soil texture
Consolidation settlement of soil
Subsidence of soil
Secondary structure development (cracks, lenses, etc.)
Soils that shrink or swell (expansive clays)
Soils susceptible to freezing and thawing (silts and clays)
Frost heave and ice lens formation
Clogged (or worse no) underdrains
Improperly sized underdrains
Underdrains with blocked inlets and outlets
Gas generation from subsoil
Anchor trench failure
Failure around appurtenances

It should also be noted that no liner system, even if perfectly designed and constructed, will result in zero seepage losses, i.e., "nothing is absolute." Kays[11] presents an interesting list of seepage rates (see Table 8.5) based on his experiences, the rates varying from maximum values for soil liners to minimum values for synthetic membranes. Note that these values are for a single primary liner only. If extreme concern is warranted, an underdrain system can be placed beneath the primary liner to collect any leakage passing through and either treat it or recirculate it back to the containment area. Additionally, a secondary liner can be placed beneath the underdrain system, a practice currently recommended for hazardous waste landfills.

Table 8.5 Estimated seepage rates from various liner systems[†]

Material	Thickness, in	Minimum expected seepage rate at 20 ft of water depth after 1 year of service, in/day
Open sand and gravel		96
Loose earth		48
Loose earth plus chemical treatment[‡]		12
Loose earth plus bentonite[‡]		10
Earth in cut		12
Soil cement (continuously wetted)	4	4
Gunite	1.5	3
Asphalt concrete	4	1.5
Unreinforced concrete	4	1.5
Compacted earth	36	0.3
Exposed prefabricated asphalt panels	0.5	0.03
Exposed synthetic membranes	0.045	0.001

[†]After Kays, Ref. 11. The data are based on actual installation experience.

[‡]The chemical and bentonite treatments depend on pretreatment seepage rates, and in the table loose-earth values are assumed.

8.5.1 Rigid Liners

1a Shotcrete or gunite liners For this type of liner a mixture of cement, sand, and water is blown, under pressures of 20 to 100 lb/in^2, onto the prepared bottom and sides of the waste pit. Usual mix proportions are 1 part cement to 4 or 5 parts sand, with water/cement ratios in the range of 0.4 to 0.6. Nearly vertical sides can be treated. Many swimming pools are constructed by this method. To prevent cracking during the curing period, as well as afterwards should settlement occur, wire mesh is often placed before shotcreting. For landfills, the technique is not often used because of its high cracking potential for all but the most rigid of foundations, e.g., a landfill built on rock.

1b Concrete liners These liners are constructed in much the same way as is rigid highway pavement. The aggregate is stone and sand, which is then mixed with cement to which water is added to make the final product. Concrete is usually brought to the job site from a concrete plant ready for placement on a prepared subgrade. It can be reinforced with wire mesh or reinforcement bars, depending on conditions of the subgrade. Its strength is far greater than that of shotcrete, but it suffers from a drawback in that construction and expansion joints must be incorporated in the pour, both of which are subject to leaks. Waterstops are used for this purpose, but they are expensive and might not be chemically compatible with the leachate that will eventually develop in the landfill. Further, the paving of side slopes with angles greater than 30° to the horizontal presents major construction problems when using concrete, and the material's high cost greatly limits its use.

1*c* Soil cement liners Soil cement liners are meant to duplicate the qualities of concrete as closely as possible, but more cheaply by using on site materials and construction methods. If an ample supply of stone and sand is available at the site, cement can be added to the blended aggregates and mixed in place using road graders or disk harrows, water can be added, and then it all can be graded and compacted. Soil cement that has few fines (less than 20 percent silts and clays) in the aggregate but sufficient cement (varying from 3 to 12 percent by weight) can perform adequately as a liner. A major problem, however, is the final homogeneity of the material, i.e., the very real danger of its having soft areas that contain insufficient cement. This is avoided by using a batch process to make it, in which aggregates are mixed, then blended with the cement in exact proportions, and water added at the plant or after the dry mix is spread on the prepared subgrade. This, of course, closely resembles the way concrete liners are installed, and costs obviously increase proportionately. And even more than was the case with concrete, lining side slopes with soil cement presents difficulties. The Portland Cement Association (PCA) of Skokie, IL, should be contacted for additional information on this topic.

1*d* Bituminous concrete (asphalt) Centrally mixed bituminous concrete, or asphalt, has been used often for landfill liners. The technology follows asphalt pavement procedures as used in highway engineering and construction, but places greater emphasis on permeability considerations. Compacted thicknesses range from 1.5 to 6.0 in, and the material is often placed in more than one layer. The uppermost layer should be similar to a road topping, with the aggregate being finer than $\frac{3}{8}$-in-sized material. Base layers can use larger-sized aggregate, depending upon their thicknesses. Asphalts of 40 to 70 penetration grade are usually used, the amount used depending upon the location of the lining. Material having high asphaltic content is generally avoided on slopes in warm climates. The viscous nature of the material causes it to creep in such cases. Compaction is critical in limiting seepage losses and values near 100 percent Marshall stability should be targeted for.

While many liners have been (and will continue to be) made from bituminous products (their cost is very competitive), there are several areas of concern connected with their use. A major consideration is the chemical compatibility between the landfill contents and the bituminous material used in the liner. Of lesser concern, but still realistically problematic, are the accelerated aging of bituminous liners and the weed growth that can occur through them, producing preferential seepage paths for leachate and other landfill contents. The Asphalt Institute of College Park, MD, should be consulted about proper design and construction methods.

1*e* Bituminous panels Beginning in the early 1950s, $\frac{1}{2}$-in-thick asphalt panels were developed for lining water reservoirs. These relatively thin (in comparison with concrete or bituminous concrete, the major other materials in use at the time) liners had low-strength, high-deformation characteristics but, with a good subbase, could serve as an effective seepage control system. A number of joint systems became available (most of which were heat-welded using lap joints, or using butt joints with an added strip covering the ends to be joined), but the panels were heavy and awkward to handle. Later, $\frac{1}{4}$-in-panels were developed; these were placed on top of one another in a staggered pattern and bonded together with a cold-applied liquid adhesive.

Asphalt panels can follow ground contours to a remarkable degree and will bond to concrete surfaces of appurtenant structures very well.

8.5.2 Flexible Liners (Natural Materials)

2a Compacted soil liners Under advantageous conditions, naturally occurring fine-grained soils may themselves be used for the liner. The key parameter is, of course, the soil's permeability, more properly called its *hydraulic conductivity,* which should be 10^{-5} to 10^{-7} cm/s depending upon the contained materials. This automatically puts the class of required soils into the fine-grained soils category, where the montmorollinite clay family possesses the lowest values. This includes bentonite clays. See Fig. 8.4 for typical values and the influences of consolidation pressures.

At this point it should be noted that a soil having a permeability of 10^{-7} cm/s indeed has a very low flow rate. The governing equation is Darcy's law, which is as follows:

$$q = kiA \tag{8.1}$$

where q = flow rate, cm³/s

k = hydraulic conductivity, cm/s (commonly called the *coefficient of permeability* or simply *permeability*)

A = total area through which flow is occurring, cm²

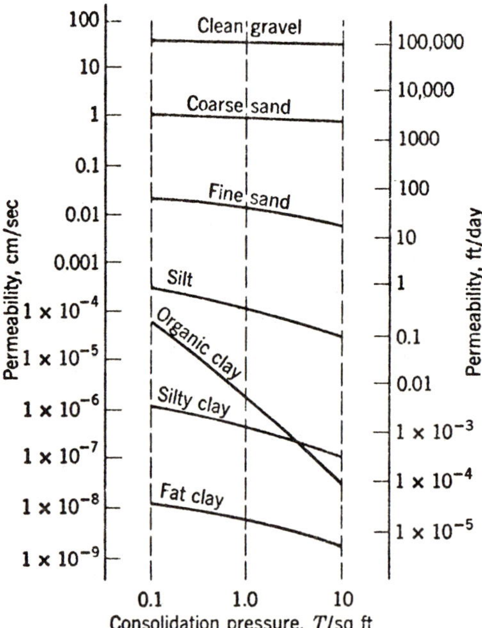

Figure 8.4 Typical soil permeability (hydraulic conductivity) values of different soil types showing influence of consolidation pressure. *(After Cedegren, Ref. 12.)*

i = hydraulic gradient (dimensionless) = $\Delta h / \Delta \ell$, where Δh is the increment of hydraulic head causing flow and $\Delta \ell$ the distance through which flow is occurring

Furthermore,

$$\frac{q}{A} = v_s = ki \tag{8.2}$$

where v_s = (superficial) velocity cm/s

This value of velocity is not the true value since the area is a combined solid and void area. Since water can only flow within the soil voids, the superficial velocity must be divided by the soil's porosity n to obtain the real velocity, i.e.,

$$v_t = \frac{v_s}{n} \tag{8.3}$$

where v_t = true velocity, cm/s

v_s = superficial velocity, cm/s

n = soil porosity (varies from 0.3 to 0.6)

$\quad = \dfrac{V_v}{V}$

V_v = void volume, cm³

V = total volume, cm³

Example 8.1 illustrates the situation.

Example 8.1 Given a fine-grained clay soil having a permeability of 10^{-7} cm/s (0.000283 feet per day), determine the true water velocity and the time needed for water to move 1 ft through it, based on a soil porosity of 0.5, for a series of hydraulic gradient values; graph the results.

SOLUTION Using Darcy's law,

$$q = kiA$$

$$\frac{q}{A} = v_s = ki$$

Thus if $i = 1.0$,

$$v_s = 0.000283 \text{ feet per day}$$

$$v_t = \frac{v_s}{n}$$

$$= \frac{0.000283}{0.5}$$

$$= 0.000566 \text{ feet per day}$$

$$t = 1767 \text{ days per foot}$$

As shown in the accompanying table, typical hydraulic gradients for storage of liquids can be as high as 50 and, for hazardous materials landfills, as low as 0.05. The true velocities and times are given for those extremes, as well as for gradients in between; see the accompanying graph.

Hydraulic gradient i	True velocity, ft/day	Time to travel one foot
0.05	0.0000283	35,335 days = 96.8 years
0.1	0.0000566	17,667 days = 48.4 years
0.5	0.000283	3,533 days = 9.68 years
1.0	0.000566	1,767 days = 4.84 years
5	0.00283	353 days = 0.968 years
10	0.00566	177 days = 0.48 years
50	0.0283	35.3 days = 0.097 years

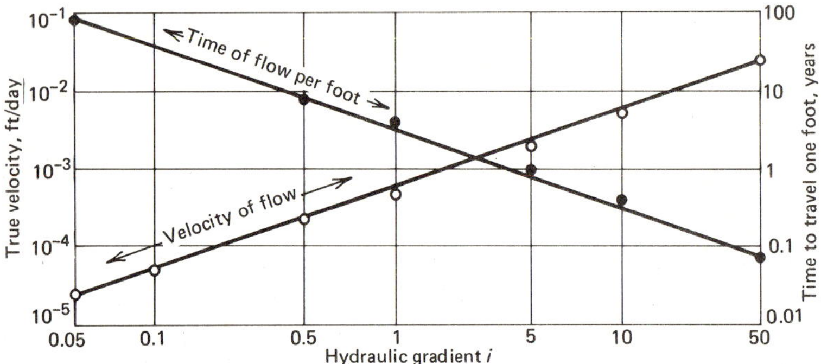

Two separate situations can be envisioned using on site fine-grained soils for liners. The first is when the soil at the bottom of the proposed landfill is itself sufficiently impermeable to serve as the barrier. This is an ideal situation and only in unusual circumstances would there be any problems connected with it, e.g., if leachates consisting of organic solvents were present. The problem of how to line the side slopes however, would have to be considered. The second, and more likely, situation is when the soil at, or near, the site is of the proper type but is not at the proper elevation or is nonhomogeneous in its makeup. In this case, the soil must be blended or homogenized, brought to the previously prepared site (if necessary), and placed and compacted in lifts. Typical lifts of clay soils are 6 to 9 in and a towed or self-propelled sheepsfoot roller is usually used for compaction. The clay is usually kept slightly moister than optimum as long as that moisture content does not seriously affect construction operations. Side slopes of the landfill can be prepared in a manner similar to the base, although it is generally necessary to use lighter rollers traveling perpendicularly to the slope.

More detail as to what constitutes the optimum type of soil for use as liners and the construction procedures needed to handle them will be given in section 2c below, on bentonite clays.

2*b* Chemically treated soils Obviously the addition of chemicals to soils can drastically alter their behavior. Chapter 6 was concerned with chemical grouting, which besides adding strength to and reducing the compressibility of soils, also decreases their permeability. In general, the process is only cost-effective when zero-, or negative-, cost chemicals are used. Kays[11] mentions a number of possibilities including chrome lignin (a byproduct of the paper industry), sodium brine (salt from numerous industries which can make calcium clays quite impermeable), and numerous types of polyphosphates. Soils can also be treated chemically by spraying. Asphalt emulsions can be used, as can chemicals like polyvinyl alcohol.[13] These materials either infiltrate into the voids of a soil or form a crust on its surface, thereby blocking off the voids.

While the idea of chemically treating soils is technically sound, the practical construction problems of doing so are usually formidable. In particular, it is difficult to achieve uniformity of treatment; there is no guarantee against irregularly treated zones having high permeabilities. Furthermore, how permanent the various types of chemical treatments will prove to be has not been documented.

2*c* Bentonite clays The neighboring states of Wyoming and Montana are particularly rich in sodium montmorillonite clay, commonly called *bentonite* or, more suitably for our purposes, *sodium bentonite*. It is a hydrated aluminum silicate mineral with the following structural formula:

$$(A\ell_{1.63}Fe_{0.17}Mg_{0.25})(A\ell_{0.07}Si_{3.93})O_{10}(OH)_2(Na_{0.24}Ca_{20.04})$$

It is obviously a very complex material, but most important, the resulting particles are flakelike, with their lengths and widths much greater than their thicknesses. This comes about because of the structural sharing of oxygen atoms forming silica tetrahedral and aluminum octahedral sheets; see Fig. 8.5. Water can easily penetrate between the sheets (it is then called *chemically bound, crystalline,* or *structural water*), or it can easily be adsorbed onto their surfaces. This latter situation, i.e., surface adsorption, is very important because bentonites (all clays for that matter) are strongly anionic, i.e., they carry a negative electrical charge due to substitutions which occur within their crystalline structure. When water in the soil voids is confronted with this situation, at least two things occur:

The water, being a polar liquid, aligns itself in an orderly fashion, to help satisfy the charge of the bentonite sheets

Cations in the water, which are usually only partly hydrated, also move to the clay's surface; if these cations are predominantly sodium, the bentonite is then called a sodium bentonite.

These changes result in an adsorbed water layer of highly viscous water molecules and entrapped cations which is bound to the surface of the bentonite sheets; see Fig. 8.6. This forms the adsorbed water layer which protrudes into the soil's voids, drastically affecting the soil's permeability. Since only the unbound, or free, water left in the soil voids flows in accord with hydrodynamic principles, e.g., according to Darcy's

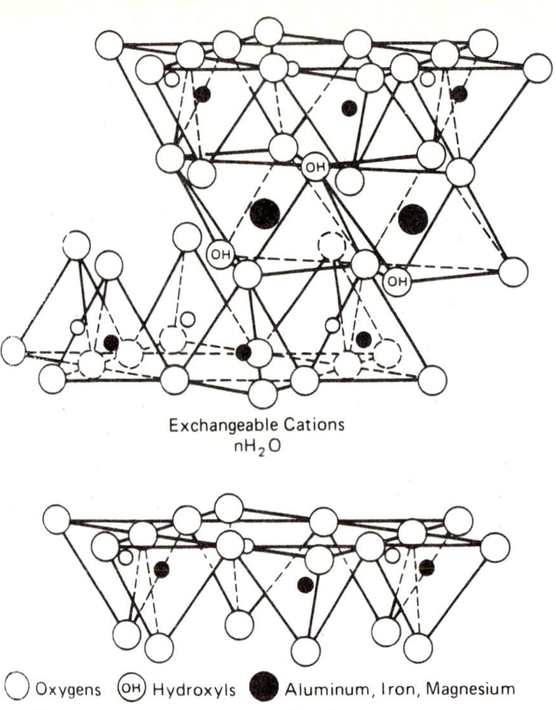

Figure 8.5 Sketch of silica tetrahedra and aluminum octahedra and their formation into sheet structures typical of bentonite clay. *(After Grim, Ref. 14.)*

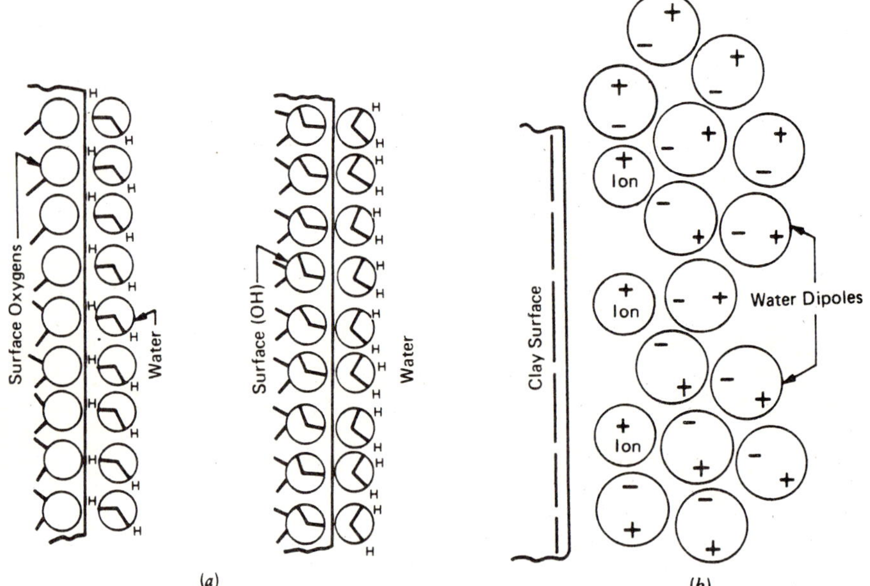

Figure 8.6 Possible mechanisms of water adsorption of clay surfaces. (*a*) Hydrogen bonding, (*b*) ion hydration, (*c*) attraction by osmosis, (*d*) dipole attraction. *(After Mitchell, Ref. 15.)*

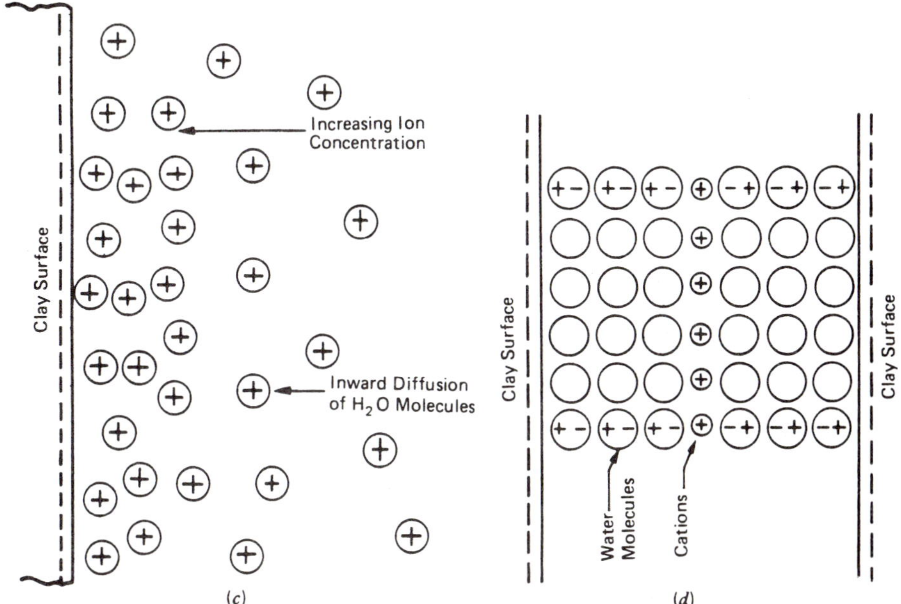

Figure 8.6 (*Continued*)

law, it should be understandable why clays have such low permeabilities when compared with other soil types of comparable void size; recall Fig. 8.4.

This situation has long been recognized by engineers and contractors looking for liners to retard seepage from landfills of all types. Example 8.2 shows how to determine required clay liner thickness for a landfill.

Example 8.2 Determine the thickness t required for a bentonite clay liner whose permeability k is 10^{-7} cm/s when the anticipated seepage rate q of the leachate through the liner is 0.01 gallons per day per square foot of liner if the leachate level in the landfill is 10 ft deep and could potentially fill the entire containment area.

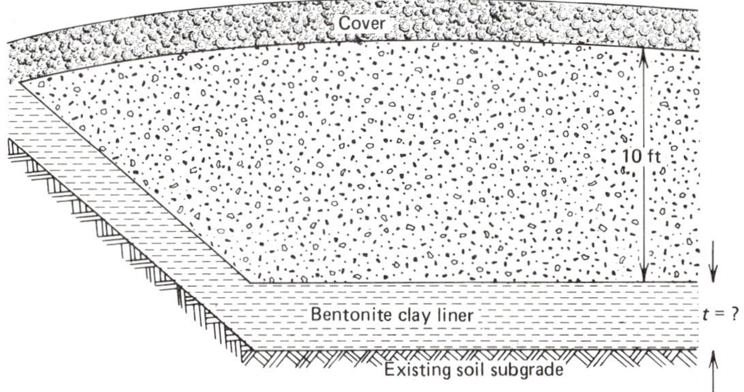

SOLUTION Convert the permeability to gallons per day per square foot using 1 foot per day $\simeq$ 7.5 gallons per day per square foot. Thus

$$k = 10^{-7}\ \text{cm/s}$$

$$= 2.83 \times 10^{-4}\ \text{ft/day}$$

$$= 21.2 \times 10^{-4}\ \text{gal/day per square foot of liner}$$

Now, using Darcy's law,

$$q = kiA$$

$$= k\frac{\Delta h}{\Delta \ell}A$$

or

$$(0.01) = 0.00212\frac{10 + t}{t}1.0$$

Therefore

$$4.72t = 10 + t$$

and

$$t = \frac{10}{3.72}$$

$$= 2.69\ \text{ft}$$

The use of clay liners in industrial and municipal solid waste landfills has progressed to the point where many companies are mining, processing, and supplying clays for specialty purposes. Additionally, construction companies are specializing in the construction of bentonite clay liners. Certain leachates, however, have been particularly troublesome for clay liners (including bentonites): those consisting of organic solvents, and those containing high levels of dissolved salts, acids, or alkalies. Alarming results were revealed by a systematic study undertaken by Anderson et al.[16,17] into how the permeabilities of four types of clay were affected by the seepage through them of several organic liquids as compared with the seepage of water. Figure 8.7 illustrates that behavior, showing that the passage of as little as one pore volume of 100% aniline, methanol, and xylene increased the permeabilities of the clays from the base line value of about 10^{-8} cm/s by as much as two orders of magnitude. There were obvious structural changes within the clay soils themselves, with actual piping channels and cracks within the soil masses occurring.

For sodium bentonites the problem is lessened by the exchange of the adsorbed sodium ions on the clay's surface with ions that are present in the contaminated leachate. Also, the industry has responded by introducing inorganic chemicals with polymers that decrease the sensitivity of the clay to potential contaminants.

Much research into the problem is going on within the industry at this time (see

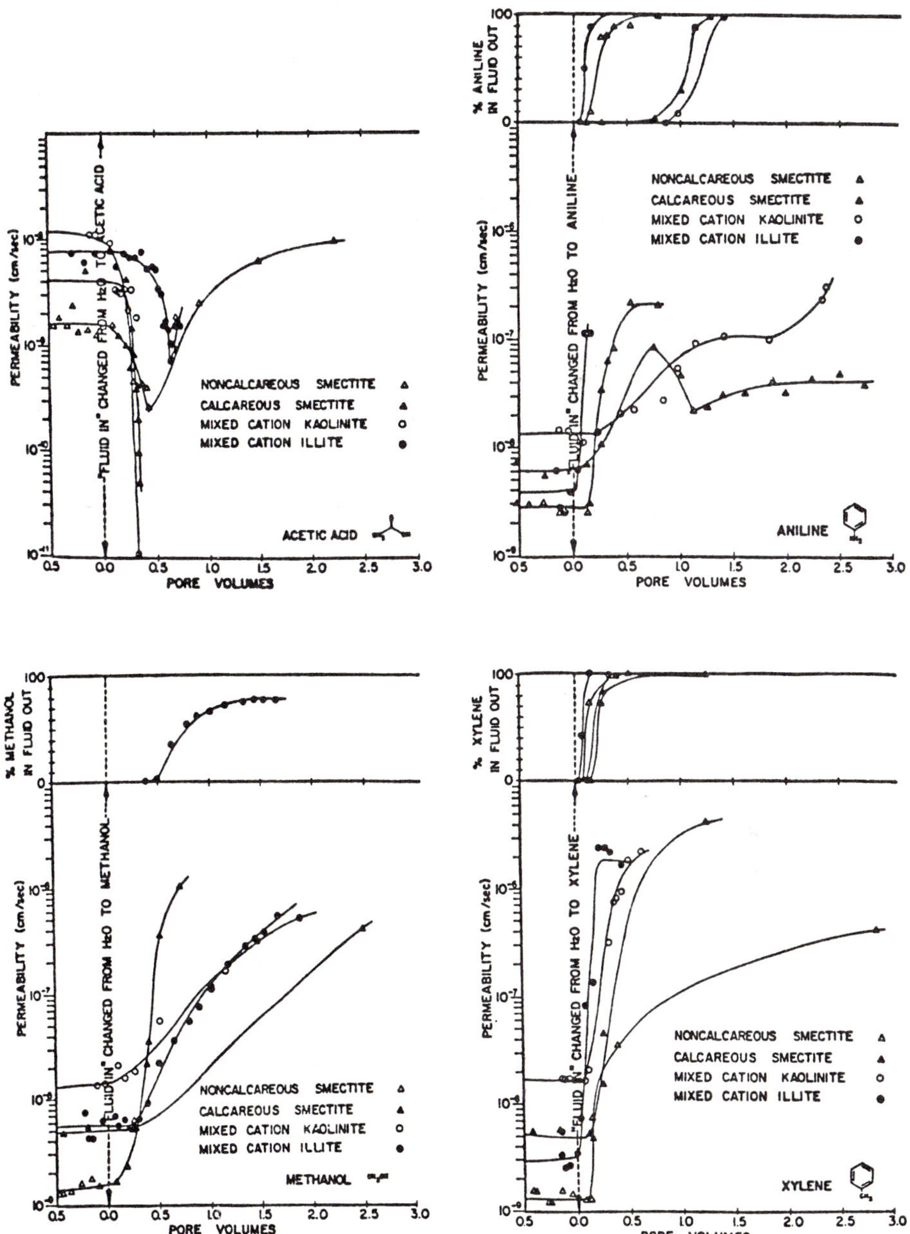

Figure 8.7 Drastic increases in permeabilities of several clay soils during seepage through them by 100% organic liquids. *(After Anderson, Ref. 17.)*

Appendix F for a list of the organizations involved). However, the effectiveness of clay liners in containing many hazardous wastes is currently being questioned. This situation, along with the EPA regulations concerning land disposal facilities for hazardous wastes (as reported in the *Federal Register* on July 26, 1982[18]) has recently produced a rapid growth in the use of synthetic materials, i.e., plastic and elastomeric geomembranes, as landfill liners. These will be discussed next, in Sec. 8.5.3.

8.5.3 Flexible Liners (Synthetic Materials)

3*a* Geomembranes (plastics and elastomers) The use of flexible liners, or geomembranes, made from synthetic materials (plastics and rubber) began with the introduction of polyvinyl chloride (PVC) and butyl rubber and has emerged over the past 15 to 20 years as an important method used in some very impressive projects. Significant in this category are a 110-acre, 50-ft-deep, water storage reservoir in Hawaii which was lined with a nylon-reinforced butyl rubber,[19] and a 278-acre, 20-ft-deep, hydroelectric power water storage reservoir in Colorado which was lined with a polyester-reinforced chlorinated polyethylene.[20]

As impressive as these statistics are, the main catalyst for the increased use of geomembranes seems to be governmental regulations. The Environmental Protection Agency's rules require that landfills, surface impoundments, and waste piles be designed so as both to prevent the migration of leachate out of them and to remove any leachate they generate during the active lifetime of the site. As per Ref. 18, dated July 26, 1982, some relevant parts of the EPA code are as follows:

> When a synthetic membrane liner, for example, is installed in a landfill, the leachate collection and removal system installed above the liner (as required by today's regulations for landfills and piles) can achieve virtually a 100% removal efficiency. In contrast, if a clay liner is used, some leachate will seep into the liner rather than be removed through the drainage layer. This leachate will remain in the soil after closure and will likely migrate to the ground water at some future time.
>
> Prevention, rather than minimization, of leachate migration similarly produces better environmental results in the case of surface impoundments used to dispose of hazardous wastes. A liner that prevents rather than minimizes leachate migration provides added assurance that environmental contamination will not occur.

Many hazardous materials fall into the category of special provisions requiring the use of double-liner systems of geomembranes. The EPA regulations state:

> These special standards require that there be two liners underlying the unit and a leak detection system between the two liners. The two liners must be designed and constructed in a manner that prevents the migration of liquids into or out of the space between the liners. This can be achieved by lapping or sealing the edges of two synthetic membrane liners at the surface.

The requirements for both clay liners and geomembranes are addressed in the following manner:

> First, the liner must be constructed of materials that will resist degradation. Synthetic liners can be degraded by exposure to incompatible wastes or leachate and in some cases by excessive expo-

sure to sunlight. Clay liners can develop highly increased permeabilities (sometimes by several orders of magnitude) when exposed to certain types of chemicals. The regulation thus requires appropriate materials to be used to avoid such problems. When the permit is issued, the appropriateness of the liner material will be considered in the specific context of the wastes to be placed in the impoundment.

Second, liner materials must be of sufficient strength and thickness to prevent failure due to physical stresses (e.g., earth-moving equipment, dredging equipment, and the weight of large volumes of liquid wastes). While this requirement applies to all liners, it is especially crucial for synthetic liners, which can rupture if they are mishandled or are too thin. Most synthetic liners need to be at least 30 mils (thousandths of an inch) thick to assure that this requirement is complied with.

Third, the foundation underneath the liner must be capable of supporting the liner and resisting pressure gradients. If the support system settles, compresses, or uplifts, the liner may rupture or crack.

Finally, the liner must cover all surrounding earth likely to be in contact with the waste or leachate. This assures that liners will be placed not only underneath the wastes but also on the sides of the wastes. Thus, lateral as well as vertical migration will be prevented.

Thus, because of both the cost-effectiveness of geomembranes and EPA regulations, the use of flexible synthetic liners for landfills will continue to experience a strong and steady growth at least for the foreseeable future. It is easy to see that geotechnical and construction engineers will require a considerable amount of exposure to and knowledge of this technology. Some general liner schematics are shown in Fig. 8.8.

Geomembranes are made from plastic or rubber-based materials by a chemical reaction called *polymerization*. In regard to plastic geomembranes, the reaction occurs in a pressure vessel, called a *polymerizer* or *reactor*, which is temperature-controlled and constantly stirred. Small molecular materials (monomers) are fed into the polymerizer to join with themselves or with other kinds of molecules to form long-chain polymers. Examples are vinyl chloride monomer forming polyvinyl chloride resin and ethylene monomer forming polyethylene resin. Obviously, other materials can be added, e.g., stabilizers, during the process. The combinations that can be produced are virtually unlimited.

It is of interest to note that the above resin and polymer manufacturing is done by chemical companies which then sell the product to separate (and usually much smaller) processing companies. Here the material is mixed (compounded) with other ingredients and processed into its final sheet form by extrusion and/or calendering. An *extruder* is a relatively small device where the resin is heated (and often further blended) and forced into a worm gear system that has a die on its outlet side. The die determines the final shape of the finished product, which can be sheet, solid tube, or hollow tube. On the other hand, a calender is a large piece of equipment having counterrotating rolls through which the heated resin is forced. The width and thickness of the resulting sheet are carefully controlled. From this point, the sheets can be factory-seamed into large widths and prefabricated to meet exact field dimensions. This is important, for it is basic to the philosophy of using geomembranes to use as few field seams as possible.

The manufacturing of rubber-based geomembranes follows a processing route similar to that just described except that rubber must be cured (vulcanized) before being formed into sheets. Here, crosslinking with sulphur or sulphur compounds

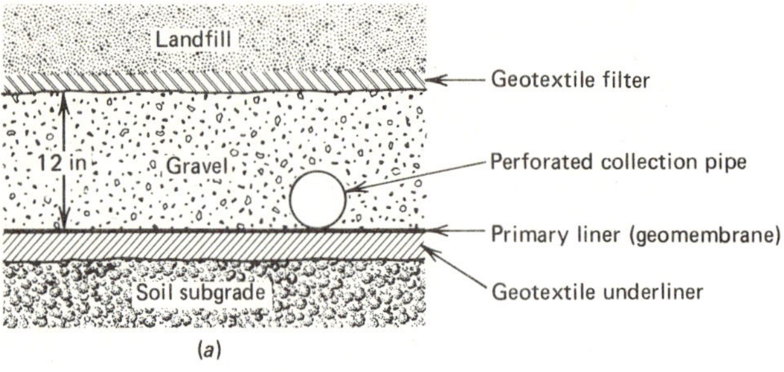

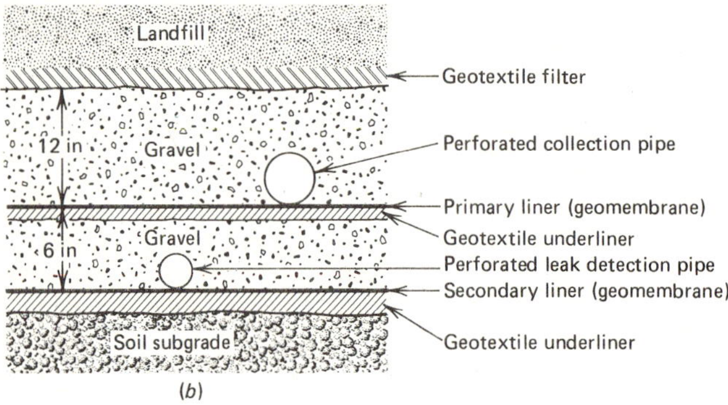

Figure 8.8 Typical cross sections of geomembrane liner systems. (*a*) Single-liner system. (*b*) Double-liner system.

under heat and pressure is necessary to reduce the tendency of natural rubbers toward ozone degradation. It is easy to visualize that plastic and rubber sheets of the types under discussion have relatively low moduli, low strengths, and extremely high elongation at failure. Figures 8.9 and 8.10 show the results of wide-width tensile tests for three types each of CPE (plastic) and EPDM (rubber) geomembrane sheets. The unreinforced membranes illustrate the above-described behavior. When a reinforcing scrim is added, however, the stress-strain behavior changes markedly. The data are self-explanatory. Not shown are the beneficial effects of scrim reinforcement with respect to puncture and tear resistance and shrinkage stability.

Reinforcing scrims are usually nylon, polyester, or polypropylene fabrics in a 5- to 20-thread-per-inch grid pattern (they are loosely woven geotextiles by our definition) and are made by calendering the resin into the scrim or by laminating the scrim between two resin sheets.[21]

Problems with delamination can arise, proper seams can be difficult to make, and a wicking action by the scrim can occur unless the ends of the sheets are properly sealed by means of a process called *selvaging*.

There are many variants of the above production methods, e.g., blown film extru-

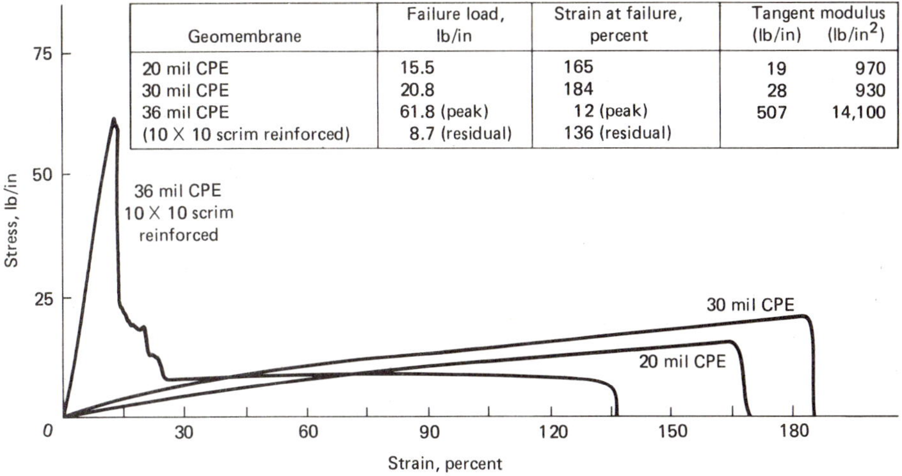

Geomembrane	Failure load, lb/in	Strain at failure, percent	Tangent modulus (lb/in)	(lb/in²)
20 mil CPE	15.5	165	19	970
30 mil CPE	20.8	184	28	930
36 mil CPE	61.8 (peak)	12 (peak)	507	14,100
(10 × 10 scrim reinforced)	8.7 (residual)	136 (residual)		

Figure 8.9 Stress-versus-strain behavior of 20-mil CPE, 30-mil CPE, and 36-mil (10 × 10) reinforced CPE plastic geomembranes during wide-width tensile tests.

sion can make sheets up to 20 ft wide in one operation, but the above discussion should suffice as an introduction here. For further information see Kays[11] or any one of a number of handbooks and texts on producing plastics and rubber liners. A directory of companies that make geomembrane systems can be found in Appendix F.

Table 8.6, in describing the most common geomembranes, presents a number of relevant features including the usual thickness of each type of sheet and the method

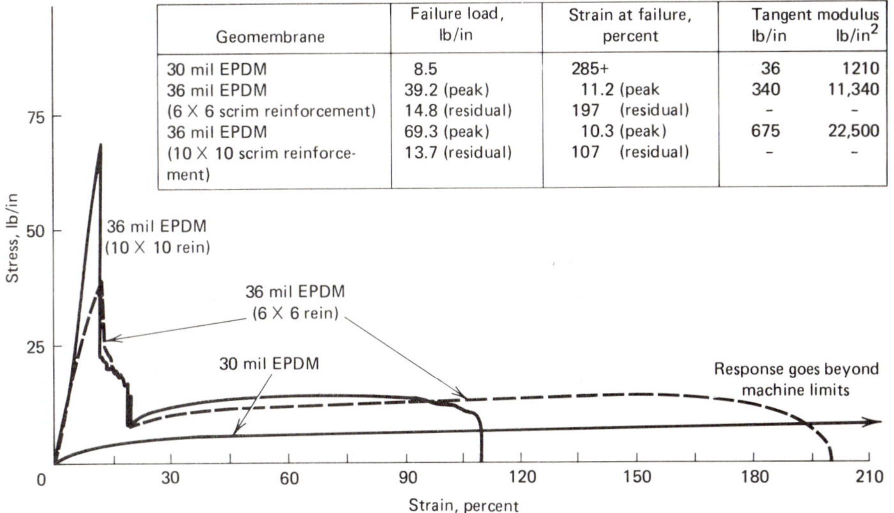

Geomembrane	Failure load, lb/in	Strain at failure, percent	Tangent modulus lb/in	lb/in²
30 mil EPDM	8.5	285+	36	1210
36 mil EPDM	39.2 (peak)	11.2 (peak	340	11,340
(6 × 6 scrim reinforcement)	14.8 (residual)	197 (residual)	–	–
36 mil EPDM	69.3 (peak)	10.3 (peak)	675	22,500
(10 × 10 scrim reinforcement)	13.7 (residual)	107 (residual)	–	–

Figure 8.10 Stress-versus-strain behavior of 30-mil EPDM, 36-mil (6 × 6) reinforced EPDM, and 36-mil (10 × 10) reinforced EPDM rubber geomembranes during wide-width tensile tests.

Table 8.6 Commonly used geomembranes

Common designation	Chemical name	Polymer class	Thickness mils	Field seaming method
PVC	Polyvinyl chloride	Plastic	10–30	Adhesive
PE	Polyethylene†	Plastic	20–100	Welding
CPE	Chlorinated polyethylene	Plastic	30–50	Adhesive
CSPE	Chlorosulfonated polyethylene	Plastic	30–50	Adhesive
3110	Elasticized polyolefin	Plastic		Adhesive
Butyl	Butyl rubber	Rubber	40–60	Gum tape
EPT	Ethylene propylene terpolymer	Rubber	30–50	Gum tape
EPDM	Ethylene propylene diene monomer	Rubber	20–40	Gum tape
Neoprene	Polychloroprene	Rubber	20–60	Adhesive

†Polyethylene is covered by ASTM D1248. This specification provides for three types:
Type 1 is polyethylene in a density range of 0.910 to 0.925 g/cm³. This is termed *low-density* or *branched polyethylene*, or *LDPE*.
Type 2 is polyethylene in a density range of 0.926 to 0.949 g/cm³. This is known as *medium-density polethylene*, or *MDPE*
Type 3 is polyethylene in a density range of 0.941 to 0.965 g/cm³. This is termed *high-density* or *linear polyethylene*, or *HDPE*.

used for each to make field seams. See Fig. 8.11 for sketches of these field joint details. Regarding factory seams for splicing sheets of various widths together, dielectric and thermal methods are often used. These produce seams that are superior to those produced by adhesives and tapes but are very expensive and not conducive to field operations.[22] Thus there is a definite advantage to prefabricating sheets in large sizes whenever possible. Except for PVC (which can be produced in thin sheets, then accordion-pleated for shipment in boxes), all other geomembranes are shipped to the job site in rolls, spread out by a team of laborers, and joined as per Fig. 8.11. Major problems often occur at appurtenances (protrusions through the geomembrane for pipes, fittings, weirs, etc.) and at the points of anchorage. Figure 8.12 illustrates some of the common ways of handling such situations.

Field seams can of course be visually inspected, but only major defects and flaws in them can be observed by the unaided eye. An option that always exists is to cut out a sample of the seam and test it for shear or peel resistance. However, this is only useful for assessing the adequacy of the particular type of seam being tested and not the adequacy of skill going into the seaming process. The latter can be inspected by a number of nondestructive testing methods:

Air lance method. Here a high-pressure air lance is placed at the edge of the sealed overlap joint and defects often become obvious.
Vacuum test. A vacuum is created at the overlap interface and a loss of vacuum indicates a leaking seam.
Pressure pocket test. Here a double seam is made with an unbonded section in between. Air pressure is then pumped into the inner section and a pressure drop indicates a leaking section.

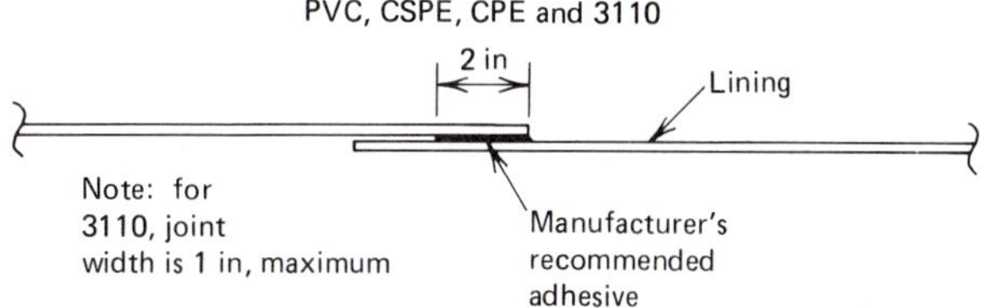

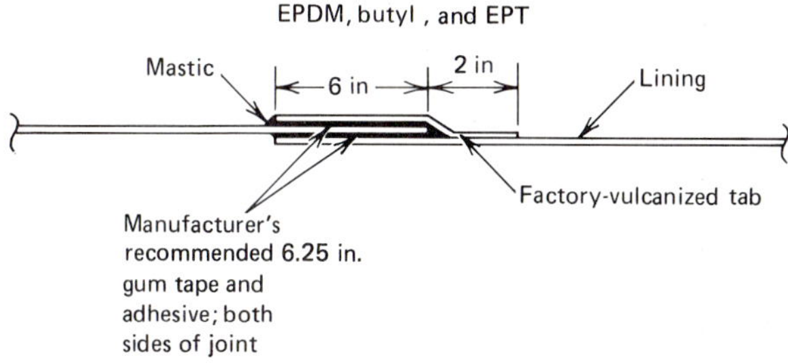

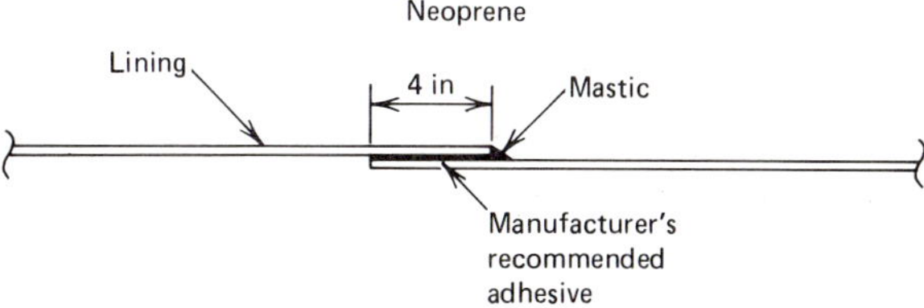

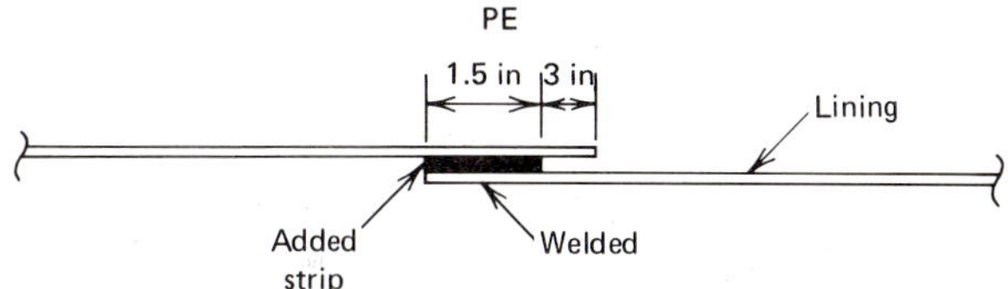

Figure 8.11 Typical methods of making field seams for various geomembranes. *(Adapted from Kays, Ref. 11.)*

Dead load test. An in situ placed tensile load is applied to the seam to assess its load-carrying ability. Due to its non-continuous nature, this is probably the least attractive of the seam inspection methods.

Besides problems relating to seams, the extremely important aspect of *chemical compatibility* between waste material and geomembrane must also be addressed. The first source of information should be the producer of the polymer. Most of the firms that make them are large, having full staffs of chemists and chemical engineers, and can provide excellent brochures containing information on a vast number of chemicals and how their specific geomembrane behaves with respect to them. Unfortu-

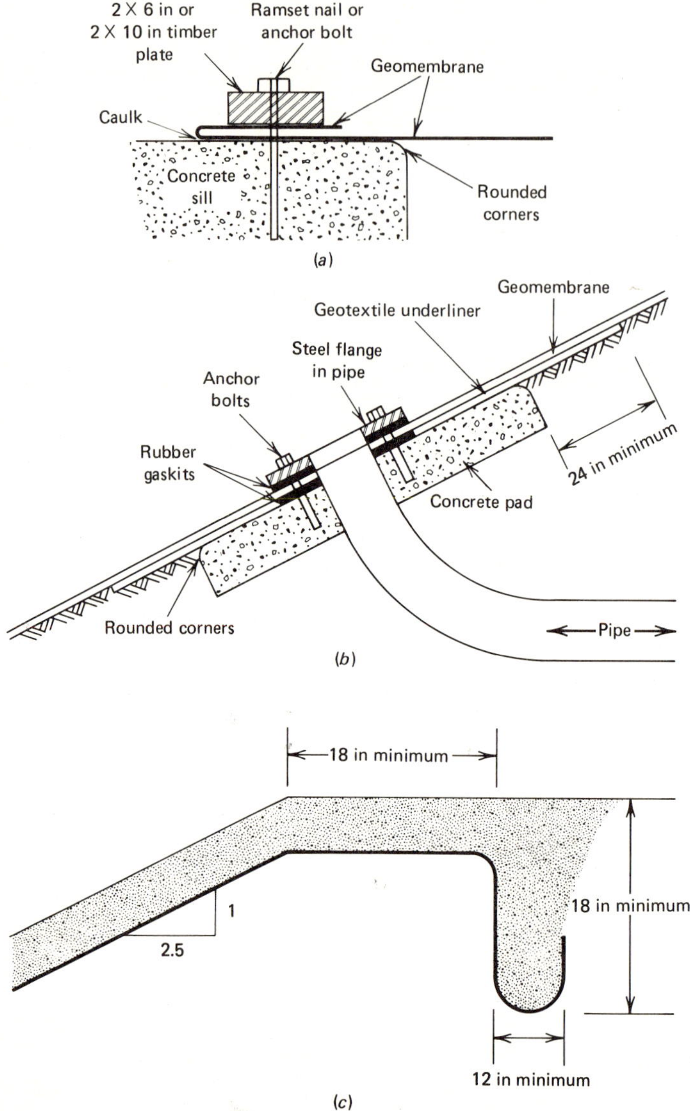

Figure 8.12 Details of geomembrane attachments to appurtenances and methods for anchorage. (*a*) Suggested detail at concrete apron or sill. (*b*) Suggested detail for inlet or outlet pipe. (*c*) Suggested anchorage detail in soil.

nately, many wastes are extremely complex mixtures containing organics, toxics, heavy metals, etc. Therefore a chemical assay of the material to be contained by a landfill is required to determine the individual chemicals that will be present and their concentrations. In the absence of published information, a testing laboratory should be engaged to test the candidate geomembranes with the particular waste leachate one has, or expects to have, at the site.[23,24] Test specimens of the geomembrane should first be evaluated in their as-received condition (see Table 8.7), and then immersed or exposed in the leachate, removed, and tested again. The following information about the exhumed samples is of particular interest:

Changes in thickness
Changes in weight
Changes in puncture resistance
Changes in tear resistance
Changes in stress-strain response in regard to modulus, strength, and elongation at
    failure

Testing should be performed for a 90-day period and behavioral trends noted. See Figs. 8.13 and 8.14 for the type of data reporting that is recommended. Unfortunately, there are no absolute limits on what types of changes can be tolerated, and a relative ranking of the candidate geomembranes is necessary.

In closing, it should be mentioned that geomembrane liners are not infallible—for example, see Montague.[26] The amount of acceptable water reservoir leakage has often been addressed and at least one organization has attempted to quantify the

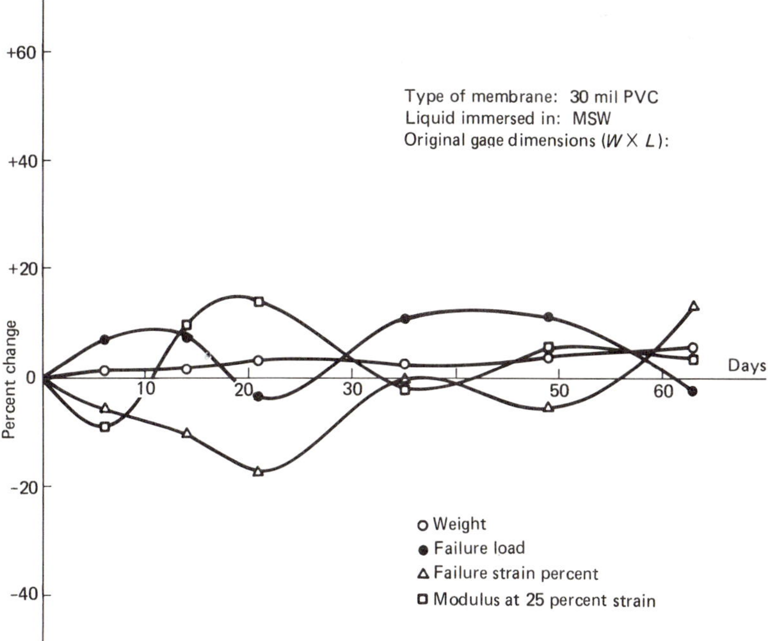

Figure 8.13 Tests for evaluating chemical compatibility between candidate geomembranes and anticipated leachates at waste sites. Example shown is for 30-mil PVC and municipal solid waste leachate.

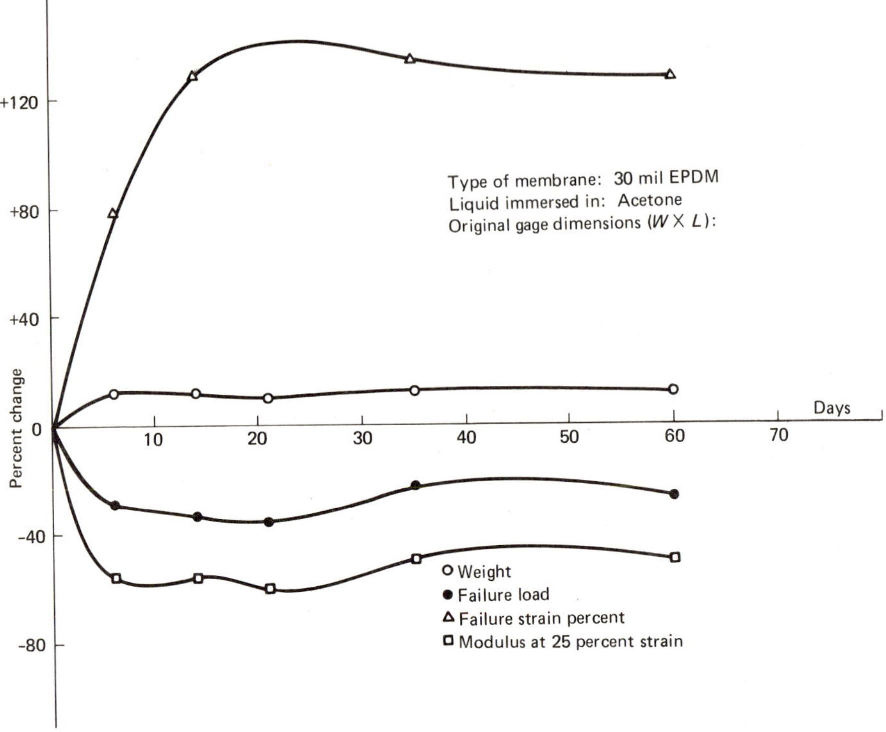

Figure 8.14 Tests for evaluating chemical compatibility between candidate geomembranes and anticipated leachates at waste sites. Example shown is for 30-mil EPDM and acetone.

limiting rate. The East Bay Water Company of Oakland, CA, offers the following
definition of the amount of leakage that can be tolerated at a site:

$$Q \leq \frac{A\sqrt{H}}{80} \tag{8.4}$$

where Q = leakage tolerance, gal/min
$\quad A$ = reservoir area, 1000 ft^2
$\quad H$ = maximum water depth, ft

The closest that one can come to a no-leakage situation is the double-liner system described near the beginning of this section and shown in Fig. 8.8b. The primary
liner of such a system is the upper geomembrane, below which is located a perforated
pipe collector system bedded within a crushed-stone drainage layer. Below this stone
layer is located a secondary geomembrane liner, which only need function if leakage
from the primary liner shows up in the underdrain system. If leakage from the primary liner does occur a downstream well monitoring system must be deployed to
check for possible leakage from the secondary liner. Obviously expensive, this type
of double liner system is targeted for landfills containing hazardous wastes.

Table 8.7 Properties of commonly used lining materials†

Property‡	Polyethylene		Polyvinyl chloride	Chlorinated polyethylene	Polypropylene	Nylon	Butyl rubber	Natural rubber	Hypalon
	Low density	High density							
Specific gravity	0.92–0.94	0.94–0.96	1.20–1.5	1.35–1.39	0.9–0.91	1.08–1.4	0.92–1.25	0.91–1.25	
Tensile strength, lb/in^2	1300–2500	2400–4800	3500–10,000	1800 min	4000–32,000	9000–11,000	1000–4000	1000–3500	1000–2000
Elongation, %	200–800	10–650	60–200	375–575	40–400	250–550			
Shore "A" hardness							15–90	20–100	55–95
Operating temperature range, °F	−70 to 180	−70 to 240	−60 to 200	−40 to 200	−60 to 220	−60 to 380	−50 to 325	−70 to 250	−45 to 250
Resistance to acids	Poor to good	Good	Good to excellent	Good to excellent	Good to excellent	Poor			Good
Resistance to bases	Good to excellent	Good to excellent	Good to excellent	Good to excellent	Good to excellent	Excellent			Good to excellent
Resistance to oxygenated solvents	Poor to good	Poor to good	Good	Poor					Good
Resistance to aromatic and halogenated solvents	Fair	Fair	Good	Poor	Good	Good	Poor	Poor	Fair
Resistance to aliphatic (petroleum) solvents	Poor to fair	Fair to good	Good	Good	Good	Excellent	Poor	Poor	Good
Water vapor permeability, perm mils	3–14	1.8–2.2	3–18	0.040–0.048	0.25–1	0.09–1.0	0.15		2.0
Weatherability	Poor	Poor	Good	Excellent	Poor	Fair	Good	Fair	Excellent
Time to crack, h	900	300	No crack till 2500 h	No effect to 4000 h	100	1200			
Time to chalk, h	No effect till 2500 h	600	300	No effect to 4000 h	600	No effect till 2500 h			
Time to fade, h	300	300	100	No effect to 4000 h	900	200			

†After Kumar and Jedlicka, Ref. 25.

‡ASTM test methods for various properties: specific gravity, D751; tensile strength, D97; elongation, D412 or D882; Shore "A" hardness, D676; water vapor permeability, E96.

8.6 LANDFILL CAPS (CLOSURES)

Whenever a landfill has ceased to be usable (or when it is judged to be a hazard in its current condition), it should be capped in an environmentally safe and secure manner. EPA's Office of Solid Waste estimates that there are as many as 32,000 hazardous waste dump sites (landfills) throughout the United States. Of those, 1200 to 2000 may present significant health or environmental problems.[27] The *Comprehensive Environmental Response, Compensation, and Liability Act of 1980,* or "Superfund" for short, with $1.6 billion at its disposal, is aimed at correcting this situation.

The EPA regulations of July 26, 1982, suggest the fundamental design philosophy meant to guide the capping of hazardous waste landfills. Note that these closures are for landfills that do not impact on local groundwater sources (closures that do so will be considered in Sec. 8.7). The recommendations are as follows[18]:

> Closure of Land Disposal Units—A cap is a top liner, placed on the unit at closure. Caps, like bottom liners cannot be expected to last forever. However, a properly designed and maintained cap can prevent the entry of liquids into the closed unit, and thus the formation and migration of leachate, for many years and can minimize it thereafter in the absence of damage. Therefore, EPA requires that the cap be designed and constructed to provide long-term minimization of the movement of liquids into the closed unit. Because clays will generally last longer than synthetic materials, clay caps rather than synthetic caps should usually be the materials chosen to provide long-term minimization.
>
> To avoid the build-up of liquids in the closed landfill or impoundment (the "bathtub effect"), EPA requires that the cap be as impermeable as the bottom liner. This will require the installation of a synthetic membrane cap whenever the bottom liner is synthetic. Thus, many units will be required to have two-layer caps consisting of a synthetic layer to avoid the bathtub effect and provide short-term prevention of infiltration, and a clay layer to provide long-term minimization of precipitation infiltration and leachate generation. EPA believes that this will provide maximum short-term and long-term protection of human health and the environment.

The above philosophy leads to two different design methods, which are shown schematically in Fig. 8.15.

For the single-layer cap shown in Fig. 8.15*a*, the landfill must be reasonably stable with unconfined compressive strengths of 1500 lb/ft² (10.4 lb/in² = 71.6 kPa ≃ 2.4 CBR) or greater. In this case, the clay seal provides an impermeable barrier preventing surface water from entering the landfill and thereby eliminating the formation of additional leachate. The design of this layer in regard to thickness follows the procedures discussed in Sec. 8.5.2, paragraph 2*c,* for bentonite clays. Obviously, the clay used must be chemically compatible with the contents of the landfill. The overlying sand-gravel drainage layer, or bulky, needle-punched geotextile,[28] is meant to divert surface water from ever interfacing with the clay seal. This drainage layer must be adequately graded and provided with underdrains at its boundaries to convey accumulated water from the landfill area. In cases of heavy infiltration, a composite geotextile may be necessary; see Fig. 7.2*f.* The uppermost soil and topsoil layers serve the dual roles of protecting underlying layers and providing the means to grow a vegetative cover.

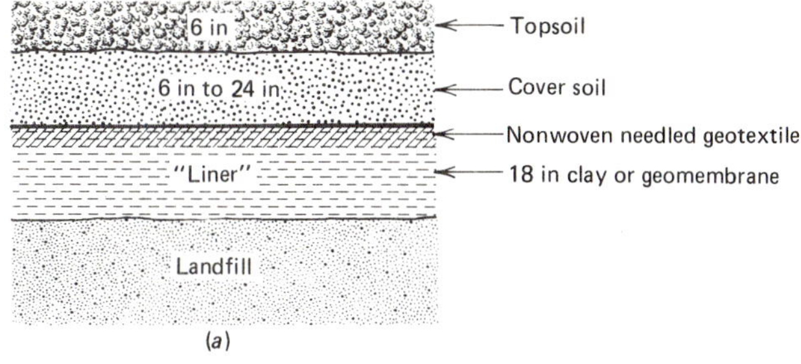

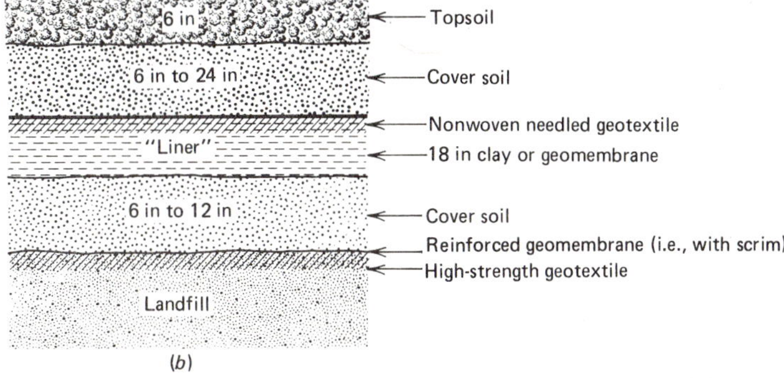

Figure 8.15 Two types of landfill cap (closure) systems; which is used depends primarily on the stability of the landfill being capped. (*a*) Single-liner closure system (for reasonable to good landfill stability). (*b*) Double-liner closure system (for poor to marginal landfill stability).

The double-layer closure system shown in Fig. 8.15*b* is considerably more intricate in both its design and construction. This is used for landfills that are less stable (those having unconfined compressive strengths of less than 1500 lb/ft^2 down to levels exhibited by the viscous liquids that make up the contents of some landfills), where if a single-layer closure were attempted either failure of the clay seal and overlying soil would occur or very large surface deformations, the "bathtub effect," would result.

The primary consideration in dealing with closures of soft landfills should be to stabilize the landfills. That can be done by mixing their contents with more stable in situ or off site materials—gravel, sandy or silty soils, fly ash, incinerator residue, lime, cement, etc. When a minimum unconfined compressive strength of 500 lb/ft^2 is reached, a geotextile should be used to "span" irregularities in mixing at the site (unless the site is very small). This geotextile should have low creep characteristics[29,30] and a strength and modulus compatible with the landfill characteristics and geometry. It must be securely anchored around the perimeter of the site and its joints must be sewn so as to achieve at least 80% of the strength of the geotextile fabric itself. As it interfaces directly with the landfill, it must be chemically compatible with it. A geomembrane should be placed over the geotextile to provide a secondary seal protecting against any infiltration of surface liquids. It can be tied into the same anchorage trench as the lower geotextile layer. Depending on the strength of the geotextile vis-à-vis the landfill, the geomembrane may require a reinforcing scrim. The clay seal, placed over a soil "working layer" that should go above the geomembrane, provides the primary liner, but it must be as thin as possible so as not to overload the geotextile-geomembrane-reinforced landfill. Over the clay seal should be placed a second geotextile (in preference to a sand-gravel drainage layer, which would weigh more). This geotextile should be needle-punched and nonwoven or a composite system, with high transmissivity characteristics.[28] It also should be sewn at its joints, rather than overlapped, so it can add additional reinforcement to the entire system. The uppermost soil layer and topsoil serve the same functions as they do in the single-layer closure system.

Certainly these closure systems are not inexpensive, but it seems as though their costs are the price we must pay for the careless and reckless manner in which we have disposed of our waste products in the past.

8.7 SEEPAGE CUTOFF WALLS

Existing landfills that actively generate leachates which flow into the groundwater create a particularly perplexing problem. The first line of defense against such a situation (assuming that the landfill cannot be removed, relocated, or treated in place) is to cover it so that surface water cannot enter it and generate additional leachate. The landfill caps described in Sec. 8.6 addressed this part of the problem. Leachate generated within capped landfills and escaping laterally or into the soil below are addressed in this section.

Regarding lateral flow of leachates out of a landfill, the construction of a seepage cutoff wall is an economical and technically feasible solution. Such cutoff walls come in a number of forms:

Interconnected sheet piling (wood, concrete, or steel)
Overlapping columns of grouted soil (usually using chemical grouts)
Relatively thin chemical walls formed by jet grouting
Slurry-constructed walls filled with soil-bentonite, cement-bentonite, or a
    geomembrane

The construction methods for these were all discussed in Chaps. 3 and 6, and only a few aspects need further elaboration here.

Most seepage cutoff walls seem to be constructed by the slurry method and are backfilled with a soil-bentonite mixture. The obvious question arises as to what permeability results when bentonite is added to the differing types of soil that might be available for use. And the obvious answer is that testing is required; see Refs. 31

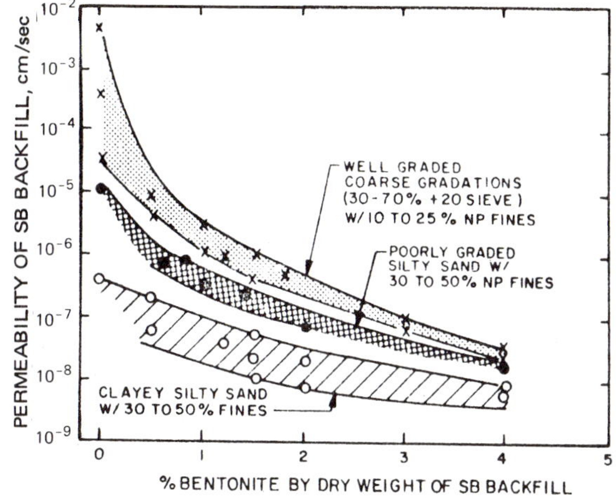

Figure 8.16 Relationship between weight percent of bentonite added to various soils and their resulting permeabilities. *(After D'Appolonia, Ref. 32.)*

and 32 for information about such testing. Figure 8.16 shows that additions of bentonite to soil amounting to 4 percent by weight are usually sufficient to reduce the permeability of the soil to less than 10^{-7} cm/s.

The method of mixing the bentonite and soil is critically important so that a homogeneous blend is produced for placement in the trench that will contain no high-permeability zones. And certainly the chemical compatibility of the leachate with the bentonite and soil must be established. Tests of the type described by Anderson et al.[16,17] are required in this regard. Adaptations of those same tests can be used to determine the hydraulic gradient at which failure will occur,[31,32] although that is usually only a problem in regard to very deep cutoff walls.

The depth of the cutoff trench must be considered, and in those cases where an impermeable stratum (called an *aquaclude*) exists, an adequate key can usually be established; see Fig. 8.17a. When rock is encountered and it is sound (Fig. 8.17b), packages of dry bentonite can be placed at the bottom of the trench which, when exposed to water, will swell considerably. Alternatively, and if the rock is fractured, post-seepage-wall-construction grouting may be necessary to obtain an adequate seal; see Fig. 8.17c. Finally, the case of no available aquaclude for a considerable distance (and an obviously low water table) requires a very deep cutoff wall; see Fig. 8.17d. Here flow net analyses are required to estimate the depth at which seepage loss reaches a minimum; see Ref. 33.

Regarding leaking landfills whose locales have no impermeable layers beneath them but do have high water tables, the situation is bleak. One possible remedy is to create a bottom using a chemical grout and the jet grouting method described in Chap. 6. Here a grout pipe with a horizontal hole near its bottom is inserted into the earth at predesignated centers throughout the landfill. When the pipe reaches the bottom of the landfill, it is rotated 360°, thus forming horizontal disks of chemically stabilized material. The elevation control of adjacent disks is critical so that no open space is left between them, and a considerable overlap is required. A major disad-

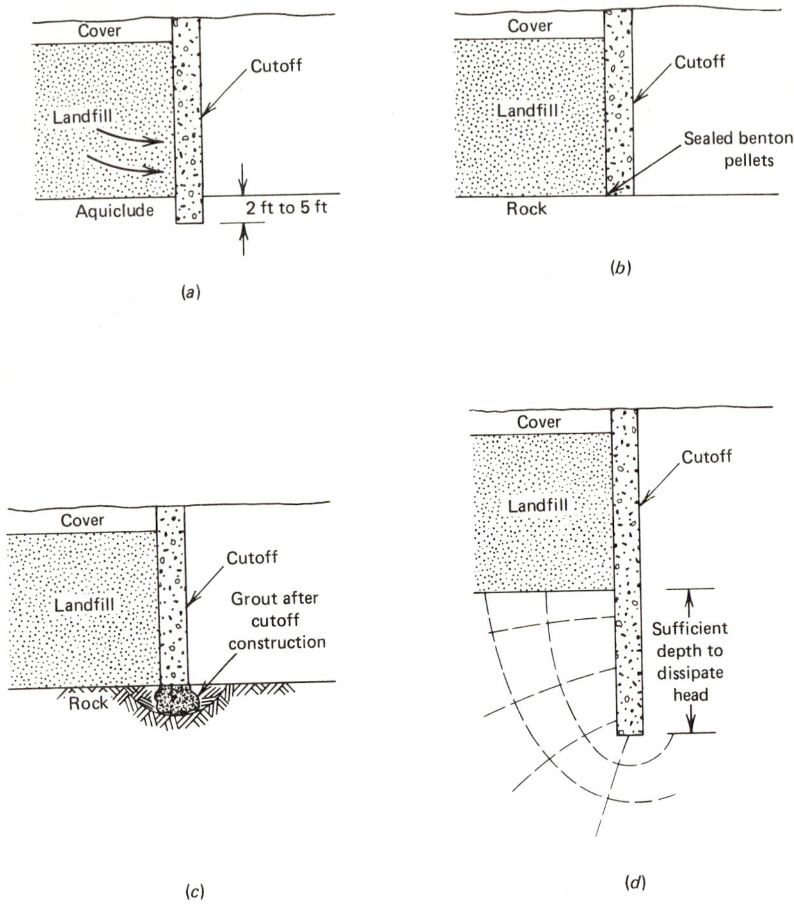

Figure 8.17 Various types of slurry trench keys and suggested depths. (*a*) Key in aquaclude. (*b*) Key in firm rock. (*c*) Key in fractured rock. (*d*) No available aquaclude.

vantage of this method is that the landfill itself must be pierced by the grout pipes, which can pose a hazard to workers and the environment alike. But clearly this particular area of in situ formation of seepage cutoff barriers beneath existing landfills is a fruitful subject for research and development; see, for example, Ref. 34.

PROBLEMS

8.1. Investigate the method used for solid waste disposal in your neighborhood and write a brief report about it. In particular, discuss the method in terms of the categories given in Table 8.1.

8.2. Investigate the method of sewage treatment in your neighborhood and in particular how the residual solid material (i.e., the sludge) is disposed of.

8.3. Investigate the heavy industries in your neighborhood and determine if they have waste treatment facilities, if they store liquid wastes, if they ship out bulk waste, etc. Pay particular attention to the industries mentioned in Table 8.2.

8.4. If a waste is classified "hazardous," as per Fig. 8.1, it must be properly disposed of in a safe and environmentally secure manner. Determine for your neighborhood where the closest such facility is located. (*Note:* This will require outside investigation and perhaps a call to your regional EPA office.)

8.5. Landfill liners can fail through a number of possible mechanisms. The most common of those mechanisms were discussed in Sec. 8.5. Show sketches of each illustrating the phenomena involved.

8.6. Calculate the anticipated seepage loss in gallons per day and in gallons per day per square foot of liner from a 10-acre reservoir that is 40 ft deep using Kays' data shown in Table 8.5 for liners made out of the following materials:
   (*a*) Loose earth plus bentonite
   (*b*) Compacted earth
   (*c*) Geomembrane

8.7. Recalculate the answer to Example 8.1 Sec. 8.5 for a sandy silt having a permeability 3×10^{-3} cm/s, and comment on the velocity and flow time results.

8.8. (*a*) Calculate the required thickness t of a clay liner whose permeability $k = 10^{-7}$ cm/s for a landfill whose leachate will have an anticipated seepage rate of q through the liner of 0.01 gallons per day per square foot of liner as the leachate heads (depths) vary from 1 to 20 ft. Plot the thicknesses against leachate heads in a graph, as shown in Fig. P8.8.
   (*b*) Repeat the problem for clay liners with permeabilities of 10^{-8} and 10^{-9} cm/s, plotting those results as well.

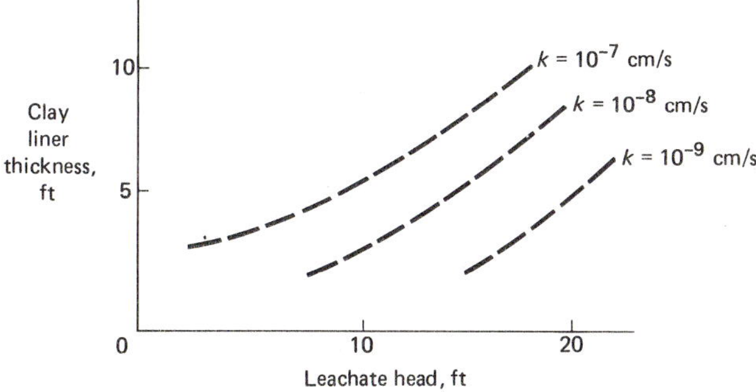

Figure P8.8

   (*c*) Redo your calculations for a clay liner having a permeability of 10^{-7} cm/s and leachate with a head of 10 ft, but now vary the anticipated seepage rate through the liner from 10 to 0.001 gallons per day per square foot of liner (recall Prob. 8.6) and plot the results on a semi-log axis. Each seepage rate plotted should equal one-tenth of the previous one.

8.9. What are the reasons for placing a geotextile beneath a geomembrane when one is dealing with double-liner systems as shown in Fig. 8.8*b*?

8.10. (*a*) If the foundation soil beneath a geomembrane-lined landfill is compressible and the long-term settlement is expected to be about 5 ft, as shown in Fig. P8.10, what strain will be mobilized in a firmly anchored geomembrane (neglect the friction generated between geomembrane and soil).

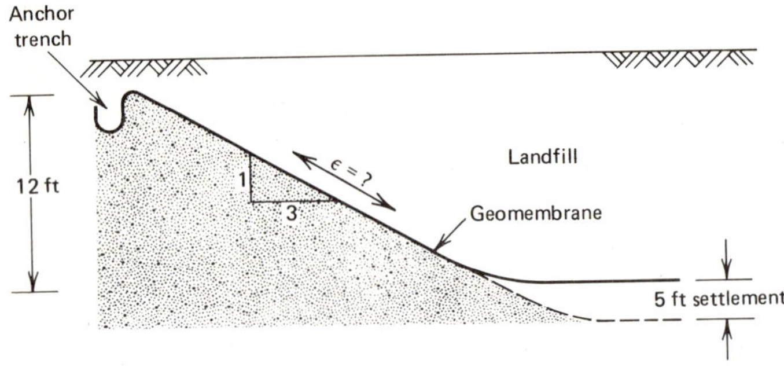

Figure P8.10

(*b*) Could the 10 × 10 reinforced 36-mil CPE membrane whose stress-versus-strain response is shown in Fig. 8.9 be used in this situation without failing, or would the unreinforced 30-mil CPE have to be used?

(*c*) Considering the relatively large anticipated settlement in this case, what are the implications of using the reinforced rather than the unreinforced lining?

8.11 (*a*) When testing the strength of geomembrane materials, the stress is usually expressed in units of pounds per inch width of material. How do you convert this to a true stress unit of proper dimensions, i.e., pounds per square inch?

(*b*) What is the difference between a tangent modulus of elasticity and a secant modulus of elasticity?

(*c*) Check the tangent modulus values for the six geomembranes shown in Figs. 8.9 and 8.10.

(*d*) Compare those tangent modulus values to the secant moduli at 10 percent strain for the reinforced membranes and at 30 percent strain for the unreinforced membranes.

(*e*) Why are the stress-versus-strain response curves of the scrim-reinforced geomembranes irregular (i.e., why do they jump up and down) beyond their peak strength and then eventually smooth out?

8.12. A series of accelerated aging tests have been run on a group of candidate geomembranes for a period of 1 year. The control test being run is a diffusion test whereby gas, under high pressure, is forced through the geomembrane. The resulting diffusion data (in units of square centimeters per second) is then divided by the geomembrane thickness to get the usual "average" permeability value. Based on a minimally acceptable threshold permeability value of 10^{-7} cm/s, what is the anticipated service life of each geomembrane listed in the accompanying table?

Geomembrane	Thickness, mils	Diffusion cm^2/s				
		As received	10 days	32 days	90 days	350 days
HDPE	20	0.10×10^{-12}	0.32×10^{-12}	0.053×10^{-11}	0.105×10^{-11}	0.34×10^{-10}
CPE-OR	40	0.62×10^{-13}	0.21×10^{-12}	0.25×10^{-11}	0.80×10^{-11}	0.85×10^{-10}
EPDM	30	0.11×10^{-10}	0.35×10^{-10}	0.55×10^{-10}	0.084×10^{-9}	0.42×10^{-9}
PVC	40	0.68×10^{-12}	0.35×10^{-11}	0.62×10^{-10}	0.29×10^{-9}	0.81×10^{-8}

8.13. Describe in detail, and using sketches, a diffusion test which can be used to determine an "average" coefficient of permeability of a geomembrane. Note that in the as received condition geomembranes have permeabilities on the order of 10^{-9} to 10^{-12} cm/s, which is much too low to detect using standard geotechnical engineering variable-head permeability tests. (*Hint:* Look in the literature on polymers and plastics for comparable tests.)

8.14. (*a*) Using Eq. (8.4), calculate the maximum reservoir leakage in gallons per day that can be tolerated from a 10-acre water reservoir that averages 40 ft deep.

(*b*) Compare your result with those found in Prob. 8.6 and determine if a compacted earth liner is acceptable.

(*c*) Would a geomembrane liner be acceptable under the same conditions?

8.15. (*a*) For a soil-bentonite slurry-constructed seepage cutoff wall, as shown in Fig. P8.15 as well as in 8.17*d*, what is the flow beneath the wall assuming saturated conditions throughout. (Although analytic formulas are available for this solution, the drawing of a seepage flow net will produce reasonably accurate results.)

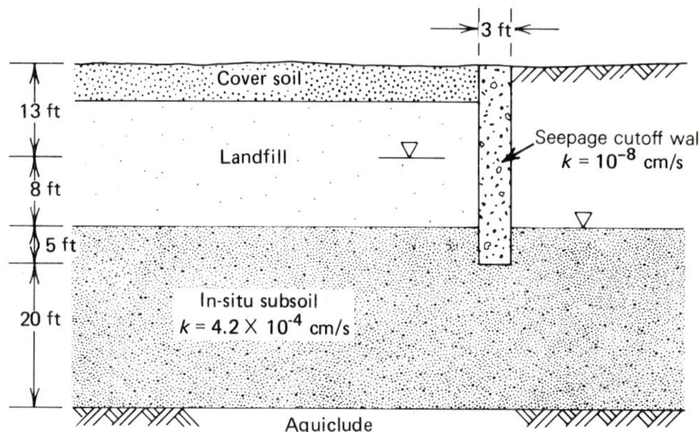

Figure P8.15

(*b*) Recalculate your answer for cutoff wall depths of 10 and 15 ft.

(*c*) What would be the flow if the seepage cutoff wall went down into the aquaclude?

(*d*) Plot on a graph the cutoff wall depth (beneath the bottom of the landfill) versus your calculated values of flow.

REFERENCES

1. Franklin Assoc., Ltd., *Post-Consumer Solid Waste and Resource Recovery Baseline,* Resource Conservation Committee, Washington, DC, 1979.
2. Newton, M., *Model State Hazardous Waste Management Act,* U.S. EPAOSW Report No. SW-635, Washington, DC, 1977.
3. ———, *State Decision Makers Guide to Hazardous Waste Management,* U. S. EPAOSW Report No. SW-612, Washington, DC, 1977.
4. ———, *Final Report, Program for the Management of Hazardous Wastes,* Batelle Memorial Institute, Richland, WA, 1973.
5. Tchobanoglous, G., H. Theisen, and R. Eliassen, *Solid Waste,* McGraw-Hill, New York, 1977.
6. Vesilind, P. A., and A.E. Rimer, *Unit Operations in Resource Recovery Engineering,* Prentice-Hall, Englewood Cliffs, NJ, 1981.
7. Brunner, D.R., and D. J. Keller, *Sanitary Landfill Design and Operation,* U.S. EPAOSWMP Report No. SW-65 ts, Washington, DC, 1972.
8. Eliassen, R., "Refuse Collection and Disposal in American Civil Engineering Practice," in *American Civil Engineering Practice,* R. W. Abbett (ed.), J. Wiley and Sons, New York, 1956.

9. Chian, E. S. K., and F. B. DeWalle, "Sanitary Landfill Leachates and Their Treatment", *J. Environ. Eng. Div., ASCE,* Vol. 102, No. EE2, April 1976, pp 411–431.

10. Hwang, J., and R. M. Koerner, "Location of Groundwater Pollution Sources from Limited Field Data," in *Proc. Nat. Conf. on Mgmt. of Uncontrolled Hazardous Waste Substances,* Hazardous Materials Control Research Institute, Washington, D.C., Oct 28–30, 1981, pp. 317–321.

11. Kays, W. B., *Construction of Lining for Reservoirs, Tanks, and Pollution Control Facilities,* Wiley, New York, 1977.

12. Cedegren, H. R., *Seepage, Drainage and Flow Nets,* Wiley, New York, 1967.

13. Koerner, R. M. and T. A. Okrasinski, "Erosion Control of Granular Soils Using PVA," *J. Constn. Div., ASCE,* Vol. 104, No. CO3. September 1978, pp. 279–294.

14. Grim, R. E., *Clay Mineralogy,* 2d ed., McGraw-Hill, New York, 1968.

15. Mitchell, J. K., *Fundamentals of Soil Behavior,* Wiley, New York, 1976.

16. Anderson, D. C., K. W. Brown, and J. Green, "Organic Leachate Effects on the Permeability of Clay Liners," in Ref. 10, pp. 223–229.

17. Anderson, D., "Does Landfill Leachate Make Clay Liners More Permeable?" *Civil Eng., ASCE,* September 1982, pp. 66–69.

18. EPA, "Hazardous Waste Management System; Permitting Requirements for Land Disposal Facilities, Part 11," *Federal Register,* July 26, 1982, Vol. 47, No. 143, pp. 32,274–32,388.

19. Chuck, R. T., "Largest Butyl Rubber Lined Reservoir," *Civil Eng. ASCE,* May 1970, pp. 44–47.

20. "Membrane Liner Seals 278—Acre Reservoir," *Public Works Mag.,* February 1981.

21. Ewald, G. W. "Stretching the Lifespan of Synthetic Pond-Linings," *Chem. Eng.,* October 1, 1973, pp. 67–70.

22. Schmidt, R. K., and P. W. Barker, "Selection Criteria for the Use of Synthetic Membrane Liners in the Containment of Hazardous Wastes," in *Proc. Fifty-Fourth Water Pollution Control Federation Conf.,* October 1981.

23. Haxo, H. E., Jr., "Effects on Liner Materials of Long-Term Exposure in Waste Environment," in *Proc. Eighth Symp. on Land Disposal of Hazardous Waste,* Cincinnati, Ohio, EPA Publication No. 600/9-82-002, March 1982, pp. 191–211.

24. Haxo, H. E., Jr., et al., *Liner Materials Exposed to Municipal Solid Waste Leachate,* Final report, and *Liner Materials Exposed to Hazardous and Toxic Wastes,* Final report, both to U.S. EPA, Cincinnati, in preparation.

25. Kumar, J., and J. A. Jedlicka, "Selecting and Installing Synthetic Pond Linings," *Chem. Eng.,* February 5, 1973, p. 67.

26. Montague, P., "Hazardous Waste Landfills: Some Lessons from New Jersey," *Civil Eng., ASCE,* September 1982, pp. 53–56.

27. Touhill, C. J., A. J. Shockrow, and A. P. Pajak, "Hazardous Waste Management of Abandoned Dump Sites—Evolving Perspectives," *J. Hazardous Mater.,* Vol. 6, 1982, pp. 261–265.

28. Koerner, R. M., and J. A. Bove, "In-Plane Hydraulic Properties of Geotextiles," accepted by ASTM *J. Geotech. Test.,* April, 1983.

29. Shrestha, S. C., and J. R. Bell, "Creep Behavior of Geotextiles Under Sustained Loads," in *Proc. Second Int. Conf. on Geotextiles,* Vol. 3, Las Vegas, August 1–6, 1982, pp. 769–774.

30. Hwu, B. L., "Sustained Load (Creep) Characteristics of a Wide Range of Geotextiles and Geomembranes," Master's thesis, Drexel University, Philadelphia, June 1983.

31. Xanthakos, P. P., *Slurry Walls,* McGraw-Hill, New York 1979.

32. D'Appolonia, D. J., "Soil-Bentonite Slurry Trench Cutoffs," *J. Geotech. Eng. Div., ASCE,* Vol. 106, No. GT4, April 1980, pp. 399–417.

33. Stefan, H., and J. Meyer, "Seepage Study Using Electrical Analog," *J. Hydraulics Div., ASCE,* Vol. 96, No. HY2, February 1970, pp. 411–416.

34. Brunsing, T. P., W. E. Grube, Jr., and W. J. Davis-Hoover, "A Block Displacement Technique to Isolate Uncontrolled Hazardous Waste Sites," in *Proc. Conf. on Mgmt of Uncontrolled Hazardous Waste Sites,* Hazardous Materials Control Research Inst. Washington, D.C., November 29–December 1, 1982, pp. 249–253.

FOUNDATION CONTROL TECHNIQUES

9.1 OVERVIEW

For any construction project, irrespective of its size, a certain amount of knowledge about the details of subsurface soil conditions is required. Traditionally, this knowledge has come from information gathered during a subsurface soil boring and/or rock coring test program. The general types of exploratory methods were presented in Table 2.16, where the procedures they required, their uses, and their limitations were described. The goal there was to illustrate how one develops a "fence diagram," which is actually a section or profile of the subsurface soil and rock at a given site. See, for example, Figs. 2.14 through 2.16, which show such diagrams for projects of gradually increasing size and significance.

Irrespective of the number, size, and location of borings taken for a project, however, a certain amount of conjecture and speculation about subsurface detail cannot be avoided. Indeed, if the design and construction of foundation systems were so straightforward and simplistic as to have no uncertainty, any untrained novice could do the work. That not being the case, however, the professionals involved must have as adequate a knowledge as possible of factors that can affect design and construction techniques, and such knowledge usually comes from the testing and assessment of previously completed projects and the project under current consideration. Certainly this type of feedback produces the stepping stones from which future progress can be made.

This chapter will cover three distinct areas of testing: laboratory methods, traditional field testing methods, and nondestructive testing (NDT) methods. A discussion of these topics should help us understand the strengths and weaknesses of various techniques described earlier in the book, and thus also to gain a perspective on the reliability and sensitivity of those techniques.

This chapter will also cover the instrumentation needed for performance monitoring in relation to three time frames: during construction, in the short term after construction, and in the long term after construction. This is a particularly strong area of current activity, and should see considerable growth in the future.

9.2 LABORATORY TESTING

The heart of geotechnical engineering since its very inception has been the attempt to test representative soil samples from project sites and use the resulting information in the design of those projects. Obviously the reconstitution of the soil after receiving the samples in the laboratory should simulate in situ and future stresses as closely as possible. The importance of laboratory testing is so universally accepted that every institution which teaches geotechnical engineering has a soil testing laboratory associated with it, as does almost every geotechnical consulting firm. The results obtained from laboratory tests generate the basic design information in terms of the following categories:

1. Physical properties:
    a. Mineral identification
    b. Density
    c. Density limits
    d. Water content
    e. Gradation characteristics
    f. Particle shape characteristics
    g. Consistency limits
2. Hydraulic properties:
    a. Permeability (hydraulic conductivity)
    b. Piping potential
    c. Characteristics of pore fluid (chemistry)
3. Mechanical properties:
    a. Compaction limits
    b. Consolidation properties
    c. Elastic constants
    d. Shear strength (in all of its many varieties and as shown by the different ways of testing it)

Out of this maze of information determined from laboratory testing, it is possible to characterize and identify the different soils and their properties at a given site so as to take them into account when designing a project. (Much of the soil information presented in Appendix B was obtained by carefully planned sequential laboratory testing programs.) Many manuals and texts on laboratory testing are available—see, for example, Refs. 1 through 3. Since most of the tests are standardized, the appropriate ASTM procedures for carrying them out should be used.

Questions of isotropy and nonhomogeneity constantly creep into any discussion about the value and validity of laboratory soil testing. In other words, how representative of the full-scale field situation is the specimen (or are the specimens) being tested? The entire methodology of undisturbed sampling, encapsulation, and capping of the specimens, their transportation to the laboratory, their withdrawal and retrieval from the specimen containers, preparing them for testing, and finally reconstituting the original (soil stresses and pore water pressures) conditions at the site from which they were removed, must be taken into account in the context of the test values obtained. In areas of complex stratigraphy the situation becomes even more problematic since the number of samples and test specimens that can be removed from a site is always limited. Thus it is necessary to average the test results to determine the final design values.

All of this might seem to present a quite bleak picture of laboratory testing and its usefulness, yet it is not meant to serve in any way as a total condemnation of the method. It has served the profession well over the years and certainly will continue to do so within its limits for many more years to come.

What will also happen, however, is that other forms of soil testing will emerge to either replace those laboratory tests which are not satisfactory, or to augment the data obtained through such methods. And those new forms of soil testing fall into the category of field testing, both traditional and nondestructive, which are the subjects of Secs. 9.3 and 9.4.

9.3 TRADITIONAL FIELD TESTING

Field testing of soils and rocks for their intrinsic compressibility and strength parameters has a number of inherent advantages over the laboratory testing methods just discussed, but also some disadvantages. The advantages:

No sample need be removed from its natural state; thus the soil or rock stresses and
 pore water pressures involved should be representative of in situ conditions.
Total time needed for testing and evaluating the test results is short relative to that
 required for the laboratory testing sequence.
There is an immediate feedback of test results, a particular advantage in construction
 control monitoring.
Changes in test programs, or in the number or location of tests, can be made on the
 basis of information gained while the field work is in progress. This can be very
 cost-effective.

The disadvantages:

The advancement of boreholes creates some soil and rock disturbance, although
 nowhere near the amount created by the standard sampling methods required
 for laboratory testing.
Generally, no sample is recovered for visual identification or classification.
Flexibility of stress state imposition is limited to the particular device being used.

Test operators must be trained in regard to the particular test being conducted. These are generally new skills which must be taught to the field personnel.

The main field tests used to determine intrinsic compressibility and strength parameters are the standard penetration, vane shear, cone penetration, and pressuremeter tests.

9.3.1 Standard Penetration Test (SPT)

This is the bulwark of soil field testing. So much has been written about it, its advantages, and disadvantages, that a concise review borders on the impossible. See, for example, Ref. 4 and the number of discussions that this one paper generated. The test itself is a dynamic one in that a sampling spoon is placed in an open or cased hole and driven into the soil under investigation. The number of blows required to drive the casing 1 ft using a standard-size hammer falling a standard distance (called the *blow count,* or simply *N*) has been empirically correlated to different soil types. Table 9.1 presents common variations on this testing method and the soil properties determined from their results. The second procedure listed, using a 140-lb hammer falling 30 in, is quite commonly used, and the equipment it requires is standard for most North American drillers. In fact, blow counting on the sampling spoon, as well as on the casing, is standard operating procedure for drillers, and *the information it produces about soil at the site being drilled comes absolutely free* as part of a typical drilling contract. This, as well as the fact that the sampling spoon brings up a soil sample from the subsurface location at which the blow count was obtained, is a powerful impetus to continue using the method and to examine its further capabilities. This research is already in progress, as a number of papers presented during a conference at North Carolina State University in 1973 indicates.[6] Blow count has been correlated to many engineering properties, such as modulus of elasticity, cohesion in fine-grained soils, and friction angle in granular soils; see Appendix B for some typical response curves.

9.3.2 Vane Shear Test

This is used in evaluating the shearing resistance of cohesive soils. It was originally developed for use in sensitive clays which are difficult to sample. The procedure is to drive a casing down to the desired subsurface elevation, remove the soil from within the casing, and then insert a vane on a shaft into the soil below the casing. A typical vane consists of two 2-in-wide, 6-in-long rectangular plates at right angles to each other. Once the vane is pushed into the undisturbed soil below the casing, the upper end of the shaft is fitted to a calibrated torque spring setup and the entire rod then rotated by hand or motor until the soil offers no resistance to its turning (or torque). At this point, i.e., at failure, the slip surface is a right circular cylinder of soil whose height and diameter are the same as those of the vane itself. The shear strength of

Table 9.1 Variations in standard penetration tests†

Sampler	1-in extra-heavy pipe 1.315-in. OD, 0.957-in ID		Raymond 2.0-in OD, 1.375-in ID		2.50-in OD		3.00-in OD	
Hammer	± 140 lb, ± 30-in fall		140 lb, 30-in fall		300 lb, 18-in fall		300 lb, 18-in fall	
Soil	Designation	Blows per ft	Designation	Blows per ft	Designation	Blows per ft	Designation	Blows per ft
Sand and silt, relative density	Loose	Less than 9	Very loose	Less than 4	Loose	0–15	Very loose	Less than 8
	Firm	9–13	Loose	4–10	Compact	16–50	Loose	8–16
	Hard	14–49	Medium	10–30	Very compact	Over 50	Medium	16–55
	Hardpan	Over 50	Dense	30–50			Compact	55–110
			Very dense	Over 50			Very compact	Over 110
Clay, consistency	Soft	Less than 5	Very soft	Less than 2	Very soft	0–2	Very soft	Less than 8
	Medium		Soft	2–4	Soft	3–10	Soft	8–16
	Hard	11–30	Medium	4–8	Stiff	11–30	Medium stiff	16–55
			Stiff	8–15	Hard	Over 30	Stiff to medium hard	55–110
			Very stiff	15–30			Very hard	Over 110
			Hard	Over 30				

†After Hvorslev, Ref. 5.

the soil is found as follows:

$$\tau = \frac{6T}{\pi D^2(3H + D)} \tag{9.1}$$

where τ = shear strength of soil, lb/in^2
 T = limiting value of torque, in·lb
 D = width of the vane, in
 H = height of the vane, in

If this response is plotted, a curve similar to those shown on Fig. 9.1 results. In Fig. 9.1, characteristic remolded and undisturbed shear strength responses of typical clay soils are shown. In either case, the maximum shear strength equals the cohesion of the soil if the clay is undrained. If granular particles or organic material are present, however, the test value loses its technical significance, i.e., the shear plane remains undefined and there is no way to separate out c and ϕ from the shear strength determined by the test.

The vane shear test is not widely used in the United States. Many papers discussing this method and the cone penetration test are available in Ref. 7.

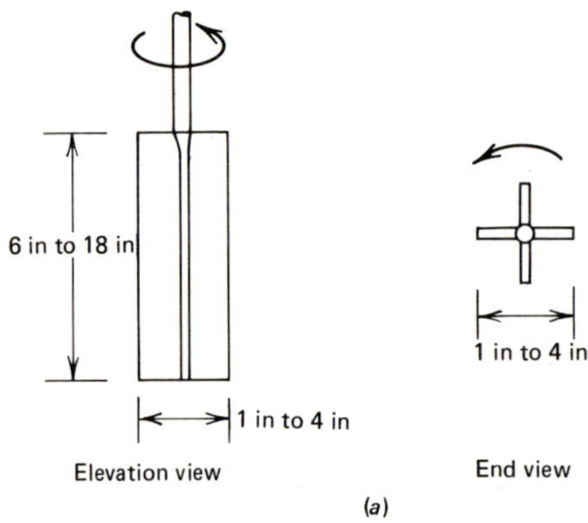

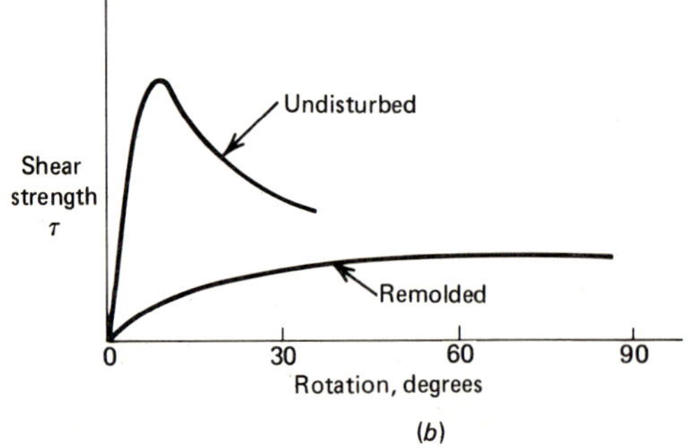

Figure 9.1 Details of vane shear test for obtaining data on in situ shear strength of cohesive soils. (*a*) Typical details of vane. (*b*) Typical response curves.

9.3.3 Cone Penetration Test

This consists of vertically advancing a pointed cone on the end of a stiff drill rod down into the soil. Advancement can be done statically or dynamically, and many kinds of hardware can be used. The cone point, however, is usually set at either 60° (for granular soils) or 45° (for fine-grained soils). Figure 9.2a to e shows some of the configurations that have been used. The data obtained can be empirically correlated to soil strength as shown in Fig. 9.2f. Strata changes are easily noted by means of this test, but identification must come from individual, and separate, borings. This type of test is very useful in field stabilization work where "soundings" of the site before and after soil modification can show the degree of success of the stabilization method being used. (Various stabilization and densification methods were described in Chaps. 4 and 5.) It is rapid, easy to deploy, and reasonably easy to interpret when used for such applications. To use this technique alone for the obtaining of soil shear strength parameters is somewhat more difficult, but it is certainly possible to do so since the stress state and failure conditions it reveals are quite clearly defined.

9.3.4 Pressuremeter Test

This is more involved than the three tests just described, but provides considerably more information. It can be used in soil, soft rock, and even hard rock, and consists of inserting a torpedolike probe into an uncased borehole or rock core boring and then inflating, or jacking, the expandable probe against the sides of the hole. Figure

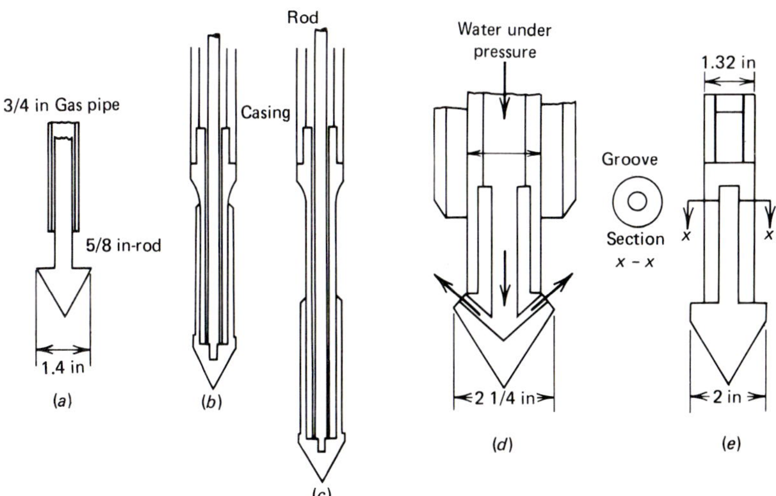

Figure 9.2 Various types of cone penetrometers, and a typical response curve. (a) Original Dutch cone. (b) and (c) Refined Dutch cone with point retracted while advancing casing and with point extended after measurement of resistance. (d) Wash-point penetrometer. (e) Conical drive point. (f) Typical cone response curve.

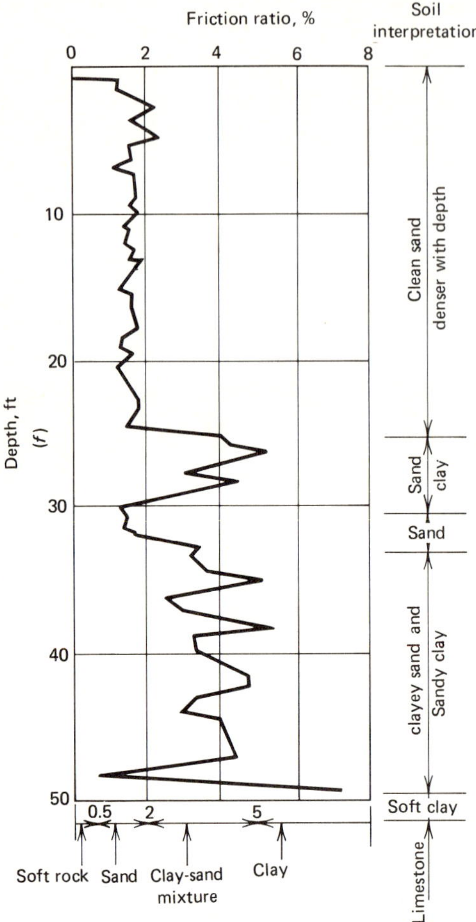

Figure 9.2 (*Continued*)

9.3*a* shows how the technique is carried out in soil using an inflatable rubber bladder as a probe, and Fig. 9.3*b* shows how it is carried out in rock using a hydraulically activated jack. While the test is being performed, pressure is monitored along with volume change (in soil) or borehole deformation (in rock). Obviously the stress is being applied through radial expansion of the borehole. Typical response data for soil and rock are shown in Fig. 9.3*c* and *d*. The moduli of elasticity E and failure pressures p_f can be obtained from those curves. This is an interesting, and somewhat challenging, test to perform, and one which will be used in the future for high-quality geotechnical work. A recent adaptation of it is known as a *self-boring pressuremeter;* see Ref. 8.

Some other types of in situ shear strength testing methods not discussed here (but in Ref. 8 and 9) are the following:

Borehole shear test
Plate-bearing-load test

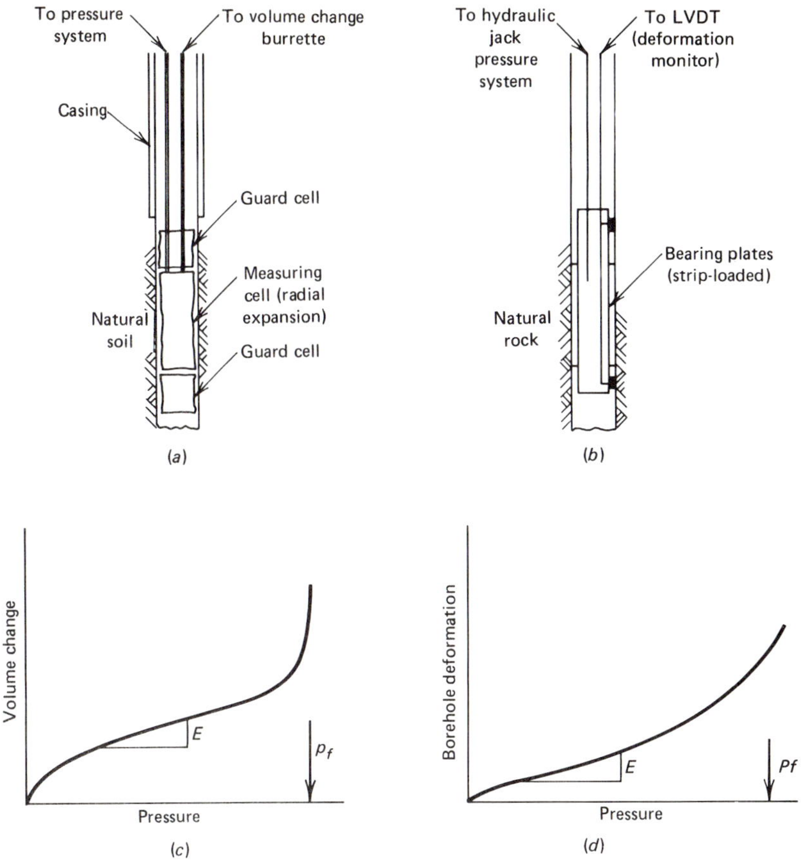

Figure 9.3 Schematic diagrams of pressuremeters and typical response curves. (*a*) Schematic of soil pressuremeter. (*b*) Schematic of rock jack. (*c*) Typical soil pressuremeter response. (*d*) Typical rock-jack response.

Field direct-shear test
Hydraulic fracturing test
Acoustic pressuremeter test
Flat-plate dilatometer

9.4 NONDESTRUCTIVE TESTING (NDT), INCLUDING GEOPHYSICS TECHNOLOGY

The long-established technique of taking borings which are then tested in situ or in a laboratory is excellent for analyzing the subsurface conditions prevailing at the specific location of the boring. However, when one wants detailed information about subsurface conditions between boring sites, problems arise. One could take additional borings, but that would be costly and time-consuming. How then is one to gain the desired information relatively easily, taking into account the money and time constraints as well as any adverse factors that may be encountered in terms of the site's depth and accessibility, the presence of surface and subsurface water, and so forth? Obviously, what is needed are suitable nondestructive testing (NDT) methods. Such methods will probably never completely replace the need for borings, but they could mean a reduction in the number of borings needed at any given site, thus producing large savings in money and time.

Unfortunately, the use of NDT methods in geotechnical and construction engineering has not grown rapidly. They are an outgrowth of the general area of geophysics technology which is used extensively by the oil, gas, and minerals exploration industries. However, the subsurface depth that such technology explores and the general lack of specific informational detail it produces, has made the transfer of geophysics technology to geotechnical and construction engineering difficult. Nevertheless such a "close-in" technology has been developed. Indeed, the candidate methods are many, and we have selected only those which have been used to some extent and are beyond the stage of laboratory feasibility studies in their development. Because of space limitations, we will discuss each method only briefly. For further information, books by Griffith and King,[10] Grand and West,[11] Sharma,[12] and Dobrin[13] can be consulted; those works give a general background to the use of the more traditional geophysical probing techniques. The geophysical techniques are covered in Secs. 9.3.1 through 9.3.4. In regard to the newer NDT methods, covered in Secs. 9.3.5 through 9.3.11, there are two organizations taking a leadership role in standardizing and exploring their capabilities—the American Society for Testing and Materials (ASTM) and the American Society for Nondestructive Testing (ASNT). References to work being done on those newer methods will be cited at the appropriate points in our discussions of them.

9.4.1 Seismic Reflection

This standard geophysical method utilizes an impulse (a hammer blow, an explosion, etc.) which is generated in the ground at a given point, either at the surface or below it. The elastic (seismic) waves emanating from this source are picked up via a series of transducers, called *geophones,* on the surface. If a well-defined layer of soil or rock (or layers) is (are) present, there will be a characteristic series of return echoes. The travel time t of the wave (down into the soil interface and back up) along with the velocity of propagation v of the wave through the soil gives the depth d of the interface, i.e., $d = vt/2$. It is also possible that buried objects could be picked up in this way by analyzing the echo response received as the transducers are placed at different locations. However, the technique has not been widely used for shallow-depth exploration.

9.4.2 Seismic Refraction

In a related geophysical method called refraction surveying, an impulse is generated in the ground and the time it takes the echoes to reach a transducer is measured for varying transducer-to-impact distances. The time is plotted as a function of the distance, and if a well-defined layer exists beneath the ground's surface, there will be a characteristic break in the response curve from which the depth of the layer can be determined. The method is good for determining subsurface stratigraphy but does need a relatively sharp boundary between the layers being surveyed for distinct defintion. Top-of-rock surveying, where a sharp interface between the rock surface and overburden exists, is an excellent example of such a situation.

9.4.3 Electrical Resistivity

This geophysical method applies current to the ground through electrodes and depends on the fact that any subsurface variation in conductivity alters the form of the current flow within the earth. This means the distribution of electric potential is affected. The degree to which the potential at the surface is affected depends on the size, shape, location, and electrical resistivity of the subsurface mass affecting it. It is therefore possible to obtain information about the subsurface distribution of various bodies from measurements of electric potential made at the surface. Four electrodes are used, two in an outer set and two in an inner set, and they are deployed in a straight line on the ground's surface. The outer set transmits the current, the inner set receives it. As the four electrodes are marched along the surface, resistivity readings are obtained, plotted, and interpreted on the basis of known responses. This interpretation is straightforward for simple topography, but can be difficult for complex subsurface conditions. The method has been used to great advantage in plotting out subsurface seepage plumes of contaminated liquids since the conductivity of many pollutants differs from that of normal groundwater.

9.4.4 Magnetometry

A *magnetometer* is a device which measures very minute changes in the earth's magnetic field. Any magnetic object, e.g., an iron ore deposit, buried steel, etc., will alter the earth's magnetic field locally and thus potentially can be detected. The magnetometer most commonly used today involves proton nuclear magnetic resonance. The nuclear spin of the proton can be made to flip a radio frequency (RF) field. When the radio frequency field and the dc magnetic field have the appropriate values, there is a maximum flip rate and, hence, maximum energy absorption. The frequency of the proton magnetometer is set in such a way that maximum absorption occurs under the normal magnetic field of the earth. If a magnetic object is encountered, the frequency goes out of tune, and the change needed to bring it back in tune is a very precise measure of the change in magnetic field caused by the magnetic object. Several types of magnetometers are commercially available.

The technique has been used for subsurface profiling, but is only effective when the magnetic properties of the various layers being probed are different, which is not usually the case. It can, however, be used in top-of-rock surveying or for locating buried magnetic objects, e.g., metal drums.

9.4.5 Ground-Probing Radar (GPR)

This method is similar in principle to seismic reflection except that it is electromagnetic waves (which are much faster than seismic waves) that are propagated into the soil.[14,15] The bulk of subsurface probing at shallow depths has been based on the transmitting of pulsed RF waves in the frequency range from about 1 to 900 MHz. A schematic diagram of the method is shown in Fig. 9.4. The transmitted pulse travels through soil, or water, until it meets up with an object or material having dissimilar electrical characteristics. Part of it is then reflected back to the ground's surface where it is received and its time of travel is measured. The depth d at which the interface is located is then calculated from $d = vt/2$, where v is the wave velocity (which is equal to $c/\sqrt{\epsilon_r}$, where c is the velocity of light and ϵ_r is the relative dielectric constant of the material in which the wave is propagating) and t is the pulse's travel time. The relative dielectric constants of numerous soils having different water contents have been evaluated by many investigators, including Okrasinski et al.[17]

By using a system which transmits RF pulses in the 100- to 200-MHz frequency range and then presenting the reflected signal in a real-time printout, a visual subsurface profile can be developed. See Fig. 9.5a for an example of such a profile, in which a region of sandy soil was probed and the area's water table was monitored

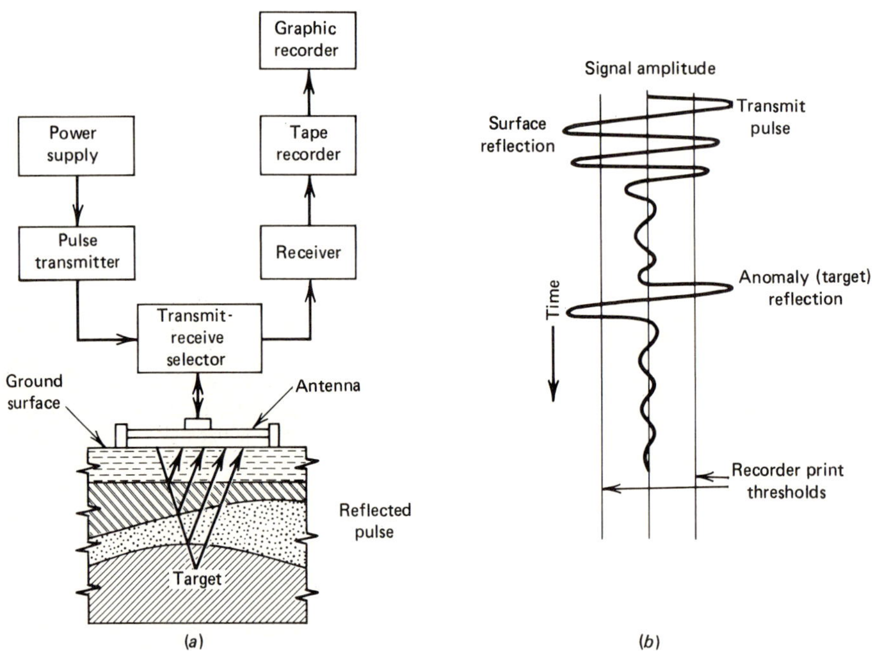

Figure 9.4 Schematic diagram and output of ground-probing radar system used to investigate subsurface anomalies. (*a*) General instrumentation schematic. (*b*) Oscilloscope output of waveform. (*After Morey, Ref. 16.*)

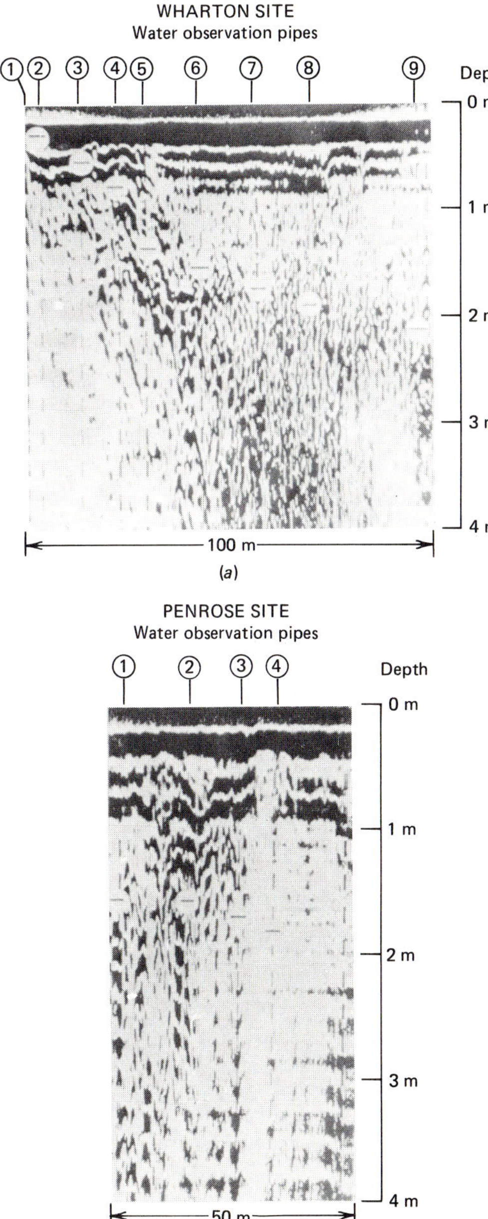

Figure 9.5 Examples of ground-probing radar printouts showing different subsurface conditions. (*a*) Pulsed RF survey of Wharton site with actual water table elevations indicated by horizontal line in circle. (*b*) Pulsed RF survey of Penrose site with actual water table elevations indicated by horizontal line in circle. (*c*) GPR printout showing steel drums of various sizes with 3.5 ft (1.1 m) of soil cover above. (*After Koerner et al., Ref. 14.*)

and verified; Fig. 9.5*b* shows a profile in which changes in soil stratigraphy are identified; and Fig. 9.5*c* shows a GPR printout indicating the location of buried steel drums of various sizes.[14,18]

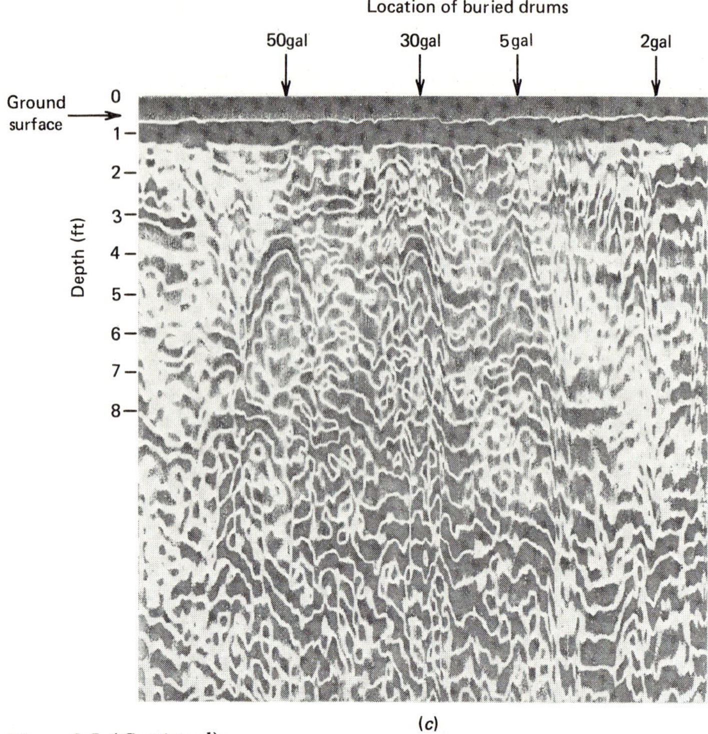

Figure 9.5 (*Continued*)

9.4.6 Continuous-Wave (CW) Microwave

The CW microwave technique is related to the pulsed GPR method except that a continuous wave is transmitted into the soil of the area being examined. The wave is swept in frequency over a relatively narrow range, sent into the soil by means of a transmitting antenna, and the wave reflection from the ground's surface and the wave from a subsurface reflection plane interfere with each other as sensed by the receiving antenna. The depth d to the reflecting interface can be determined from the spacing (in frequency) between interference maxima (or minima) as the frequency is swept:

$$d = \frac{c}{2n}\left(1 - \frac{1}{n^2}\cos^2\theta\right)^{-1/2}\left(\frac{x}{f_2 - f_1}\right) \tag{9.2}$$

where d = distance to reflecting surface, m

c = velocity of light, m/s

n = index of refraction of the material ($= \sqrt{\epsilon_r}$, the relative dielectric constant)

θ = angle of incidence, deg

x = number of peaks separating f_1 and f_2

f_1 = arbitrarily chosen initial frequency peak, Hz
f_2 = arbitrarily chosen final frequency peak, Hz

The strength of the interference is presented as a function of the lateral position of the antennas in order to map out subsurface detail. The research into some systems of this type is at an advanced stage, and they may be available for use shortly; see Koerner et al.[19]

9.4.7 Eddy Current Methods (Metal Detector and Inductance Methods)

These utilize a search coil with an alternating current running through it. This produces an ac magnetic field which induces eddy currents in subsurface metallic objects. These eddy currents in turn produce a magnetic moment in the metal which then interacts back (via mutual inductance) on the search coil. The size of this mutual inductance is read out, which gives an indication of the location and size of the metallic object(s) or layer if a traverse is being performed. This is the principle by which most metal detectors work. It is a good technique to use in the search for buried metal containers and has a penetration depth of up to 10 ft in dry soils.

The same principle has recently been used to construct a device for subsurface profiling, as shown in Fig. 9.6a. Called an *inductance meter,* it operates in a manner similar to instruments that utilize electrical resistivity, except that the profiling can be performed much more quickly and it is much easier to interpret the data it produces. The theory is explained by McNeill,[20] and typical results for mapping out a seepage plume and for locating buried metal drums are shown in Fig. 9.6b and c, respectively.

9.4.8 Pulse-Echo Ultrasonics

In this NDT method, a pulse of elastic energy is beamed into the material being investigated. The pulse is typically a few microseconds long and has a carrier frequency of about 1 MHz. The elastic wave is reflected back from anomalies and discontinuities within the material, and the nature of the reflected pattern gives an indication of the depth and the spatial extent of those anomalies and discontinuities. The method is well developed and there is a great deal of equipment available on the market. However, it is oriented to the study of metals, polymers, ceramics, and composites materials, and is not particularly well suited to the geotechnical and construction fields.

9.4.9 Acoustic Emission (AE)

This method (which has also been called a *passive ultrasonics technique*) makes use of the fact that a material under stress, which subsequently deforms, generates noise (called *acoustic emissions*). The noises are monitored by means of a transducer, which senses them and then amplifies, filters, counts, and/or records them as the investigator wishes. The total acoustic emission counts, or count rate, is then related to the stress causing the emissions in order to assess the material's integrity and

(a)

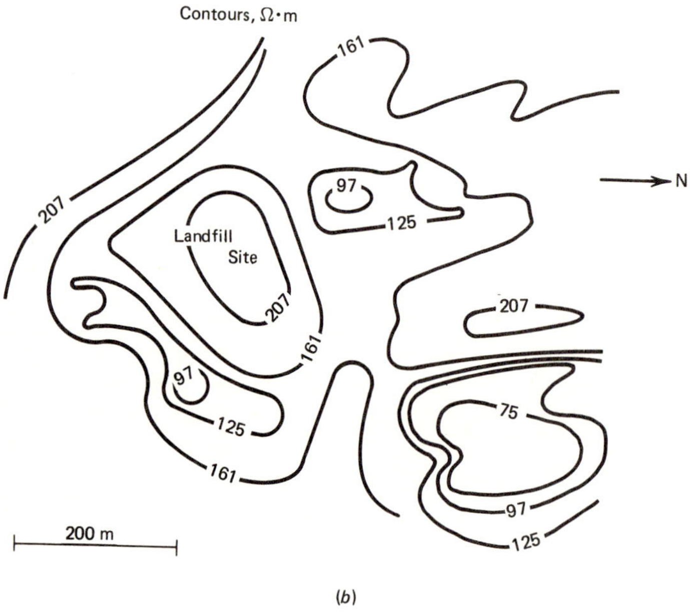

(b)

Figure 9.6 (*a*) Photograph of inductance meter of 12-ft length between sending and receiving coils. (*b*) Results of inductance survey of subsurface seepage plume. (*c*) Results of inductance survey carried out to locate buried metal containers showing the influence of results to various offset distances.

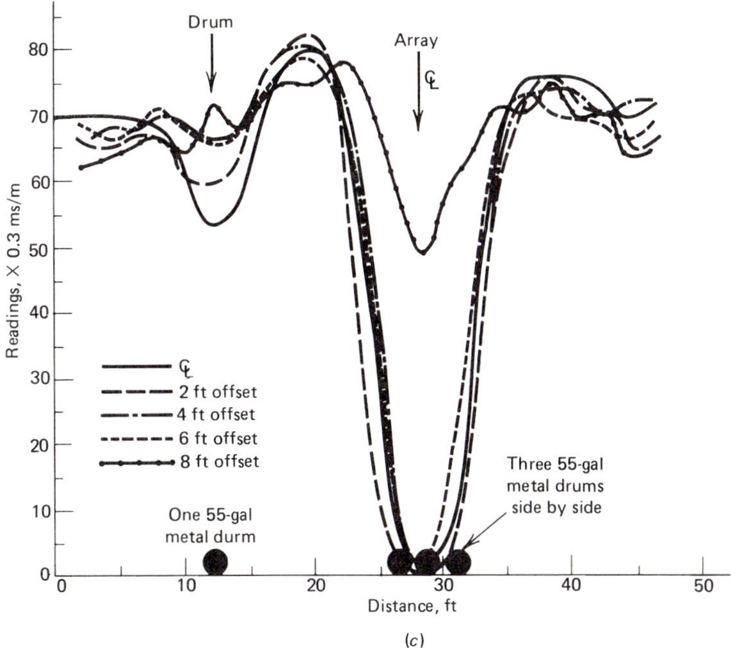

Figure 9.6 (*Continued*)

stability. In geotechnical engineering the method has been used for assessing earth dam and embankment stability (see Fig. 9.7a)[21] and for determining seepage location and movement (see Fig. 9.7b),[22] and is currently being investigated for use in determining soil and rock prestress by means of an acoustic pressuremeter.

9.4.10 Heat Pulse

The generation, or use, of heat for testing or evaluation purposes in soils is an interesting concept which has not yet been utilized to any great extent. Certainly deforming soil, or rock, masses generate heat at their points of contact, and temperature-sensing equipment is available to monitor these small changes in temperature; yet no systems are available for monitoring deformation processes. What has been developed, however, is a heat probe for insertion into moving groundwater to sense its direction and flow rate.[23] The probe has a central pin which sends out a pulse of heat on a regular basis. A set of 10 receiving probes, centrally located around the transmitting probe, are sequentially monitored for their temperatures (see Fig. 9.8a). The results are then plotted on a polar diagram, in vector form, and the resultant vector gives the direction and magnitude (i.e., the flow rate) of the groundwater flow (see Fig. 9.8b). The probe can be deployed at different locations around a site to determine its general groundwater regime (see Fig. 9.8c). The system has been used to advantage for mapping out polluted seepage plumes.

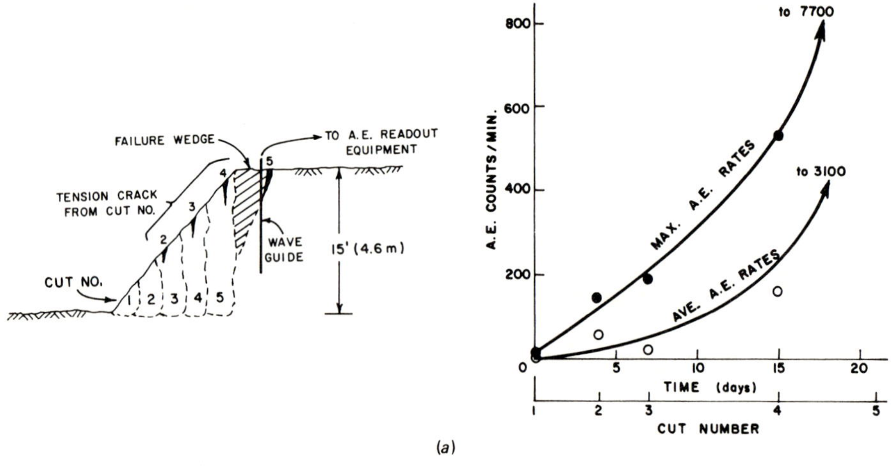

(a)

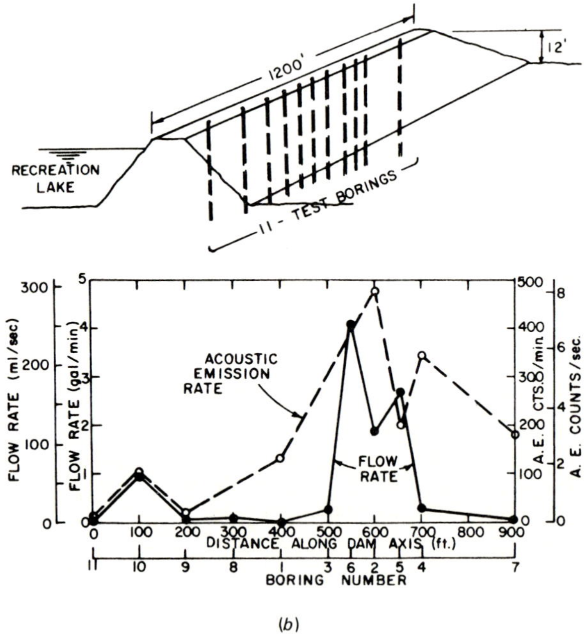

(b)

Figure 9.7 Examples of acoustic emission monitoring of stability and seepage in earth masses. (*a*) Cross section and AE response of stability monitoring of a sequentially excavated earth embankment. (*After Koerner et al., Ref. 21.*) (*b*) Schematic diagram and flow and AE rate response to seepage monitoring from within casing placed along axis of leaking earth dam. (*After Koerner et al., Ref. 22.*)

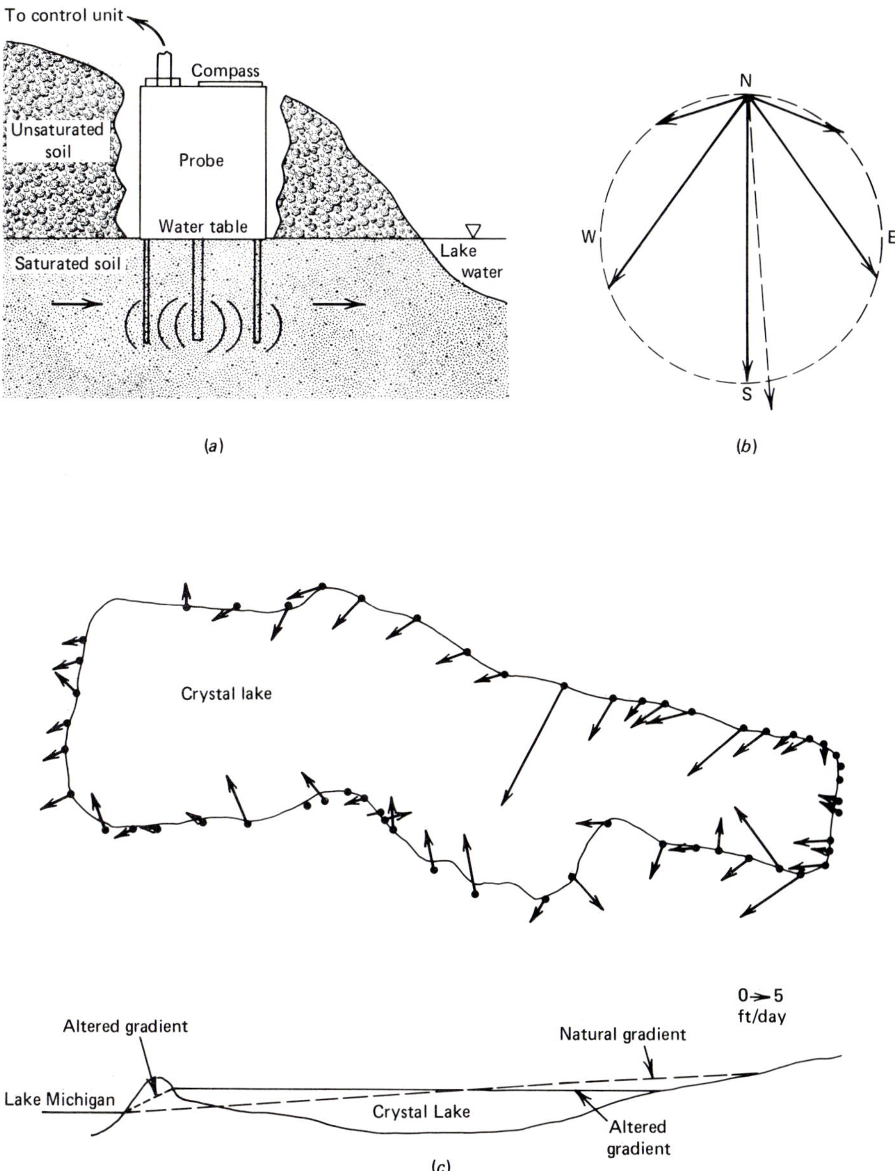

Figure 9.8 Temperature sensing, via heat pulse, to monitor direction and rate of groundwater flow. (*a*) Schematic of temperature probe. (*b*) Resolution of heat vectors. (*c*) Results of groundwater movement around lake. (*After Kerfoot and Skinner, Ref. 23.*)

9.4.11 Nuclear Methods

After describing seismic, magnetic, electromagnetic, acoustic, and thermal NDT methods, it seems appropriate to take a quick look at nuclear methods. Indeed, nuclear density and moisture meters are commercially available and are a common feature of geotechnical and construction engineering activities. Used mainly to monitor fills as they are being placed, the units provide an immediate feedback of data about in situ conditions in the vicinity being probed.

Nuclear density and moisture meters consist of a sealed isotope material or a neutron generator which transmits a radioactive pulse. A detector and a readout device (a suitable sealer) monitor its reflection back. Calibration curves of the materials in question are necessary to determine actual densities or water contents of the soils being evaluated. Two options are available:

Backscatter: The source is transmitted into the soil from the ground's surface, reflects off the subsurface soil, or water, as it penetrates the ground, and is received back at the ground's surface by the detector. (Both source and detector are housed in the same unit.)

Transmission: A probe containing the source (measuring about 9 to 12 in) is pressed into the ground and transmits the pulse to the detector at the ground surface.

Both methods are fully described in ASTM literature and are widely accepted throughout the geotechnical and construction industries.

9.4.12 Borehole Monitoring Methods

Each of the above described methods is usually deployed from (or near) the ground's surface, and the energy involved passes vertically through the ground until an anomaly is met and it is reflected back. It is then received by detectors at the ground's surface and appropriately analyzed. However, if boreholes are available, many of the techniques discussed can be adapted so that the energy is transmitted horizontally through the earth. The transmission is then received at an adjacent borehole, or reflected back to the same borehole, and analyzed.

In the transmission of seismic waves from a borehole, the crosshole method is generally used,[24] although echo reflection has also been used. In the crosshole method, determination of soil or rock properties (usually the moduli) is often the goal of the study. Velocities of both the compressional (P) wave and dilational or shear (S) wave can be measured and used to calculate the modulus of elasticity E or the shear modulus G by means of the following equations:

$$v_p = \left(\frac{K + \frac{4}{3}G}{\rho}\right)^{1/2} \tag{9.3}$$

$$v_p = \frac{E(1 - \mu)}{\rho(1 + \mu)(1 - 2\mu)} \tag{9.4}$$

$$v_s = \left(\frac{G}{\rho}\right)^{1/2} \qquad (9.5)$$

$$v_s = \frac{E}{2\rho(1 + \mu)} \qquad (9.6)$$

where v_p = compressional wave velocity, ft/s
  v_s = shear wave velocity, ft/s
  E = modulus of elasticity, lb/in^2
  G = shear modulus of elasticity, lb/in^2
  K = bulk modulus (or compressibility) of elasticity, lb/in^2
  ρ = density of the material, lb/in^3
  μ = Poisson's ratio

The technique uses a hammer impulse to generate the seismic wave and a velocity transducer to receive it. A storage oscilloscope is used to analyze the arrival times of the various waves involved. The completeness of soil grouting has also been assessed using this method by comparing propagation velocities before and after the grouting process.[25]

Electromagnetic waves of high frequency (1 MHz to 1 GHz) have also been used to assess grouting completeness by comparing graphic outputs of their transmission through the soil before and after grouting. Here the transmitting and receiving antennas are torpedolike probes which are sequentially and systematically lowered into the boreholes to assess all orientations of possible paths between them. The process has been called *geotomography*[26] or *crosshole radar.*[27]

Nuclear probes, too, have been lowered down into individual rock coreborings to determine subsurface stratigraphy and to obtain data on the physical properties of soils such as their water content, density, and porosity. This technology is common in petroleum exploration (where it is called *nuclear well logging*) and is being used in combination with electrical resistivity logs to provide in situ data about sites where foundation structures are to be built. See Wright and Harrell[28] and Crosby et al.[29] for additional information.

9.5 FOUNDATION INSTRUMENTATION

After all is said and done, one can validly ask, Is the foundation that has been built really working as intended? Obviously, if it fails, the answer is no. But even if the foundation has not failed, can it be termed "successful?" Perhaps there were alternative schemes that could have been used which were less expensive and equally safe? Perhaps the factor of safety for the system is ridiculously high? Perhaps it has not failed because of circumstances not intrinsic to its own design, e.g., added superstructure support or favorably changed conditions at the system's location. When a foundation fails, the fault lies with the designer or contractor or both, and the situation clearly is unacceptable. In the case of an extremely high factor of safety, the fault usually lies with the designer, and the situation is also unacceptable. No doubt there are many foundations whose factors of safety are too high. But while we can learn from foundation failures (if the legal profession lets us publish information about them!), we do not learn anything from ultraconservative designs.

The point of these remarks is that foundations and their supporting soil or rock should be monitored after construction much more carefully than is currently done— and not just those with designs considered marginally safe to begin with, but even those with routine designs. Only then can progress toward better design be made. With good field performance data as a basis for design, failures should become less frequent, designs can become more cost-effective, and new and innovative foundation techniques can be confidently developed.

This section is intended to serve as an introduction to the methods and devices that can be used for monitoring the performance of many of the types of foundations described earlier in this book. For additional information on the subject, see ASTM publication STP 554[30] and STP 584,[31] and works by Hanna[32] and Dunnicliff.[33]

9.5.1 Load Measurement

In a number of foundation situations, the actual load carried by a particular structural member is of critical concern. Examples of such situations are pile loads, anchor loads, soil and rock tieback loads, temporary sheeting strut and bracing loads, etc.

The most direct way of measuring loads in these cases is by means of a *proving ring* incorporated into the loaded system. Shown in Fig. 9.9*a*, a proving ring is made of hardened steel and has a high modulus of elasticity. As the ring is stressed by the load whose magnitude is being determined, a dial gauge or LVDT (electronic deformation device) is used to measure the deformation it is undergoing. The use of a calibration curve allows one to calculate the load itself, and that value, when divided by the cross-sectional area of the member, also gives the stress being exerted by the load. Although proving rings are simple and inexpensive, they are generally not rugged enough for use in the field and so are used predominantly in laboratory work.

Field values of loads are more commonly obtained by means of *load cells*. There are five types currently available:[33]

Tell-tale
Mechanical
Hydraulic
Vibrating wire strain gauge
Electrical resistance strain gauge

The strain gauge types are by far the most sensitive over a wide range of load capacities, having good temperature stability, and can be shielded from moisture and adverse environmental conditions. It should be noted that they are also the most expensive type of load measuring device.

In the vibrating wire load cell, the deformation of the member being tested is sensed by a set of parallel-housed vibrating wires; see Fig. 9.9*b*. The basic principle is that a change in the tension of the wires indicates a change in the cell's natural

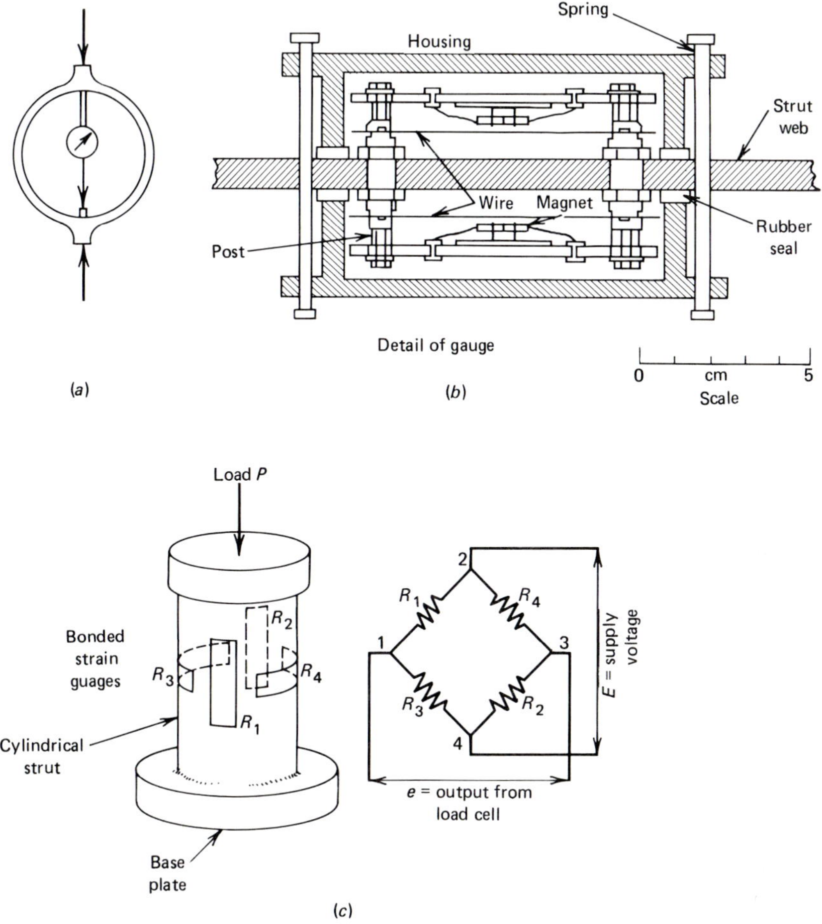

Figure 9.9 Various devices for measuring load directly and indirectly. (*a*) Direct-reading proving ring. (*b*) Vibrating wire strain gauge type of load cell. (*After Bjerrum et al., Ref. 34.*) (*c*) Electrical resistance strain gauge type of load cell and circuit diagram. (*After Hanna, Ref. 32.*)

frequency caused by its deformation. The vibrational outputs from each wire are measured, averaged, converted to strain, and finally to load by means of the following equations:

$$\epsilon = c_1 \frac{4L\rho^2}{Eg}\left(f_1^2 - f_0^2\right)$$

$$P = c_2(f_1^2 - f_0^2)$$

where ϵ = strain of the tensioned wire
$\quad L$ = length of wire, in
$\quad \rho$ = density of wire material, lb/in^3

E = modulus of wire, lb/in
g = gravitation constant, in/s^2
c_i = gauge constants (depends on geometry and design)[32,34]
f_0 = frequency at zero strain of gauge, Hz
f_1 = frequency at subsequent strain of gauge, Hz
P = load causing deformation, lb

A number of physical configurations are available, with capacities up to 1000 tons and accuracies of better than ± 1 percent.[33]

An equally popular type of load cell is one constructed of electrical strain gauges. Here a series of strain gauges or a set of strain rosettes is placed on a steel or aluminum cylinder as shown in Fig. 9.9c. The strain on the cylinder (caused by the unknown load) is measured by the strain gauges, converted to stress, and then converted to load using straightforward principles of mechanics; see most texts on strength of materials, e.g., Timoshenko and MacCollough.[35]

9.5.2 Earth Pressure Measurements

Adaptations of the load-measuring devices just described can also be used to measure earth pressures. An almost infinite variety of field situations requires this type of information. The systems that have been devised fall into the following categories:

Pneumatic
Hydraulic
Vibrating wire strain gauges
Bonded electrical resistance strain gauges
Unbonded electrical resistance strain gauges

However, a few cautions are in order before beginning to describe the different types. One must recall that the strength of soil is based on effective stress, the basic relationship for saturated soils being:

$$\bar{\sigma} = \sigma - u_w$$

where $\bar{\sigma}$ = effective (intergranular) stress, lb/ft^2
σ = total stress, lb/ft^2
u_w = pore water pressure, lb/ft^2

Here we are concerned with monitoring total stress. Later, in Sec. 9.5.4 we will discuss pore water pressure monitoring. The difference between the two values, at a specific locale in the soil, gives the effective stress. The other aspect of earth pressure monitoring that is of concern (as it is in deformation and pore water pressure monitoring also) is the disturbance of the soil caused by the introduction of the monitoring device itself. Attempts at including a correction factor in the calculations have been made,[32] but the wide variety of devices and methods used to place them make it difficult to treat the subject analytically. Monitoring results must take this problem into account.

Earth pressures can be measured by devices analogous to the load cells used in measuring loads, and Fig. 9.10*a* and *b* shows the electrical resistance and vibrating wire types. Figure 9.10*c* shows a hydraulic cell, known as the *Glotzl cell,* which is oil-filled and has a high area-to-thickness ratio. It is mechanically equilibrated in such a way that the line pressure becomes equal to the cell pressure. The earth pressure is equal to this bypass gauge pressure reading corrected for elevation minus the zero gauge reading. This same general principle can be used in pneumatic devices, as shown in Fig. 9.10*d* where the equilibration interface is the flexible diaphragm shown in the illustration.

While systems such as these are indeed clever, their accuracy has been seriously challenged. Weiler and Kulhawy[36] are particularly skeptical of the results they give, in part because of the need to calibrate the cells in the soil in which they will be used. The calibration process is difficult and, even if done extremely carefully, depends greatly for its success on how the cells are placed in the field.

9.5.3 Deformation Monitoring

The problems inherent to measuring earth pressures just described leave the door open to deformation monitoring. But even though such monitoring is much easier and more straightforward to carry out, we must still ask once the soil deformations are determined, What do they mean? Since foundation engineering is a "stress-oriented" profession and its practitioners feel somewhat comfortable about designing in terms of limiting stresses on foundation members or soil masses, there is less confidence in relying on deformation data for design or "calling a project quits" on the basis of such data. Even if the soil deformation determined is converted to strain, the stress-versus-strain response of the soil must also be determined. That response can be obtained from laboratory tests, but how does that curve relate to deformation and design decisions in the first place? Indeed, these are difficult questions, but a review of deformation monitoring technology is warranted because the insights its use can provide into ground and/or foundation movements are so very important.

To begin with, there are the conventional methods of surveying by means of benchmark leveling (for elevation changes) and offset measurements (for lateral displacement). Standard levels and transits can certainly be used for this, but more sensitive electronic distance measurement devices and lasers give greater accuracy. If information on subsurface deformations is desired, a suitable rod or pipe, adequately anchored, can be utilized, which extends through the soil from the point being monitored to the ground's surface; see Fig. 9.11 for typical configurations.

Remote settlement gauges serve the same purpose as the devices just described, but avoid the need for vertical risers coming up through the soil. Since such risers seem to be targets for trucks, dozers, scrapers, and every other piece of construction equipment required on a project, eliminating them is indeed a noble aim. A number of manometer-type setups are shown in Ref. 33.

To measure subsurface movements in a nonvertical orientation, the use of an extensometer is quite common. These units are placed within boreholes and anchored

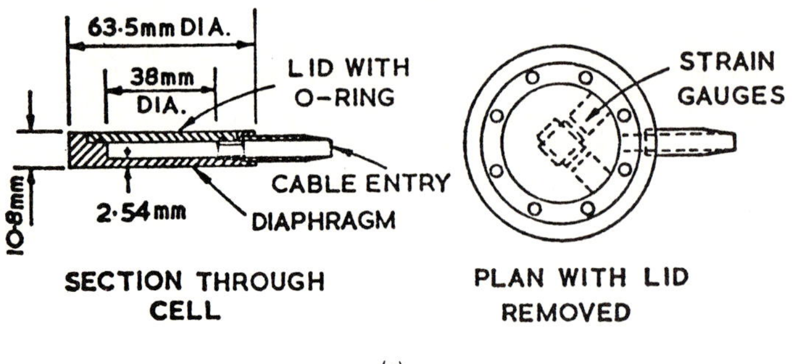

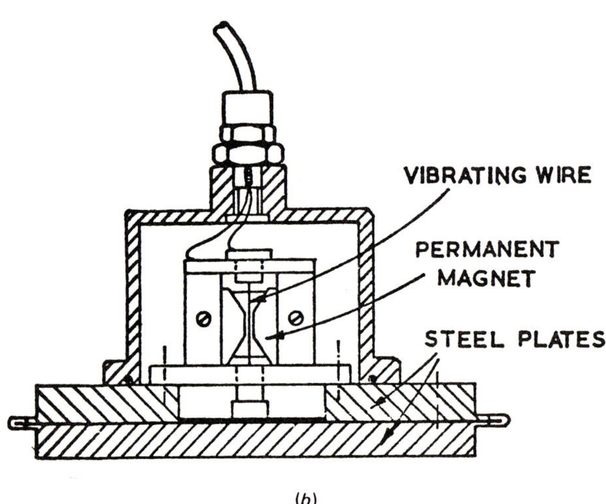

Figure 9.10 Various types of earth-pressure-measuring cells. (*a*) Electrical resistance strain gauge cell. (*b*) Vibrating wire cell. (*c*) Hydraulic cell. (*d*) Pneumatic cell. *(After Hanna, Ref. 32, and Dunnicliff, Ref. 33.)*

at prescribed locations, and they measure axial change in the longitudinal borehole dimensions. Rods or wires extend from the anchor to the head as shown in Fig. 9.12*a*. Either a dial indicator or an LVDT can serve for the readout. When wires are used, the borehole is sufficiently wide to bring a series of them to the ground's surface, thus creating a multiple-point extensometer; see Fig. 9.12*b* and *c*. The individual points can be mechanically or magnetically anchored, or grouted into place.

Rather than using a ridiculously large number of extensometer points to obtain a continuous profile of subsurface movements, an inclinometer can be used. Although several types of these devices are available, one common instrument uses a closed servo-accelerometer as a tilt sensor. The housing is a wheeled probe riding in a grooved plastic or aluminum guide pipe; see Fig. 9.13. Readings are taken at closely

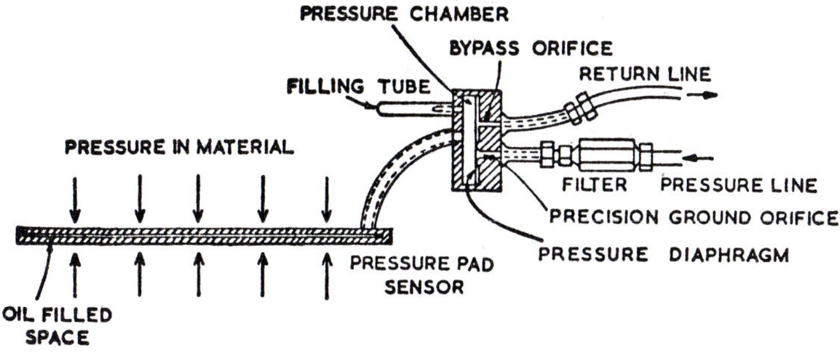

(c)

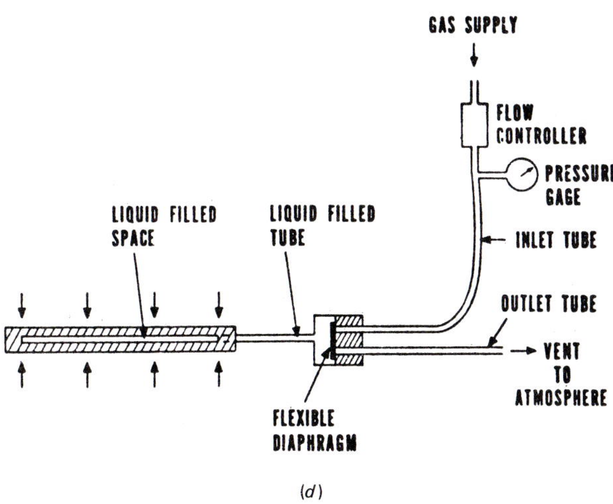

(d)

Figure 9.10 (*Continued*)

spaced intervals and the horizontal movement from one set of readings to the next is plotted and compared. Cumulative movement trends can easily be interpreted to provide data for investigations into slope stability movements, retaining walls deformations, tunnel convergence, etc.

A totally different way of measuring soil strain is to embed two, matched, free-floating coils in the soil at a known distance apart. One coil is connected by cables to an oscillator (in the instrument package at the ground's surface) to produce an electromagnetic (EM) field in the soil. By mutual inductance a field is developed in the other coil, and its magnitude is a function of the distance separating the two coils. A calibration curve gives the actual distance, and strain can be calculated in the customary manner. Both static and dynamic strains can be sensed because of the rapid rise time of the system; see Selig for further details.[37]

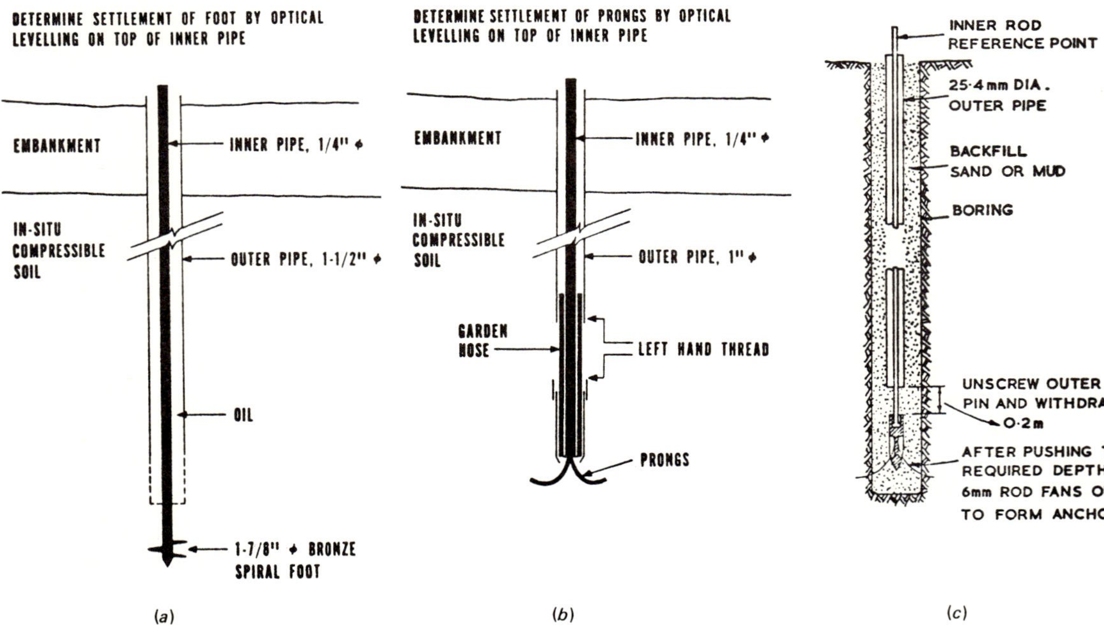

Figure 9.11 Anchored settlement platforms for monitoring subsurface movements. (*a*) Spiral foot gauge. (*b*) Borros anchor. (*c*) Geonor probe. (*After Dunnicliff, Ref. 33, and Hanna, Ref. 32.*)

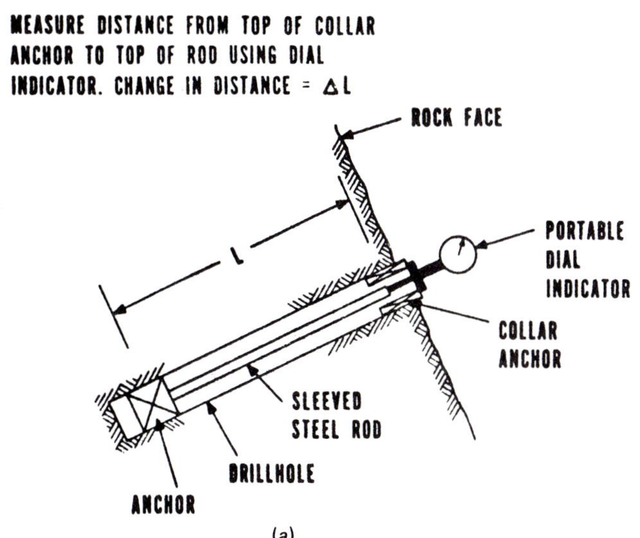

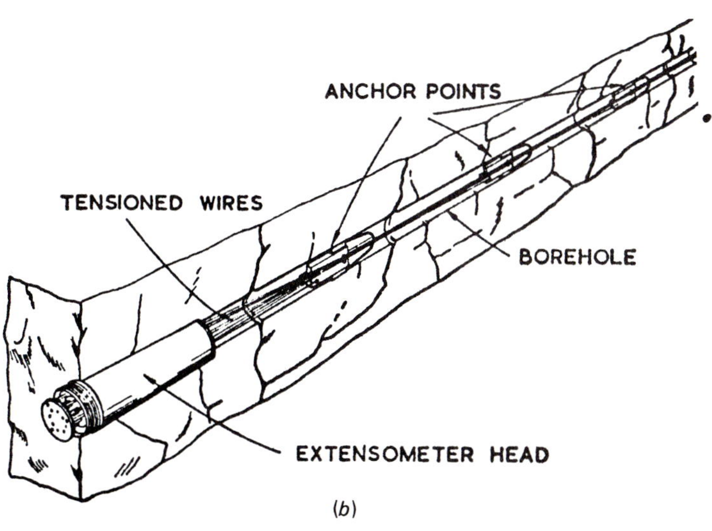

Figure 9.12 Various types of borehole extensometers. (*a*) Single-point extensometer. *(After Dunnicliff, Ref. 33.)* (*b*) Multiple-point extensometer *(After Terrametrics Inc.).* (*c*) Triple-rod extensometer with alarm device *(After Inerfels Inc.).*

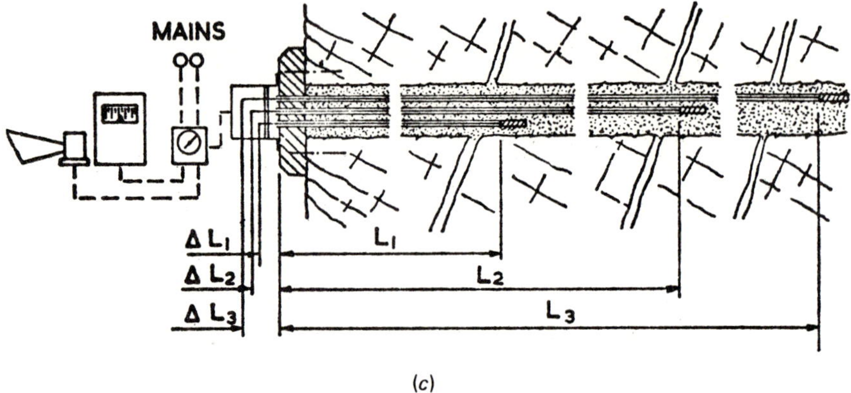

Figure 9.12 (*Continued*)

9.5.4 Pore Water Pressure Measurement

Two topics will be covered here: water observation pipes and piezometers. Water observation pipes are used to determine the elevation of the underground water table. As shown in Fig. 9.14*a*, they are merely open standpipes with slots or a porous stone at their bottoms; they are placed into boreholes in the ground, and water entering them rises to its equilibrium position. The simplicity of this device belies its importance. It should be the first item used on every boring job and should be monitored until the very end of the project, and even beyond if possible. Piezometers are used to measure pore water pressure at a point beneath the underground water table level. They too are slotted or have porous stones at their bottoms, but their base areas must be sealed off from the part of their boreholes above the point being monitored. This allows the pore water pressure at the bottom of the piezometer to be measured. Two conditions can result:

The water in the piezometer standpipe can rise to the same level as shown in a water observation pipe, i.e., to the underground water table level.

The water in the piezometer can rise above that level in a water observation pipe at the same location. The latter case indicates excess pore water pressure, meaning that the soil is undergoing a consolidation process and is not in equilibrium.

As seen in Fig. 9.14*b* through *d*, there are various types of piezometers available. These different devices have been nicely contrasted by Dunnicliff[33]; see Table 9.2.

Figure 9.13 Photograph and schematic drawing showing inclinometer system for measuring soil and rock movements. (*After Slope Indicator Co.*)

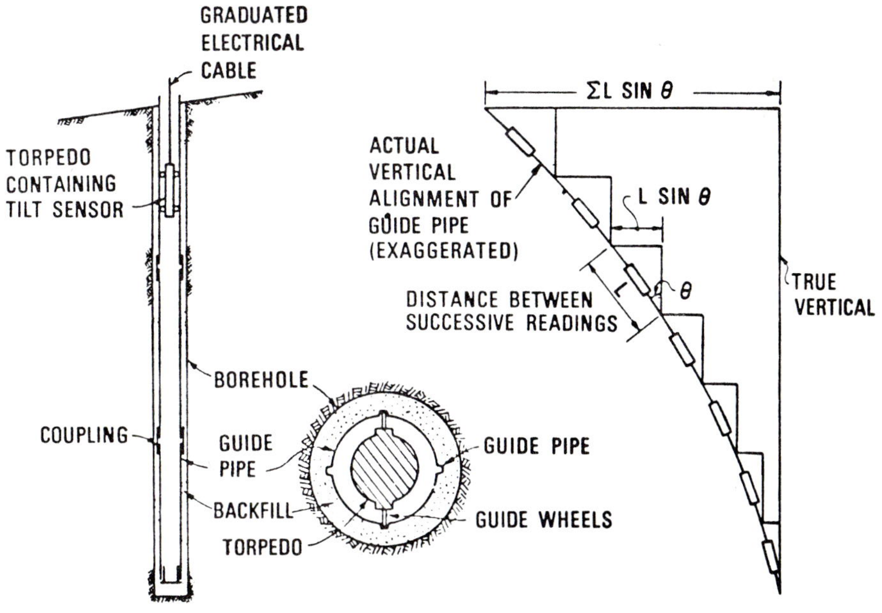
GRADUATED
ELECTRICAL
CABLE
TORPEDO
CONTAINING
TILT SENSOR
ACTUAL
VERTICAL
ALIGNMENT OF
GUIDE PIPE
(EXAGGERATED)
DISTANCE BETWEEN
SUCCESSIVE READINGS
ΣL SIN θ
L SIN θ
L
θ
TRUE
VERTICAL
BOREHOLE
COUPLING
GUIDE
PIPE
BACKFILL
TORPEDO
GUIDE PIPE
GUIDE WHEELS

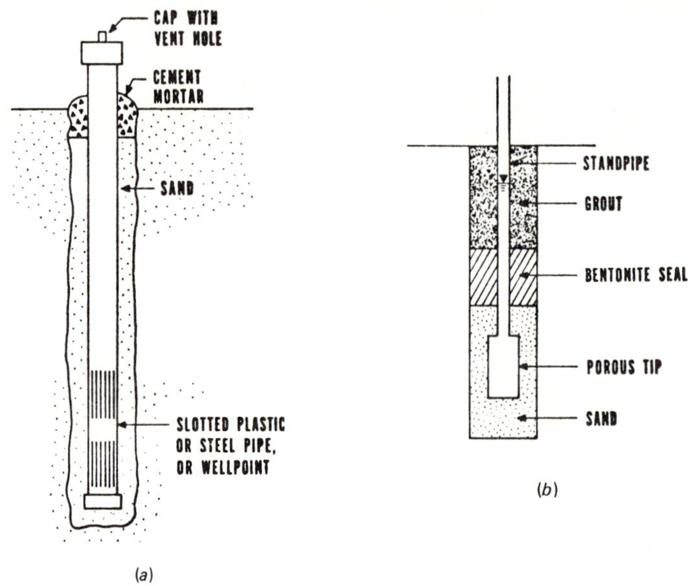

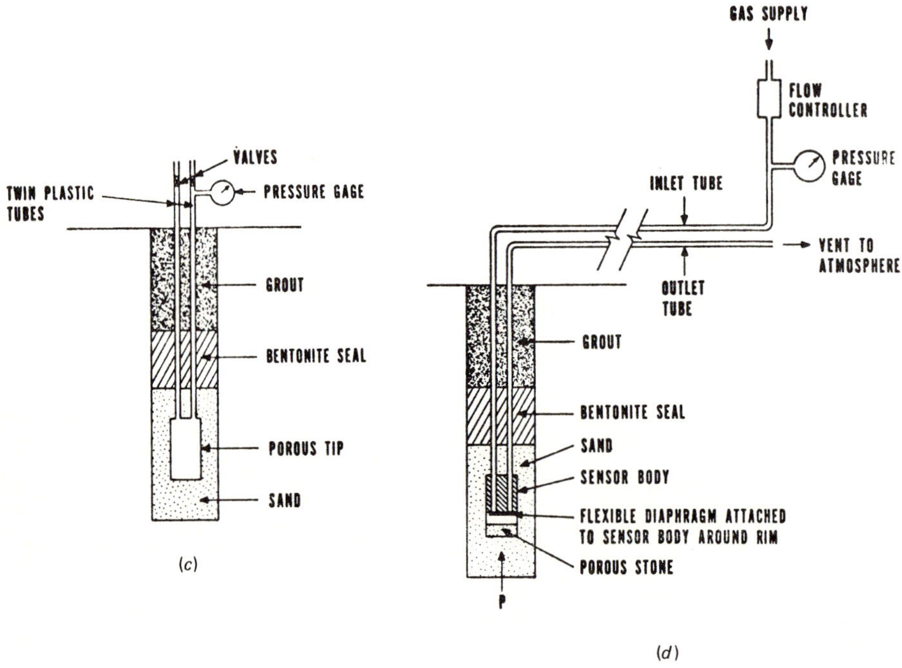

Figure 9.14 Various types of configurations of water observation pipes and piezometers. (*a*) Observation well. (*b*) Open standpipe piezometer. (*c*) Twin-tube hydraulic piezometer with local readout. (*d*) Typical pneumatic piezometer (similar to vibrating wire or electrical resistivity systems except for point and readout hardware). *(After Dunnicliff, Ref. 33.)*

430

Type	Advantages	Limitations and precautions
Observation well	Simple, inexpensive, universally available.	Not applicable if perched or artesian water tables present. Metal elements may corrode. Long time lag.
Open standpipe piezometer	Simple, inexpensive, reliable. Long performance record. Self-de-airing if inside diameter of standpipe is greater than 0.4 in.	Porous filter can plug because of repeated water inflow and outflow. Long time lag. Tubing must be raised nearly vertical. Cannot be used if piezometric level is above top of standpipe. Freezing problems. Subject to damage by construction equipment and consolidation of soil around standpipe.
Twin-tube hydraulic piezometer	Simple, reliable. Long experience record. Less time lag and less prone to damage than open standpipe piezometer.	Freezing problems. Tubing must not be significantly above piezometric elevation. Periodic de-airing required.
Pneumatic piezometer	Stable. Short time lag. Capability of purging lines. Minimum interference to construction. Level of tubes and readout independent of level of tip. No freezing problems.	Dry gas should be used. Check valve displacement should be minimal (can be as low as 0.002 cm^3). Models with tube to soil side of diaphragm may increase time lag. More expensive than above types.
Vibrating-wire strain gauge piezometer	Easy to read. Can be used to read negative pore pressures. Short time lag. Minimum interference to construction work. Level of wires and readout independent of level of tip. Suitable for automatic recording. Frequency signal permits data transmission over long distances. No freezing problems.	Not suitable for dynamic readings. Overvoltage protection and grounding system necessary in regions of thunderstorm activity. More expensive than pneumatic piezometers.
Bonded resistance strain gauge piezometer	Easy to read. Can be used to read negative pore pressures. Short time lag. Suitable for dynamic measurements and automatic recording. Minimum interference to construction work. Level of wires and readout independent of level of tip. No freezing problems.	Some versions sensitive to temperature. Long-term stability not yet verified. False data can result from transmission over long distances. Overvoltage protection and grounding system necessary in regions of thunderstorm activity. More expensive than pneumatic piezometers.

†After Dunnicliff, Ref. 33.

Table 9.3 Advantages and limitations of various contractual arrangements for instrumentation installation†

Method	Advantages	Limitations
1. Installation by owner's personnel	Owner has direct control over cost and quality. Flexible to accommodate changes.	Potential problems with contractor cooperation if instrumentation work interferes with other work. Owner must plan for work load well in advance. Assumes owner has necessary in-house skills. Cannot always be financed by construction funds.
2. Bid items in prime contract, with no prequalification	Initial cost will tend to be least. Least effort required to write specifications. Financed by construction funds.	Generally contractor will shop for lowest-price subcontractor, with risk of lowest quality and invalid measurement data. Requires strong and experienced supervision by owner's representative. Not flexible to accommodate changes.
3. Bid items in prime contract, with prequalification	Initial cost will tend to be low. Excludes inexperienced instrumentation subcontractors. Small specifications writing effort. Financed by construction funds.	Generally contractor will shop for lowest-price "qualified" subcontractor with risk of subcontract having inadequate price, cutting of corners, and thus invalid measurement data.

		Often difficult to substantiate desire to reject questionably qualified subcontractor. Usually requires strong and experienced supervision by owner's representative. Not flexible to accommodate changes.
4. Instrumentation specialist contracting with owner	Owner has direct control over cost and quality. Flexible to accommodate changes. Instrumentation specialist can, if retained early enough, assist with design of monitoring program.	Potential problems with contractor cooperation if instrumentation work interferes with other work. Cannot always be financed by construction funds.
5. Instrumentation specialist contracting with contractor	In selecting the specialist, owner has some control over cost and quality. Facilitates cooperation and scheduling with contractor. Flexible to accommodate changes. Financed by construction funds.	Assumes "professionalism" on part of instrumentation specialist, who has negotiated with the owner but contracted with the contractor. Not permitted under some public agency regulations.

†After Dunnicliff, Ref. 33.

9.5.5 Summary

In summarizing this section on foundation instrumentation, it is probably obvious that we are active proponents of field monitoring. The information gained by knowing how a foundation is performing in terms of its loads, the stresses, deformations, and strains of the soil on which it is built, and the water (and air) pore pressures being mobilized is invaluable. All too often our designs are decided on intuition and a judgment guided by precious few documented case histories or personal experiences. On the one hand this way of doing things can be costly (when it results in overdesigning), while on the other it can be dangerous (when it results in underdesigning).

Once such information is obtained, *it must be published*. Time must be set aside, after a project is completed, to accomplish this vital task. Indeed, this kind of paper work is the last part of a project. However, we may ask, Who should pay for such instrumentation? Again Dunnicliff[33] has discussed various funding possibilities, and his suggestions are summarized in Table 9.3. The point to be made is that if the desire to obtain the data and publish it is there, the task can be accomplished. Our sincere hope, as a parting shot if you will, is that documented case histories of foundations become the wave of the future. We all—engineers, contractors, owners, the public in general—will be the winners.

PROBLEMS

9.1 In testing soil or rock in the laboratory, what are the main purposes of the following:
    (*a*) Testing physical properties?
    (*b*) Testing hydraulic properties?
    (*c*) Testing mechanical properties?

9.2 How would the quality of the pore fluid, e.g., gasoline mixed in with the water, etc., affect:
    (*a*) The testing of physical properties?
    (*b*) The testing of hydraulic properties?
    (*c*) The testing of mechanical properties?

9.3 List the advantages and disadvantages of:
    (*a*) Testing soils in the laboratory versus testing them in the field.
    (*b*) Testing rock in the laboratory versus testing it in the field.

9.4 (*a*) Give examples (with sketches) of isotropic and anisotropic soil and rock profiles.
    (*b*) Give examples (with sketches) of homogeneous and nonhomogeneous soil and rock profiles.

9.5 Prepare a table listing the advantages and disadvantages of the following traditional field test methods (they are all ASTM testing methods):
    (*a*) Standard penetration test
    (*b*) Vane shear test
    (*c*) Cone penetration test
    (*d*) Pressuremeter test

9.6 In the accompanying table, rate the ability of the field tests listed to provide information about the following conditions and problems: (*a*) soil densification results using the vibroflotation, Terra-Probe, or deep dynamic compaction methods (refer back to Chap. 4); (*b*) soil consolidation results using the preloading or drain wick methods (refer back to Chap. 5); (*c*) adequacy of compaction grouting for soil densification (refer back to Chap. 6); (*d*) identifying subsurface soil and rock stratigraphy; (*e*) evaluating the uniformity of strength within a thick soil stratum. Use a scale of A (most able) to F (least able), and consider each test method strictly on its own merits, i.e., do not compare methods.

	Conditions and problems				
Field test method	(a)	(b)	(c)	(d)	(e)
1. Standard penetration					
2. Vane shear					
3. Core penetration					
4. Pressuremeter					

9.7 Look up information about and write short descriptions of the following field test methods:
   (a) Borehole shear test
   (b) Plate-bearing-load test
   (c) Field direct-shear test
   (d) Hydraulic fracturing test
   (e) Acoustic pressuremeter test

9.8 Compare and contrast traditional field test methods to NDT methods.

9.9 In the accompanying table, rate the ability of the NDT methods listed to solve the following problems: (a) locating the boundary between soil and rock which is 20 to 40 ft beneath the ground's surface; (b) locating a water table level in granular soil which is 5 to 20 ft beneath the ground's surface; (c) defining a shallow seepage plume of spilled hazardous liquid waste; (d) locating utility lines below a city street; (e) detecting 55-gal steel drums buried 2 to 15 ft beneath the ground's surface; (f) locating voids beneath concrete pavement or behind brick sewers; (g) monitoring the stability of earth dams; (h) determining the adequacy of chemical grouting for seepage control (refer back to Chap. 6). Use a scale of A (most able) to F (least able), and consider each method strictly on its own merits, i.e., do not compare methods.

	Problems							
NDT method	(a)	(b)	(c)	(d)	(e)	(f)	(g)	(h)
1. Seismic reflection								
2. Seismic refraction								
3. Electrical resistivity								
4. Magnetometry								
5. Ground probing radar								
6. Continuous wave microwave								
7. Eddy current methods								
8. Pulse-echo ultrasonics								
9. Acoustic emission								
10. Heat pulse								
11. Nuclear methods								

9.10 (a) Describe the ASTM nuclear backscatter and transmission methods used to obtain the in situ density of an embankment as it is being placed.

   (b) Compare these methods to others used for this purpose, such as the ASTM sand cone and rubber balloon methods.

   (c) Describe the two nuclear methods with respect to measuring in situ moisture content (also an ASTM procedure).

9.11 What type of field instrumentation would you use to measure strut loads in a braced excavation if the struts were 10 × 10-in wood beams? If the struts were 6-in steel beams? Illustrate your method by sketches. (Refer to Sec. 3.3.)

9.12 What type of field instrumentation would you use to measure the tieback or anchor loads of a temporary wall? What would you change in their design if this were a permanent wall and creep deformation were a potential problem? (Refer to Sec. 3.5.)

9.13 For a 110-ft vertical, sided, unbraced cut in rock, what type of horizontal deformation measurement scheme would you suggest? (Assume a duration of the cut opening of about 2 years.)

9.14 Design a preliminary scheme for monitoring an existing 20-ft-high earth dam containing a hazardous waste liquid. (The dam is suspected of being unstable.)

9.15 Design a monitoring scheme for a 100-ft-high earth dam that is itself only in the design stage, which is to be founded on somewhat compressible soil.

9.16 Design a monitoring scheme for a 100-ft-high masonry dam that is itself only in the design stage, which is to be founded on highly fractured rock. (The rock will be cement-grouted at the start of construction.)

9.17 Design a monitoring scheme for the following subsidence situations (refer to Secs. 1.2.7 and 1.2.9):

(*a*) Local Karst topography (i.e., decomposing limestone) near the shallow foundations of five-story masonry buildings. The rock surface is 10 to 20 ft beneath the footings.

(*b*) Withdrawal of groundwater due to temporary construction dewatering. Residential houses are within the zone of drawdown influence.

(*c*) Withdrawal of groundwater by a municipality for use in drinking *or* withdrawal of oil by conventional oil field methods.

9.18 Design a monitoring scheme to assess the adequacy of a slurry trench cutoff wall recently constructed around a leaking hazardous materials waste landfill, as shown in Fig. P9.18. (Refer also to Sec. 3.4.3.)

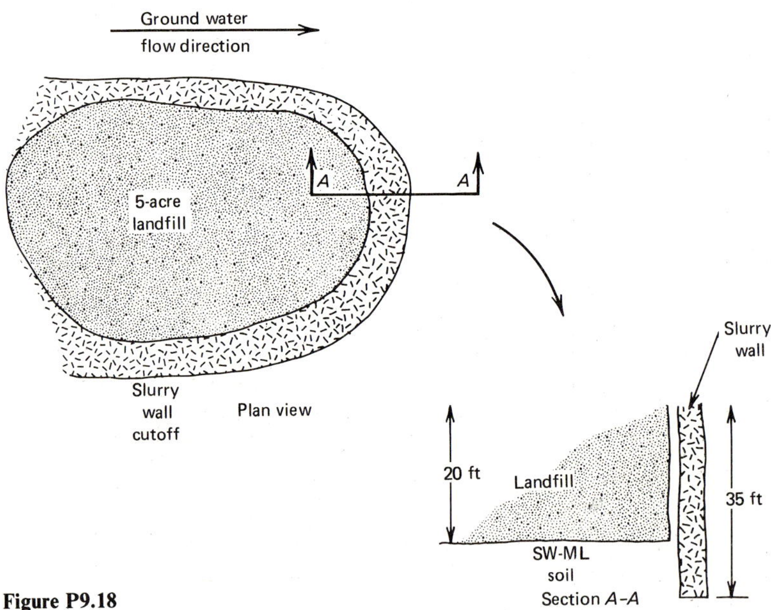

Figure P9.18

9.19 Discuss ideas, based on NDT methods, of how you would determine if a geomembrane (pond liner) were leaking, what the leakage rate was, and where the leak was located. Refer to Fig. 8.8*a* for a typical single-liner system.

9.20 Discuss ideas, based on NDT methods, of how you would determine if a clay or geomembrane cap, in a closure like those shown in Fig. 8.15*a* and *b,* were allowing surface water to infiltrate into a completed landfill.

REFERENCES

1. Lambe, T. W., *Soil Testing for Engineers,* Wiley, New York, 1951.
2. Bowles, J. E., *Engineering Properties of Soils and Their Measurement,* 2d ed., McGraw-Hill, New York, 1978.
3. Rosenfarb, J. L., and R. M. Koerner, *Experimental Soil Mechanics,* Vols. 1 and 2, 2d ed., Drexel University Press, Philadelphia, 1978.
4. Fletcher, G. F. A., "The Standard Penetration Test; Its Uses and Abuses," *J. Soil Mech. Found. Eng. Div., ASCE,* Vo. 91, No. SM4, July 1965, pp. 67–76.
5. Hvorslev, J. M., *Subsurface Exploration and Sampling of Soils for Civil Engineering Purposes,* U.S. Army Corps of Engineers, Vicksburg, MS, 1949.
6. *Proc. Conf. on In-Situ Measurement of Soil Properties* (Vols. 1 and 2), Spec. Conf. of Geotech. Eng. Div., ASCE, North Carolina State University, Raleigh, NC, June 1–4, 1975.
7. *Proc. Conf. on Vane Shear and Cone Penetration Resistance Testing on In-Situ Soils,* Seattle, ASTM Publication No. STP 399, 1965.
8. Schmertmann, J. H., "Measurement of In-Situ Shear Strength," in Ref. 6, Vol. 2, pp. 57–138.
9. Lord, A. E., Jr., and R. M. Koerner, "An Acoustic Pressuremeter for Use in Soil and Rock Stress History Determination," submitted to *J. Acoust. Emiss.,* January 1983.
10. Griffith, D. H., and R. F. King, *Applied Geophysics,* Pergamon, Oxford, 1965.
11. Grand, F. S., and G. F. West, *Interpretation Theory in Applied Geophysics,* McGraw-Hill, New York, 1965.
12. Sharma, P. V., *Geophysical Methods in Geology,* Elsevier, Amsterdam, 1976.
13. Dobrin, M. B., *Introduction to Geophysical Prospecting,* McGraw-Hill, New York, 1976.
14. Koerner, R. M., A. E. Lord, Jr., and J. J. Bowders, "Utilization and Assessment of a Pulsed RF System to Monitor Subsurface Liquids," in *Proc. Nat. Conf. on Mgmt. of Uncontrolled Hazardous Waste Substances,* Hazardous Materials Control Research Institute, Washington, D.C., October 28–30, 1981, pp. 165–170.
15. Lord, A. E., and R. M. Koerner, "Electromagnetic Methods in Subsurface Investigations," *Proc. Eng. Found. Conf. on Subsurface Exploration Methods,* ASCE, Santa Barbara, December 1981, in press.
16. Morey, R. M., "Continuous Subsurface Profiling by Impulse Radar," in *Proc. Subsurface Exploration for Underground Excavation and Heavy Construction,* ASCE, New York, 1974, pp. 213–232.
17. Okrasinski, T. A., R. M. Koerner, and A. E. Lord, Jr., "Dielectric Constant Determination at L-Band Microwave Frequencies," *Geotech. Eng. J., ASTM,* Vol. 1, No. 3, 1979, pp. 134–140.
18. Bowders, J. J., R. M. Koerner, and A. E. Lord, Jr., "Buried Container Detection Using Ground Probing Radar," *J. Hazardous Mater.,* Vol. 7, 1982, pp. 1–17.
19. Koerner, R. M., J. S. Reif, and M. J. Burlingame, "Detection Methods for Location of Subsurface Water and Seepage," *J. Geotech. Eng. Div., ASCE,* Vol. 105, No. GT11, November 1979, pp. 1301–1316.
20. McNeill, D. J., "Electromagnetic Resistivity Mapping of Contaminated Plumes," in *Proc. Mgmt. of Uncontrolled Hazardous Waste Sites,* Washington, DC, HMCRI, November 29–December 2, 1982, pp. 1–6.
21. Koerner, R. M., A. E. Lord, Jr., and W. M. McCabe, "Acoustic Emission Monitoring of Soil Stability," *J. Geotech. Eng. Div., ASCE,* Vol. 104, No. GT5, May 1978, pp. 571–582.
22. Koerner, R. M., W. M. McCabe, and A. E. Lord, Jr., "Acoustic Emission Behavior and Monitoring of Soils," in *Acoustic Emissions in Geotechnical Engineering Practice,* ASTM Publication No. STP 750, V. P. Drnevich and R. E. Gray (eds.), ASTM, New York, 1981, pp. 93–141.
23. Kerfoot, W. B., and S. M. Skinner, Jr., "Direct Groundwater Flow Measurement," in *Proc. Fifth Nat. Groundwater Quality Symposium,* Las Vegas, October 8, 1980, available through K-V Assoc., Falmouth, MA 02540.
24. Stokoe, K. H., and R. D. Woods, "In-Situ Shear Wave Velocity by Cross-Hole Method," *J. Soil Mech. Found. Eng. Div., ASCE,* Vol. 98, No. SM5, May 1972, pp. 443–460.
25. Partos, A., R. D. Woods, and J. P. Welsh, "Soil Modification for Relocation of Die Forging Oper-

ations," in *Proc. Conf. on Grouting in Geotech. Eng.,* ASCE, W. H. Baker (ed.), New Orleans, February 1982, pp. 938–958.

26. Laine, E. F., R. J. Lytle, and J. T. Okada, "Cross Bore-Hole Observation of Soil Grouting," *J. Geotech. Eng. Div., ASCE,* Vol. 106, No. GT8, August 1980, pp. 871–875.

27. Huck, P. J., and M. J. Waller, "Quality Control for Grouting," in Ref. 25, pp. 781–791.

28. Wright, J. W., and H. C. Harrell, "Current Techniques in Engineering Geophysical Logging," in K. H. Stokoe (ed.), *Geophysical Methods in Geotechnical Engineering,* ASCE, New York, 1979, pp. 51–89.

29. Crosby, J. W., B. Konstantinidis, and P. Davis, "Geotechnical Applications of Borehole Geophysics," *J. Geotech. Eng. Div., ASCE,* Vol. 107, No. GT10, October 1981, pp. 1255–1267.

30. *Field Testing and Instrumentation of Rock,* G. B. Clark (ed.), ASTM Publication No. STP 554, 1974.

31. *Performance Monitoring for Geotechnical Construction,* L. I. Stern and C. J. Dunnicliff (eds.), ASTM Publication No. STP 584, 1975.

32. Hanna, T. H., *Foundation Instrumentation,* Trans. Tech. Publ., Clausthal, W. Germany, 1973.

33. Dunnicliff, J., *Geotechnical Instrumentation for Monitoring Field Performance,* NCHRP Publication No. 89, Transportation Research Board, Washington, DC, 1982.

34. Bjerrum, L., T. C. Kenny, and B. Kjarensli, "Measuring Instruments for Strutted Excavations," *J. Soil Mech. Found. Eng. Div., ASCE,* Vol. 91, No. SM1, pp. 111–141.

35. Timoshenko, S., and G. H. MacCullough, *Elements of Strength of Materials,* 3d ed., Van Nostrand, New York, 1949.

36. Weiler, W. A., and F. H. Kulhawy, *Behavior of Stress Cells in Soils,* Geotech. Eng. Report No. 78-2, School of Civil and Environmental Engineering, Cornell University, Ithaca, 1978.

37. Selig, E. T., "Soil Strain Measurement Using Inductance Coil Method," in Ref. 31, pp. 141–158.

SOIL CLASSIFICATION SYSTEMS, SOIL COMPONENTS AND FRACTIONS, AND ENGINEERING USE CHART

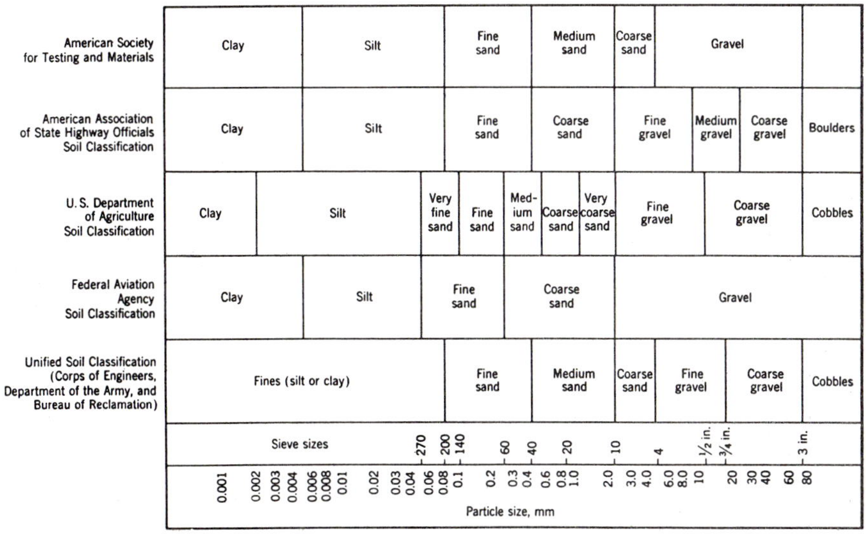

Classification system									
American Society for Testing and Materials	Clay	Silt	Fine sand	Medium sand	Coarse sand	Gravel			
American Association of State Highway Officials Soil Classification	Clay	Silt	Fine sand	Coarse sand	Fine gravel	Medium gravel	Coarse gravel	Boulders	
U. S. Department of Agriculture Soil Classification	Clay	Silt	Very fine sand / Fine sand / Medium sand / Coarse sand / Very coarse sand	Fine gravel	Coarse gravel	Cobbles			
Federal Aviation Agency Soil Classification	Clay	Silt	Fine sand	Coarse sand	Gravel				
Unified Soil Classification (Corps of Engineers, Department of the Army, and Bureau of Reclamation)	Fines (silt or clay)	Fine sand	Medium sand	Coarse sand	Fine gravel	Coarse gravel	Cobbles		

Sieve sizes: 270 | 200 | 140 | 60 | 40 | 20 | 10 | 4 | ½ in. | ¾ in. | 3 in.

Particle size, mm: 0.001 | 0.002 | 0.003 | 0.004 | 0.006 | 0.008 | 0.01 | 0.02 | 0.03 | 0.04 | 0.06 | 0.08 | 0.1 | 0.2 | 0.3 | 0.4 | 0.6 | 0.8 | 1.0 | 2.0 | 3.0 | 4.0 | 6.0 | 8.0 | 10 | 20 | 30 | 40 | 60 | 80

Soil-separation size limits of ASTM, AASHO, USDA, FAA, and Corps of Engineers and USBR. *(From R. D. Krebs and R. D. Walker, Highway Materials, McGraw-Hill, New York, 1971.)*

Unified soil classification†

Major divisions		Field identification procedures (Excluding particles larger than 3 in and basing fractions on estimated weights)			Group symbols‡	Typical names
Coarse-grained soils—more than half of material is *larger* than No. 200 sieve size§ (The No. 200 sieve size is about the smallest particle visible to naked eye)	Gravels—more than half of coarse fraction is larger than No. 7 sieve size	Clean gravels (little or no fines)		Wide range in grain size and substantial amounts of all intermediate particle sizes	GW	Well-graded gravels, gravel-sand mixtures, little or no fines
		Clean gravels (little or no fines)		Predominantly one size or a range of sizes with some intermediate sizes missing	GP	Poorly graded gravels, gravel-sand mixtures, little or no fines
		Gravels with fines (appreciable amount of fines)		Nonplastic fines (for identification procedures, see *ML* below)	GM	Silty gravels, poorly graded gravel-sand-silt mixtures
		Gravels with fines (appreciable amount of fines)		Plastic fines (for identification procedures, see *CL* below)	GC	Clayey gravels, poorly graded gravel-sand-clay mixtures
	Sands—more than half of coarse fraction is smaller than No. 7 sieve size (For visual classification the ¼-in size may be used as equivalent to the No. 7 sieve size)	Clean sands (little or no fines)		Wide range in grain sizes and substantial amounts of all intermediate particle sizes	SW	Well-graded sands, gravelly sands, little or no fines
		Clean sands (little or no fines)		Predominantly one size or a range of sizes with some intermediate sizes missing	SP	Poorly graded sands, gravelly sands, little or no fines
		Sands with fines (appreciable amount of fines)		Nonplastic fines (for identification procedures, see *ML* below)	SM	Silty sands, poorly graded sand-silt mixtures
		Sands with fines (appreciable amount of fines)		Plastic fines (for identification procedures, see *CL* below)	SC	Clayey sands, poorly graded sand-clay mixtures

Identification procedures on fraction smaller than No. 40 sieve size

Major divisions		Dry strength (crushing characteristics)	Dilatancy (reaction to shaking)	Toughness (consistency near plastic limit)	Group symbols‡	Typical names
Fine-grained soils—more than half of material is *smaller* than No. 200 sieve size¶	Silts and clays liquid limit less than 50	None to slight	Quick to slow	None	ML	Inorganic silts and very fine sands, rock flour, silty or clayey fine sands with slight plasticity
		Medium to high	None to very slow	Medium	CL	Inorganic clays of low to medium plasticity, gravelly clays, sandy clays, silty clays, lean clays
		Slight to medium	Slow	Slight	OL	Organic silts and organic silt-clays of low plasticity
	Silts and clays liquid limit greater than 50	Slight to medium	Slow to none	Slight to medium	MH	Inorganic silts, micaceous or diatomaceous fine sandy or silty soils, elastic silts
		High to very high	None	High	CH	Inorganic clays of high plasticity, fat clays
		Medium to high	None to very slow	Slight to medium	OH	Organic clays of medium to high plasticity
Highly organic soils		Readily identified by color, odor, spongy feel, and frequently by fibrous texture			Pt	Peat and other highly organic soils

Information required for describing soils	Laboratory classification criteria		
Give typical name; indicate approximate percentages of sand and gravel; maximum size; angularity, surface condition, and hardness of the coarse grains; local or geologic name and other pertinent descriptive information; and symbols in parentheses For undisturbed soils add information on stratification, degree of compactness, cementation, moisture conditions, and drainage characteristics Example: *Silty sand*, gravelly; about 20% hard, angular gravel particles $\frac{1}{2}$-in maximum size; rounded and subangular sand grains coarse to fine, about 15% nonplastic fines with low dry strength; well compacted and moist in place; alluvial sand; (*SM*)	$C_U = \dfrac{D_{60}}{D_{10}}$ Greater than 4 $C_C = \dfrac{(D_{30})^2}{D_{10} \times D_{60}}$ Between 1 and 3		
	Not meeting all gradation requirements for *GW*		
	Atterberg limits below A line, or *PI* less than 4	Above A line with *PI* between 4 and 7 are *borderline cases* requiring use of dual symbols	
	Atterberg limits above A line, with *PI* greater than 7		
	$C_U = \dfrac{D_{60}}{D_{10}}$ Greater than 6 $C_C = \dfrac{(D_{30})^2}{D_{10} \times D_{60}}$ Between 1 and 3		
	Not meeting all gradation requirements for *SW*		
	Atterberg limits below A line or *PI* less than 5	Above A line with *PI* between *borderline cases* requiring use of dual symbols	
	Atterberg limits below A line with *PI* greater than 7		

| Give typical name; indicate degree and character of plasticity, amount and maximum size of coarse grains; color in wet condition, odor if any, local or geological name, and other pertinent descriptive information, and symbol in parentheses

For undisturbed soils add information on structure, stratification, consistency in undisturbed and remoulded states, moisture and drainage conditions

Example:
Clayey silt, brown; slightly plastic; small percentage of fine sand; numerous vertical root holes; firm and dry in place; loess; (*ML*) |

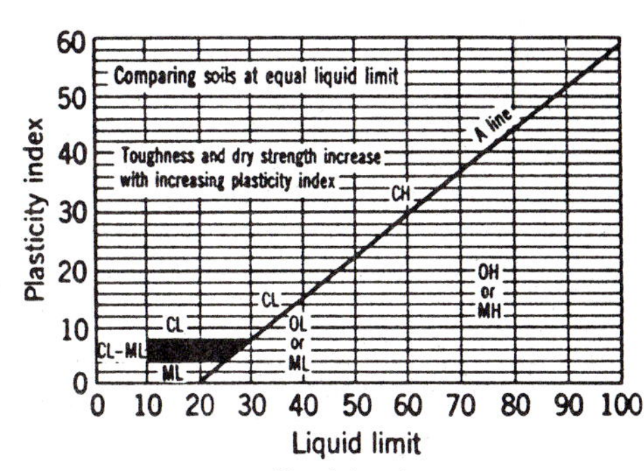

Plasticity chart
for laboratory classification of fine grained soils

†From A. A. Wagner, "The Use of the Unified Soil Classification System by the Bureau of Reclamation," in *Proc. Fourth Int. Conf. on Soil Mech. and Found. Eng.,* Vol. 1, London, 1957, pp. 125–128.

‡*Boundary classifications.* Soils possessing characteristics of two groups are designated by combinations of group symbols. For example *GW-GC*, well-graded gravel-sand mixture with clay binder.

§All sieve sizes on this chart are U.S. standard; see ASTM for details.

¶*Field identification procedure for fine-grained soils or fractions:* These procedures are to be performed on the minus No. 40 sieve size particles, approximately $\frac{1}{64}$ in. For field classification purposes, screening is not intended, simply remove by hand the coarse particles that interfere with the tests.

Dilatancy (reaction to shaking):

> After removing particles larger than No. 40 sieve size, prepare a pat of moist soil with a volume of about $\frac{1}{2}$ in³. Add enough water if necessary to make the soil soft but not sticky. Place the pat in the open palm of one hand and shake horizontally, striking vigorously against the other hand several times. A positive reaction consists of the appearance of water on the surface of the pat, which changes to a livery consistency and becomes glossy. When the sample is squeezed between the fingers, the water and gloss disappear from the surface, the pat stiffens and finally it cracks or crumbles. The rapidity of appearance of water during shaking and of its disappearance during squeezing assist in identifying the character of the fines in a soil.

> Very fine clean sands give the quickest and most distinct reaction whereas a plastic clay has no reaction. Inorganic silts, such as a typical rock flour, show a moderately quick reaction.

Dry Strength (crushing characteristics):

> After removing particles larger than No. 40 sieve size, mould a pat of soil to the consistency of putty, adding water if necessary. Allow the pat to dry completely by oven, sun, or air drying, and then test its strength by breaking and crumbling between the fingers. This strength is a measure of the character and quantity of the colloidal fraction contained in the soil. The dry strength increases with increasing plasticity.

> High dry strength is characteristic for clays of the *CH* group. A typical inorganic silt possesses only very slight dry strength. Silty fine sands and silts have about the same slight dry strength but can be distinguished by the feel when powdering the dried specimen. Fine sand feels gritty whereas a typical silt has the smooth feel of flour.

Toughness (consistency near plastic limit):

> After removing particles larger than the No. 40 sieve size, a specimen of soil about $\frac{1}{2}$ in³ size, is moulded to the consistency of putty. If too dry, water must be added, and if sticky, the specimen should be spread out in a thin layer and allowed to lose some moisture by evaporation. Then the specimen is rolled out by hand on a smooth surface or between the palms into a thread about $\frac{1}{8}$ in in diameter. The thread is then folded and rerolled repeatedly. During this manipulation the moisture content is gradually reduced and the specimen stiffens, finally loses its plasticity, and crumbles when the plastic limit is reached.

> After the thread crumbles, the pieces should be lumped together and a slight kneading action continued until the lump crumbles.

> The tougher the thread near the plastic limit and the stiffer the lump when it finally crumbles, the more potent is the colloidal clay fraction in the soil. Weakness of the thread at the plastic limit and quick loss of coherence of the lump below the plastic limit indicate either inorganic clay of low plasticity, or materials such as a kaolin-type clays and organic clays which occur below the A line.

> Highly organic clays have a very weak and spongy feel at the plastic limit.

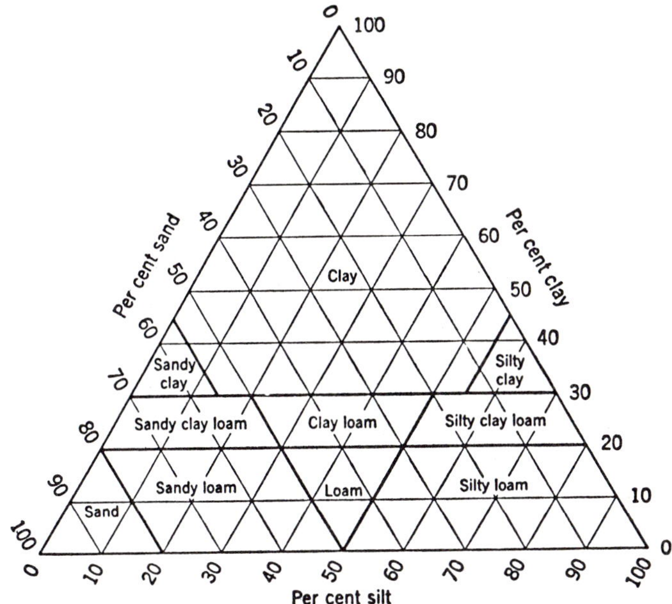

U.S. Bureau of Soils triangular classification chart. *(From D. S. Taylor, Fundamentals of Soil Mechanics, Wiley, New York, 1948.)*

Soil components and fractions†

Soil	Soil component	Symbol‡	Grain size range and description‡	Significant properties
Fine-grained components	Boulder	None	Rounded to angular, bulky, hard, rock particle, average diameter more than 12 in	Boulders and cobbles are very stable components, used for fills, ballast, and to stabilize slopes (rip-rap). Because of size and weight, their occurrence in natural deposits tends to improve the stability of foundations. Angularity of particles increases stability.
	Cobble	None	Rounded to angular, bulky, hard, rock particle, average diameter smaller than 12 in but larger than 6 in	
	Gravel Coarse Fine	G	Rounded to angular, bulky, hard, rock particle, passing 3-in sieve (76.2 mm) retained on No. 4 sieve (4.76 mm) 3 to $\frac{3}{4}$ in $\frac{3}{4}$ in to No. 4	Gravel and sand have essentially same engineering properties differing in degree. The No. 4 sieve is arbitrary division, and does not correspond to significant change in properties. They are easy to compact, little affected by moisture, not subject to frost action. Gravels are generally more perviously stable,
	Sand	S	Rounded to angular, bulky, hard, rock particle, passing No. 4 sieve (4.76 mm) retained on No. 200 Sieve (0.74 mm)	and resistant to erosion and piping than are sands. The well-graded sands and gravels are generally less pervious and more stable than those which are poorly graded (uniform gradation). Irregularity of particles increases the stability slightly. Finer uniform sand approaches the characteristics of silt: i.e., a decrease in its permeability and a reduction in its stability is apparent with an increase in moisture content.
	Coarse		No. 4 to 10 sieves	
	Medium		No. 10 to 40 sieves	
	Fine		No. 40 to 200 sieves	

	Silt	*M*	Particles smaller than No. 200 sieve (0.74 mm) identified by behavior: that is, slightly or nonplastic regardless of moisture and exhibits little or no strength when air-dried	Silt is inherently unstable, particularly when moisture is increased, with a tendency to become quick when saturated. It is relatively impervious, difficult to compact, highly susceptible to frost heave, easily erodible, and subject to piping and boiling. Bulky grains reduce compressibility; flaky grains, i.e., mica, diatoms, increase compressibility, produce an "elastic" silt.
Coarse-grained components	Clay	*C*	Particles smaller than No. 200 sieve (0.074 mm) identified by behavior: that is, can be made to exhibit plastic properties within a certain range of moisture and exhibits considerable strength when air-dried	The distinguishing characteristic of clay is cohesion or cohesive strength, which increases with decrease in moisture. The permeability of clay is very low; it is difficult to compact when wet and impossible to drain by ordinary means; when compacted it is resistant to erosion and piping, is not susceptible to frost heave, and is subject to expansion and shrinkage with changes in moisture. The properties are influenced not only by the size and shape (flat, platelike particles) but also by their mineral composition; i.e., the type of clay-mineral, and chemical environment or base exchange capacity. In general, the montmorillonite clay mineral has greatest, illite and kaolinite the least, adverse effect on the properties.
	Organic matter	*O*	Organic matter in various sizes and stages of decomposition	Organic matter present even in moderate amounts increases the compressibility and reduces the stability of the fine-grained components. It may decay causing voids or by chemical alteration change the properties of a soil; hence organic soils are not desirable for engineering uses.

†From A. A. Wagner, "The Use of the Unified Soil Classification System by the Bureau of Reclamation," in *Proc. Fourth Int. Conf. on Soil Mech. and Found. Eng.,* Vol. 1, London, 1957, pp. 125–134.

‡*Note:* The symbols and fractions were developed for the Unified Classification System. For field identification, $\frac{1}{4}$ in is assumed equivalent to the No. 4, and the No. 200 is defined as "about the smallest particle visible to the unaided eye." The sand fractions are not equal divisions on a logarithmic plot; the No. 10 was selected because of the significance attached to that size by some investigators. The No. 40 was chosen because the "Atterberg limits" tests are performed on the fraction of soil finer than the No. 40.

Engineering use chart†   (Note: comparisons range from 1 to 14; where 1 is excellent, 14 is unacceptable)

Typical names of soil groups	Group symbols	Important properties				Relative desirability for various uses									
						Rolled earth dams			Canal sections		Foundations		Roadways		
													Fills		
		Permeability when compacted	Shearing strength when compacted and saturated	Compressibility when compacted and saturated	Workability as a construction material	Homogeneous embankment	Core	Shell	Erosion resistance	Compacted earth lining	Seepage important	Seepage not important	Frost heave not possible	Frost heave possible	Surfacing
Well-graded gravels, gravel-sand mixtures, little or no fines	GW	Pervious	Excellent	Negligible	Excellent	...	...	1	1	...	...	1	1	1	3
Poorly graded gravels, gravel-sand mixtures, little or no fines	GP	Very pervious	Good	Negligible	Good	...	...	2	2	...	...	3	3	3	
Silty gravels, poorly graded gravel-sand-silt mixtures	GM	Semipervious to impervious	Good	Negligible	Good	2	4	...	4	4	1	4	4	9	5
Clayey gravels, poorly graded gravel-sand-clay mixtures	GC	Impervious	Good to fair	Very low	Good	1	1	...	3	1	2	6	5	5	1
Well-graded sands, gravelly sands, little or no fines	SW	Pervious	Excellent	Negligible	Excellent	...	...	3 (if gravelly) 4 (if gravelly)	6	...	...	2	2	2	4
Poorly graded sands, gravelly sands, little or no fines	SP	Pervious	Good	Very low	Fair	...	...	...	7 (if gravelly) 8 (if gravelly)	...	...	5	6	4	
Silty sands, poorly graded sand-silt mixtures	SM	Semipervious to impervious	Good	Low	Fair	4	5	...		5 (erosion critical)	3	7	8	10	6

Clayey sands, poorly graded sand-clay mixtures	*SC*	Impervious	Good to fair	Low	Good	3	2	· · ·	5	2	4	8	7	6	2
Inorganic silts and very fine sands, rock flour, silty or clayey fine sands with slight plasticity	*ML*	Semipervious to impervious	Fair	Medium	Fair	6	6	· · ·	· · ·	6 (erosion critical)	6	9	10	11	
Inorganic clays of low to medium plasticity, gravelly clays, sandy clays, silty clays, lean clays	*CL*	Impervious	Fair	Medium	Good to fair	5	3	· · ·	9	3	5	10	9	7	7
Organic silts and organic silt-clays of low plasticity	*OL*	Semipervious to impervious	Poor	Medium	Fair	8	8	· · ·	· · ·	7 (erosion critical)	7	11	11	12	
Inorganic silts, micaceous or diatomaceous fine sandy or silty soils, elastic silts	*MH*	Semipervious to impervious	Fair to poor	High	Poor	9	9	· · ·	· · ·	· · ·	8	12	12	13	
Inorganic clays of high plasticity, fat clays	*CH*	Impervious	Poor	High	Poor	7	7	· · ·	10	8 (volume change critical)	9	13	13	8	
Organic clays of medium to high plasticity	*OH*	Impervious	Poor	High	Poor	10	10	· · ·	· · ·	· · ·	10	14	14	14	
Peat and other highly organic soils	*Pt*														

†From A. A. Wagner, "The Use of the Unified Soil Classification System by the Bureau of Reclamation," in *Proc. Fourth Int. Conf. on Soil Mech. and Found. Eng.*, Vol. 1, London, 1957, pp. 125–134.

B

A COLLECTION OF TYPICAL SOIL PROPERTIES AND REPRESENTATIVE LABORATORY TEST VALUES[†]

Porosity, void ratio, and unit weight of typical soils in natural state[†]

Description	Poros-ity n, %	Void ratio e	Water content w,[‡] %	Unit weight[§]			
				γ_d, g/cm^3	γ, g/cm^3	γ_d, lb/ft^3	γ, lb/ft^3
Uniform sand, loose	46	0.85	32	1.43	1.89	90	118
Uniform sand, dense	34	0.51	19	1.75	2.09	109	130
Mixed-grained sand, loose	40	0.67	25	1.59	1.99	99	124
Mixed-grained sand, dense	30	0.43	16	1.86	2.16	116	135
Glacial till, very mixed-grained	20	0.25	9	2.12	2.32	132	145
Soft glacial clay	55	1.2	45	· · ·	1.77	· · ·	110
Stiff glacial clay	37	0.6	22	· · ·	2.07	· · ·	129
Soft slightly organic clay	66	1.9	70	· · ·	1.58	· · ·	98
Soft very organic clay	75	3.0	110	· · ·	1.43	· · ·	89
Soft bentonite	84	5.2	194	· · ·	1.27	· · ·	80

[†]*Source:* K. Terzaghi and R. B. Peck, *Soil Mechanics in Engineering Practice,* 2d ed., J. Wiley and Sons, New York, 1968.

[‡]w = water content when saturated, in percent of dry weight.

[§]$_d$ = unit weight in dry state; γ = unit weight in saturated state.

[†]*Note:* The values are not meant to replace representative laboratory or in situ testing of samples from a particular site. They are given for rough estimates and guides to more extensive investigation as required.

Typical values of soil index properties[a]

	Particle size and gradation				Voids[b]							Unit weight[c], lb/ft³						
	Approx. size range, mm		Approx. D_{10}, mm	Approx. range unif. coef., C_u	Void ratio			Porosity, %			Dry wt., γ_{dry}			Wet wt., γ_{wet}		Sub. wt., γ_{sub}		
	D_{max}	D_{min}			e_{max} (loose)	e_{cr}	e_{min} (dense)	n_{max} (loose)	n_{min} (dense)	Min. (loose)	100% Mod. AASHO	Max. (dense)	Min. (loose)	Max. (dense)	Min. (loose)	Max. (dense)		
Granular materials																		
1. Uniform Materials																		
a. Equal spheres (theoretical values)	· · ·	· · ·	· · ·	1.0	0.92	· · ·	0.35	47.6	26.0									
b. Standard Ottawa *sand*	0.84	0.59	0.67	1.1	0.80	0.75	0.50	44	33	92	· · ·	110	93	131	57	69		
c. Clean, uniform *sand* (fine or medium)	· · ·	· · ·	· · ·	1.2–2.0	1.0	0.80	0.40	50	29	83	115	118	84	136	52	73		
d. Uniform, inorganic *silt*	0.05	0.005	0.012	1.2–2.0	1.1	· · ·	0.40	52	29	80	· · ·	118	81	136	51	73		
2. Well-graded materials																		
a. Silty *sand*	2.0	0.005	0.02	5–10	0.90	· · ·	0.30	47	23	87	122	127	88	142	54	79		
b. Clean, fine to coarse *sand*	2.0	0.05	0.09	4–6	0.95	0.70	0.20	49	17	85	132	138	86	148	53	86		
c Micaceous *sand*	· · ·	· · ·	· · ·	· · ·	1.2	· · ·	0.40	55	29	76	· · ·	120	77	138	48	76		
d. Silty *sand* and *gravel*	100	0.005	0.02	15–300	0.85	· · ·	0.14	46	12	89	· · ·	146[d]	90	155[d]	56	92		
Mixed soils																		
1. Sandy or silty *clay*	2.0	0.001	0.003	10–30	1.8	· · ·	0.25	64	20	60	130	135	100	147	38	85		
2. Skip-graded silty *clay* with stones or rock fragments	250	0.001	· · ·	· · ·	1.0	· · ·	0.20	50	17	84	· · ·	140	115	151	53	89		
3. Well-graded *gravel, sand, silt,* and *clay* mixture	250	0.001	0.002	25–1000	0.70	· · ·	0.13	41	11	100	140	148[e]	125	156[e]	62	94		

449

Typical values of soil index properties[a] (*Continued*)

	Particle size and gradation				Voids[b]						Unit weight[c], lb/ft^3						
	Approx. size range, mm		Approx. D_{10}, mm	Approx. range unif. coef., C_u	Void ratio			Porosity, %		Dry wt., γ_{dry}			Wet wt., γ_{wet}		Sub. wt., γ_{sub}	
	D_{max}	D_{min}			e_{max} (loose)	e_{cr}	e_{min} (dense)	n_{max} (loose)	n_{min} (dense)	Min. (loose)	100% Mod. AASHO	Max. (dense)	Min. (loose)	Max. (dense)	Min. (loose)	Max. (dense)
Clay soils																
1. *Clay* (30 to 50% clay sizes)	0.05	0.5μ	0.001	· · ·	2.4	· · ·	0.50	71	33	50	105	112	94	133	31	71
2. Colloidal *clay* (−0.002 mm ≥ 50%)	0.01	10 Å	· · ·	· · ·	12	· · ·	0.60	92	37	13	90	106	71	128	8	66
Organic soils																
1. Organic *silt*	· · ·	· · ·	· · ·	· · ·	3.0	· · ·	0.55	75	35	40	· · ·	110	87	131	25	69
2. Organic *clay* (30 to 50% clay sizes)	· · ·	· · ·	· · ·	· · ·	4.4	· · ·	0.70	81	41	30	· · ·	100	81	125	18	62

[a]*Source:* B. K. Hough, *Basic Soils Engineer,* 2d ed., Ronald, New York, 1969.

[b]Granular materials may reach e_{max} when dry or only slightly moist. Clays can reach e_{max} only when fully saturated.

[c]Granular materials reach minimum unit weight when at e_{max} and with hygroscopic moisture only. Clays reach minimum unit wet weight when fully saturated at e_{max}. The unit submerged weight of any saturated soil is the unit wet weight minus the unit weight of water.

[d]Applicable for very compact glacial till. Unusually high unit weight values for tills are sometimes due not only to an extremely compact condition but to unusually high specific gravity values.

[e]Applicable for hardpan.

General note: Tabulation is based on $G = 2.65$ for granular soil, $G = 2.7$ for clays, and $G = 2.6$ for organic soils.

Permeability and drainage characteristics of soils†

Coefficient of permeability k in cm per sec (log scale)

10^2	10^1	1.0	10^{-1}	10^{-2}	10^{-3}	10^{-4}	10^{-5}	10^{-6}	10^{-7}	10^{-8}	10^{-9}

Drainage

Good	Poor	Practically impervious

Soil types

Clean gravel	Clean sands, clean sand and gravel mixtures	Very fine sands, organic and inorganic silts, mixtures of sand silt and clay, glacial till, stratified clay deposits, etc.	"Impervious" soils, e.g., homogeneous clays below zone of weathering

"Impervious" soils modified by effects of vegetation and weathering

Direct determination of k

Direct testing of soil in its original position—pumping tests; reliable if properly conducted; considerable experience required

Constant-head permeameter; little experience required

Indirect determination of k

Falling-head permeameter; reliable; little experience required	Falling-head permeameter; unreliable; much experience required	Falling-head permeameter; fairly reliable; considerable experience necessary

Computation from grain size distribution; applicable only to clean cohesionless sands and gravels		Computation based on results of consolidation tests; reliable; considerable experience required

†*Source:* H. R. Cedergren, *Seepage, Drainage and Flow Nets*, J. Wiley and Sons, New York, 1967.

Typical values for compression index C_c[†]

Soil	Liquid limit	Plastic limit	Virgin compression index		References
			Undisturbed	Remolded	
Boston blue clay	41	20	0.35	0.21	Lambe and Whitman, 1969
Chicago clay	58	21	0.42	0.22	Lambe and Whitman, 1969
Louisiana clay	74	26	0.33	0.29	Lambe and Whitman, 1969
New Orleans clay	79	26	0.29	0.26	Lambe and Whitman, 1969
Fort Union clay	89	20	0.26		Lambe and Whitman, 1969
Mississippi loess	23–43	17–29	0.09–0.23		Sheeler, 1968
Delaware organic silty clay	84	46	0.95		Schmidt & Gould, 1968
Indiana silty clay	36	20	0.21	0.12	
Marine sediment, BC Canada	130	74	2.3		Finn et al., 1971

[†]*Source:* H. F. Winterkorn and H.-Y. Fang, "Soil Technology and Engineering Properties of Soils," Chap. 2 in *Foundation Engineering Handbook,* H. F. Winterkorn and H.-Y. Fang (eds.), Van Nostrand Reinhold, New York, 1975, pp. 67–120.

Angle of shearing resistance (internal friction) of cohesionless soils†

No.	General description	Grain shape	D_{10}, mm	C_u	Loose		Dense	
					e	ϕ, deg	e	ϕ, deg
1	Ottawa standard sand	Well rounded	0.56	1.2	0.70	28	0.53	35
2	Sand from St. Peter sandstone	Rounded	0.16	1.7	0.69	31	0.47	37‡
3	Beach sand from Plymouth, MA	Rounded	0.18	1.5	0.89	29		
4	Silty sand from Franklin Falls Dam site, NH	Subrounded	0.03	2.1	0.85	33	0.65	37
5	Silty sand from vicinity of John Martin Dam, CO	Subangular to subrounded	0.04	4.1	0.65	36	0.45	40
6	Slightly silty sand from the shoulders of Ft. Peck Dam, MT	Subangular to subrounded	0.13	1.8	0.84	34	0.54	42
7	Screened glacial sand, Manchester, NH	Subangular	0.22	1.4	0.85	33	0.60	43
8§	Sand from beach of hydraulic fill dam, Quabbin Project, MA	Subangular	0.07	2.7	0.81	35	0.54	46
9	Artificial, well-graded mixture of gravel with sands No. 7 and No. 3	Subrounded to subangular	0.16	68	0.41	42	0.12	57
10	Sand for Great Salt Lake fill (dust gritty)	Angular	0.07	4.5	0.82	38	0.53	47
11	Well-graded, compacted crushed rock	Angular	. . .	. . .	. . .	. . .	0.18	60

†*Source:* R. D. Holtz and W. D. Kovacs, *An Introduction to Geotechnical Engineering,* Prentice-Hall, Englewood Cliffs, NJ, 1981.

‡The angle of internal friction of the undisturbed St. Peter sandstone is larger than 60° and its cohesion so small that slight finger pressure or rubbing, or even stiff blowing at a specimen by mouth, will destroy it.

§Angle of internal friction measured by direct shear test for No. 8, by triaxial tests for all others.

Values of the shear strength parameters c' and ϕ' from undrained tests with pore-pressure measurement†

Material‡	Clay fraction <0.002 mm, %	Optimum water content, %	Optimum dry density, lb/ft³	Water content as tested, %	Cohesion c', lb/ft²	Angle of shearing resistance ϕ', deg
Moraine						
	<1	8	131	6.8	890	42
	<1	8.8	131	8.8	750	44
	<1	8.8	131	10.2	115	41
Moraine						
	2	12.2	120	9.6	1440	36
	2	12.2	120	12.2	980	37
	2	12.2	120	14.8	0	36
Boulder clay	4	7.8	135	6.8	1300	38
	4	7.8	135	7.8	980	38
	4	7.8	135	8.8	270	37
Boulder clay	10	9.5	130	9.5	1440	33
Boulder clay	19	10.7	126	8.8	2350	27
	19	10.7	126	10.7	1660	24
				12.4	490	28
Residual clay*	25	47.0	72	48.2	1200	31
Residual clay	44	23.0	98	20.0	2590	24
	44	23.0	98	23.0	2260	21
	44	23.0	98	25.0	620	24

†*Source:* A. W. Bishop and D. J. Henkel, *The Measurement of Soil Properties in the Triaxial Test,* Arnold, London, 1957.

‡The compactive effort used in sample preparation approximated in all cases to that used in the *standard compaction test.* Samples 4 in in diameter and 8 in in height were used except where indicated by an asterisk*.

Representative values for angle of shearing resistance (internal friction) †

Soil	Type of test‡		
	Unconsolidated-undrained UU, deg	Consolidated-undrained CU, deg	Consolidated-drained CD, deg
Gravel			
Medium size	40–55		40–55
Sandy	35–50		35–50
Sand			
Loose dry	28.5–34		
Loose saturated	28.5–34		
Dense dry	35–46		43–50
Dense saturated	1–2 less than dense dry		43–50
Silt or silty sand			
Loose	20–22		27–30
Dense	25–30		30–35
Clay	0 if saturated	14–20	20–42

†J. E. Bowles, *Foundation Analysis and Design,* 2d ed., McGraw-Hill, New York, 1977.

‡See a laboratory manual on soil testing for a complete description of these tests, e.g., chap. 9, Ref. 1, 2, or 3

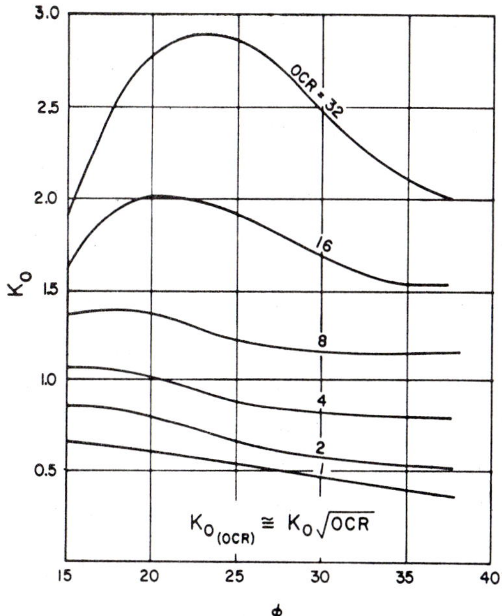

Relation between overconsolidation ratio (OCR), K_0, and drained angle of shearing resistance (internal friction). *(From J. E. Bowles, Foundation Analysis and Design, 2d ed., McGraw-Hill, New York, 1977.)*

Coefficient of earth pressure at rest, K_0: average values on first loading[†]

Material	Clay fraction <0.002, %	Activity $\dfrac{PI}{\%\ \text{clay}}$	Type of test	Value of K_0
Moraine, compacted	1	· · ·	Undrained	0.36
Boulder clay, compacted	10	0.44	Undrained	0.43
Residual clay, compacted	21	0.44	Undrained	0.42
Boulder clay, compacted	19	0.59	Undrained	0.56
Residual clay, compacted	20	1.55	Undrained	0.66
Sandy clay, undisturbed	15	0.90	Drained	0.43
Clay, remoulded	40	0.60	Drained	0.70
Loose sand, saturated	· · ·	· · ·	Drained	0.46
Dense sand, saturated	· · ·	· · ·	Drained	0.37
Raw sugar, dry	· · ·	· · ·	Drained	0.50
Loose lead shot, dry	· · ·	· · ·	Drained	0.44

[†]*Source:* A. W. Bishop and D. J. Henkel, *The Measurement of Soil Properties in the Triaxial Test,* Arnold, London, 1957.

Typical values of the coefficient of consolidation c_v from dissipation tests[†]
Total stress increment 0–30 lb/in^2

Type of soil	Clay fraction <0.002 mm, %	Optimum water content, %	Optimum dry density, lb/ft^3	Water content as tested, %	Value of c_v, cm^2/s
Compacted[‡]:					
Moraine	<1	9.5	129	9.1	1.2
				10.4	1.8×10^{-2}
Boulder clay	4	7.8	134	8.4	1.9×10^{-2}
				9.3	3.7×10^{-3}
Boulder clay	10	9.5	130	10.4	2.9×10^{-3}
Boulder clay	19	10.7	126	10.7	7.6×10^{-4}
				13.5	4.3×10^{-4}
Residual clay	20	27.5	94	28.2	6.6×10^{-4}
				33.0	8.6×10^{-5}
Residual clay	21	16.0	111	16.1	2.7×10^{-2}
				19.8	1.9×10^{-4}
Undisturbed:					
Boulder clay	~19	· · ·	· · ·	14.9	3.6×10^{-4}
Residual clay	31	· · ·	· · ·	26.3	2.9×10^{-3}

[†]*Source:* S. W. Bishop and D. J. Henkel, *The Measurement of Soil Properties in the Triaxial Test,* Arnold, London, 1957.

[‡]The compactive effort used in sample preparation approximated in all cases to that used in the *standard compaction test.*

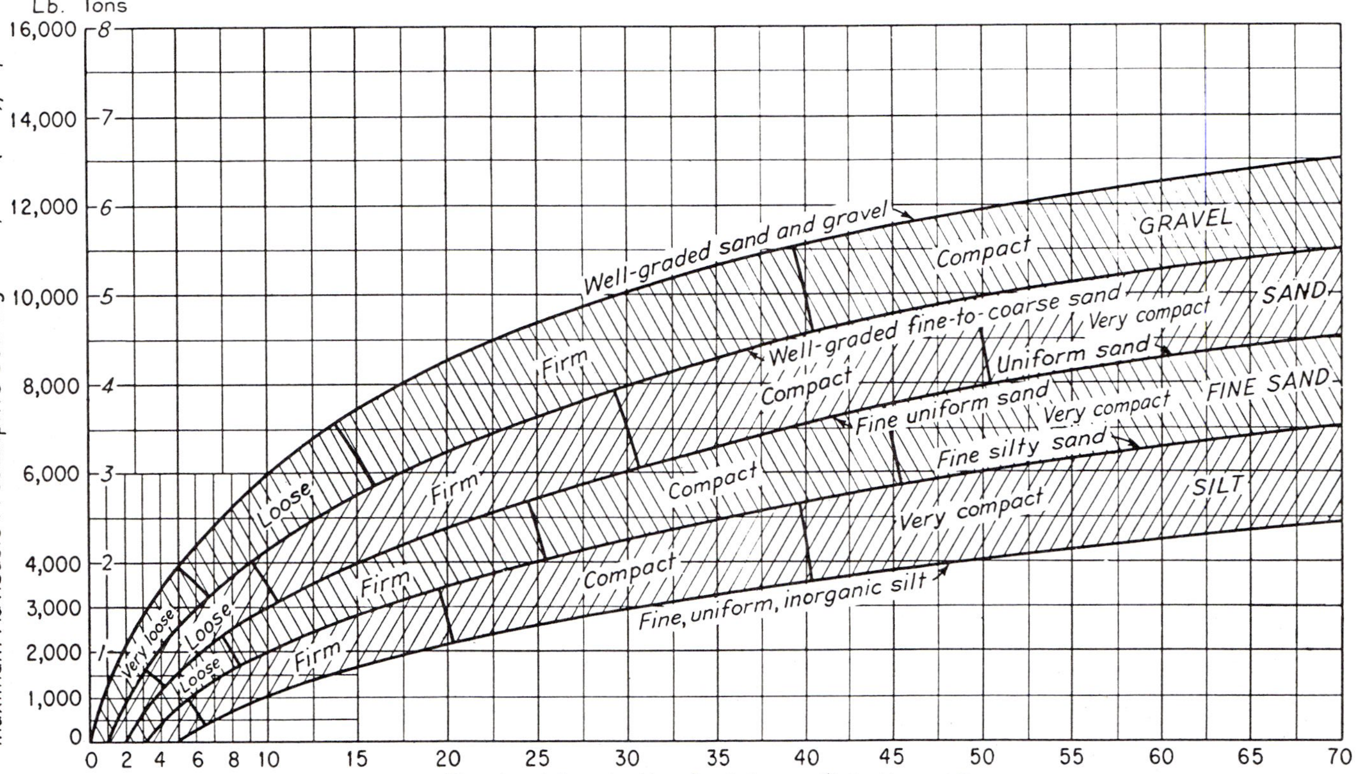

*Number of blows of 140-lb pin-guided drive weight falling 30 in per blow required to drive a split-barrel spoon with a 2 in. outside diameter 12 in.

†Values must be corrected for effect of weak substrata, high ground water, and surcharge.

Presumptive bearing capacity values, granular soils. (*From B. K. Hough, Basic Soils Engineering, Ronald, New York, 2d ed., 1969.*)

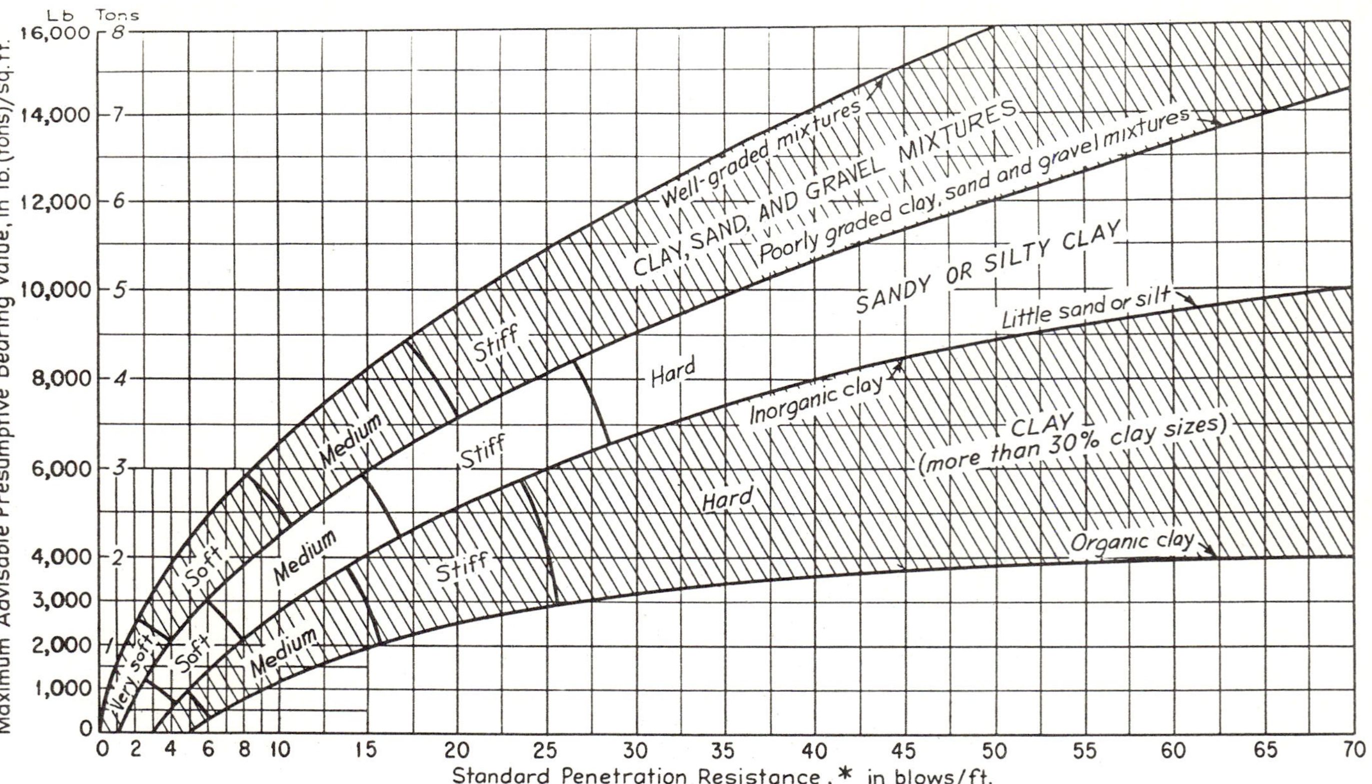

*Number of blows of 140-lb. pin-guided weight falling 30 in. per blow required to drive a split-barrel sample spoon with a 2-in. outside diameter 12 in.

†Higher values may be used for precompressed (or compacted) clays of low sensitivity than for normally loaded or extra-sensitive clays.

Presumptive bearing capacity values, clay and mixed soils. (*From B. K. Hough, Basic Soils Engineering, 2d ed., Ronald, New York, 1969.*)

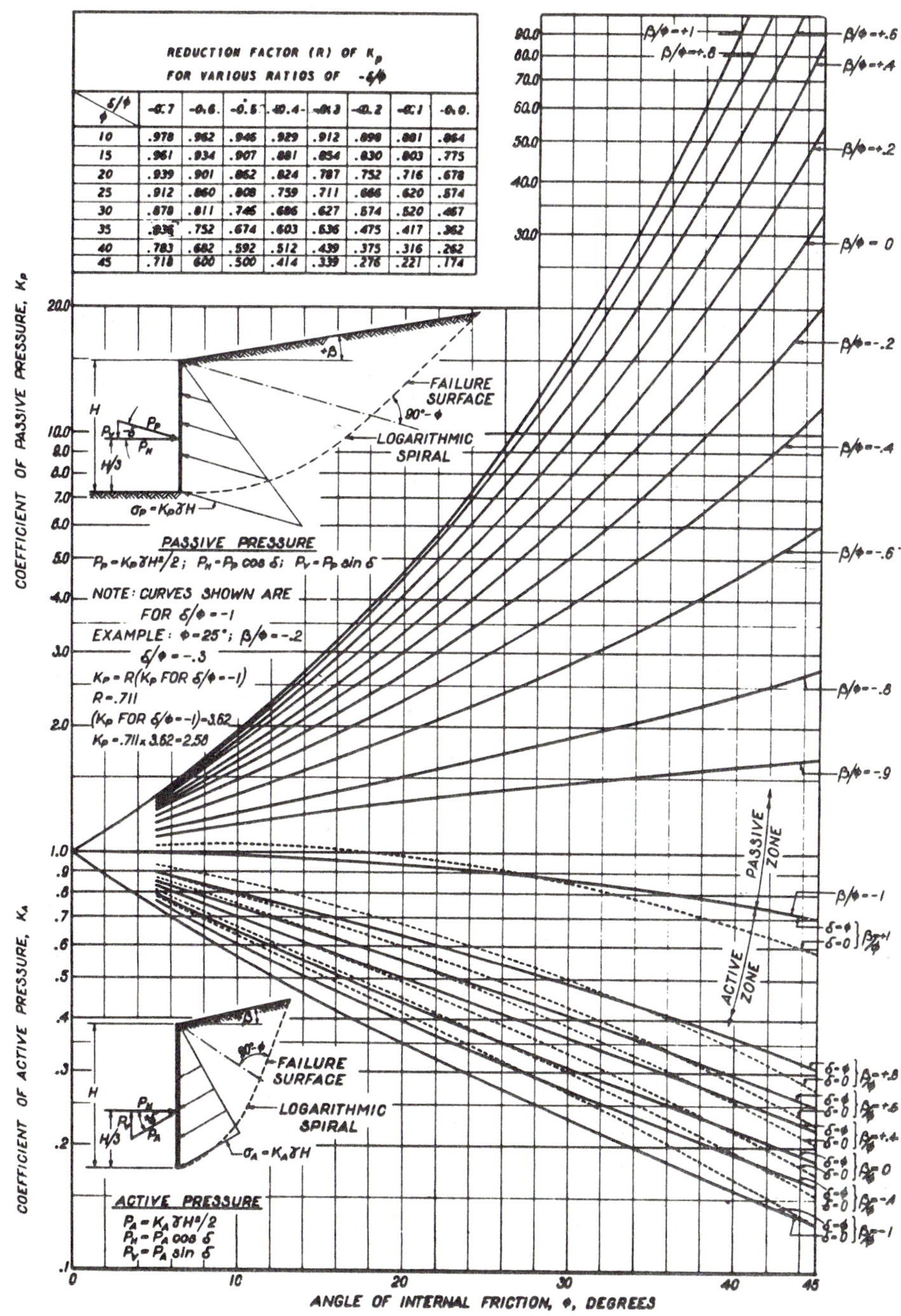

Active and passive earth pressure coefficients by the logarithmic spiral method. (*After Caquot and Kensel, 1948; Refer to Design Manual: Soil Mechanics, Foundations, and Earth Structures,* NAVDOCK DM-7, *Department of the Navy, Bureau of Yards and Docks, Washington, DC, 1967.*)

TECHNICAL DETAILS OF VARIOUS IMPACT PILE HAMMERS

C.1

    Conmaco, Inc.
    820 Kansas Avenue
    P.O. Box 5097
    Kansas City, Kansas 66119
    (913) 371-3930

C.2

    L. B. Foster Company
    4825 N. Scott Street—Suite 35
    Schiller Park, Illinois 60176
    (312) 671-0380

C.3

    MKT Geotechnical Systems
    Box 793
    Dover, New Jersey 07801
    (201) 361-2800

C.4

    Raymond International, Inc.
    P.O. Box 27456
    Houston, Texas 77027
    (713) 623-1500

C.5

    Vulcan Iron Works, Inc.
    2909 Riverside Drive
    Chattanooga, Tennessee 37406
    (615) 698-1581

Conmaco: Single-acting, cable-type pile hammer specifications

Specifications	Hammer model									
	50	65	80	100	115	140	160	200	300	5300
Rated striking energy, ft·lb	15,000	19,500	26,000	32,500	37,375	42,000	48,750	60,000	90,000	150,000
Blows/min—normal stroke and no set†	60	60	50	50	50	60	60	60	55	46
Steam pressure at hammer, lb/in^2	80	100	85	100	120	110	120	120	150	160
Air pressure at hammer, lb/in^2	80	100	85	100	120	110	120	120	150	
Steam consumption, lb/h	2794	3230	2580‡	2945‡	3425‡	3844	4182	5563	6944	8064
Volume of free air (adiabatic), ft^3/min	565	625	730	820	910	829	899	1195	1471	1711
Recommended compressor size	600	750	1050	1050	1200	1500	1800			
Normal stroke, in	36	36	39	39	39	36	36	36	36	60
Boiler horsepower—ASME rating§	81	94	75	85	99	111	121	161	201	234
$\sqrt{EW}$ rating (square root of energy × striking weight)	8660	11,258	14,442	18,028	20,732	24,248	28,145	34,641	51,961	67,082
Diameter of piston, in	13½	13½	16½	16½	16½	19	19	21¾	21¾	21¾
Length of hammer, ft-in	13'0"	13'0"	14'2"	14'2"	14'2"	13'10¼"	13'10¼"	15'0"	16'5"	20'9½"
Inlet connection, diameter in inches	2	2	2½	2½	2½	3	3	3	3	4
Size of hoisting line, diameter in inches	¾	¾	⅞	⅞	⅞	1	1	1	1	1
Weight of striking parts, lb	5000	6500	8000	10,000	11,500	14,000	16,250	20,000	30,000	30,000
Net weight, lb (female jaws)	10,600	12,100	17,280	19,280	20,780	30,750	33,200	44,560	55,390	57,130
Shipping weight, lb (female jaws)	11,000	12,500	18,000	20,000	21,500	31,500	33,950	45,560	56,390	58,330

†Hammer operating speed is a function of many variables, including penetration resistance, cushion material, pressure, stroke, etc. Speeds given are for hard driving under ideal conditions. Energy is not a function of speed.

‡Special reduced cylinders.

§Boiler horsepower based on 6 ft^2 of heating surface.

Kobe (via L. B. Foster Co.): Diesel hammer specifications

Specification	Model		
	KC25	KC35	KC45
Overall length, mm (ft-in)	5101 (16'9″)	5125 (16'10″)	5460 (17'11″)
Diameter of anvil, mm (in)	590 (23.23)	700 (27.56)	800 (31.50)
Distance between center of hammer and lead pipe, mm (in)	430 (16.93)	490 (19.29)	580 (22.83)
Depth of hammer, mm (in)	900 (35.43)	1000 (39.37)	1190 (46.85)
Max. width of hammer, mm (in)	780 (30.71)	900 (35.43)	1000 (39.37)
Total weight, kg (lb)	5500 (12,100)	7900 (17,400)	11,200 (24,700)
Weight of ram, kg (lb)	2500 (5510)	3500 (7720)	4500 (9920)
No. of blows, blows/min	39–60	39–60	39–60
Energy output per blow, kg-m (ft-lb)	7500 (54,200)	10,500 (75,900)	13,500 (97,600)
Explosion pressure on pile, kg (lb)	96,000 (211,600)	133,000 (293,200)	170,000 (374,800)
Fuel consumption (light oil), l/h (U.S. gal/h)	9–12 (2.38–3.17)	12–16 (3.17–4.23)	17–21 (4.49–5.55)
Lube oil consumption, l/h (U.S. gal/h)	1.0 (0.26)	1.4 (0.37)	2.0 (0.53)
Fuel tank capacity, l (U.S. gal)	40 (10.6)	41.5 (11.0)	65 (17.2)
Lube oil chamber capacity, l (U.S. gal)	30 (7.9)	30 (7.9)	30 (7.9)
Cooling water tank capacity, l (U.S. gal)	100 (26.4)	150 (39.6)	250 (66.0)
Lubricant for ram and anvil	Motor oil SAE 40–50	Motor oil SAE 40–50	Motor oil SAE 40–50

MKT: pile hammer specifications

Variable-stroke steam or air hammers

Model	Energy rating, ft·lb (kg·m)	Ram weight, lb (kg)	Strokes/ minute	Compressor size, ft³/min (m³/min)	Net weight, lb (kg)
MS-350	15,000–31,000 (2074–4287)	7716 (3500)	40–50	900 (25.2)	10,000 (4540)
MS-500	22,000–44,000 (3043–6085)	11,000 (5000)	40–50	1200 (34)	15,250 (6908)

Double-acting hammers

	Model		
Specifications	9B3	10B3	11B3
Energy rating, ft·lb (kg·m)	8750 (1210)	13,100 (1812)	19,150 (2648)
Ram weight, lb (kg)	1600 (726)	3000 (1361)	5000 (2268)
Strokes/min	145	105	95
Air consumption, ft³/min (m³/ min)	600 (16.99)	750 (21.24)	900 (25.49)
Compressor size, ft³/min (m³/ min)	900 (25.49)	1200 (33.98)	1200 (33.98)
Net weight, lb (kg)	7000 (3175)	10,850 (4922)	14,000 (6350)
Shipping weight with fittings, lb (kg)	7100 (3221)	11,000 (4990)	14,200 (6441)

Double-acting sheet pile hammers

	Model		
Specifications	5	6	7
Energy rating, ft·lb (kg·m)	1000 (138)	2500 (346)	4150 (574)
Ram weight, lb (kg)	200 (91)	400 (181)	800 (363)
Strokes/min	300	275	225
Air consumption, ft³/min (m³/min)	250 (7.08)	400 (11.33)	450 (12.74)
Compressor size, ft³/min (m³/min)	365 (10.34)	600 (16.99)	750 (21.24)
Net weight, lb (kg)	1500 (680)	2900 (1315)	5000 (2268)
Shipping weight with fittings, lb (kg)	1560 (708)	2970 (1347)	5075 (2302)

MKT: pile hammer specifications (*Continued*)

Single-acting with interchangable ram sizes

Specification	Model				
	DE-20B/30B			DE-50B/70B	
	DE-20B	DE-30B	DE-50B	DE-50B	DE-70B
Energy rating, ft·lb (kg·m)	12,000–17,000 (1660–2351)	16,800–23,800 (2323–3292)	30,000–42,500 (4149–5878)	30,000–42,500 (4149–5878)	42,000–59,500 (5809–8229)
Ram weight, lb (kg)	2000 (907)	2800 (1270)	5000 (2268)	5000 (2268)	7000 (3175)
Strokes/min	40–50	40–50	40–50	40–50	40–50
Width overall, in (mm)	20 (508)	20 (508)	26 (660)	26 (660)	26 (660)
Length overall, ft-in (mm)	15'4" (4673)	15'4" (4673)	14'9" (4496)	15'10" (4826)	15'10" (4826)
Shipping weight with drive cap, lb (kg)	7400 (3357)	8200 (3720)	13,335 (6050)	14,035 (6366)	16,035 (7275)

Double-acting—convertible to single-acting

Specification	Model					
	DA-15C		DA-35C		DA-55B	
	Double-acting	Single-acting	Double-acting	Single-acting	Double-acting	Single-acting
Energy rating, ft·lb (kg·m)	6000–8200 (835–1134)	6600–9350 (913–1293)	15,600–21,000 (2157–2904)	16,800–23,800 (2323–3291)	31,200–38,200 (4315–5283)	30,000–42,500 (4149–5878)
Ram weight, lb (kg)	1100 (500)	1100 (500)	2800 (1270)	2800 (1270)	5000 (2268)	5000 (2268)
Strokes/min	86–90	40–50	78–82	40–50	78–82	40–50
Width overall, in (mm)	20 (508)	20 (508)	20 (508)	20 (508)	26 (660)	26 (660)
Length overall, ft-in (mm)	13'11" (4242)	13'11" (4242)	18 (5486)	18 (5486)	17'4" (5283)	17'4" (5283)
Shipping weight with drive cap, lb (kg)	5700 (2585)	5700 (2585)	11,750 (5330)	11,750 (5330)	18,325 (8312)	18,325 (8312)

Raymond: Pile hammer specifications

Specification	Single-acting Model number										
	1	1S	0	2/0	3/0	4/0	5/0	22X	30X	8/0	40X
Rated energy (ft·lb)	15,000	19,500	24,375	32,500	40,625	48,750	56,875	56,900	75,000	81,250	100,000
Ram weight, lb	5000	6500	7500	10,000	12,500	15,000	17,500	22,050	30,000	25,000	40,000
Normal stroke, in	36	36	39	39	39	39	39	31	30	39	30
Operating speed, blows/min[a]	60	58	52	50	48	46	44	58	70	33–38	64
Diameter of piston, in	$13\frac{1}{2}$	$13\frac{1}{2}$	18	18	18	18	18	18	26	18	26
Total weight, lb	11,000	12,500	16,100	18,550	21,225	23,800	26,450	31,750	52,000	34,000	62,000
Overall height, ft-in	12′9″	12′9″	15′9″	15′0″	15′7″	16′1″	16′9″	17′8″	19″1″	19′4″	19′1″
Steam pressure at hammer, lb/in^{2b}	80	104	110	110	120	120	135	135	135	135	135
Minimum boiler horsepower, hp[c]	40	40	50	55	70	85	100	100	200	140	250
Minimum-size steam hose, in[d]	$1\frac{1}{2}$	$1\frac{1}{2}$	2	2	$2\frac{1}{2}$	$2\frac{1}{2}$	3	3	3	3	3

Raymond: Pile hammer specifications (*Continued*)

	Differential					Hydraulic	
	Model Number						
	Steam					Hydraulic	
Specification	15M[e]	65C	80C	125CX	150C	65CH	80CH
Rated energy, ft·lb	15,000	19,500	24,450	40,625	48,750	19,500	24,450
Ram weight, lb	5000	6500	8000	15,000	15,000	6500	8000
Normal stroke, in	18	16	$16\frac{1}{2}$	15	18	16	$16\frac{1}{2}$
Equivalent stroke, ft	3.0	3.0	3.05	2.71	3.25	3.0	3.05
Operating speed, blows/min[a]	75–90	100–110	95–105	110–120	95–105	128–136	110–120
Diameter of large piston, in	$13\frac{7}{16}$	16	17	$23\frac{1}{4}$	$23\frac{1}{4}$	$2\frac{1}{2}$	$2\frac{21}{32}$
Diameter of small piston, in	$7\frac{5}{16}$	$9\frac{1}{4}$	$10\frac{3}{16}$	$13\frac{5}{8}$	$13\frac{5}{8}$	$1\frac{7}{16}$	$1\frac{19}{32}$[f]
Total weight, lb	10,305	14,675	17,885	32,800	32,500	14,615	17,780
Overall height, ft-in	20'10"	11'8"	12'2"	15'9"	15'9"	12'1"	11'10"
Operating pressure at hammer, lb/in^2	120	120	120	120	120	5,000	5,100
Minimum boiler horsepower, hp[c]	60	70	80	150	150	NA	NA
Recommended size steam hose, in[d]	2	2	$2\frac{1}{2}$	3	3	NA	NA

[a]Hammer operating speed is function of many variables including penetration resistance, capblock, steam pressure, stroke, etc. Speeds given are for hard driving under ideal conditions. Delivered energy is not a function of operating speed.

[b]Provide minimum 15 lb/in^2 higher pressure at boiler to allow for losses in pipe and hose.

[c]Boiler horsepower based on 10 ft^2 of heating surface per horsepower.

[d]Steam piping should be next size larger.

[e]Internal hammer used with special Raymond Step-Taper mandrel.

[f]Piston rod diameter—hammer has no lower piston.

Vulcan: pile hammer specifications

Specifications		Onshore (land) hammers—single-acting and differential-acting Model numbers												
	1	50C	65C	06	80C	08	010	140C	012	014	016	200C	020	030
Operating data:														
Rated striking energy, ft·lb	15,000	15,100	19,200	19,500	24,450	26,000	32,500	36,000	39,000	42,000	48,750	50,200	60,000	90,000
(kJ)	(20.3)	(20.5)	(26.0)	(26.4)	(33.1)	(35.3)	(44.1)	(48.8)	(52.9)	(57.6)	(66.1)	(68.1)	(81.3)	(122.0)
Operating cycle[a]	S	D	D	S	D	S	S	D	S	S	S	D	S	S
Blows/min—normal stroke, no set	60	117	115	60	109	50	50	101	50	59	58	95	59	54
Nominal stroke, in	36	15.5	15.5	36	16.5	39	39	15.5	39	36	36	15.5	36	36
(mm)	(914)	(394)	(394)	(914)	(419)	(991)	(991)	(394)	(991)	(914)	(914)	(394)	(914)	(914)
Rated operating pressure at hammer, lb/in^2	80	120	150	100	120	83	105	140	125	110	120	142	120	150
(bar)	(5.52)	(8.27)	(10.34)	(6.89)	(8.27)	(5.72)	(7.24)	(9.65)	(8.62)	(7.58)	(8.27)	(9.79)	(8.27)	(10.34)
Steam consumption, lb/h	2794	4312	5244	3230	6210	4158	5022	7279	5835	6111	6814	8970	7500	8550
(kg/h)	(1267)	(1956)	(2370)	(1465)	(2817)	(1886)	(2278)	(3302)	(2647)	(2772)	(3091)	(4069)	(3402)	(3878)
Required boiler horsepower, hp	81	125	152	94	180	125	150	211	175	200	210	260	250	300
Air consumption (adiabatic), ft^3/min	565	880	991	625	1245	880	1002	1425	1075	1161	1275	1746	1634	1833
(m^3/min)	(160)	(24.9)	(28.1)	(17.7)	(35.3)	(24.9)	(28.4)	(40.4)	(30.4)	(32.9)	(36.1)	(49.4)	(46.3)	(51.9)
Required air compressor size, ft^3/min	600	900	1050	750	1300	900	1050	1600	1200	1300	1300	1800	1800	2000
(m^3/min)	(17.0)	(25.5)	(29.7)	(21.2)	(36.8)	(25.5)	(29.7)	(45.3)	(34.0)	(36.8)	(36.8)	(51.0)	(51.0)	(56.6)
Dimensional data:														
Bore, in	13.5	14	14	13.5	17	16.5	16.5	19	16.5	19	19	21¾	21¾	21¾
(mm)	(343)	(356)	(356)	(343)	(432)	(419)	(419)	(483)	(419)	(483)	(483)	(552)	(552)	(552)
Net area of piston,[b] in^2	133.51	55.91	55.91	133.51	81.51	197.92	197.92	99.40	197.92	255.26	255.26	132.73	335.75	335.75
(cm^2)	(861.4)	(360.7)	(360.7)	(861.4)	(525.9)	(1276.9)	(1276.9)	(641.3)	(1276.9)	(1646.8)	(1646.8)	(856.3)	(2166.1)	(2166.1)
Length of hammer (overall), ft-in	12′9″	11′0″	12′1″	12′9″	12′7″	14′10″	15′7″	12′3″	15′7″	13′11″	13′11″	13′11″	15′7″	16′4″
(m)	(3.88)	(3.35)	(3.68)	(3.88)	(3.84)	(4.52)	(4.75)	(3.73)	(4.75)	(4.24)	(4.24)	(4.24)	(4.75)	(4.98)
Distance across female jaws, in	20	20	20	20	26	26	26	32	26	32	32	37	37	37
(mm)	(508)	(508)	(508)	(508)	(660)	(660)	(660)	(813)	(660)	(813)	(813)	(940)	(940)	(940)
Width of female jaws, in	8¼	8¼	8¼	8¼	9¼	9¼	9¼	11¼	9¼	11¼	11¼	11¼	11¼	11¼
(mm)	(210)	(210)	(210)	(210)	(235)	(235)	(235)	(286)	(235)	(286)	(286)	(286)	(286)	(286)
Largest outside diameter of pile,[c] in	18	18	18	18	24	24	24	29	24	29	29	34	34	34
(mm)	(457)	(457)	(457)	(457)	(610)	(610)	(610)	(737)	(610)	(737)	(737)	(864)	(864)	(864)
Size of hose, in	2	2	2	2	2½	2½	2½	3	2½	3	3	4	3	3
(mm)	(51)	(51)	(51)	(51)	(64)	(64)	(64)	(76)	(64)	(76)	(76)	(102)	(76)	(76)

Vulcan: pile hammer specifications (*Continued*)

Onshore (land) hammers—single-acting and differential-acting
Model numbers

Specifications	1	50C	65C	06	80C	08	010	140C	012	014	016	200C	020	030
Weight data:														
Weight of striking parts, lb	5000	5000	6500	6500	8000	8000	10,000	14,000	12,000	14,000	16,250	20,000	20,000	30,000
(kg)	(2268)	(2268)	(2948)	(2948)	(3629)	(3629)	(4536)	(6350)	(5443)	(6350)	(7371)	(9072)	(9072)	(13,608)
Net weight of hammer, lb	9700	11,782	14,886	11,200	17,885	16,750	18,750	27,984	20,750	27,500	30,250	39,000	41,670	53,420
(kg)	(4400)	(5344)	(6752)	(5080)	(8112)	(7598)	(8505)	(12,693)	(4412)	12,474	13,721	17,690	18,942	24,231
Shipping weight of hammer, lb	10,100	15,279	15,279	12,100	18,635	17,500	19,500	28,724	21,500	28,250	32,000	39,833	42,520	54,020
(kg)	(4581)	(6930)	(6930)	(5488)	(8453)	(7938)	(8845)	(13,029)	(9752)	(13,041)	(14,515)	(18,068)	(19,287)	(24,503)

Offshore hammers—single-acting

Model number

Specifications	016	020	030	340	530	360	540	560	5100	5150	5250	6300
Operating data:												
Rated striking energy, ft·lb	48,750	60,000	90,000	120,000	150,000	180,000	204,500	312,500	500,000	750,000	1,250,000	1,800,000
(kJ)	(66)	(82)	(122)	(163)	(204)	(245)	(278)	(425)	(679)	(1019)	(1695)	(2446)
Blows/min—normal stroke, no set	58	59	54	62	42	60	48	47	48	46	38	42
Nominal stroke, in	36	36	36	36	60	36	60	60	60	60	60	72
(mm)	(914)	(914)	(914)	(914)	(1524)	(914)	(1524)	(1524)	(1524)	(1524)	(1524)	(1829)
Rated operating pressure at hammer, lb/in^2	120	120	150	120	150	130	130	150	150	175	200	235
(bar)	(8.27)	(8.27)	(10.34)	(8.27)	(10.34)	(8.96)	(8.96)	(10.34)	(10.34)	(12.07)	(13.79)	(16.20)
Steam consumption, from and at 212°F, lb/h	4182	5563	6944	12,230	8064	17,460	14,126	20,897	35,977	45,426	56,537	96,723
(100°C, kg/h)	(1897)	(2523)	(3150)	(5547)	(3658)	(7920)	(6407)	(9479)	(16,319)	(20,605)	(25,640)	(43,873)
Boiler horsepower, from and at 212°F, (100°C), hp	121	161	201	354	234	506	409	606	1043	1317	1639	2804
Required boiler horsepower, 60°F (16°C) feedwater, hp	150	200	250	500	300	650	500	750	1300	1700	2000	3400
Air consumption (adiabatic), ft^3/min	1275	1634	1883	3400	2076	4652	3755	5410	9326	11,399	13,846	22,536
(m^3/min)	(36.1)	(46.3)	(51.9)	(96.3)	(58.8)	(131.7)	(106.3)	(153.2)	(264.1)	(322.8)	(392.1)	(638.2)

Dimensional data:

Bore, in	19	21.75	21.75	32	21.75	36	32	36	48	49	49	62
(mm)	(483)	(552)	(552)	(813)	(552)	(914)	(813)	(914)	(1219)	(1245)	(1245)	(1575)
Net area of piston, in^2	255.26	335.75	335.75	740.63	335.75	943.22	740.63	943.22	1696.46	1772.64	1772.64	2819.59
cm^2	(1646.84)	(2166.12)	(2166.12)	(4778.25)	(2166.12)	(6085.28)	(4778.25)	(6085.28)	(10,944.88)	(11,436.36)	(11,436.36)	(18,190.87)
Length of hammer (overall), ft-in	13'7⅜"	14'8"	16'4⅞"	18'7"	20'4⅞"	19'0"	22'7"	23'0"	27'4"	26'3½"	28'3"	30'0½"
(m)	(4.15)	(4.47)	(5.00)	(5.66)	(6.22)	(5.79)	(6.88)	(7.01)	(8.33)	(8.01)	(8.61)	(9.15)
Distance across male jaws, in	54	54	54	80	54^2	88	80	88	120	120^d	144^e	144
(mm)	(1372)	(1372)	(1372)	(2032)	(1372)	(2235)	(2032)	(2235)	(3048)	(3048)	(3638)	(3658)
Width of male jaws, in	10¼	10¼	10½	14	$10¼^f$	18¾	14	18¾	22	22^d	22^e	22
(mm)	(260)	(260)	(260)	(356)	(260)	(476)	(356)	(476)	(559)	(559)	(559)	(559)
Largest outside diameter of pile, ^{c} in	48	48	72	48^2	72	72	72	72	96	96	120	120
(mm)	(1219)	(1219)	(1829)	(1219)	(1829)	(1829)	(1829)	(1829)	(2438)	(2438)	(3048)	(3048)
Size of Hose, in	3	3	3	3(2)	3	4(2)	4(3)	4(3)	4(3)	6(2)	6(2)	6(2)
[mm]	[76]	[76]	[76]	[76(2)]	[76]	[102(2)]	[102(3)]	[102(3)]	[102(3)]	[152(2)]	[152(2)]	[152(2)]

Weight data:

Weight of striking parts, lb	16,250	20,000	30,000	40,000	30,000	60,000	40,900	62,500	100,000	150,000	250,000	300,000
(kg)	(7371)	(9072)	(13,608)	(18,144)	(13,608)	(27,216)	(18,552)	(28,350)	(45,359)	(68,039)	(133,398)	(136,078)
Net weight of hammer, lb	33,340	43,785	55,410	85,819	57,680	124,830	94,595	134,060	219,000	275,000	470,000	575,000
(kg)	(15,123)	(19,861)	(25,134)	(38,920)	(26,163)	(56,622)	(42,900)	(60,809)	(99,337)	(124,738)	(143,256)	(260,816)
Shipping weight of hammer, lb	34,500	45,000	57,500	88,639	60,000	129,000	98,000	138,430	228,000	283,000	490,000	593,700
(kg)	(15,649)	(20,412)	(26,082)	(40,199)	(27,216)	(58,513)	(44,000)	(62,791)	(103,419)	(128,367)	(149,352)	(269,298)
Weight of pipe cap, lb	9600	10,000	10,000	31,300	10,000	42,600	32,800	43,100	69,500	75,000	88,000	139,285
(kg)	(4354)	(4536)	(4536)	(14,197)	(4536)	(19,323)	(14,878)	(19,550)	(31,525)	(34,014)	(26,822)	(63,179)
Weight of leaders, lb	16,850	16,850	16,850	30,000	16,850	64,750	33,700	68,000	110,000	110,000	137,000	137,000
(kg)	(7643)	(7643)	(7643)	(13,608)	(7643)	(9370)	(15,286)	(30,844)	(49,895)	(49,895)	(41,758)	(62,142)
Assembled weight, lb	59,790	70,635	82,260	147,119	84,530	232,180	164,500	245,160	398,500	460,000	695,000	851,285
(kg)	(27,120)	(32,039)	(37,313)	(66,721)	(38,342)	(105,315)	(74,603)	(111,203)	(180,757)	(208,627)	(211,836)	(386,136)

Vulcan: pile hammer specifications (*Continued*)

Offshore pile hammer—differential-acting

Specifications	200C	Specifications	200C
Operating data:			
Rated striking energy, ft·lb	50,200	Equivalent stroke, in	30.1
(kJ)	(69)	(mm)	(765)
Blows/min—normal stroke, no set	98	Striking velocity at impact, ft/s	12.75
Nominal stroke, in	15.5	(m/s)	(3.89)
(mm)	(394)		
Rated operating pressure at hammer, lb/in^2	142	Largest diameter of pile,^{c} in	48
(bar)	(9.79)	(mm)	(1219)
Steam consumption, from and at 212°F, lb/h	8970	Size of hose, in	4
(kg/h)	(4069)	(mm)	(102)
Boiler horsepower, hp	260	*Weight data:*	
Air consumption (adiabatic), ft^3/min	1746	Weight of striking parts, lb	20,000
(m^3/min)	(49.4)	(kg)	(9072)
		Net weight of hammer, lb	43,315
Dimensional data:		(kg)	(19,194)
Diameter of small piston, in	13	Shipping weight of hammer, lb	43,530
(mm)	(330)	(kg)	(19,745)
Area of small piston, in^2	132.73	Weight of pipe cap, lb	10,000
(cm^2)	(856.3)	(kg)	(4536)
Length of hammer (overall), ft·in	12′11″	Weight of leaders, lb	16,850
(m)	(3.94)	(kg)	(7643)
Distance across male jaws, in	54	Assembled weight lb	69,165
(mm)	(1372)	(kg)	(31,373)
Width of male jaws, in	10¼		
(mm)	(260)		

^{a}S—single-acting hammers; D—differential-acting hammers.

^{b}Single-acting hammers—surface area of rod end of piston; differential-acting hammers—area of small piston.

^{c}Permitted by standard jaws.

^{d}44 × 22 in (3652 × 559 mm) male jaws available for 120-in (3048-mm) pile.

^{e}120 × 22 in (3048 × 559 mm) male jaws available for 96-in (2438-mm) pile.

^{f}80 × 14 in (2032 × 356 mm) male jaws available for 72-in (1829-mm) pile.

TECHNICAL DETAILS OF VARIOUS VIBRATORY PILE DRIVERS

D.1

L. B. Foster Company
4825 N. Scott Street—Suite 35
Schiller Park, Illinois 60176
(312) 671-0380

D.2

MKT Geotechnical Systems
Box 793
Dover, New Jersey 07801
(201) 361-2800

Foster vibratory pile drivers-extractors†

Specification	Model		
	1200	1700	4000
Type	Hydraulic	Hydraulic	Hydraulic
Eccentric moment, in·lb (kg·cm)	1200	1740	4000
	(1410)	(2000)	(4810)
Dynamic force, tons (t)	40.3	49	111
	(35.0)	(44.5)	(101.1)
Frequency, vibrations/min	0–1425	0–1400	0–1400
Pile amplitude, in (mm)	$\frac{1}{4}$–$\frac{1}{2}$	$\frac{5}{16}$–$\frac{3}{4}$	$\frac{5}{16}$–$1\frac{1}{4}$
	(6–13)	(8–19)	(8–32)
Maximum output of hydraulic motor, hp (kW)	85	147	200
	(63)	(110)	(226)
Maximum operating pressure, lb/in^2 (bar)	4000	5000	5000
	(275)	(350)	(350)
Maximum flow, gal/min (L/min)	40	60	120
	(151)	(230)	(450)
Working line pull for extraction, tons (t)	20	30	45
	(18)	(27)	(41)
Pile clamping force, tons (t)	58/118	100/200	100/200
	(53/107)	(90/180)	(90/180)
Suspended weight, lb‡ (kg)	8200	12,900	18,800
	(3795)	(5850)	(8530)
Shipping weight, lb§ (kg)	13,600	20,000	32,300
	(6169)	(9072)	(14,700)
Length, in (cm)	59	83	118
	(150)	(210)	(300)
Width, in (cm)	22	22	22
	(56)	(56)	(81)
Throat width, in (cm)	12	12	12
	(30)	(30)	(30)
Hammer height, in (cm)	$82\frac{2}{3}$	75	$76\frac{1}{2}$
	(210)	(190)	(194)
Head height,¶ in (cm)	$29\frac{1}{2}$	36	36
	(75)	(91)	(91)

†Electric series also available.

‡Suspended weight includes driver-extractor, typical driving-extracting head, hydraulic hoses.

§Shipping weight includes driver-extractor, power pack, driving-extracting head, hydraulic hoses.

¶Head height. For typical driving-extracting head. Some units can be tandemized with same unit.

L. B. Foster Company reserves the right to make revisions to the specifications without notification.

MKT vibratory pile drivers-extractors

Specification	Model			
	V-5	V-16	V-20	V-36
Driving force, tons (t)	30	78.3	100.5	182
	(27.2)	(71)	(91.1)	(165.2)
Hydraulic motor output, hp	59	161	295	550
Steady-state frequency, cycles/min	1450	1750	1650	1600
Driving amplitude, in (mm)	0.5	0.47	0.58	0.75
	(12.7)	(12)	(14.7)	(19)
Maximum line pull, tons (t)	20	50	50	80
	(18.1)	(45.4)	(45.4)	(72.6)

GEOTEXTILE (FILTER FABRIC) DIRECTORY*

All North American manufacturers and distributors of geotextiles were asked recently to participate in a survey listing their products. Those companies that responded to the survey are listed below, along with the products they manufacture or supply to the industry.

E. I. du Pont de Nemours and
   Co., Inc.
Explosives Products Division
Wilmington, DE 19898
302-774-2421

Texel, Inc.
485, des Erables
St-Elzéar
Beauce-Nord
P. Quebec
GOS 2J0
1-800-463-8866 in Canada
1-418-387-5910 out of
   Canada
Contact M. Jacques Côté

Nicolon Corp./Baycor Div.
Suite 1990 Peachtree Corners
   Plaza
Norcross, GA 30073
404-447-6272
Contact Joe L. Burchfield

American Enka Co.
Enka, NC 28728
704-667-9476

Synflex Northwest Inc.
1840—130 St., #13
Bellevue, WA 98005
206-885-2524
Contact Rod Newton

Exxon Chemical Americas
380 Interstate North, Suite
   375
Atlanta, GA 30339
404-955-2300
Contact Harold Hartley or Ted
   Gailer

Mirafi, Inc.
P.O. Box 240967
Charlotte, NC 28224
(704)523-7477   or   (800)438-
   1855
Contact R.G. Carroll

Hoechst Fibers Ind.
1515 Broadway
New York, NY 10036
212-790-9459
Contact Alan Sterling

Crown Zellerbach
3720 Grant Street
Washougal WA 98671
206-835-9547 or 1-800-426-
   0700
Contact T. G. Collins

Dominion Textile Inc.
1950 Sherbrooke St. West
Montreal. Que. H3H 1E7
514-937-5711
Contact Peer Moldvar

Amoco Fabrics Co.
550 Interstate North Pkwy.
Atlanta, GA 30099
404-955-0935
Contact Larry Hale

*Modified from *Industrial Fabrics Products Review*, vol. 58, no. 3, July 1981, pp. 24–26.

Collins and Aikman Corp.
1803 North Main St.
Roxboro, NC 27573
919-599-1111
Contact Leigh C. Woodall, Jr.

Phillips Fibers Corp.
P.O. Box 66
Greenville, SC 29602
803-242-6600
Contact L. T. Gray

Carthage Mills Erosion Control
  Co.
125 West 66th St.
Cincinnati, OH 45216
513-242-2740
Contact Polly Toennis

Rhone Poulenc/Nonwoven
  Dept.
69 rue Casimir Perier
95870 Bezons—France
(3)GHF.33.40
Contact G. Cazier

Foss Manufacturing Co., Inc.
One Whitney Ave., P.O. Box
  175
Haverhill, MA 01830
617-374-0121
Contact Brian Jeffrey

Burlington Industrial Fabrics
  Co.
Link Dr.
Rockleigh, NJ 07647
201-767-4586
Contact Peter E. Stevenson

Chicopee
Peachtree Corners Plaza, Suite
  1950
Norcross, GA 30071
404-449-4960 or 1-800-241-
  7566
Contact Marshall Scott

Manufacturer or distributor	Fabric trade name	Structure	Fiber	Width	Weight
Du Pont	Typar 3201, 3341, 3401, 3601	Nonwoven	Polypropylene	151 to 197 in	2 oz/yd^2, 3.4 oz/yd^2, 4 oz/yd^2, 6 oz/yd^2
Du Pont	Reepav	Nonwoven	Polyester	148 in	3 oz/yd^2
Texel	Texel 7607, 7609, 7612, 7618, 7643	Nonwoven	Polyester	Up to 40 ft	240 g/m^2 to 2000 g/m^2
Nicolon	40/30	Woven	Polypropylene	12 ft and multiples of 12 ft	
Nicolon	70/06, 70/20, 100/08	Woven	Polypropylene	6-ft multiples	
Nicolon	100/05	Woven	Polyvinylidene chloride	6-ft multiples	
Nicolon	Geolon 200	Woven	Polypropylene	12.5 and 16.5 ft	
Nicolon	Geolon 400	Woven	Polypropylene	12-ft multiples	
Nicolon	Geolon 1250	Woven	Polypropylene	16.7 ft	
Enka	Enkamat		Nylon	39 in	265 g/m^2 and 405 g/m^2
Enka	Enkadrain		Nylon and polyester	39 in	735 g/m^2
Enka	Stabilenka	Nonwoven	Polyester	3.5 and 7 ft	2 oz/yd^2 3 oz/yd^2 4 oz/yd^2
Synflex	Permealiner	Woven and nonwoven	Polypropylene		
Exxon	Industrial Fabrics Style 49862	Woven	Polypropylene	12.5 and 15.5ft	3.4 oz/yd^2
Celanese	Mirafi 500X	Woven	Polypropylene	12.5 and 17.5 ft	
Celanese	Mirafi 140N	Nonwoven	Polypropylene	15 ft	
Celanese	Mirafi 600X	Woven	Polypropylene	12.5ft	
Celanese	Mirafi 700X	Woven	Polypropylene	6 ft and 12 ft	

Hoechst	Trevira	Nonwoven	Polyester	Up to 16 ft	
Crown Zellerbach	Fibretex 150, 200 300, 400, 600	Nonwoven	Polypropylene	13 and 17.5 ft	150 g/m^2 to 600 g/m^2
Crown Zellerbach	Fibretex Ten-1	Woven	Polypropylene	12.5 and 15.5 ft	3.4 oz/yd^2
Dominion Textile	Penroad 50, 150, 250, 350, 450, 550	Nonwoven	Polyester	3.65 m	200 g/m^2, 270 g/m^2, 340 g/m^2, 400 g/m^2, 540 g/m^2, 680 g/m^2
Dominion Textile	Penroad 100, 500	Woven	Polypropylene	3.82 m	125 g/m^2 and 150 g/m^2
Amoco	ProPex 4545, 4551, 4553	Nonwoven	Polypropylene	14.5 ft	4.5 oz/yd^2, 6 oz/yd^2, 8 oz/yd^2
Amoco	ProPex 1325, 1369	Nonwoven	Polypropylene	14.5 ft	6.5 oz/yd^2
Amoco	ProPex 2002	Woven	Polypropylene	14.5 ft	4.5 oz/yd^2
Collins & Aikman	Diamond 8	Woven	Nylon or polyester	60 in	11.5 oz/yd^2
Collins & Aikman	Contain 4	Woven	Polyester	60 in	13.75 oz/yd^2
Collins & Aikman	Retain	Woven	Nylon	72 in	9 oz/yd^2
Collins & Aikman	SB-375	Stitchbound	Polyester	56 to 144 in	3.75 oz/yd^2 and 8 oz/yd^2
Phillips	Petromat	Nonwoven	Polypropylene	75 and 150 in	4.3 oz/yd^2
Phillips	Supac	Nonwoven	Polypropylene	Up to 180 in	4 oz/yd^2, 5 oz/yd^2, 16 oz/yd^2
Carthage Mills	Poly-Filter X	Woven	Polypropylene	6 to 84 ft (custom fab.)	7.2 oz/yd^2
Carthage Mills	Poly-Filter GB	Woven	Polypropylene		6.6 oz/yd^2
Carthage Mills	Filter-X	Woven	Polyvinylidene chloride	6 to 72 ft (custom fab.)	11.6 oz/yd^2
Rhone-Poulenc	Bidim	Nonwoven	Polyester	2.7 m, 4.2 m, 5.3 m	50 g/m^2 to 1000 g/m^2

Manufacturer or distributor	Fabric trade name	Structure	Fiber	Width	Weight
Foss	F-55	Nonwoven	Polyester	36 in, 75 in, 150 in, 186 in (other widths up to 186 in available upon request)	5.5 oz/yd^2
Foss	F-65	Nonwoven	Polyester	36 in, 75 in, 150 in, 186 in (other widths up to 186 in available upon request)	6.5 oz/yd^2
Foss	F-90	Nonwoven	Polyester	36 in, 75 in, 150 in, 186 in (other widths up to 186 in available upon request)	9 oz/yd^2
Foss	F-180	Nonwoven	Polyester	36 in, 75 in, 150 in, 186 in (other widths up to 186 in available upon request)	18 oz/yd^2

GEOMEMBRANE (FLEXIBLE MEMBRANE), BENTONITE, AND OTHER LINER SYSTEMS DIRECTORY†

A. Polymeric membrane liners
   1. Polymer producers
   2. Manufacturers of polymeric membrane sheeting
   3. Fabricators of liners
   4. Installing contractors
B. Bentonite producers and suppliers
C. Other liner materials
D. Miscellaneous organizations in the liner industry

A. Polymeric membrane liners
   1. Polymer Producers

Dow Chemical Company
2040 Dow Center
P.O. Box 1847
Midland, MI 48640
Contact: John L. Hartman, product sales manager, Chlorinated Polyethylene Designed Products Department
Phone: 517-636-1000

E. I. du Pont de Nemours and Co., Inc.
Elastomer Chemicals Dept.
Wilmington, DE 19898
Contact: Richard J. Arhart
Phone: 302-999-3160
Contact: Gerald E. Fisher
3245 Sunnyside Ave.
Brookfield, IL 66513
Phone: 312-485-6881

E. I. du Pont de Nemours and Co., Inc.
Explosive Products Div.
1007 Market St.
Wilmington, DE 19898
Contact: Mr. T. J. Enright, geotextile product manager
Phone: 302-774-1000

†From U.S. EPA Report "Lining of Waste Impoundment and Disposal Facilities" by Metrecon, Inc., Oakland, Calif. 94623, R. Landreth, EPA Project Officer, 1983.

Exxon Chemical Co.
Elastomer Technology Div.
P.O. Box 45
Linden, NJ 07036
Phone: 201-474-0100

Hercules Incorporated
910 Market St.
Wilmington, DE 19899
Contact: Norman C. Mac-
Arthur, product manager,
Elastomers
Phone: 302-575-6293

Monsanto Industrial Chemicals
Co.
260 Springside Dr.

Akron, OH 44313
Contact: Gary E. O'Connor,
project manager, Com-
mercial Development
Dept., Rubber Chemicals
Div.
Phone: 216-666-4111

Pantasote, Inc.
26 Jefferson St.
Passaic, NJ 07055
Contact: Larry Kamp
Phone: 201-777-8500

Polysar, Ltd.
Technical Development Div.
Vidal St.

Sarnia, Ontario
Canada N7T 7M2
Contact: Carl Hancock
Phone: 519-337-8251

Shell Chemical Company
605 N. Main St.
Altamont, IL 62411
Contact: Larry Watkins
Phone: 618-483-6517

Uniroyal Chemical Company
Spencer St.
Naugatuck, CT 06488
Contact: Allen Crepeau
Phone: 203-723-3825

A. 2. Manufacturers of Polymeric Membrane Sheeting

Burke Rubber Company
2250 S. Tenth St.
San Jose, CA 95112
Contact: D. Kutnewsky, man-
ager, Flexible Membranes
Phone: 408-297-3500

Carlisle Syntech Systems
Division of Carlisle Corporation
P.O. Box 7000
Carlisle, PA 17013
Contact: William Witherow,
product manager, Liners
and Waterproofing
Phone: 717-245-7000

Columbus Coated Fabrics
1280 N. Grant St.
Columbus, OH 43216
Contact: Lee Fishbein
Phone: 614-225-6069

Cooley, Inc.
50 Esten Ave.
Pawtucket, RI 02862
Contact: Paul Eagleston, vice
president
Phone: 401-724-9000

Gaco
P.O. Box 88698
Seattle, WA 98188

Contact: Earle Johnson,
San Jose, CA
Phone: 415-341-5661

B. F. Goodrich Company
Engineered Rubber Products
Div.
500 S. Main St.
Akron, OH 44318
Contact: R. D. Cunningham,
sales manager, Environ-
mental Products
Phone: 216-379-2226

Gundle Lining Systems, Inc.
1340 East Richey Rd.
Houston, TX 77073
Contact: Richard K. Schmidt,
president
Phone: 713-433-8564

Harte & Company
16 E. 34th St.
New York, NY 10016
Contact: R. H. Dickinson
Phone: 212-481-1210

Mainline, Inc.
3292 S. Highway 97
Redmond, OR 97756
Contact: DeWitt Maine
Phone: 503-548-4207

Oxford, Inc.
220 Rainbow Blvd. North
Niagara Falls, NY 14303
Contact: Albert Hooper,
vice president
Phone: 716-283-6900

Pantasote, Inc.
26 Jefferson St.
Passaic, NJ 07055
Contact: Larry Kamp
Phone: 201-777-8500

Plymouth Rubber
Company
104 Revere St.
Canton, MA 02021
Contact: Charles Neese
Phone: 617-828-0220

Protective Coatings, Inc.
1602 Birchwood Ave.
Ft. Wayne, IN 46803
Contact: Elmo Murrell, presi-
dent
Phone: 219-422-7503

Sarnafil (U.S.), Inc.
Canton Commerce Center
Canton, MA 02021
Contact: Clark Gunness
Phone: 617-828-5400

Schlegel Lining Technology, Inc.
P.O. Box 7730
The Woodlands, TX 77380
Contact: James M. Price, president
Phone: 713-273-3066 (Conroe)
713-350-1813 (Houston)

Shelter-Rite, Inc.
Division of Seaman Corp.
P.O. Box 331

Millersburg, OH 44654
Contact: Dr. Bala Venktaraman, vice president, Research and Development
Phone: 216-674-2015

Stevens Elastomeric & Plastics Products, Inc.
27 Payson Ave.
Easthampton, MA 01073
Contact: Arnold G. Peterson
Phone: 413-527-0700

Tenneco Chemicals, Inc.
P.O. Box 189
Piscataway, NJ 08805
Contact: Bob Hayes/Kent Turner
Phone: 201-356-2550

Uniroyal, Inc.
312 N. Hill St.
Mishawaka, IN 46544
Contact: D. L. Zimmerman
Phone: 212-256-8181

A. 3. Fabricators of Liners

Advance Construction Specialties
P.O. Box 17212
Memphis, TN 38117
Contact: H. M. VanNieuwenhuyze, president
Phone: 901-362-0980

Brown and Brown
P.O. Drawer 269
Mobile, AL 36601
Contact: Sam Brown, president
Phone: 205-479-6581

Electra Tarp, Inc.
Park Centre
7241 Whipple Ave., N.W.
North Canton, OH 44720
Contact: Bob Fulmer, president
Phone: 216-497-1496

Environetics, Inc.
9824 Industrial Dr.
Bridgeview, IL 60455
Contact: Ray Winters, president
Phone: 312-585-6000

Fabrico Manufacturing Corp.
1300 W. Exchange Ave.
Chicago, IL 60609
Contact: Jay Sabath, sales manager
Phone: 312-254-4211

McKittrick Mud Co.
P.O. Box 3343

Bakersfield, CA 93305
Contact: Bill Wheeler, president
Phone: 805-325-5013

Palco Linings, Inc.
7571 Santa Rita Circle
P.O. Box 919
Stanton, CA 90680
Contact: Richard Cain, president
Phone: 714-898-0867

Poly-Plastics
P.O. Box 299
Springfield, OH 45501
Contact: Roland Harmer, president
Phone: 513-323-4625

M. Putterman & Co.
2221 West 43rd St.
Chicago, IL 60609
Contact: A. Berman, president
Phone: 312-927-4120

Revere Plastics
16 Industrial Ave.
Little Ferry, NJ 07643
Contact: Larry Smith, president
Phone: 201-641-0777

Southwest Canvas Mfg. Co.
Oklahoma City, OK
Contact: Richard C. Nelson, manager
Phone: 405-672-3355

Staff Industries
240 Chene St.
Detroit, MI 48207
Contact: Charles E. Staff, president
Phone: 313-259-1820
800-526-1368

Synflex Industries, Inc.
2004-750 Jervis St.
Vancouver, B.C.
Canada V6E 2A9
Contact: Gerald W. Salberg, president
Phone: 604-682-3621

Watersaver Company, Inc.
5890 E. 56th Ave.
Commerce City, CO
P.O. Box 16465
Denver, CO 80216
Contact: Bill Slifer, vice president
Phone: 303-623-4111

Manufacturers who also fabricate: (see section A.2.)

Burke Rubber Company
Carlisle Tire and Rubber Co.
B. F. Goodrich
Schlegel Lining Technology, Inc.
Gundle Lining Systems

A. 4. Installing Contractors

Crestline Supply Corp.
2987 South 300 West
Salt Lake City, UT 84115
Contact: Guy Woodward
Phone: 801-487-2233

Enviroclear, Inc.
P.O. Box 242
Falls Village, CT 06031
Contact: Don Thompson, president
Phone: 212-997-0100
      518-325-3332

Gaston Containment Systems, Inc.
1853 N. Main St.
P.O. Box 1157
El Dorado, KS 67042
Contact: John Saenz
Phone: 316-321-5140

Globe Linings, Inc.
1901 East Wardlow Rd.
Long Beach, CA 90807
Contact: William Kays
Phone: 213-426-2587
      213-636-6315

Gulf Seal Corporation
Suite 275
700 Regency Square Blvd.
Houston, TX 77036
Contact: William J. Way, vice president and general manager
Phone: 713-782-9220

Key Enterprises, Inc.
P.O. Box 6606
Odessa, TX 79760
Contact: Ken Stewart, president
Phone: 915-362-2368

MWM Contracting Corp.
347 N. Main St.
Milford, MI 48042
Contact: Joe McCullough
Phone: 313-685-9350
      313-685-1201

National Seal Co., Inc.
7701 E. Kellogg
Wichita, KS 67202
Contact: John W. Owen
Phone: 316-681-1931

Pacific Linings, Inc.
P.O. Drawer GGGG
Indio, CA 92201
Contact: John Blatt, president
Phone: 714-347-0828

Plasti-Steel, Inc.
3588 W. 13th St.
Vickers-KSB&T Building
Wichita, KS 67203
Contact: M. C. Green, president
Phone: 316-262-6861

Sta-Flex Corporation
16 Post Rd.
Greenland, NH 03840
Contact: Lou Peloquin
4917 New Ramsey Ct.

San Jose, CA 95136
Phone: 408-224-0604

The Thurston Wallace Co.
5470 E. Evans Ave.
Denver, CO 80222
Contact: Cliff Heller, vice president
Phone: 303-758-2232

Tri State Construction
959 108th Ave., N.E.
Belleview, WA 98004
Contact: Joe Agostino
Phone: 206-455-2570

Unit Liner Co.
P.O. Box 789
Shawnee, OK 74884
Contact: J. A. Hendershot, president
Phone: 405-275-4600

Fabricators who also install:
   (see Section A.3.)
McKittrick Mud Co.
Synflex Industries, Inc.

Manufacturers who also install:
   (see Section A.2.)
B. F. Goodrich Company
Gundle Lining Systems, Inc.
Schlegel Lining Technology, Inc.

B. Bentonite Producers and Suppliers

American Colloid Company
5100 Suffield Ct.
Skokie, IL 60077
Contact: Nick Kesslering
Ryan F. McKendrick
Phone: 312-583-0400

Archer-Daniels-Midland
P.O. Box 15166
Commerce Station

Minneapolis, MN 55415
Phone: 612-371-3400

Ashland Chemical
9450 Midwest Ave.
Cleveland, OH 44125
Phone: 216-587-2230

Dowell
1150 N. Utica St.

P.O. Box 21
Tulsa, OK 74102
Contact: Chris Parks, Mining and Construction Technical Services
Phone: 918-560-2972

Dresser Minerals
P.O. Box 6504
Houston, TX 77005
Phone: 713-972-2670

International Minerals &
 Chemical Corp.
IMC Foundry Products
17350 Ryan Road
Detroit, MI 48212
Contact: G. Alther
Phone: 313-368-6000

Charles Pfizer & Co.
235 E. 42nd St.
New York, NY 10017
Phone: 212-573-2323

Wilbur Ellis Co.
P.O. Box 1286

Fresno, CA 93715
Phone: 209-226-1934

Wyo-Ben Products, Inc.
P.O. Box 1979
Billings, MT 59103
Phone: 406-252-6351

C. Other Liner Materials

Arizona Refining Company
P.O. Box 1453
Phoenix, AZ 85001
Contact: J. R. Bagley, president
Phone: 602-258-4843

The Asphalt Institute
Asphalt Institute Building
College Park, MD 20740
Phone: 301-277-2458

Chevron, USA, Inc.
Asphalt Div.
P.O. Box 7643
San Francisco, CA 94120

Contact: Kenneth Coleman,
 K. D. Coleman Co.
P. O. Box 414
Santa Clara, CA 95052
Phone: 408-244-8948

I.U. Management Systems, Inc.
1500 Walnut St.
Philadelphia, PA 19102
Phone: 215-985-660

Michelle Corporation
Division of Weychem Canada
 Limited
P.O. Box 4794

Charleston Heights, SC 29405
Contact: F. Weyrich, president
Phone: 803-554-4033

Phillips Petroleum Company
Commercial Development Div.
Bartlesville, OK 74004
Contact: Floyd H. Holland
Phone: 918-661-6428

Portland Cement Association
Old Orchard Rd.
Skokie, IL 60076
Phone: 312-066-6200

D. Miscellaneous Organizations in Liner Industry

Hovater-Way Engineers, Inc.
23011 Moulton Parkway, Suite
 F-5
Laguna Hills, CA 92653
Contact: Louis R. Hovater
Phone: 714-830-4050
Type of service: Design engi-
 neering, specializing in
 membrane liners.

Inquip Associates, Inc.
P.O. Box 2182
Santa Barbara, CA 93120
Contact: O. E. Hensgen
Phone: 805-963-6785
Type of service: Consulting,
 engineering, and contract-

ing of various lining materials and slurry cutoff
 walls.

Lining Materials
23011 Moulton Parkway, Suite
 F-4
Laguna Hills, CA 92653
Contact: George J. Miller, gen-
 eral manager
Phone: 714-581-9292
Type of service: Supplier of a
 variety of lining materials

Liquid Containment Systems
P.O. Box 324
South Holland, IL 60473

Contact: Jack Moreland, president
Phone: 312-468-2500
Type of service: Design, planning, engineering, and installation of liners.

National Sanitation Foundation
NSF Building
Ann Arbor, MI 48105
Contact: Gary W. Sherlaw,
 director standards devel-
 opment
Phone: 313-769-8010
Type of service: Developing na-
 tional specifications for
 membrane liners.

INDEX